LuCaNT: Databases, Algorithms, and Computational Number Theory

CONTEMPORARY MATHEMATICS

840

LuCaNT: Databases, Algorithms, and Computational Number Theory

Conference
LMFDB, Computation, and Number Theory (LuCaNT) 2025
July 7–11, 2025
Institute for Computational and Experimental Research in Mathematics (ICERM)
Providence, Rhode Island

John W. Jones
Jennifer Paulhus
Andrew V. Sutherland
John Voight
Editors

AMS AMERICAN MATHEMATICAL SOCIETY

2020 *Mathematics Subject Classification.* Primary 11-06, 14-06, 11-04, 11-11, 08-04.

Library of Congress Cataloging-in-Publication Data

Cataloging-in-Publication Data has been applied for by the AMS.
See http://www.loc.gov/publish/cip/.

Contemporary Mathematics ISSN: 0271-4132 (print); ISSN: 1098-3627 (online)
DOI: https://doi.org/10.1090/conm/840

∞ The paper used in this book is acid-free and falls within the guidelines
established to ensure permanence and durability.
Visit the AMS home page at `https://www.ams.org/`

10 9 8 7 6 5 4 3 2 1 31 30 29 28 27 26

Contents

Preface

The LuCaNT (LMFDB, Computation, and Number Theory) conference took place from July 7 to 11, 2025, at the Institute for Computational and Experimental Research in Mathematics (ICERM) located in Providence, Rhode Island, and affiliated with Brown University. It was the second in what will likely be a series of conferences on the topics of mathematical databases, computation, and algorithms, particularly in the areas of number theory and arithmetic geometry.

The conference series aims to provide a venue to highlight the value and importance of mathematical databases in the modern mathematical research landscape. The field of mathematical databases has emerged as an important area of research at the intersection of computer science and mathematics. It seeks to address questions that arise when organizing, storing, and providing access to mathematical knowledge in a structured manner. These databases are intended to be easily searchable and navigable, providing researchers, educators, and students with a convenient way to access mathematical content. There are many challenges in developing and maintaining mathematical databases, ranging from data quality and integration, to standardization, search design, updating and maintenance, and usability. In addition, many algorithmic and theoretical questions emerge naturally when considering systematic computations that must be done on a large scale. Of special concern in this regard at LuCaNT, the LMFDB (the L-Functions and Modular Forms Database, `http://lmfdb.org`) is an online resource for mathematical objects arising in the Langlands program and the connections between them.

This proceeding features papers on databases, related algorithms and computations, and applications of the LMFDB or other databases to new research. Many of the papers were presented at the conference. A new feature in this proceeding is a collection of short papers (about 5 pages each) that provide concise reports on the themes of the conference. The goal in providing a short paper option is to give researchers space to announce a significant computation which may only require a few details to explain, present a novel proof of a known result, give an outline of a new algorithm with a link to the implementation, or elaborate upon some aspect of creating or maintaining a database. The conference also featured four 50-minute invited talks.

- David Roe (MIT) began the conference with *Incubating Mathematical Databases* where he talked about two different database projects he is working on. The first, MathBases, is a database of mathematical databases. The second is a nascent database of Diophantine equations which will compile and collate in an easily searchable way known results about a large swath of such equations. The intended audience for this database would be researchers in mathematics and related fields who may stumble across an equation they need to solve, but who may not necessarily have the training to know how to solve such equations.

- In *Do Mathematicians Dream of Data?*, Katja Berčič (University of Ljubljana) gave a thoughtful and wide ranging talk on data and mathematics. She asked us to think about what data is to mathematicians, talked about ancient versions of databases, and discussed current practices of data management. She ended by talking about how to engender more trust among users in the data we present.
- The third plenary talk was given by Jennifer Balakrishnan (Boston University) entitled *Shadow Line Distributions*. Shadow lines are 1-dimensional $\mathbb{Q}_p$ vector spaces attached to certain elliptic curves with trivial p-torsion (and with some additional conditions on p). She and her co-authors used the LMFDB in an intriguing way to study the distribution of these lines.
- Finally, Alexander Kasprzyk (University of Warwick) delivered his talk *The Graded Ring Database*. He described this database, one of the older mathematical databases around, and reflected back on the pros and cons of the many design decisions that were made over the years.

Additionally, two lightning talk sessions provided an opportunity for early-career researchers to advertise their work. A software demo session highlighted computations on modular curves and abelian varieties in the LMFDB, as well as an explanation of how "portraits" for each number field in the LMFDB were created, and a way to quickly and efficiently enumerate low genus hyperelliptic curves.

Various associate editors discussed recent or upcoming improvements to sections of the LMFDB, and several managing editors gave a behind-the-scenes overview of how the data stored in the LMFDB gets onto the webpages which researchers use. This presentation was followed by a robust session where many new participants worked on small coding projects to fix or improve pieces of the LMFDB. There were also many active informal research conversations and collaborations which took place during open times in the schedule.

As with the previous volume, Jeroen Hanselman (TU Kaiserslautern) in his role with the Mathematical Research Data Initiative (MaRDI) provided a code review for papers with associated software implementations. Authors were asked to place code in publicly available repositories. The review then provided authors with a report giving suggestions for improving the code to address issues such as clarity of instruction, ease of use, and correctness, all with the purpose of making the computational results of each paper reproducible.

The organizers are deeply grateful to the staff at ICERM for all their technical and administrative work for the conference, as well as to the ICERM director Brendan Hassett for his support. We are grateful to the LMFDB associate editors for their time and effort in handling the peer review process for the papers submitted to this volume and to the referees for providing efficient and valuable feedback. Finally, we thank the AMS staff for their assistance in preparing this volume, and Alex Scorpan for his help with EditFlow.

John W. Jones
Jennifer Paulhus
Andrew V. Sutherland
John Voight

Contemporary Mathematics
Volume **840**, 2026
https://doi.org/10.1090/conm/840/16803

Families of p-adic fields

Jordi Guàrdia-Rubies, John W. Jones, Kevin Keating, Sebastian Pauli,
David P. Roberts, and David Roe

ABSTRACT. We improve the database of p-adic fields in the LMFDB by systematically using Krasner-Monge polynomials and working relatively as well as absolutely. These improvements organize p-adic fields into families. They thereby make long lists of fields more manageable and various theoretical structures more evident. In particular, the database now includes all degree n extensions of $\mathbb{Q}_p$, for $p < 200$ and $n \leq 23$.

1. Introduction

In arithmetic parts of mathematics, it is often useful to work one prime at a time. When working at a single prime p, the field $\mathbb{Q}_p$ of p-adic numbers commonly plays a central role. Also important are finite-degree field extensions of $\mathbb{Q}_p$. The number of isomorphism classes of such extensions of a given degree n is finite and given by a formula due to Monge [**Mon11**, Thm 1]. Some cardinalities are given in Table 1.1.

n	1	2	3	4	5	6	7	8	9	10	11	12	13	14	15	16
$p=2$	1	7	2	59	2	47	2	1823	3	158	2	5493	2	590	4	890111
$p=3$	1	3	10	5	2	75	2	8	795	6	2	785	2	6	1172	13
$p=5$	1	3	2	7	26	7	2	11	3	258	2	17	2	6	1012	17
$p=2$	1	3	2	10	2	8	2	49	3	10	2	43	2	12	1	389
$p=3$	1	2	4	3	2	10	2	4	28	1	2	20	2	1	16	5
$p=5$	1	2	2	3	6	4	2	4	3	16	2	6	2	1	20	5

TABLE 1.1. Top: The number of isomorphism classes of degree n field extensions of $\mathbb{Q}_p$. Bottom: The number of degree n families over $\mathbb{Q}_p$. Cases where all ramification is tame are in gray.

An online database giving defining polynomials and various invariants of p-adic fields appeared in 2006, in connection with the paper [**JR06**], which built on [**Ama71**, **Pan95**, **PR01**]. The original database quickly expanded via various works [**JR04**, **JR08**, **Awt12**, **ABM+15**], to include complete detailed tables for many (p, n), including all those listed on Table 1.1 except $(2, 16)$. Another improvement was a migration in 2011 from an *ad hoc* platform to the LMFDB [**LMFDB**], so that the data can be more easily inspected from a wider variety of perspectives.

The purpose of this paper is to describe a substantial improvement we have recently made to the database in the LMFDB. The starting point for the improvement is the systematic use of certain Eisenstein polynomials. These polynomials were introduced long ago by Krasner [**Kra37**] and their theory was brought into modern form by Monge [**Mon14**]. In this approach, fields are naturally organized into families. The number of families for small (p, n) is also given in Table 1.1.

The improved database is at

`https://lmfdb.org/padicField/`

and has a page for each of the 1,335,301 fields with $p < 200$ and $n \leq 23$. The new case on Table 1.1, $(p, n) = (2, 16)$, by itself accounts for about 64% of the fields in this range. The next two largest contributors are $(2, 20)$ and $(3, 18)$, and they contain 314,543 and 130,647 fields respectively. These two new cases account for about 23% and 9% of the total, so that all the other (p, n) together contribute about 4%.

The improved database also now has a page for each of the 19585 families with $p < 200$ and $n \leq 47$. It is not reasonable to populate all of these families with fields, as there are in total about 897 billion. However, if a need arises for including all the fields belonging to a specific reasonably-sized family, doing so will be straightforward. Some supporting code for the database is at `https://github.com/roed314/padic_db`.

The subsections of this introduction give a first idea of the previous structure of the database and how the systematic introduction of families improves it. Sections 2 and 3 then present the theory necessary for the improvement, with some of the more subtle details and various algorithmic issues deferred to the companion paper [**GRJK+**]. Sections 4 and 5 encourage the reader to explore and appreciate the database, first by focusing on sample families and then by focusing on connections to various theorems in the Galois theory of p-adic fields.

1.1. The previous field-by-field approach. We begin by describing some aspects of the database as it stood before our recent improvements. Degree n fields were presented as $L = \mathbb{Q}_p[x]/f(x)$ with $f(x) \in \mathbb{Z}[x]$ a degree n polynomial obtained from a search over possibilities. Detailed attention was not paid to the choice of $f(x)$. Rather the focus was on the most important invariants of L.

The extension $L/\mathbb{Q}_p$ has a normal closure $L^{\text{gal}}/\mathbb{Q}_p$ and hence a Galois group $\text{Gal}(L^{\text{gal}}/\mathbb{Q}_p)$. The general theory of p-adic fields gives a decreasing filtration of this group by normal subgroups. The successive subquotients Q^s each have a size $|Q^s|$ and also an associated slope s. Here slopes of -1, 0, and positive rational numbers correspond to no ramification, tame ramification, and wild ramification.

The database focused on the filtered group $\text{Gal}(L^{\text{gal}}/\mathbb{Q}_p)$. Filtered groups are somewhat unwieldy objects, so the database gave only associated numerical invariants. To represent the group, it gave its standard label nTj in the list of conjugacy classes of transitive subgroups of the symmetric group S_n [**BM83, CHM98, Hul05**]. To represent the filtration, it gave the *Galois slope content* W_t^u. Here $u = |Q^{-1}|$, $t = |Q^0|$, and a wild subquotient Q^s of size p^ρ contributes ρ copies of s to the weakly increasing list W of wild slopes. So the word "content" is in the spirit of "Jordan-Hölder content."

As an example, one of the 795 nonic 3-adic fields L was represented by the polynomial

$$f(x) = 21 + 18x + 18x^2 + 21x^3 + 9x^4 + 18x^5 + x^9. \tag{1.1}$$

The Galois group has $324 = 2^2 3^4$ elements. On the standard list from $9T1 = C_9$ to $9T45 = S_9$, it is $9T24$. The Galois slope content is $[\frac{1}{2}, \frac{1}{2}, \frac{2}{3}, \frac{3}{2}]_2^2$. In general, Galois groups were determined in the above-cited papers by computing and factoring many resolvents over $\mathbb{Q}_p$. The Galois slope content was determined by studying the ramification in the fields defined by these factors.

Previous to our current improvements, the database emphasized Artin slopes $\hat{s}_k = s_k+1$ rather than our current slopes s_k, displaying e.g. $[\frac{3}{2}, \frac{3}{2}, \frac{5}{3}, \frac{5}{2}]_2^2$ in the above example. Artin slopes $\hat{s}_k$ are indeed often more convenient in global applications. We are now emphasizing the smaller slopes s_k, often called Swan slopes, because they are more natural in detailed local analyses.

1.2. The new family approach. The Galois slope content of any finite extension $L/\mathbb{Q}_p$ splits cleanly into the *visible slope content* $[s_1, \ldots, s_w]_\epsilon^f$, with associated degree $[L : \mathbb{Q}_p] = f\epsilon p^w$, and the rest, called *hidden slope content*, with associated degree $[L^{\text{gal}} : L]$. In example (1.1), the visible slope content is $[\frac{1}{2}, \frac{3}{2}]_1^1$ and the hidden slope content is $[\frac{1}{2}, \frac{2}{3}]_2^2$. The visible slope content is enormously easier to compute than the hidden slope content. As we will explain, no Galois-theoretic concepts are needed.

We say that two extensions of $\mathbb{Q}_p$ belong to the same *absolute family* if their visible slope contents are the same. The key idea underlying this paper is that one can find defining polynomials for all the extensions in a family by suitably specializing a single *generic polynomial* belonging to the family. These specializations are the previously mentioned Krasner-Monge polynomials.

Continuing the example begun in (1.1), consider the family of 3-adic fields with visible slope content $[\frac{1}{2}, \frac{3}{2}]_1^1$. Following the general recipe we will present, the generic polynomial is

$$f(a_3, a_{10}, b_{11}, b_{13}, \pi, x) = \pi \left(1 + a_{10}\pi x + b_{11}\pi x^2 + a_3 x^3 + b_{13}\pi x^4\right) + x^9. \tag{1.2}$$

Specializing via $\pi = 3$, $a_\sigma \in \{1, 2\}$, and $b_\sigma \in \{0, 1, 2\}$ gives thirty-six 3-Eisenstein polynomials in $\mathbb{Z}[x]$. They represent bijectively the thirty-six entries on the list of 795 nonic 3-adic fields which have visible slope content $[\frac{1}{2}, \frac{3}{2}]_1^1$. It couldn't be easier! Moreover, as an important bonus, the coordinates provided by generic polynomials often give rise to clean descriptions of the hidden invariants. In the case of (1.2) there are ten possibilities for the pair consisting of the Galois group and the hidden slope content, one pair being the above $(9T24, [\frac{1}{2}, \frac{2}{3}]_2^2)$. Three fields have this pair, namely the ones with $(a_3, a_{10}, b_{11}) = (1, 2, 2)$. The particular field defined by (1.1) comes from $b_{13} = 2$. The other nine subsets likewise have very elementary descriptions, as can be seen in Table 4.1.

Our sample visible slope content $[\frac{1}{2}, \frac{3}{2}]_1^1$ has two simplifying features: its unramified degree f is 1 and the other data in the visible slope content measuring ramification is *rigid*, as we will explain in §2.3. We broadly describe the general case in this paper, but defer a full treatment of the complications associated with $f > 1$ and nonrigidity to [**GRJK+**].

Using symbols like $[\frac{1}{2}, \frac{3}{2}]_1^1$ in a naming scheme for families would be unwieldy as part of a URL. The database instead uses labels in the form $p.f.e.cL$, as in

3.1.9.18b for the example. Here c is the common discriminant-exponent of all the fields in the family and the letter L resolves ambiguity. Similarly, the mathematically ideal Eisenstein coefficients do not work well as labels identifying fields within a family. The database instead appends a subfamily number ℓ followed by a counter j, so that the example field (1.1) becomes 3.1.9.18b2.9. In (1.2), two fields are in the same subfamily if they have the same a_{10}. The general notion of subfamily involves residual polynomials and is given in [**GRJK+**].

1.3. The new relative context. The paper [**JR06**] and the subsequent papers extending the database were aggressively absolute: p-adic fields of degree N were always given by a degree N polynomial with coefficients in $\mathbb{Z}$. However it is often better to build fields in towers and the theory of generic polynomials fits perfectly into this paradigm. The theory of slopes generalizes to this relative context and we say that two extensions L_1/K and L_2/K belong to the same family over K if they have the same relative visible slope content $I = [s_1, \dots, s_w]_\epsilon^f$. We denote this family, viewed simply as a finite set of isomorphism classes of extensions, by I/K.

Henceforth, we call I a *Herbrand invariant*, because we have other ways of expressing the data in I that do not directly involve slopes, as we will be explaining in the next section. In fact our viewpoint is that p-adic Herbrand invariants are simple combinatorial objects that could be described independently of p-adic fields. To get a family, one combines two objects of different nature, I and K, subject to a numerical compatibility condition.

To continue the example begun in (1.1) yet further, take any finite extension K of $\mathbb{Q}_3$ as ground field, with residual cardinality denoted by q. Consider all extensions L/K with Herbrand invariant $I = [\frac{1}{2}, \frac{3}{2}]_1^1$. This family I/K is again bijectively indexed by certain specializations of the exact same generic polynomial (1.2). The difference is that π now is specialized to a uniformizer of K, rather than to the uniformizer 3 of $\mathbb{Q}_3$, and the a_σ and b_σ to elements of K with distinct reductions modulo π. Thus the real purpose of (1.2) is to get all relative extensions L/K of type $[\frac{1}{2}, \frac{3}{2}]_1^1$, for any fixed 3-adic base field K. Directly generalizing the case $K = \mathbb{Q}_3$, this family has cardinality $|I/K| = (q-1)^2 q^2$.

While we work relatively throughout this paper, the improved database keeps the original context of p-adic fields L as one of its two focal points. It has a basic bipartite structure. On the one hand, each p-adic field L within range has, as before, a homepage. On the other hand, each family I/K now also has a homepage. The two homepages are linked if K can be realized as a subfield of L such that L is in the family I/K.

In the case of the example family $[\frac{1}{2}, \frac{3}{2}]_1^1/\mathbb{Q}_3$, each field L in it has exactly one cubic subfield K' having discriminant-exponent 3. The cubic extension L/K' then appears in the relative family $[\frac{7}{2}]_1^1/K'$. There are two possibilities for K', namely 3.1.3.3a1.1 or 3.1.3.3a1.2, each of which occurs for half of the L. So the fields in the absolute family $[\frac{1}{2}, \frac{3}{2}]_1^1/\mathbb{Q}_3$ come half each from the relative families 3.1.3.3a1.1-1.3.9a and 3.1.3.3a1.2-1.3.9a. Here, the syntax for relative families is (base field)-$f.e.c$(tiebreaker). Two fields L in $[\frac{1}{2}, \frac{3}{2}]_1^1/\mathbb{Q}_3$ also have three more cubic subfields, now with discriminant-exponent 5. The index page lmfdb.org/padicField/families/?label_absolute=3.1.9.18b gives an overview of all possibilities.

1.4. Notation. We gather and comment on our most basic notations for the reader's convenience. As already indicated, a prime p is fixed, the symbol K is reserved for a finite extension of the field $\mathbb{Q}_p$ of p-adic numbers, and the symbol L is reserved for a finite extension of K. We let $\mathcal{O}$ be the ring of integers of K, Π its maximal ideal, $\kappa = \mathcal{O}/\Pi$ its residue field, and $q = |\kappa|$ its residual cardinality. We usually view K as fixed and L as varying.

Many fields and numbers are associated to a given L/K. The most basic come from its standard tower

$$K \overset{f}{\subseteq} L_{\mathrm{ur}} \overset{\epsilon}{\subseteq} L_0 \overset{p^w}{\subseteq} L. \tag{1.3}$$

Here L_{ur} is the maximal unramified subextension and L_0 is the maximal tamely ramified subextension. The superscripts indicate relative degrees so that the entire degree $n = [L : K]$ comes with a canonical factorization, $f\epsilon p^w$. In many circumstances it is enough to work with the ramification index $e = \epsilon p^w$, but in this paper it is usually best to separately emphasize its tame factor ϵ and its wild factor p^w.

The most familiar quantity capturing ramification in L/K is the discriminant $\mathrm{disc}(L/K) = \Pi^c$. We are emphasizing the importance of the discriminant-exponent c in our LMFDB labeling scheme. But for deeper analysis we prefer to switch to the *mean* m via the transformation equation

$$c = f(e - 1 + em). \tag{1.4}$$

So in the continuing nonic 3-adic example, the focus on $c = 18$ is shifted onto $m = 10/9$.

Most of our attention is focused on a discrete invariant W measuring the wild ramification present in the extension L/L_{ur}. One way to describe W is by the vector $[s_1, \ldots, s_w]$ already emphasized. Two similar ways are given in (2.4) and (2.5) and relations are summarized either by (2.6) and (2.7) or, in a different manner, by (3.6) and (3.7). The invariant W is the wild part of the Herbrand invariant $I = \mathrm{Inv}(L/K) = W_\epsilon^f$.

The set of all extensions L/K sharing a given I is denoted I/K. We think of I as a discrete invariant of L/K. In contrast, we view the invariants necessary to distinguish fields inside of I/K as continuous. They have their own complicated notation, introduced in Section 3.

All the notation for L/K has its analog for $K/\mathbb{Q}_p$, starting with $\mathbb{Q}_p \subseteq K_{\mathrm{ur}} \subseteq K_0 \subseteq K$. If we were to name everything, we would have to take care to avoid notation clashes. Fortunately, our considerations here make very little use of the internal structure of K, as we are focused mainly on extensions of K. The invariants of K which enter our considerations most often are its absolute ramification index $e_K = [K : K_{\mathrm{ur}}] = \mathrm{ord}_\Pi(p)$ and the above-mentioned residual cardinality $q = |\kappa|$.

In Sections 2 and 3 we work completely constructively, making no mention of any separable, i.e. algebraic, closure of $\mathbb{Q}_p$. In Sections 4 and 5 the attention shifts to Galois theory. There we fix a separable closure $\mathbb{Q}_p^{\mathrm{sep}}$ of $\mathbb{Q}_p$ and view the ground field K as in $\mathbb{Q}_p^{\mathrm{sep}}$. Given an abstract extension L/K of degree n, we let L^{gal} be the compositum of the images of the n different K-linear embeddings of L into $\mathbb{Q}_p^{\mathrm{sep}}$. So the Galois group $\mathrm{Gal}(L^{\mathrm{gal}}/K)$ is a quotient of $\mathrm{Gal}(\mathbb{Q}_p^{\mathrm{sep}}/K)$.

1.5. The Laurent series case. There is an extremely parallel case where one replaces the absolute ground field $\mathbb{Q}_p$ by the field $\mathbb{F}_p((t))$ of Laurent series over $\mathbb{F}_p$. Here a separable closure is smaller than an algebraic closure. It is the former which is the setting for the analog, which accounts for our notation $\mathbb{Q}_p^{\text{sep}}$ in this paper. The direct connections between the two cases are of great current interest [**Sch14**, §2]. As the database contains only extensions of $\mathbb{Q}_p$, we will limit our discussion of the Laurent series case to occasional brief remarks that clarify the p-adic case.

2. Herbrand invariants

Herbrand invariants are fundamental to this paper because they index families. This section explains what they are and how to work with them explicitly.

2.1. An elementary approach emphasizing canonical subfields. Serre's standard text [**Ser79**, IV] associates a Herbrand function $\phi : \mathbb{R}_{\geq 0} \to \mathbb{R}_{\geq 0}$ to any Galois extension L/K of p-adic fields. The construction involves filtering $\text{Gal}(L/K)$ by a descending family of subgroups indexed in two ways, by lower and upper numbering. Remark 1 of [**Ser79**, IV.3] extends the definition of ϕ to general extensions L/K by reduction to the Galois case. Details of this reduction were first explained in print by Deligne in a six-page appendix to [**Del84**].

It is less widely known that the Herbrand function of a general extension L/K can be understood without any reference to group theory whatsoever. In this subsection, we use the method of *slope polygons* to get relevant numerical quantities. We are following [**JR06**, §3.4], except that the Artin slopes there are translated to Swan slopes here. Figure 2.1 starts our second continuing example, based on

$$f(x) = 1 + 6x^4 + x^8. \tag{2.1}$$

This particular polynomial is chosen because it defines a Galois extension over both $\mathbb{Q}$ and $\mathbb{Q}_2$, with Galois group D_4; note that $f(x+1)$ is 2-Eisenstein. The reader can then follow along, using number field software to confirm our statements, e.g. *Pari*'s `nfsubfields` to get subfields and `smalldiscf` to get their discriminants and ultimately their means via (1.4). The fact that this example is Galois is irrelevant for the method we are describing.

Consider towers $L/L'/L_{\text{un}}/K$. For each, one has the degree $e' = [L' : L_{\text{un}}]$, the mean m' of L'/L_{un}, and thus a point $(e', e'm')$ in a Cartesian plane. The *slope polygon* S is the lower boundary of the convex hull of all such points. Over each interval $[\epsilon p^{k-1}, \epsilon p^k]$ the polygon S is just a segment with some slope s_k. The symbol $[s_1, \ldots, s_w]_\epsilon^f$ is then the visible slope content emphasized in the introduction.

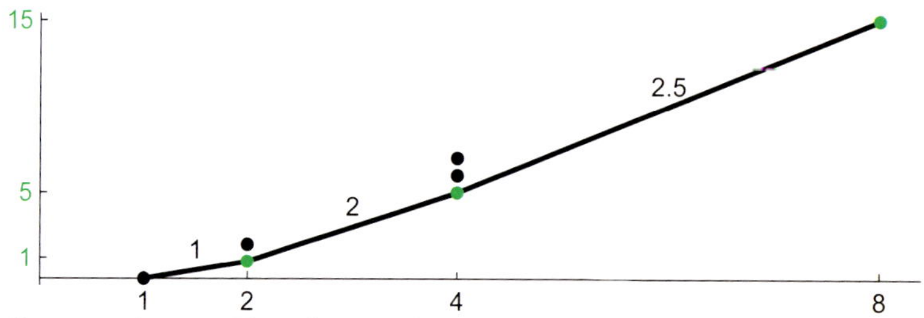

FIGURE 2.1. The slope polygon S associated to the octic extension $L/\mathbb{Q}_2$ defined by (2.1). The slopes $[1, 2, 2.5]$ and means $\langle \frac{1}{2}, \frac{5}{4}, \frac{15}{8} \rangle = \langle 0.5, 1.25, 1.875 \rangle$ are indicated.

Define h to be the function on $[1, e]$ having graph S. For $k = 1, \ldots, w$, define $m_k = \frac{h(\epsilon p^k)}{\epsilon p^k}$. Clearly, the data $\langle m_1, \ldots, m_w \rangle_\epsilon^f$ contains the exact same information as $[s_1, \ldots, s_w]_\epsilon^f$. If the point $P_k = (\epsilon p^k, \epsilon p^k m_k)$ is a turning point or the right endpoint of S then we say that the index k is final. Then P_k comes from exactly one tower $L/L_k/L_{\text{un}}/K$. The standard chain (1.3) can be extended to a more refined chain from K to $L = L_w$ by including all the other *canonical subfields* L_k. Note that for a final index k, the extension L_k/L_{un} has mean m_k. For a non-final index k, there may or may not be a extension mapping to P_k. In the example, the indices $k = 1$, 2, and 3 are all final.

2.2. Four perspectives on Herbrand invariants. The *slopes* s_k just introduced are usually called *breaks in the upper numbering*. There is similarly a dual polygon called the ramification polygon in which certain elementary quantities r_k appear along with the means m_k again. These *rams* r_k coincide with the *breaks in the lower numbering*. We will emphasize our elementary and non-Galois-theoretical viewpoint by using these two terms systematically, and not making further reference to breaks or numbering. Even when we bring back Galois groups in the last two sections, we will use the briefer terminology of slopes.

We write the Herbrand invariant of a p-adic extension L/K in four ways, the first classical but the others advantageous in various situations:

$$\text{Inv}(L/K) = (\phi, f) \quad (\phi \text{ is the Herbrand function}) \tag{2.2}$$

$$= [s_1, \ldots, s_w]_\epsilon^f \quad (\text{the } s_k \text{ are the slopes}) \tag{2.3}$$

$$= \langle m_1, \ldots, m_w \rangle_\epsilon^f \quad (\text{the } m_k \text{ are the means}) \tag{2.4}$$

$$= (r_1, \ldots, r_w)_\epsilon^f \quad (\text{the } r_k \text{ are the rams}). \tag{2.5}$$

As a matter of notation, the subscript ϵ and superscript f are allowed to be omitted when they are 1. Our key reference [**Mon14**] had different aims that did not require emphasis on Herbrand invariants. However it makes essential use of all the quantities s_k, m_k and r_k.

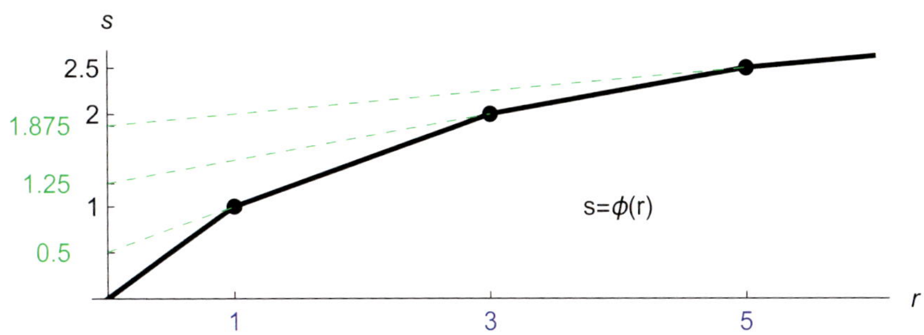

FIGURE 2.2. The Herbrand function ϕ for the octic extension $L/\mathbb{Q}_2$ defined by (2.1). It takes the rams $(r_1, r_2, r_3) = (1, 3, 5)$ of this extension to its slopes $[s_1, s_2, s_3] = [1, 2, 2.5]$. The means $\langle 0.5, 1.25, 1.875 \rangle$ are obtained by the indicated extensions of segments.

As a convention, we put $s_0 = m_0 = r_0 = 0$. Then (2.3)–(2.5) are related via

$$s_k = \frac{pm_k - m_{k-1}}{p-1}, \qquad m_k = \sum_{j=1}^{k} \frac{p-1}{p^{k+1-j}} s_j, \tag{2.6}$$

$$r_k = \epsilon p^k \frac{m_k - m_{k-1}}{p-1}, \qquad m_k = \sum_{j=1}^{k} \frac{p-1}{\epsilon p^j} r_j. \tag{2.7}$$

Formulas (2.6) reflect the perspective of slope polygons: each slope s_k is a certain rise-over-run and each mean m_k is a weighted average of slopes, with the formal slope $s_0 = 0$ having coefficient $1/p^w$, so that the coefficients sum to one. Formulas (2.7) reflect the dual perspective of ramification polygons. Direct transformation formulas between slopes and rams are a little more complicated, and one can just compose two transformations with means in the middle.

Figure 2.2 uses our octic 2-adic continuing example to illustrate how the classical version (2.2) is related to the more numerical versions (2.3)–(2.5). In general, the graph of ϕ starts at $(r_0, s_0) = (0, 0)$ and goes linearly to the (r_k, s_k) in order, with an actual step being taken only if k is a final index, as otherwise we have $(a_k, b_k) = (a_{k+1}, b_{k+1})$. It ends with a ray emanating from (r_w, s_w). If k is 0 or a final index then the slope of the segment with left endpoint (r_k, s_k) is $1/(\epsilon p^k)$. With this strong condition on slopes of the Herbrand segments, just ϕ determines all the s_k, m_k, and t_k.

2.3. Automorphisms, mass, and rigidity. An automorphism in $\mathrm{Aut}(L/K)$ necessarily stabilizes all of the canonical subextensions of L/K. Because of this fact, the Herbrand invariant I alone constrains the size of $\mathrm{Aut}(L/K)$. A key input is that if k is a final index, with the ram r_k having been repeated ρ times, then the inertia group associated to $L_k/L_{k-\rho}$ has the form $C_p^\rho \rtimes C_d$, where d is the denominator of the ram r_k. The action is such that the step can only have nontrivial automorphisms if $d = 1$, i.e. if r_k is integral.

Define the *ambiguity number* of a p-adic Herbrand invariant $(r_1, \ldots, r_w)_\epsilon^f$ to be $\mathrm{Amb}(I) = f\epsilon p^i$, where i is the number of integral rams (see §3.5 for the definition of $\mathrm{Amb}(I/K)$, the ambiguity of I over K). Then an extension L/K in any I/K has $|\mathrm{Aut}(L/K)|$ dividing $\mathrm{Amb}(I)$. The *mass* of L/K is by definition $1/|\mathrm{Aut}(L/K)|$ and L/K is called *rigid* if its mass is 1. We say that I is *rigid* if $\mathrm{Amb}(I) = 1$. So all extensions L/K in a family I/K with a rigid I are rigid.

2.4. Classification of Herbrand invariants. To classify all Herbrand invariants and say which actually occur over a given ground field K, it is best to use rams as was done in different language in [**PS17**, Prop 3.10]. First consider totally wildly ramified extensions L/K of degree p^ρ having just a single ram r repeated ρ times. The possibilities for r depend only on the absolute ramification index e_K of the ground field K.

Define first $\mathcal{R}_\rho^\infty$ to be the set of positive rational numbers with denominator dividing $p^\rho - 1$ and numerator not divisible by p. Then $\mathcal{R}_\rho^\infty$ is the correct collection of r in the parallel case of Laurent series ground fields $\mathbb{F}_q((t))$. For the characteristic zero fields K on which this paper is focused, the set of possibilities $\mathcal{R}_\rho^{e_K}$ contains all the elements of $\mathcal{R}_\rho^\infty$ which are less than $pe_K/(p-1)$. These elements exhaust $\mathcal{R}_\rho^{e_K}$ except that $\mathcal{R}_1^{e_K}$ also contains $pe_K/(p-1)$. We call this last ram *arithmetic* and all the smaller rams *geometric*.

Now consider all extensions L/K with canonical factorization $f\epsilon p^w$. The Herbrand invariants that can arise are $(r_1, \ldots, r_w)^f_\epsilon$ where the r_k are weakly increasing and each in their allowed set, from the above considerations. For example, to get all strictly increasing sequences, each r_k is chosen from $\mathcal{R}_1^{e_K \epsilon p^{k-1}}$. This description lets one produce any entry on the lower half of Table 1.1. If p is relatively prime to n, then the wild considerations are all vacuous and the possible Herbrand invariants are simply $(\ )^{n/\epsilon}_\epsilon$ as ϵ runs over divisors of n.

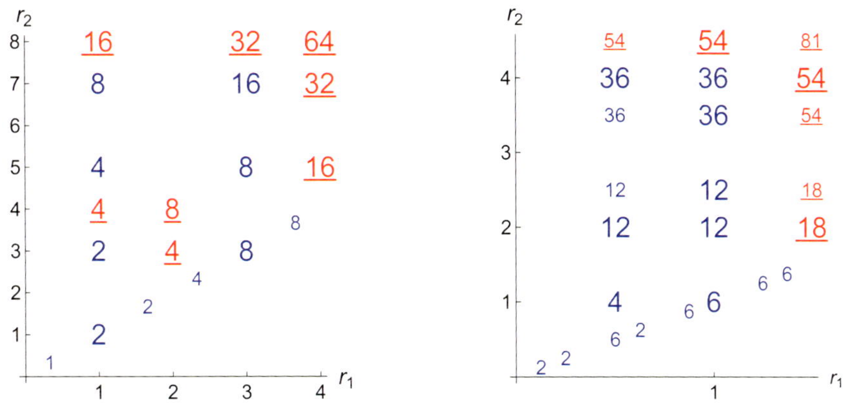

FIGURE 2.3. p-adic Herbrand invariants $I = (r_1, r_2)$ with $r_j \leq kp^j/(p-1)$. Left: $p = 2$ and $k = 2$. Right: $p = 3$ and $k = 1$

Figure 2.3 illustrates the case of Herbrand invariants (r_1, r_2) over 2- and 3-adic fields, with an eye towards giving a visual understanding of the general case. Rather than represent a Herbrand invariant I by a point at (r_1, r_2), we represent I by an integer. This integer is the mass of the family I/K, where K is any ground field compatible with I and having residue field of size p, as explained around (3.5). Families for which both rams are geometric are presented in blue, and families which have an arithmetic ram are presented in underlined red. Making an independent distinction, rigid Hurwitz invariants are indicated by a smaller font.

One should imagine each half of Figure 2.3 extended to the entire first quadrant of its (r_1, r_2) plane, so that for each positive integer k there is an underlined red hook H_k of entries with upper right corner at $(kp/(p-1), kp^2/(p-1))$. The Herbrand invariants (r_1, r_2) compatible with a given p-adic field K are exactly the arithmetic ones on H_{e_K} and the geometric ones under H_{e_K}. In other words, they are the ones on H_{e_K}, which are all underlined red, and the blue ones beneath H_{e_K}.

As a numerical example, consider the total number of families over $\mathbb{Q}_p$ with degree p^2. There is just one family with ramification index $e = 1$, the unramified family $(\ )^{p^2}$. There are p families with $e = p$, the geometric families $(1)^p$, $(2)^p$, ..., $(p-1)^p$ and the arithmetic family $(p)^p$. Guided by Figure 2.3, one can check that there are $p^3 - \frac{p^2}{2} + \frac{p}{2}$ families with $e = p^2$. For $p = 2$ and 3 respectively, the total counts are $1 + 2 + 7 = 10$ and $1 + 3 + 24 = 28$, with 7 and 24 being the numbers of families on or under H_1 on Figure 2.3 and 10 and 28 appearing as entries in the lower half of Table 1.1.

To see the seven and twenty-four families $p.1.p^2.C$ listed as an indexing table, one can search for absolute families with residual characteristic p, residual field degree $f = 1$, and ramification degree $e = p^2$. To see a sixteen-line table corresponding to the $e_K = 2$ case of the left side of Figure 2.3, one can search analogously for relative families. While there are nineteen numbers in the left side of the figure, the three in the underlined red hook do not correspond to families, because they would involve an arithmetic ram, prohibited by the above rules. One needs to input an appropriate ground field by its label, say $K = \mathbb{Q}_2(i)$ by `2.1.2.2a1.1`, as well as $f = 1$ and $e = 4$ again. The three tables produced in this paragraph contain summary information, such as the masses appearing in Figure 2.3.

2.5. Explicit composition via sorting. Let $L/K/Q/\mathbb{Q}_p$ be a tower with

$$\text{Inv}(K/Q) = (t_1, \ldots, t_{w'})^{f'}_{\epsilon'} \quad \text{and} \quad \text{Inv}(L/K) = (t_{w'+1}, \ldots, t_{w'+w''})^{f''}_{\epsilon''}. \tag{2.8}$$

A great virtue of the traditional functional presentation of Herbrand invariants is that one has the simple formula $\phi_{L/Q}(r) = \phi_{K/Q}(\phi_{L/K}(r))$. Here we translate this formula into the more explicit language of rams, so as to be able to go directly from (2.8) to $\text{Inv}(L/Q) = (r_1, \ldots, r_w)^f_\epsilon$

Of course, $f = f'f''$, $\epsilon = \epsilon'\epsilon''$, and $w = w' + w''$. To do the nontrivial part, we first define $T_k = \epsilon'' t_k$ and formally write

$$\text{Inv}(L/Q) = (T_1, \ldots, T_{w'}, t_{w'+1}, \ldots, t_w)^f_\epsilon. \tag{2.9}$$

Then we change the entries of the w-vector by iterating the following replacement in any order. Whenever there are two adjacent entries (a, b) with $a > b$, replace them with $(b, b + p(a - b))$. When the w-vector becomes weakly increasing, it is the desired ram vector $(r_1, \ldots, r_w)$.

The database uses this procedure to pass from the Herbrand invariant of a general relative extension L/K to the Herbrand invariant of the corresponding absolute extension $L/\mathbb{Q}_p$. If K is a canonical subfield of L, both steps of the process are trivial and, very simply, $r_k = t_k$ for all k.

To see the duality between rams and slopes, one can consider a modification of this process. Step 1 yielding (2.9) is exactly the same. In the modified Step 2, one iteratively replaces adjacent increasing (a, b) by $(b, b - (b - a)/p)$ until the w-vector becomes weakly decreasing. As a new final Step 3, one reverses the w-vector obtained to make it weakly increasing. The kth entry of this final vector is ϵs_k, where s_k is the kth slope of the extension L/Q.

3. Eisenstein polynomials

This section pictorially describes a well-behaved nonempty finite subset $\text{Eis}(L/K)$ of the infinite set $\text{Eisen}(L/K)$ of all Eisenstein polynomials defining any given totally ramified extension L/K. Thus $f = 1$ for the entirety of this section.

3.1. Set-up. Let $\text{Eisen}(e/K)$ be the space of Eisenstein polynomials of degree e over K. We write an element of $\text{Eisen}(e/K)$ as

$$f(x) = F_0 + F_1 x + \cdots + F_{e-1} x^{e-1} + x^e. \tag{3.1}$$

So the F_i run over the maximal ideal Π of the ring of integers $\mathcal{O}$ of K, except that F_0 is not in Π^2. One has decompositions into finitely many parts,

$$\text{Eisen}(e/K) = \sqcup_I \text{Eisen}(I/K), \qquad \text{Eisen}(I/K) = \sqcup_L \text{Eisen}(L/K). \tag{3.2}$$

On the left, I runs over totally ramified degree e Herbrand invariants that are compatible with K. On the right, L runs over isomorphism classes of extensions of K which have Herbrand invariant I.

To present things as concretely as possible, we choose a generator π of Π. We choose also a set of representatives $\tilde{\kappa}$ containing 0 and 1 for the q-element residue field κ. Rather than work with the F_i, we will work with π-adic expansions, writing each F_i as $\pi \sum_{j=0}^{\infty} f_{i,j}\pi^j$, with $f_{i,j} \in \tilde{\kappa}$. If K is unramified over $\mathbb{Q}_p$ we always take the uniformizer π to just be p. A theoretically natural choice would be to take the nonzero elements of $\tilde{\kappa}$ to be the $(q-1)^{\text{st}}$ roots of unity in K. However, we make the computationally more convenient choice of $\tilde{\kappa} = \{0, \ldots, q-1\}$ when q is prime.

It will also be convenient to use single-indexing simultaneously with double-indexing, with σ and (i, j) determining each other via $\sigma = je + i$, $j = \lfloor \sigma/e \rfloor$, and $i \in \{0, \ldots, e-1\}$. With this convention, (3.1) is written

$$F(x) = \pi \left(\sum_{\sigma=0}^{\infty} f_\sigma \pi^j x^i \right) + x^e. \tag{3.3}$$

The last term x^e will not be mentioned much in our narrative because it does not contain an unspecified coefficient.

A problem with the right part of (3.2) is that the sets involved are infinite. We will be replacing these sets by finite nonempty sets $\mathrm{Eis}(I/K) = \sqcup_L \mathrm{Eis}(L/K)$. In the case when I is rigid, each of the $\mathrm{Eis}(L/K)$ contains just one polynomial.

3.2. Eisenstein diagrams. Fix a totally ramified Herbrand invariant $I = [s_1, \ldots, s_w]_e$. To explicitly describe the sets $\mathrm{Eis}(I/K)$ for all compatible K at once, we consider the *Eisenstein diagram of I* in the vertical strip R with horizontal coordinate $i \in [0, e)$ and vertical coordinate $s \in [0, \infty)$. As indicated by the name, this diagram depends on I only, not on K. Always we draw only $[0, e) \times [0, s_w]$ as there is no useful information associated to the rest of the strip.

Figures 3.1 and 3.2 each show the Eisenstein diagram for a family discussed previously, as described in their captions. Most of our discussion of Eisenstein diagrams is supported by one or both of these figures. We will give links to the database for illustrations of other phenomena.

A spiral with points representing terms. We think of the rectangle R as a cut and unrolled cylinder, with each horizontal line segment of constant level s coming from a circle. We draw the spiral that starts at $(0, 0)$ and goes up with slope $1/e$. This spiral wraps every time it meets the right edge of R. The part of the spiral that goes from $(0, j)$ to $(e, j+1)$ is called the jth ramp.

We place conditions on the term $\pi f_\sigma \pi^j x^i$ by referring to the unique point P_σ on ramp j with horizontal coordinate i. Equivalently P_σ is the unique point on the spiral with $s = \sigma/e$. Thus $P_0 = (0, 0)$ is the starting point of the spiral, and as one goes up one encounters the points $P_1, \ldots, P_{\lfloor e s_w \rfloor}$ in order.

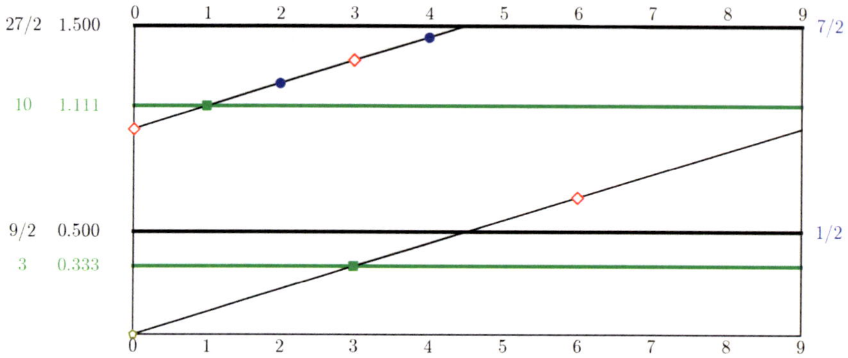

FIGURE 3.1. The Eisenstein diagram of the introductory nonic 3-adic Herbrand invariant with means $\langle \frac{1}{3}, \frac{10}{9} \rangle$, slopes $[\frac{1}{2}, \frac{3}{2}]$, rams $(\frac{1}{2}, \frac{7}{2})$, and generic polynomial (1.2). Various general facts are relatively easy to see here because the two bands do not overlap.

Bands. For $k = 1, \ldots, w$, the band B_k is defined to be the set of points in R satisfying $m_k \le s < s_k$. As k increases, the m_k strictly increase and so the bottom edges of bands go up. However the s_k only weakly increase. So the top edges of the bands B_k for k in a segment of common s_k agree, even as the widths of these bands successively decrease by a factor of p, as in `2.1.16.30a`, the sample family summarized in Table 4.3. It will also be convenient to let B_0 denote the bottom edge of the rectangle, i.e. the points with $s = 0$.

We shade bands by gray and indicate overlaps by darkening the gray. We color bottom edges of bands green and the bounding top edges black. In the rare cases where an upper edge agrees with a lower edge we dash this boundary using black and green, as in `2.1.16.79a`, the family highlighted in §5.5. These conventions assist in visually identifying the bands, even when they overlap.

Index and types of points. We say that the *index* of an integer $i \in [0, e)$ is $k = w - \text{ord}_p(\gcd(i, p^w)) \in \{0, 1, \ldots, w\}$. We partition the points P_σ into five types, using colors and shapes to distinguish the types.

The point $P_0 = (0, 0)$ plays a special role. We call it the *D-point* and we mark it by an olive pentagon. A D-point is *critical* and drawn solid if $\gcd(\epsilon, q-1) > 1$. Otherwise it is *negligible* and drawn hollow. Here, like with the red diamonds below, the solidness indicates complications in the process of choosing unique distinguished polynomials for fields L/K in the family I/K.

For $\sigma > 0$, we use the band B_k to classify the points P_σ of index k as follows.

- *Z-points* are points beneath B_k. Their associated color is *clear,* meaning we don't draw them.
- *A-points* are the unique points P_σ which are at the bottom edge of geometric bands B_k for which k is final in its segment, meaning that either $k = w$ or $s_{k+1} > s_k$. We draw them as *solid green squares.*
- *B-points* are points in their governing band B_k which are not A-points. They are represented by *solid blue disks.*
- *C-points* are points above their governing band. They are represented by *red diamonds.* A C-point is *critical* and drawn solid if it is on the top border of any band. Otherwise it is *negligible* and drawn hollow.

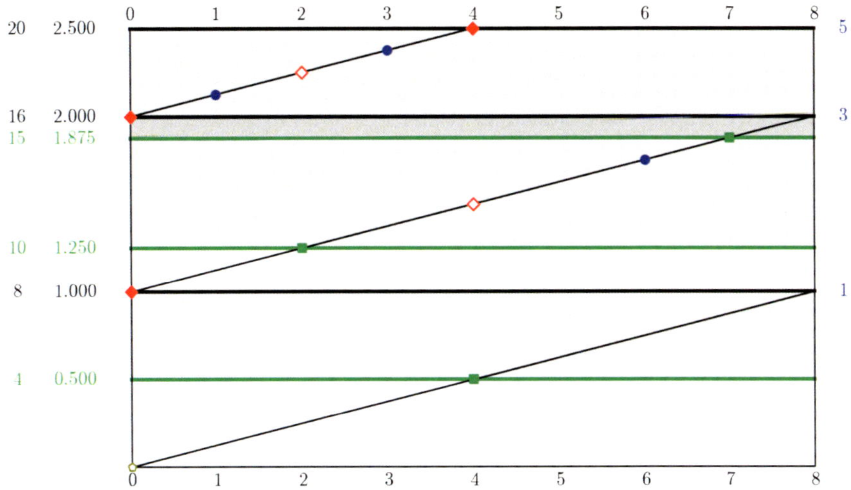

FIGURE 3.2. Eisenstein diagram for the octic 2-adic Herbrand invariant with means $m = \langle \frac{1}{2}, \frac{5}{4}, \frac{15}{8} \rangle$, slopes $s = [1, 2, \frac{5}{2}]$, and rams $r = (1, 3, 5)$.

The infinite set $\mathrm{Eisen}(I/K)$ now has the following explicit description. It is the subset of $\mathrm{Eisen}(e/K)$ where $f_\sigma = 0$ for Z-points and $f_\sigma \neq 0$ for A-points. The key finite set $\mathrm{Eis}(I/K)$ is the subset of $\mathrm{Eisen}(I/K)$ where $f_0 = 1$ if the D-point is negligible and $f_\sigma = 0$ for negligible C-points. The fact that $\mathrm{Eis}(I/K)$ still represents all extensions in I/K is not at all obvious. The proof of a considerably stronger statement is by Monge's reduction algorithm [**Mon14**, §2]. Basic aspects of this theory will be described in §3.5 below, and the full theory is described in a computational context in [**GRJK+**].

3.3. Generic polynomials and numerics. The generic polynomial for a given Herbrand invariant has the form $\left(\pi \sum_\sigma f_\sigma \pi^j x^i\right) + x^e$. Here σ runs over all nonnegative integers for which P_σ is drawn, but not negligible. To make structure more evident, we replace f_σ by d_σ, a_σ, b_σ, or c_σ according to whether P_σ is a D-point, A-point, B-point, or C-point. So when these coefficients run independently over $\tilde{\kappa}$, except for the inequalities $d_0, a_\sigma \neq 0$, one gets the set $\mathrm{Eis}(I/K)$. As an example beyond (1.2), the generic polynomial corresponding to the 2-adic Herbrand invariant $[1, 2, 2.5] = (1, 3, 5) = \langle 0.5, 1.25, 1.875 \rangle$ of Figures 2.1, 2.2 and 3.2 is

$$\begin{aligned} &f(a_4, a_{10}, a_{15}, b_{14}, b_{17}, b_{19}, c_8, c_{16}, c_{20}; \pi; x) = (\pi + \pi^2 c_8 + \pi^3 c_{16}) + \pi^3 b_{17} x + \\ &\pi^2 a_{10} x^2 + \pi^3 b_{19} x^3 + (\pi a_4 + \pi^3 c_{20}) x^4 + \pi^2 b_{14} x^6 + \pi^2 a_{15} x^7 + x^8. \end{aligned} \tag{3.4}$$

Other examples are given in the next sections.

Let $\delta \in \{0, 1\}$ be the number of critical D-points, and let α, β, and γ be the number of A-points, B-points, and critical C-points respectively. Clearly $|\mathrm{Eis}(I/K)| = (q-1)^{\delta+\alpha} q^{\beta+\gamma}$. Monge reduction (see §3.5) says that $|\mathrm{Eis}(L/K)| = (q-1)^\delta q^\gamma / |\mathrm{Aut}(L/K)|$. Accordingly, one gets a mass formula

$$M(I) := \sum_{L/K \in I/K} \frac{1}{|\mathrm{Aut}(L/K)|} = (q-1)^\alpha q^\beta. \tag{3.5}$$

In the rigid case $\delta = \gamma = 0$ where $|\mathrm{Aut}(L/K)| = 1$ is forced, (3.5) becomes a cardinality formula. As a non-rigid example, from `2.1.8.22d` there are 32 fields in the family (3.4), with (3.5) becoming $8(\frac{1}{8}) + 20(\frac{1}{4}) + 4(\frac{1}{2}) = 8$.

The numbers $M(I)$ in Figure 2.3 come from (3.5) with q set equal to either 2 or 3. A clarifying check on various numbers is to view the space $\mathrm{Eisen}(e/K)$ of degree e Eisenstein polynomials over K as an infinite product $\mathbb{F}_q^\times \times \mathbb{F}_q \times \mathbb{F}_q \times \cdots$, with the successive factors corresponding to the variables f_0, f_1, f_2, ... Giving each factor its uniform probability measure turns $\mathrm{Eisen}(e/K)$ into a probability space. Serre proved in [**Ser78**, Thm. 2] that the chance that a random polynomial in $\mathrm{Eisen}(e/K)$ is in $\mathrm{Eisen}(I/K)$ is the Serre mass $SM(I) = M(I)/q^{em_w}$. Figure 2.3 lets one see the terms in the resulting formula $\sum_{I/K} SM(I) = 1$ for three different (p, e_K). Taking the easiest case $(2,1)$ as an example, there are seven families, and one can use $em_2 = 2r_1 + r_2$. The sum is the dot product $(1,2,2,2,4,4,8) \cdot 2^{-(1,3,5,5,6,7,8)}$ and it is indeed 1.

3.4. Slopes, means, and rams on Eisenstein diagrams. When drawing Eisenstein diagrams, both here and on the family pages of the database, we indicate the slopes, means, and rams in appropriate places. There is no need to tick the vertical axis at the left of the diagram in the traditional way, because the left endpoint of the jth ramp is the integer j. Instead we give decimal approximations to lower edges m_k and upper edges s_k. In the leftmost column, we also give the scaled versions em_k and es_k. These scaled versions make the subscripts on the marked points P_σ more immediately identifiable.

The ram r_k is printed to the immediate right of the upper boundary of the band B_k. It gives two point counts as follows. First, keeping in mind that B_k includes its lower boundary but not its upper boundary, the number of drawn points in B_k of index at most k is $\lceil r_k \rceil$. Second, the total number of red diamonds on or below the top edge of B_k is $\lfloor r_k \rfloor$.

We have just given two interpretations of rams, but there are two easier ones. For these further interpretations, it is useful to scale rams to

$$(3.6) \qquad \textit{small rams } r_k^* = \frac{(p-1)r_k}{\epsilon p^k} \qquad \text{and } \textit{tiny rams } \; r_k' = \frac{r_k}{\epsilon p^k}.$$

Then

$$(3.7) \qquad r_k^* = m_k - m_{k-1} \qquad \text{and} \quad r_k' = s_k - m_k.$$

Together with the standing convention $s_0 = m_0 = r_0 = 0$, these simple equations give the complete relations between slopes, means, and rams. So the tiny ram r_k' is immediately visible as the vertical width of the kth band.

The set of Herbrand invariants $I = (r_1, \ldots, r_w)_\epsilon^f$ compatible with a given ground field K has an appealing geometric description if one uses small rams. Namely $(r_1^*, \ldots, r_w^*)$ must be in the cube $(0, e_K]^w$, with all r_k geometric if it is in the interior $(0, e_K)^w$. The hook H_j of §2.4 indicates a part of the boundary of $[0, j]^2$. In our Eisenstein diagrams, arithmetic bands, meaning the ones where $r_k^* = (p-1)r_k'$ is an integer, as discussed in §2.4, are marked by a black segment at their right end. Thus the bottom two bands in `3.1.27.99c` are marked, while the top one is not.

3.5. The Herbrand function and Monge reduction. Of the four versions (2.2)–(2.5) of totally ramified Herbrand invariants, the only one that is not immediately evident from an Eisenstein diagram is the first one, involving the classical Herbrand function ϕ. To read ϕ off from an Eisenstein diagram, one can think of a particle starting at time $r = 0$ at the point $P_0 = (0, 0)$ and moving up the spiral. The particle starts at a speed of p^w steps per second, where moving from P_σ to $P_{\sigma+1}$ counts as a step. Every time the particle crosses the top edge of a band it decreases its speed by a factor of p. This is a natural definition of speed, because the particle encounters C-points exactly at positive integral times. At any time $r \in \mathbb{R}_{\geq 0}$, the particle has traveled some number of ramps $s \in \mathbb{R}_{\geq 0}$. Then, as in Figure 2.2, $s = \phi(r)$.

The Monge reduction algorithm iteratively simplifies a given Eisenstein polynomial without changing the field it defines. Its general nature can be understood in terms of the moving particle. At the initial D-point a change of variables is made trying to make $f_0 = 1$. At each subsequent C-point P_σ a change of variables is made trying to make $f_\sigma = 0$. These changes to f_σ also change some of the $f_{\sigma'}$ for $\sigma' > \sigma$. When the particle leaves the drawn window, the complicated process can be stopped. Instead one can just turn all the f_σ with $\sigma > es_w$ to 0, as this change does not affect the field defined, by an effective version of Krasner's Lemma.

The reduction process is completely successful at negligible points but only partially successful at critical points. If one simply does not make the coordinate change at critical points, then one gets a surjection from the infinite set $\mathrm{Eisen}(L/K)$ to the $(q-1)^\delta q^\gamma$-element set $\mathrm{Eis}(L/K)$. If one makes the coordinate changes at the critical points as well, then the ambiguity from a critical D-point reduces from $q - 1$ to $\gcd(q - 1, \epsilon)$, and the ambiguity from a critical C-point at the top of ρ bands reduces from q to at most $\gcd(q, p^\rho)$. Multiplying these bounds together gives an ambiguity constant $\mathrm{Amb}(I/K)$. It depends on K only through q and is a divisor of the integer $\mathrm{Amb}(I)$ introduced in §2.3. The full reduction algorithm gives a surjection from $\mathrm{Eisen}(L/K)$ to a subset of $\mathrm{Eis}(L/K)$ of size a divisor of $\mathrm{Amb}(I/K)/|\mathrm{Aut}(L/K)|$. The sequel [**GRJK+**] implements the full algorithm, and moreover deterministically chooses a distinguished polynomial from the set of outputs.

4. Sample families

This section is aimed at facilitating the reader's exploration of the database. It summarizes the internal structure of several families, emphasizing topics which support the more theoretical discussions of the next section.

4.1. Partitions of the introductory family $[\frac{1}{2}, \frac{3}{2}]/\mathbb{Q}_3$. All absolute families are naturally partitioned in two related ways, into subfamilies and into packets. The subfamilies are determined by using the coefficients corresponding to points at the bottoms of bands; these are d_0, the a_σ, and sometimes also some b_σ, as discussed in [**GRJK+**] in the context of residual polynomials. Our naming convention for fields incorporates subfamilies, because subfamilies are both elementary and important.

a_3	a_{10}	b_{11}	b_{13}	HSC	G	a_3	a_{10}	b_{11}	b_{13}	HSC	G
1	1	1	0	$[\,]_2$	$9T5$	2	1	2	2	$[\,]_2$	$9T3$
1	1	$0,2$	$0,1,2$	$[1]_2$	$9T10$	2	1	2	$0,1$	$[\,]_2^3$	$9T10$
1	1	1	$1,2$	$[\,]_2^3$	$9T11$	2	1	0	$0,1,2$	$[1]_2$	$9T10$
1	2	1	$0,1,2$	$[1]_2^2$	$9T18$	2	1	1	$0,1,2$	$[1]_2$	$9T11$
1	2	2	$0,1,2$	$[\frac{1}{2},\frac{2}{3}]_2^2$	$9T24$	2	2	1	$0,1,2$	$[\,]_2^2$	$9T8$
1	2	0	$0,1,2$	$[\frac{1}{2},1]_2^2$	$9T24$	2	2	$0,2$	$0,1,2$	$[\frac{1}{2},1]_2^2$	$9T24$

TABLE 4.1. Partitions of the introductory family $[\frac{1}{2},\frac{3}{2}]/\mathbb{Q}_3$ = `3.1.9.18b` into two subfamilies and ten packets.

Giving a name to a concept introduced in §1.2, we say that two fields are in the same *packet* if they have the same Galois group G and the same hidden slope content HSC. The database has (G, HSC) for all its fields in degree ≤ 15. However since G and particularly HSC can be hard to compute in higher degrees, packets are not incorporated into our labeling scheme.

As the sample family of this subsection, we reconsider the family $[\frac{1}{2},\frac{3}{2}]/\mathbb{Q}_3$ of the introduction. Its generic polynomial from (1.2) is $3+9a_{10}x+9b_{11}x^2+3a_3x^3+9b_{13}x^4+x^9$. Table 4.1 breaks the 36 fields into two subfamilies of eighteen fields each, according to the value of a_{10}. The canonical cubic subfield is defined by the Eisenstein polynomial $3+3a_3y+y^3$, so the relative families discussed at the end of §1.3 correspond to the left and right halves of Table 4.1. In a more complicated way, the table breaks the family into its ten packets. Thus the packet discussed in §1.2 is given in the second from bottom line in the left half.

4.2. Easy linear packets in $[\frac{3}{2}]_5/\mathbb{Q}_3$. The family $[\frac{3}{2}]_5/\mathbb{Q}_3$ with LMFDB label `3.1.15.29a` has a very simple packet structure as follows. The generic polynomial is $3+9(b_{16}x+b_{17}x^2+b_{19}x^4+b_{20}x^5+b_{22}x^7)+x^{15}$. Specializing via $b_\sigma \in \{0,1,2\}$ gives $3^5 = 243$ polynomials bijectively representing the 243 fields of the family. If the list $(b_{16}, b_{17}, b_{19}, b_{22})$ starts with exactly 0, 1, 2, or 3 zeros, then the hidden slope content is $[j/10, j/10, j/10, j/10]_2^4$ for $j = 13$, 11, 7, 1, and the Galois group is $15T64 = C_3^4 : (S_3 \times F_5)$. If the list is simply $(0,0,0,0)$, then the hidden slope content is $[\ ]_2^4$, and the Galois group is $15T11 = S_3 \times F_5$. The fact that the coordinates b_σ render the packet structure transparent is an example of the "important bonus" mentioned after (1.2).

4.3. Complicated linear packets in $[2]_7/\mathbb{Q}_2$. The family $[2]_7/\mathbb{Q}_2$ with LMFDB label `2.1.14.27a` has generic polynomial

$$(2+8c_{28})+4\left(b_{15}x+b_{17}x^3+b_{19}x^5+b_{21}x^7+b_{23}x^9+b_{25}x^{11}+b_{27}x^{13}\right)+x^{14}.$$

There are 256 fields falling into sixteen packets (G, HSC), with the hidden slope content HSC always determining the Galois group G. The possibilities for the wild slopes are indexed by the set $\{13,5,3,\emptyset\} \times \{11,9,1,\emptyset\}$. Here an index j generically contributes $[j/7, j/7, j/7]$ but $\emptyset$ contributes the empty list. Inspecting the database shows that the list of wild slopes does not depend on b_{21} and c_{28}. Table 4.2 shows the dependence on the remaining coefficients. In the displayed vectors, each $\star$, v, or d can be independently 0 or 1, except that the d's must sum to an odd number and the v's to an even number.

	11	9	1	$\emptyset$
13	$(1,0,\star,\star,\star,\star)$	$(1,1,1,\star,\star,\star)$	$(1,1,0,d,d,d)$	$(1,1,0,v,v,v)$
5	$(0,1,\star,\star,0,\star)$	$(0,0,1,1,\star,\star)$	$(0,0,0,1,\star,1)$	$(0,0,0,1,\star,0)$
3	$(0,1,d,1,d,\star)$	$(0,0,1,0,0,\star)$	$(0,0,0,0,1,1)$	$(0,0,0,0,1,0)$
$\emptyset$	$(0,1,v,1,v,\star)$	$(0,0,1,0,0,\star)$	$(0,0,0,0,0,1)$	$(0,0,0,0,0,0)$

TABLE 4.2. The hidden slope stratification of the family $[2]_7/\mathbb{Q}_2$, with vectors $(b_{15}, b_{17}, b_{19}, b_{23}, b_{25}, b_{27})$ indicating coefficients.

4.4. Families $[1,\ldots,1]/\mathbb{Q}_p$. Families where $\epsilon = 1$ and there is just a single wild slope s repeated w times are easier than the general case, because there are no hidden wild slopes and Galois groups can be completely determined as in Theorem 8.2 of [**GP12**]. The case where $s = 1$ is particularly interesting for several reasons. This subsection focuses on the families $[1,\ldots,1]/\mathbb{Q}_p$, there being one for each prime power p^w, using the particular family $[1,1,1,1]/\mathbb{Q}_2$ as an example.

The generic polynomial for the 2-adic Herbrand invariant $[1,1,1,1]$ is

$$\pi\left((1+\pi c_{16}) + b_8x^8 + b_{12}x^{12} + b_{14}x^{14} + a_{15}x^{15}\right) + x^{16}. \tag{4.1}$$

A field $L/\mathbb{Q}_2$ in $[1,1,1,1]/\mathbb{Q}_2$ has one or two representing polynomials according to whether $b_8 + b_{12} + b_{14}$ is even or odd. In the latter case, the two polynomials differ only in the coefficient c_{16}. The polynomial with $c_{16} = 0$ is our choice of distinguished polynomial.

$\ell.i$	b_8	b_{12}	b_{14}	c_{16}	Associated polynomial	a	u	$[j_0, j_1, j_2, j_3, j_4]$
`1.1`	0	0	0	0	$(y+1)^4$	2	4	$[1, 3, 7, 15, 30]$
`1.2`	0	0	0	1	$(y+1)^4$	2	4	$[1, 3, 7, 15, 32]$
`2.1`	0	0	1	0	y^4+y+1	1	15	$[1, 3, 7, 14, 31]$
`3.1`	0	1	0	0	$\left(y^2+y+1\right)^2$	1	6	$[1, 3, 6, 12, 31]$
`4.1`	0	1	1	0	$(y+1)\left(y^3+y^2+1\right)$	2	7	$[1, 3, 6, 15, 30]$
`4.2`	0	1	1	1	$(y+1)\left(y^3+y^2+1\right)$	2	7	$[1, 3, 6, 15, 32]$
`5.1`	1	0	0	0	y^4+y^3+1	1	15	$[1, 2, 4, 8, 31]$
`6.1`	1	0	1	0	$(y+1)^2\left(y^2+y+1\right)$	2	6	$[1, 2, 4, 15, 30]$
`6.2`	1	0	1	1	$(y+1)^2\left(y^2+y+1\right)$	2	6	$[1, 2, 4, 15, 32]$
`7.1`	1	1	0	0	$(y+1)\left(y^3+y+1\right)$	2	7	$[1, 2, 7, 15, 30]$
`7.2`	1	1	0	1	$(y+1)\left(y^3+y+1\right)$	2	7	$[1, 2, 7, 15, 32]$
`8.1`	1	1	1	0	$y^4+y^3+y^2+y+1$	1	5	$[1, 2, 7, 14, 31]$

TABLE 4.3. Information on the 12 fields `2.1.16.30a`$\ell.i$ in the family `2.1.16.30a`

Table 4.3 presents information directly available on the family page for $[1,1,1,1]/\mathbb{Q}_2$, in a form slightly modified to support the discussion here. In the first column, ℓ indexes the subfamily and i indexes the field within the subfamily.

The jump sets $[j_0, j_1, j_2, j_3, j_4]$ are discussed in general in §5.2. Commonly, a family gives rise to just a very few jump sets, often just one. This family, and conjecturally all the $[1, \ldots, 1]/\mathbb{Q}_2$, have the unusual feature that the jump set determines the field.

The general case $[1, \ldots, 1]/\mathbb{Q}_p$ has a generic polynomial of a form similar to (4.1). Write $v_w = a_{p^w-1}$ and $v_j = b_{p^w-p^{w-j}}$ for $j = 1, \ldots, w-1$. Then the parameters in the generic polynomial are $v_1, \ldots, v_w$ and also c_{p^w}. Let $L/\mathbb{Q}_p$ be the field defined by the parameters $(v_1, \ldots, v_w; c_{p^w})$. Let g be an element of $GL_w(\mathbb{F}_p)$ with characteristic polynomial

$$f(y) = y^w + \sum_{j=0}^{w-1} v_{w-j} y^j. \tag{4.2}$$

Then the Galois group of $L^{\mathrm{gal}}/\mathbb{Q}_p$ is the semidirect product $\mathbb{F}_p^w \rtimes \langle g \rangle$ [**GP12**, Theorem 8.2]. Here the wild inertial group is $\mathbb{F}_p^w$, the tame quotient of inertia has order one, and the unramified quotient is $\langle g \rangle$. Thus the hidden slope content is simply $[\ ]^u$, where u is the order of g. For the case $p^w = 2^4$, Table 4.3 gives these polynomials in factored form. The residual polynomials given in the database can be obtained from $f(y)$ by replacing each y^i with z^{2^i} and dividing by z. The database gives the Galois groups in the usual way, from the smallest group $16T166 = \mathbb{F}_2^4 \rtimes C_4 = C_2 \wr C_4$ to the largest group $16T447 = \mathbb{F}_2^4 \rtimes C_{15} = F_{16}$.

The various phenomena discussed in the example of 2^4 generalize to p^w. A field $L/\mathbb{Q}_p$ has either one or p representing polynomials according to whether $f(1)$ is different from or equal to zero in $\mathbb{F}_p$. In the latter case, the polynomials again differ only in the coefficient c_{p^w} and again $c_{p^w} = 0$ gives our choice of distinguished polynomial. So, as illustrated on the table for 2^4, the automorphism number $a = |\operatorname{Aut}(L/\mathbb{Q}_p)|$ is p if $f(y)$ has a factor $(y-1)$ and 1 otherwise. Going further, the subfields of $L/\mathbb{Q}_p$ are in natural bijection with the factors of $f(y)$ in $\mathbb{F}_p[y]$, with the field coming from a degree k factor having degree p^k. So there are all together $\prod(m_j+1)$ subfields, where $\prod_j f_j(y)^{m_j}$ is the factorization of $f(y)$ into irreducibles.

5. Theoretical connections

Consider finite extensions K of $\mathbb{Q}_p$ inside of a fixed separable closure $\mathbb{Q}_p^{\mathrm{sep}}$. A natural goal, which seems a long way off or perhaps not obtainable at all, is to find a group-theoretical description of each absolute Galois group $\operatorname{Gal}(\mathbb{Q}_p^{\mathrm{sep}}/K)$ together with its descending filtration by the higher ramification groups $\operatorname{Gal}(\mathbb{Q}_p^{\mathrm{sep}}/K)^s$. There are however many deep theorems towards this goal. In this section, we describe ways in which the database interacts with these theorems, rendering them more explicit.

5.1. Extensions with a given Galois group. An overview of many powerful and explicit results on absolute Galois groups is given in [**NSW08**, VII§5]. Highlights are as follows. Let K^{nil} be the maximal extension of K for which the Galois group is a pro-p-group; here nil stands for nilpotent. Shafarevich proved in the 1940s that $\operatorname{Gal}(K^{\mathrm{nil}}/K)$ is free on $[K:\mathbb{Q}_p]+1$ generators if K does not contain pth roots of unity. Demushkin proved in the 1950s that it can always be presented with $[K:\mathbb{Q}_p]+2$ generators and one explicit relation, with an example being

$$\operatorname{Gal}(\mathbb{Q}_2^{\mathrm{nil}}/\mathbb{Q}_2) = \langle x, y, z | x^2 y^3 z^{-1} y z \rangle. \tag{5.1}$$

In the 1980s, Jannsen and Wingberg gave a complete description of $\mathrm{Gal}(\mathbb{Q}_p^{\mathrm{sep}}/K)$ for p odd, and then Diekert did the same for $p = 2$, assuming K contains fourth roots of unity. All these results are silent on the ramification filtration.

For a finite group G, let G/K be the set of Galois extensions L^{gal}/K in $\mathbb{Q}_p^{\mathrm{sep}}/K$ with $\mathrm{Gal}(L^{\mathrm{gal}}/K) \cong G$. These presentations let one compute cardinalities $|G/K|$. Many completely explicit examples are given in [**Roe19**, §4] for $K = \mathbb{Q}_p$. For example, let P_8 be the Sylow 2-subgroup of S_8. Then from (5.1) one eventually gets $|P_8/\mathbb{Q}_2| = 48$. Switching language to our context of general extensions, let nTj/K be the set of isomorphism classes of degree n extensions L/K with $\mathrm{Gal}(L^{\mathrm{gal}}/K) = nTj$. Then one can multiply by a constant associated to nTj to get the cardinality $|nTj/K|$. For example, over any field, each P_8 Galois extension comes from eight isomorphism classes of abstract $8T35$ extensions, and so $|8T35/\mathbb{Q}_2| = 8 \cdot 48 = 384$.

The database lets one search by Galois group and see ramification behavior not covered by the theorems. Continuing the example of $8T35/\mathbb{Q}_2$, the database shows that they are distributed among eight families $2.1.8.C$ as summarized in Table 5.1.

C	M	s_1	s_2	s_3	m	Hidden slopes			P	Other $8Tj$	
21a	32	1	1	2.75	16	2	2.5	2.5	2	$8T38$	
24d	16	1	2.5	2.75	16	1	2	2.5	1		
25b	32	1	2	3.25	16	1	2.5	3	3	$8T21$	$8T31$
26b	32	1	2.5	3.25	16	1	2	3	3	$8T19$	$8T29$
27a	64	1	2.5	3.5	32	1	2	2.5 or 3	7	(five groups)	
29a	64	2	2.5	3.75	32	1	3	3.25	3	$8T28$	$8T30$
30a	64	2	3	3.75	32	1	2.5	3.25	3	$8T27$	$8T30$
31a	128	2	3	4	32	1	2.5	3.25	14	(ten groups)	
					192						

TABLE 5.1. The distribution of the 384 octic extensions of $\mathbb{Q}_2$ with associated Galois group $8T35$ into eight families $2.1.8.C$

The table gives some indication of the nature of each of the families, including its total mass M and the mass m coming from $8T35$ fields. Each $8T35$ field has mass $\frac{1}{2}$ and the sum 192 of the m is indeed 384/2. The part of the family consisting of $8T35$ fields is always easy to describe. For example, it is the locus where the coefficient $b_4 = 0$ in the first-listed family $21a$. When the number P of packets is at most 3, the remaining mass is evenly split among the remaining possible groups $8Tj$.

In general, the organization of p-adic fields into families provides a framework for further investigation into ramification. There are many resolvent maps $n'Tj'/K \to nTj/K$ coming from Galois theory over arbitrary fields. The coordinates d_0, a_σ, b_σ, and c_σ can be used to describe these maps in a concise and uniform way. Returning to the example, consider an $8T35$ extension $L = K[x]/f(x)$. If $f(x)$ is generically chosen then the degree twenty-eight resolvent corresponding to the subgroup $S_6 \times S_2$ of S_8 factors into irreducibles as $f_4(x)f_8(x)f_{16}(x)$. The extension $L' = K[x]/f_8(x)$ is one of L's seven different siblings, the others then being easily obtainable either by a degree 35 resolvent construction or by certain quadratic twists [**JR08**, Figure 3.1]. The horizontal lines in Table 5.1 indicate that for

$K = \mathbb{Q}_2$ one has family interchanges $21a \leftrightarrow 24d$, $25b \leftrightarrow 26b$ and $29a \leftrightarrow 30a$ under this operation $L \leftrightarrow L'$. Note that the set of six wild slopes is preserved in each of the three family interchanges as it must be. The $8T35$ parts of the families $27a$ and $31a$ are closed under operation $L \leftrightarrow L'$.

5.2. Making cyclic cases explicit via jump sets. The natural goal of identifying the filtered group $\mathrm{Gal}(\mathbb{Q}_p^{\mathrm{sep}}/K)$ was reached nearly a century ago at the much simpler level of understanding the filtered abelianization $\mathrm{Gal}(K^{\mathrm{ab}}/K)$. Namely local class field theory identifies $\mathrm{Gal}(K^{\mathrm{ab}}/K)$ with the profinite completion of the multiplicative group $K^\times$ with the inertia group being sent to the unit group $U = \mathcal{O}^\times$. For j a positive integer and $s \in (j, j+1]$, the group $\mathrm{Gal}(K^{\mathrm{ab}}/K)^s$ is sent to the j-unit group $U_j = 1 + \Pi^j$.

However this theoretically ideal solution does not immediately answer some very basic concrete questions. One such question is, *what is the set $C_w(K)$ of Herbrand invariants $[s_1, \ldots, s_w]$ coming from cyclic extensions of K of degree p^w?* In other words, for what families $[s_1, \ldots, s_w]/K$ is $(G, HSC) = (C_{p^w}, [\ ])$ one of the packets. One certainly needs the s_k to form a strictly increasing sequence of positive integers. But to go beyond this statement, one needs to understand the filtered group U_1.

Suppose the p-primary torsion in U_1 has order p^v. Then U_1 is isomorphic to $(\mathbb{Z}/p^v) \times \mathbb{Z}_p^{n_K}$. Thus the free vs. one-relator distinction from the beginning of §5.1 is visible at this abelian level. In the free case $v = 0$, the set $C_w(K)$ depends only on e_K and is given below. In the one-relator case $v \geq 1$, the situation is much more complicated and $C_w(K)$ depends on an invariant j_K extracted from the abelianization of the relation given in (5.2) below. The database tabulates j_K, with instances having been given in Table 4.3.

The description of $C_w(K)$ involves combinatorial notions, as follows. For a prime p and and a positive integer e, define $\rho_{p,e} : \mathbb{Z}_{\geq 1} \to \mathbb{Z}_{\geq 1}$ by $\rho_{p,e}(i) = \min(pi, i+e)$. Table 5.2 draws $\rho_{p,e}$ in three cases by organizing $\mathbb{Z}_{\geq 1}$ into e columns. Always $\rho_{p,e}(i)$ is the number immediately above i in its column. Let $T_{p,e}$ be the set of non-images of $\rho_{p,e}$, thus the e numbers underlined and in bold at the bottom of columns. A *jump set* of length w is a sequence $[s_1, \ldots, s_w]$ satisfying the initial condition $s_1 \in T_{p,e}$ and an inductive condition. The inductive condition requires that for $k \geq 2$ one has $s_k \geq \rho_{p,e}(s_{k-1})$, with $s_k \in T_{p,e}$ if strict inequality holds. Let $J_w(p, e)$ be the set of jump sets of length w. If $p-1$ divides e there is also a notion of *extended jump set*. Here $T_{p,e}$ is simply replaced by $T^*_{p,e} = T_{p,e} \cup \{pe/(p-1)\}$, the extra point being indicated by bold italic in Table 5.2.

The set $J_w(p, e)$ of jump sets and the set $J^*_w(p, e)$ of extended jump sets can be understood as the set of ways of climbing a "rock wall." There are choices at the very beginning, but as soon as one reaches $e_K/(p-1)$, marked by a light band in the table, the rest of the climbing path is forced to be vertical. The diagram splits into a part beneath $e_K p/(p-1)$, marked by a dark band, and the part on or above $e_K p/(p-1)$. We call the lower part *geometric*, because it agrees with the case (p, ∞) corresponding to $\mathbb{F}_q((t))$, and the upper part *arithmetic*. Table 5.2 also gives some sample counts, for $w = 0, 1, 2, \ldots$. The count for the last printed w also holds for all subsequent w.

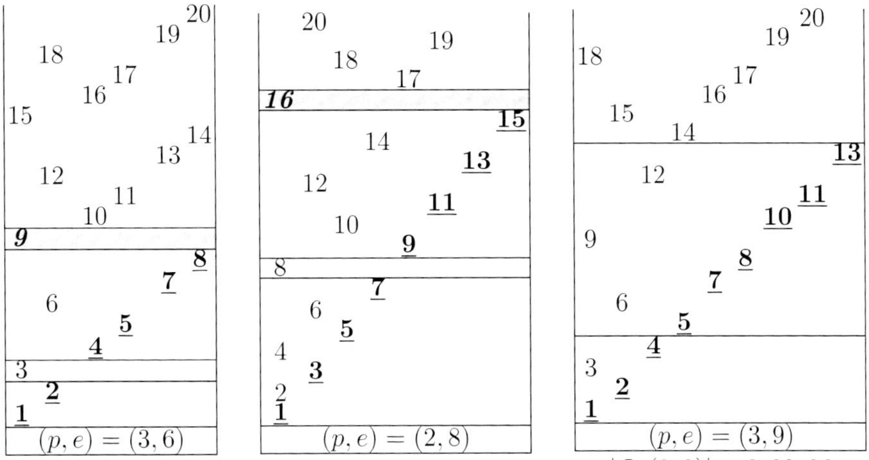

$|J_w(3,6)| = 6, 12$ $\quad$ $|J_w(2,8)| = 8, 24, 42, 53$ $\quad$ $|J_w(3,9)| = 9, 22, 26$

$|J_w^*(3,6)| = 7, 15$ $\quad$ $|J_w^*(2,8)| = 9, 23, 53, 69$

TABLE 5.2. Arrangements of the positive integers into e columns for understanding jump sets. Bands are actually just convenient thickenings of the lines at heights $e_K/(p-1)$ and $e_K p/(p-1)$, and they are drawn as lines when these heights are not integral.

The desired identification for $v = 0$ is $C_w(K) = J_w(p, e_K)$. As a trivial example, for $p > 2$ one has the familiar $C_w(\mathbb{Q}_p) = J_w(p, 1) = \{[1, \ldots, w]\}$. A partial answer for $v \geq 1$ is that $C_w(K) \subseteq J_w^*(p, e_K)$. As a simple example, $J_w^*(\mathbb{Q}_2) = \{[1, 2, \ldots, w], [2, 3, \ldots, w+1]\}$. While $C_1(\mathbb{Q}_2) = \{[1], [2]\}$ is all of $J_1^*(\mathbb{Q}_2)$, otherwise $C_w(\mathbb{Q}_2) = \{[2, 3, \ldots, w+1]\}$.

In the $v \geq 1$ case where K contains a primitive pth root of unity ζ_p, the next step towards a complete answer goes as follows. Building on Hasse [**Has02**, Ch. 15], Miki [**Mik81**, Lem 17] showed that one can write

$$\zeta_p = \alpha_0^{p^w} \alpha_1^{p^{w-1}} \cdots \alpha_{w-1}^p \alpha_w \tag{5.2}$$

with each α_i satisfying one of the following conditions:

- $v_K(\alpha_i) < pe_K/(p-1)$ and $p \nmid v_K(\alpha_i)$,
- $v_K(\alpha_i) = pe_K/(p-1)$ and α_i is not a pth power,
- $\alpha_i = 1$.

Pagano [**Pag22**, §1.2] defined the extended jump set associated to K by setting $j_i = v_K(\alpha_i - 1)$ for all $0 \leq i \leq w$ such that $\alpha_i \neq 1$; these values are independent of the choice of $\alpha_0, \alpha_1, \ldots, \alpha_w$. Undefined values in the sequence $j = [j_0, j_1, \ldots, j_w]$ are filled in using the recursion $j_{i+1} = \rho_{p,e_K}(j_i)$. The remaining steps are to use u_K to identify $C_w(K)$ as a subset of $J_w^*(p, e_K)$. This is difficult to make explicit in general, but Pagano works out many special cases.

Pagano [**Pag22**, Thm. 1.11] also gave a formula for j_K in terms of a defining Eisenstein polynomial for certain cases where $p \neq 2$ and all slopes of K are less than one. The tabulation of invariants in the database using the factorization of ζ_p suggests that j_K might always be directly expressible in terms of Eisenstein polynomials. For the case $[1, \ldots, 1]/\mathbb{Q}_2$ recall from §4.4 that the generic polynomial is written

$$2((1+2c_{2^w})+v_1x^{2^w-2^{w-1}}+\cdots+v_{w-1}x^{2^w-2^1}+v_wx^{2^w-2^0})+x^{2^w},$$

with $v_w=1$. For $0\le k\le w$ set $V_k=\sum_{i=1}^k v_i\in\mathbb{F}_2$. Then we expect

$$j_k=\begin{cases}2^{k+1}-1 & \text{if } V_k=0,\\ 2j_{k-1} & \text{if } V_k=1 \text{ and } k<w,\\ 2^{w+1}+2c_{2^w}-2 & \text{if } V_k=1 \text{ and } k=w.\end{cases}$$

We also observe the simple statement that $j_K=[j_i]$ with $j_i=\epsilon(p^i-p+1)/(p-1)$ when all slopes are greater than 1. For the case $(\epsilon,p)=(1,2)$ this formula gives $j_i=2^i-1$. In particular, if $w=4$ then the jump set is $[1,3,7,15,31]$.

5.3. Nonabelian quotients with known filtrations. The complications resolved by jump sets in the previous subsection only became serious for $w\ge 2$. Applying the easy special case of $w=1$ to tame extensions K' of a fixed field K gives a large class of extensions of K which are nonabelian and wild, but still very well understood. This situation has been studied from the point of view of primitive extensions of K by Del Corso, Dvornicich, and Monge in [**DCDM17**].

To briefly summarize [**DCDM17**] with some more specificity about slopes, let $K^{\mathrm{prim}}\subset\mathbb{Q}_p^{\mathrm{sep}}$ be the composita of all primitive extensions of K. It contains the maximal tame extension K^{tame} of K. What makes the group $\mathrm{Gal}(K^{\mathrm{prim}}/K)$ tractable is that its wild inertia subgroup $\mathrm{Gal}(K^{\mathrm{prim}}/K^{\mathrm{tame}})$ has exponent p. The wild slopes are exactly the positive rational numbers less than $e_Kp/(p-1)$ with numerator and denominator prime to p, and then $e_Kp/(p-1)$ itself. The former occur with infinite multiplicity but the latter occurs just with multiplicity 1. The quotient $\mathrm{Gal}(K^{\mathrm{tame}}/K)$ is the closure of a subgroup $\langle\tau,\sigma|\sigma\tau\sigma^{-1}=\tau^q\rangle$, where τ generates tame inertia and σ is a Frobenius element. The full group is a semidirect product $\mathrm{Gal}(K^{\mathrm{prim}}/K^{\mathrm{tame}})\rtimes\mathrm{Gal}(K^{\mathrm{tame}}/K)$ with a known action.

Consider now families $I=[s,\ldots,s]_\epsilon^f/K$ with just one visible wild slope, such as $[\frac{3}{2}]_5/\mathbb{Q}_3$, $[2]_7/\mathbb{Q}_2$, and $[1,\ldots,1]/\mathbb{Q}_p$ of §4.2, §4.3 and §4.4, respectively. The compositum $K^I\subset K^{\mathrm{prim}}$ of the Galois closures of the fields L/K in I/K is governed by the group theory just summarized. So the packets (G,HSC) that can occur can be group-theoretically calculated.

As an example of how group theory can explain otherwise mysterious patterns, consider the orbits of multiplication by p on $\mathbb{Z}/(p^\rho-1)\mathbb{Z}$, with representatives taken in $[\frac{p^\rho-1}{p-1},\frac{p(p^\rho-1)}{p-1})$. In the case of $p^\rho=2^3$, two orbits are $\{8,9,11\}$, and $\{10,12,13\}$. Corresponding to dropping to the bottom of a column in a $(p,p^\rho-1)$ table like Table 5.2, remove all factors of p. The orbits in the example then become $\{1,9,11\}$, and $\{3,5,13\}$. The two triples contain the numbers in the previously mysterious borders of Table 4.2, where one has to divide by 7 to get the hidden slopes for $[2]_7/\mathbb{Q}_2$. The general recipe involves dividing by $p^\rho-1$ at the end. In the example of $[\frac{3}{2}]_5/\mathbb{Q}_3$ from §4.2, $p^r=3^4$ and the orbit $\{56,88,104,72\}$ becomes, via the drop $72=2^33^2\to 8$, the hidden slopes $\frac{7}{10}$, $\frac{11}{10}$, $\frac{13}{10}$, and $\frac{1}{10}$ mysteriously appearing there. In general, the relatively elementary nature of $\mathrm{Gal}(K^{\mathrm{prim}}/K)$ makes us hopeful that the explicit description of packets in the families of §4.2, §4.3, and §4.4 will ultimately be specializations of the same uniform description for any $[s,\ldots,s]_\epsilon^f$.

As a second example of a nonabelian quotient with a known filtration, consider the group $\mathrm{Gal}(K^{\mathrm{nil}}/K)$ defined just before (5.1) and let $\mathrm{Gal}(K^{\mathrm{nil},p}/K)$ be its maximal quotient of nilpotency class $p-1$ and exponent p. In the strongest result of its type, Abrashkin [**Abr17**] has identified the filtration on $\mathrm{Gal}(K^{\mathrm{nil},p}/K)$ under the assumption that K contains a primitive pth root of unity. For $p \geq 3$, Abrashkin's result, when it applies, goes well past the local class field theory of the previous subsection. However it leaves a lot of cases uncovered, as for example the Sylow p-subgroup of S_{p^2} already has nilpotency class p. Thus for most of the families in the database, there is not yet a theoretical description of packets, even in principle.

5.4. Comparing families I/K for varying K. For s a real number and K a finite degree subfield of $\mathbb{Q}_p^{\mathrm{sep}}$, let $K^s \subset \mathbb{Q}_p^{\mathrm{sep}}$ be the union of subextensions L/K with all relative slopes less than s. Say that two such fields K and K' are j-close if there is an isomorphism of finite rings $\mathcal{O}_K/\Pi_K^j \to \mathcal{O}_{K'}/\Pi_{K'}^j$. As an example, take K' to have the same residual cardinality q as K, and with ramification index at least e_K. Then K and K' are e_K-close, because both finite rings are isomorphic to $\mathbb{F}_q[t]/t^{e_K}$. Deligne proved in [**Del84**, Th. 2.8] that if K and K' are j-close then $\mathrm{Gal}(K^j/K)$ and $\mathrm{Gal}(K'^j/K')$ are isomorphic as filtered groups. Thus many instances of the problem of describing filtered Galois groups have the same answer, even if we do not as yet know the answer.

To get a particular isomorphism between the Galois groups, well-defined up to conjugation, one needs to choose a particular isomorphism between the finite rings. Suppose the choices of uniformizers and residue representatives made in §3.1 are compatible with this ring isomorphism. Then the extensions L/K and L'/K' given by the same specialization of the generic polynomial correspond. While Deligne's proof is complicated, he presents its basic idea in [**Del84**, §1.3] as being this correspondence of Eisenstein polynomials. In terms of the database, the relative parts of the family pages for I/K and I/K' look extremely similar, for any I with top slope less than j.

As one of the simplest possible examples, consider the Herbrand invariant $I = [\frac{1}{3}, \frac{1}{3}]$ over 2-adic fields. Its generic polynomial is $\pi + \pi a_1 x + x^4$. Restricting attention to K with residual cardinality 2, there is just one field L/K in the family I/K, the one obtained by setting $a_1 = 1$. For $\mathbb{Q}_2$, the field $L_0 = \mathbb{Q}_2[x]/(2 + 2x + x^4)$ has associated Galois group S_4 with hidden slope content $[\;]_3^2$. So for any K, the relative Galois group is likewise S_4 and the relative hidden slope content is likewise $[\;]_3^2$. When one passes to absolute invariants, one naturally gets strong dependence on K. For example, taking K to be one of the six ramified quadratic extensions of $\mathbb{Q}_2$ yields trivial behavior: L is the compositum $K \otimes L_0$ with Herbrand invariant $[\frac{1}{3}, \frac{1}{3}, s]$, Galois group $S_4 \times C_2$, and hidden slope content $[\;]_3^2$. Here $s \in \{1, 2\}$ comes from the Herbrand invariant $[s]$ of $K/\mathbb{Q}_2$. The unique ramified cubic extension $K = \mathbb{Q}_2[x]/(2 - x^3)$ of $\mathbb{Q}_2$ yields completely different behavior. Here the unique field L is `2.1.12.12a1.1` with Galois group $C_2^6.C_9.C_6$ and slope content $[\frac{1}{9}, \frac{1}{9}, \frac{1}{9}, \frac{1}{9}, \frac{1}{9}, \frac{1}{9}]_9^6$.

As a more complicated example, take $I = [\frac{1}{2}, \frac{2}{3}]$ with generic polynomial $\pi(1 + a_3 x^3 + a_5 x^5 + c_6 x^6) + x^9$. The page for `3.1.9.13b` $= [\frac{1}{2}, \frac{2}{3}]/\mathbb{Q}_3$ says that there are eight fields, as c_6 is required to be 0 in the parameter list (a_3, a_5, c_6) whenever $a_3 \neq a_5$. The parameters $(1, 2, 0)$ and $(2, 1, 0)$ give fields having the same Galois closure, with Galois group 9T18 and Galois slope content $[\frac{1}{2}, \frac{1}{2}, \frac{2}{3}]_2^2$. The remaining six fields from (a_3, a_3, c_6) all have Galois group $9T20$ and Galois slope content $[\frac{1}{2}, \frac{1}{2}, \frac{2}{3}]_2^3$, with the splitting field depending only on a_3. Since all slopes are less than one, Deligne's comparison theorem says that all the facts just summarized also hold over any K with residual cardinality 3. Going further, one can compute via resolvent constructions that the compositum $\mathbb{Q}_3^I$ has $\mathrm{Gal}(\mathbb{Q}_3^I/\mathbb{Q}_3)$ with order $2^4 3^8$ and slope content $[\frac{1}{2}, \frac{1}{2}, \frac{1}{2}, \frac{1}{2}, \frac{5}{9}, \frac{2}{3}, \frac{2}{3}]_2^6$. So any $\mathrm{Gal}(K^I/K)$ has the same structure as a filtered group.

To see §5.2 in the light of the comparison theorem, let $C_w^s(K)$ be the subset of $C_w(K)$ consisting of Herbrand invariants $[s_1, \ldots, s_w]$ with $s_w < s$. Deligne's theorem says that $C_w^s(K) = C_w^s(\mathbb{F}_p((t)))$ holds for $s = e_K$, but in the easy case $v = 0$ the explicit descriptions says it holds for the larger integer $s = pe_K/(p-1)$. It fails for even larger integers s, as one has entered the arithmetic regime. Similarly, consider $I = [s, \ldots, s]_\epsilon^f$. Then Deligne's theorem says the filtered group $\mathrm{Gal}(K^I/K)$ is isomorphic with its geometric analog $\mathrm{Gal}(\mathbb{F}_q((t))^I/\mathbb{F}_q((t)))$ if $s < e_K$. In the case $v = 0$, the explicit description of $\mathrm{Gal}(K^I/K)$ indicated in §5.3 says that the isomorphism holds for $s < pe_K/(p-1)$.

5.5. Canonical globalization. Our complete tabulation of degree sixteen extensions of $\mathbb{Q}_2$ shows that the number of fields with automorphism group of size 1, 2, 4, 8, and 16 is respectively 9080, 833736, 44752, 2292, and 251. These newly-determined numbers sum to the previously known total 890111. The complete tabulation also gives the corresponding counts within each family. For the most ramified family `2.1.16.79a` $= [2, 3, 4, 5]$, the numbers are 0, 63488, 968, 240, and 32, for a total of 67728.

The next step in our general approach to populating the database is to determine the Galois group G and the hidden slope content HSC for each field. This is an ongoing process: we have computed almost all the Galois groups using Doris' programs [**Dor20**], but identifying HSC is harder. We describe here a completed part that is of particular interest in terms of applications to number fields.

For r an odd prime, let $\mathbb{Q}^r \subset \mathbb{C}$ be the union of all finite degree Galois extensions of $\mathbb{Q}$ with Galois group having order a power of 2 and ramification within $\{\infty, 2, r\}$. Then a special case [**Koc02**, Example 11.18] of theorems of Koch gives two related very strong results on $\mathbb{Q}^r$ for $r \equiv 3, 5$ modulo 8. Let $D_\infty = \{1, c\}$ where c is complex conjugation. Let D_r be the corresponding decomposition group at the prime r, namely $D_r = \langle \tau, \sigma | \sigma\tau\sigma^{-1} = \tau^r \rangle$ understood in the category of pro-2-groups, so that D_r has a semidirect product structure $\mathbb{Z}_2 \rtimes \mathbb{Z}_2$. The first result is that $\mathrm{Gal}(\mathbb{Q}^r/\mathbb{Q})$ is the free product $D_\infty * D_r$ in the category of pro-2-groups. The second result is that its 2-decomposition subgroup D_2 is the entire global Galois group $\mathrm{Gal}(\mathbb{Q}^r/\mathbb{Q})$.

For $r \equiv 3, 5$ (8) fixed, Koch's result says that an extension $L/\mathbb{Q}_2$ with $G = \mathrm{Gal}(L^{\mathrm{gal}}/\mathbb{Q}_2)$ a 2-group globalizes either 0 or 1 times to a number field with discriminant $\pm 2^a r^b$ and Galois group G. Typically a given field has no globalizations, as the governing local group (5.1) has three generators and one relation, while the governing global group also has three generators, but now has two relations, $c^2 = 1$ and $\sigma\tau\sigma^{-1} = \tau^r$.

For w a positive integer, consider the subfield $\mathbb{Q}^{r,w}$ generated by the subfields of $\mathbb{Q}^r$ of degree dividing 2^w. Its r-decomposition group is a quotient $D_{r,w}$ of D_r having 2^{2w} elements and depending only on r mod 2^w. So the abstract group $\mathrm{Gal}(\mathbb{Q}^{r,w}/\mathbb{Q}) = D_\infty * D_{r,w}$ depends only on r mod 2^w. Our calculations here show that the globalizing 2-adic fields for 3 and 19 agree and the globalizing fields for 13 and 29 also agree. This extends an observation made in [**JR14**, §8] for octic 2-adic fields and we do not have a proof that it holds for all r and w.

Table 5.3 illustrates the frequency of globalization. It counts the fields that globalize exactly for the extra prime r being in the indicated subset R of $\{3, 5, 11, 13\}$. There are 1131 2-adic fields that globalize to fields with discriminant $\pm 2^a$, of which 274 have $a = 79$. So the bottom right entries, corresponding to $R = \{3, 5, 11, 13\}$, include many more always-globalizing fields.

	All 677795 fields			
	$\emptyset$	$\{3\}$	$\{11\}$	$\{3,11\}$
$\emptyset$	505520	9952	9952	61158
$\{5\}$	15072	0	0	0
$\{13\}$	15072	0	0	0
$\{5,13\}$	58076	0	0	2993

	Fields in $[2,3,4,5]/\mathbb{Q}_2$			
	$\emptyset$	$\{3\}$	$\{11\}$	$\{3,11\}$
$\emptyset$	50614	1888	1888	8734
$\{5\}$	256	0	0	0
$\{13\}$	256	0	0	0
$\{5,13\}$	3384	0	0	708

TABLE 5.3. Statistics of globalization for degree sixteen extensions $L/\mathbb{Q}_2$ with Galois group $\mathrm{Gal}(L^{\mathrm{gal}}/\mathbb{Q}_2)$ having 2-power order.

The purely-local relevance of globalization is that it allows easier mechanical computation of Galois groups and it facilitates the identification of hidden slopes. Both G and HSC are in the database for all the canonically globalizing fields just discussed. Generally speaking, the database is designed so that it can present partial results. As of this writing, the database shows 185 packets inside the family $[2, 3, 4, 5]$. The numbers appearing as hidden wild slopes so far are 2, 3, 3.5, 4, 4.25, 4.5, 4.75, 5.125, 5.25, 5.375, and 5.625.

Let $\mathbb{Q}_2^{\mathrm{nil},w}$ be the subfield of $\mathbb{Q}_2^{\mathrm{nil}}$ generated by subfields of degree 2^w, so that $\mathrm{Gal}(\mathbb{Q}_2^{\mathrm{nil},w}/\mathbb{Q}_2)$ is a finite quotient of the infinite group $\mathrm{Gal}(\mathbb{Q}_2^{\mathrm{nil}}/\mathbb{Q}_2)$ of (5.1). Write its order as 2^{j_w}. A group-theoretical calculation says that $(j_1, j_2, j_3, j_4) = (3, 8, 25, 204)$. Corresponding numbers for $\mathrm{Gal}(\mathbb{Q}^{r,w}/\mathbb{Q})$ begin independently of r, being $(3, 7, 18)$. For $w = 4$, there is dependence on r mod 8, with $p = 3$ and 5 yielding 97 and 101.

We have long known the twenty-five slopes appearing in compositum $\mathbb{Q}_2^{\mathrm{nil},3}$ of all nilpotent octic extensions of $\mathbb{Q}_2$. There are three -1's coming from the unramified octic extension of $\mathbb{Q}_2$, which is famously known not to globalize. The wild slopes are then

$$1, 1, 1, 1\tfrac{1}{2}, 1\tfrac{1}{2},\ 2, 2, 2, 2\tfrac{1}{2}, 2\tfrac{1}{2}, 2\tfrac{5}{8}, 2\tfrac{3}{4},\ 3, 3, 3\tfrac{1}{4}, 3\tfrac{1}{4}, 3\tfrac{3}{8}, 3\tfrac{1}{2}, 3\tfrac{1}{2}, 3\tfrac{3}{4}, 3\tfrac{3}{4},\ 4.$$

The eighteen slopes that survive to the quotient group $\mathrm{Gal}(\mathbb{Q}^{r,3}/\mathbb{Q})$ are given for $r = 3, 5$ in [**JR14**, §8].

The basic reason for constructing a large database is that many concrete facts about ramification are not yet known and seem resistant to theoretical investigation. To underscore that much is not known, we conclude by asking a very concrete question: *what are the* 204 *slopes of* $\mathrm{Gal}(\mathbb{Q}_2^{\mathrm{nil},4}/\mathbb{Q}_2)$*?* Basic theory says that four of them are -1, and the rest are in $[1, 5]$ with all denominators being powers of 2. Analyzing high-degree composita of the new degree sixteen fields in the database will give many of these 204 numbers. We expect that a complete answer to the question may be out of reach without further theoretical advances, but computational progress can be measured by the number of slopes found.

Acknowledgements

This project arose from a SQuaRE at the American Institute of Mathematics, and we are grateful for their hospitality. Roe was supported by Simons Foundation grant 550033. We appreciate the close reading by the three anonymous referees and also Eran Assaf, as their comments have improved the quality of the paper.

References

[ABM+15] Chad Awtrey, Brett Barkley, Nicole E. Miles, Christopher Shill, and Erin Strosnider, *Degree 12 2-adic fields with automorphism group of order 4*, Rocky Mountain J. Math. **45** (2015), no. 6, 1755–1764, DOI 10.1216/RMJ-2015-45-6-1755. MR3473152

[Abr17] Victor Abrashkin, *Groups of automorphisms of local fields of period p and nilpotent class < p, II*, Internat. J. Math. **28** (2017), no. 10, 1750066, 32, DOI 10.1142/S0129167X17500665. MR3699165

[Ama71] Shigeru Amano, *Eisenstein equations of degree p in a $\mathfrak{p}$-adic field*, J. Fac. Sci. Univ. Tokyo Sect. IA Math. **18** (1971), 1–21. MR308086

[Awt12] Chad Awtrey, *Dodecic 3-adic fields*, Int. J. Number Theory **8** (2012), no. 4, 933–944, DOI 10.1142/S1793042112500558. MR2926553

[BM83] Gregory Butler and John McKay, *The transitive groups of degree up to eleven*, Comm. Algebra **11** (1983), no. 8, 863–911, DOI 10.1080/00927878308822884. MR695893

[CHM98] John H. Conway, Alexander Hulpke, and John McKay, *On transitive permutation groups*, LMS J. Comput. Math. **1** (1998), 1–8, DOI 10.1112/S1461157000000115. MR1635715

[DCDM17] Ilaria Del Corso, Roberto Dvornicich, and Maurizio Monge, *On wild extensions of a p-adic field*, J. Number Theory **174** (2017), 322–342, DOI 10.1016/j.jnt.2016.10.006. MR3597393

[Del84] P. Deligne, *Les corps locaux de caractéristique p, limites de corps locaux de caractéristique* 0 (French), Representations of reductive groups over a local field, Travaux en Cours, Hermann, Paris, 1984, pp. 119–157. MR771673

[Dor20] Christopher Doris, *Computing the Galois group of a polynomial over a p-adic field*, Int. J. Number Theory **16** (2020), no. 8, 1767–1801, DOI 10.1142/S179304212050092X. MR4143684

[GP12] Christian Greve and Sebastian Pauli, *Ramification polygons, splitting fields, and Galois groups of Eisenstein polynomials*, Int. J. Number Theory **8** (2012), no. 6, 1401–1424, DOI 10.1142/S1793042112500832. MR2965757

[GRJK+] Jordi Guàrdia-Rubies, John W. Jones, Kevin Keating, Sebastian Pauli, David P. Roberts, and David Roe, *Distinguished defining polynomials for extensions of p-adic fields*, in preparation.

[Has02] Helmut Hasse, *Number theory*, Translated from the third (1969) German edition, Classics in Mathematics, Springer-Verlag, Berlin, 2002. Reprint of the 1980 English edition [Springer, Berlin; MR0562104 (81c:12001b)]; Edited and with a preface by Horst Günter Zimmer. MR1885791

[Hul05] Alexander Hulpke, *Constructing transitive permutation groups*, J. Symbolic Comput. **39** (2005), no. 1, 1–30, DOI 10.1016/j.jsc.2004.08.002. MR2168238

[JR04] John W. Jones and David P. Roberts, *Nonic 3-adic fields*, Algorithmic number theory, Lecture Notes in Comput. Sci., vol. 3076, Springer, Berlin, 2004, pp. 293–308, DOI 10.1007/978-3-540-24847-7_22. MR2137362

[JR06] John W. Jones and David P. Roberts, *A database of local fields*, J. Symbolic Comput. **41** (2006), no. 1, 80–97, DOI 10.1016/j.jsc.2005.09.003. MR2194887

[JR08] John W. Jones and David P. Roberts, *Octic 2-adic fields*, J. Number Theory **128** (2008), no. 6, 1410–1429, DOI 10.1016/j.jnt.2007.12.014. MR2419170

[JR14] John W. Jones and David P. Roberts, *A database of number fields*, LMS J. Comput. Math. **17** (2014), no. 1, 595–618, DOI 10.1112/S1461157014000424. MR3356048

[Koc02] Helmut Koch, *Galois theory of p-extensions*, Springer Monographs in Mathematics, Springer-Verlag, Berlin, 2002. With a foreword by I. R. Shafarevich; Translated from the 1970 German original by Franz Lemmermeyer; With a postscript by the author and Lemmermeyer, DOI 10.1007/978-3-662-04967-9. MR1930372

[Kra37] Marc Krasner, *Sur la primitivité des corps p-adiques*, Mathematica (Cluj) **13** (1937), 72–191.

[LMFDB] The LMFDB Collaboration, *The L-functions and modular forms database*, `https://www.lmfdb.org`, 2025, [Online; accessed 17 January 2025].

[Mik81] Hiroo Miki, *On the ramification numbers of cyclic p-extensions over local fields*, J. Reine Angew. Math. **328** (1981), 99–115, DOI 10.1515/crll.1981.328.99. MR636198

[Mon11] Maurizio Monge, *Determination of the number of isomorphism classes of extensions of a* $\mathfrak{p}$*-adic field*, J. Number Theory **131** (2011), no. 8, 1429–1434, DOI 10.1016/j.jnt.2011.02.004. MR2793885

[Mon14] Maurizio Monge, *A family of Eisenstein polynomials generating totally ramified extensions, identification of extensions and construction of class fields*, Int. J. Number Theory **10** (2014), no. 7, 1699–1727, DOI 10.1142/S1793042114500511. MR3256847

[NSW08] Jürgen Neukirch, Alexander Schmidt, and Kay Wingberg, *Cohomology of number fields*, 2nd ed., Grundlehren der mathematischen Wissenschaften [Fundamental Principles of Mathematical Sciences], vol. 323, Springer-Verlag, Berlin, 2008, DOI 10.1007/978-3-540-37889-1. MR2392026

[Pag22] C. Pagano, *Jump sets in local fields*, J. Algebra **593** (2022), 398–476, DOI 10.1016/j.jalgebra.2021.10.038. MR4349285

[Pan95] Peter Panayi, *Computation of Leopoldt's p-adic regulator*, 1995, Thesis (Ph.D.)–University of East Anglia.

[PR01] Sebastian Pauli and Xavier-François Roblot, *On the computation of all extensions of a p-adic field of a given degree*, Math. Comp. **70** (2001), no. 236, 1641–1659, DOI 10.1090/S0025-5718-01-01306-0. MR1836924

[PS17] Sebastian Pauli and Brian Sinclair, *Enumerating extensions of* (π)*-adic fields with given invariants*, Int. J. Number Theory **13** (2017), no. 8, 2007–2038, DOI 10.1142/S1793042117501081. MR3681687

[Roe19] David Roe, *The inverse Galois problem for p-adic fields*, Proceedings of the Thirteenth Algorithmic Number Theory Symposium, Open Book Ser., vol. 2, Math. Sci. Publ., Berkeley, CA, 2019, pp. 393–409. MR3952024

[Sch14] Peter Scholze, *Perfectoid spaces and their applications*, Proceedings of the International Congress of Mathematicians—Seoul 2014. Vol. II, Kyung Moon Sa, Seoul, 2014, pp. 461–486. MR3728623

[Ser78] Jean-Pierre Serre, *Une "formule de masse" pour les extensions totalement ramifiées de degré donné d'un corps local* (French, with English summary), C. R. Acad. Sci. Paris Sér. A-B **286** (1978), no. 22, A1031–A1036. MR500361

[Ser79] Jean-Pierre Serre, *Local fields*, Graduate Texts in Mathematics, vol. 67, Springer-Verlag, New York-Berlin, 1979. Translated from the French by Marvin Jay Greenberg. MR554237

Dept. Matemàtiques, Escola Politècnica d'Enginyeria Superior de Vilanova i la Geltrú, Universitat Politècnica de Catalunya-BarcelonaTech, Av. Víctor Balaguer 1, E–08800 Vilanova i la Geltrú, Catalonia
Email address: jordi.guardia-rubies@upc.edu

School of Mathematical and Statistical Sciences, Arizona State University, P.O. Box 871804, Tempe, AZ 85287
Email address: jj@asu.edu

Department of Mathematics, University of Florida, Gainesville, FL 32611
Email address: keating@ufl.edu

Department of Mathematics and Statistics, University of North Carolina Greensboro, 116 Petty Building Greensboro, NC 27402
Email address: s_pauli@uncg.edu

Division of Science and Mathematics, University of Minnesota Morris, Morris, MN 56267
Email address: roberts@morris.umn.edu

Department of Mathematics, Massachusetts Institute of Technology, 77 Massachusetts Avenue, Cambridge, MA 02139
Email address: roed@mit.edu

Contemporary Mathematics
Volume **840**, 2026
https://doi.org/10.1090/conm/840/16804

Sampling cubic rings

Fabian Gundlach

ABSTRACT. We explain how to construct a uniformly random cubic integral domain S of given signature with $|\mathrm{disc}(S)| \leq T$ in expected time $\widetilde{\mathcal{O}}(\log T)$.

1. Introduction

In the past decades, there has been an increasing amount of interest in the statistical properties of arithmetic objects such as number fields or orders in number fields.

In [**2**], Belabas gave an algorithm that computes a list of all cubic number fields K with $|\mathrm{disc}(K)| \leq T$ in time $\widetilde{\mathcal{O}}(T)$. One can similarly enumerate all cubic integral domains S (i.e., orders in cubic number fields) with $|\mathrm{disc}(S)| \leq T$ in time $\widetilde{\mathcal{O}}(T)$. The running time is essentially optimal since the number of such fields (or rings) is $\asymp T$ as was shown by Davenport and Heilbronn in [**9**].

In this paper, we give an algorithm that constructs a uniformly random cubic integral domain S of given signature with $|\mathrm{disc}(S)| \leq T$ in expected time $\widetilde{\mathcal{O}}(\log T)$. More precisely:

THEOREM 1.1 (cf. Corollary 3.9). *There is an algorithm which for given numbers $r \in \{1,3\}$ and*

$$T \geq \begin{cases} 49, & r = 3, \\ 23, & r = 1 \end{cases}$$

computes in expected time $\widetilde{\mathcal{O}}(\log T)$ a cubic integral domain S of signature r with $|\mathrm{disc}(S)| \leq T$ and such that all such rings occur with the same probability.

Like [**2**] and [**9**], we use Levi's parametrization of cubic rings by $\mathrm{GL}_2(\mathbb{Z})$-orbits of integral binary cubic forms. (See [**12**].)

The running time $\widetilde{\mathcal{O}}(\log T)$ of the algorithm referred to in Theorem 1.1 is essentially optimal since the input has $\log T$ bits. (Moreover, the smallest possible total number of bits in the coefficients of a cubic form corresponding to such a random ring S is on average $\asymp \log T$.)

The lower bound on T in Theorem 1.1 ensures that there is at least one cubic integral domain S of signature r with $|\mathrm{disc}(S)| \leq T$.

2020 *Mathematics Subject Classification.* 11Y40, 11R16.

This work was supported by the Deutsche Forschungsgemeinschaft (DFG, German Research Foundation) — Project-ID 491392403 — TRR 358, project A4.

The most straightforward idea (used in [**2**] and [**9**]) for enumerating or counting the orbits with bounded discriminant is to construct a fundamental domain for the action of $\mathrm{GL}_2(\mathbb{Z})$ on the space of binary cubic forms and to then enumerate or count the lattice points in this domain.

The unboundedness of the fundamental domain, i.e., the presence of (long and narrow) cusps presents an inconvenience when enumerating or counting the lattice points. To deal with this issue, Bhargava introduced an elegant approach, which he calls "averaging over fundamental domains" or "thickening and cutting off the cusps". Very roughly speaking, he showed that it suffices to be able to estimate the number of points in linear transforms of a fixed region U of our choice! But it is relatively easy to count lattice points in a linear transform of a fixed (large) ball, at least as long as the linear transformation does not deform the ball too much. His method was for example used to obtain more precise counting results in [**7**].

In this paper, we adapt his method to the problem of selecting orbits uniformly at random. Very roughly speaking, it suffices to be able to pick uniformly random lattice points from linear transforms of a fixed region U of our choice and to have a good "uniform" upper bound on the number of lattice points in this transform. We use rejection sampling to ensure that all orbits occur with exactly the same probability. The quality of the uniform upper bound on the number of lattice points is important for estimating the running time of the algorithm.

Instead of using Bhargava's method, one could try to use rejection sampling to directly pick a point uniformly at random from the fundamental domain described in [**2**]. This seems possible, but not easier than the approach described in this article.

Being able to choose cubic orders of large discriminant uniformly at random could be advantageous for statistical experiments compared to completely enumerating all cubic orders of bounded discriminant when faced with error terms that decay slowly compared to the main term. For example, in [**14**, section 5], Malle experimentally studied the 2-ranks of class groups of cubic number fields. To this end, he used the algorithm of Belabas [**2**] to enumerate all number fields of small discriminant and computed the class groups of a small sample of these number fields. As Malle pointed out, his data is not entirely clear since he could (with Belabas' algorithm) only consider a relatively small range of discriminants.

An implementation of our algorithm as a standalone program is provided. (See section 4.)

Our approach can be adapted to other arithmetic objects parameterized by prehomogeneous vector spaces such as the famous parameterization of ideal classes of quadratic rings by orbits of binary quadratic forms or the parameterizations given in [**3**], [**5**]. (See section 5.)

In [**10**, section 8], to investigate the 2-ranks of class groups of cubic number fields, Dummit and Voight explained how to construct a uniformly random cubic integral domain with height at most X. (Here, the *height* $H(S)$ of a cubic integral domain S is the height of the corresponding reduced binary cubic form, i.e., the maximum absolute value of its coefficients.) They pick an integral binary cubic form of height at most X uniformly at random, accept it if it is reduced and irreducible, and start over otherwise.

Constructing cubic integral domains of bounded height is not equivalent to constructing cubic integral domains of bounded discriminant, and the latter problem appears to be more difficult: we have $|\mathrm{disc}(S)| \ll H(S)^4$, but we do not conversely have $H(S)^4 \ll |\mathrm{disc}(S)|$ for all cubic integral domains S. In fact, for any $C > 0$, as $X \to \infty$, a positive proportion of cubic integral domains with $|\mathrm{disc}(S)| \le X$ do not satisfy $H(S)^4 \le CX$. (In essence, Dummit and Voight construct points "in the main body" of the fundamental domain and ignore points "in the cusps".)

Acknowledgements. The author is grateful for helpful conversations with Noam Elkies, Jürgen Klüners, and Anne-Edgar Wilke. Moreover, the author would like to thank the anonymous referees for their careful reading and helpful suggestions and John Voight for helpful remarks.

2. Preparations

It is easy to select an integer lattice point in an axis-parallel box with integer side lengths uniformly at random. This remains true if we apply a triangular linear transformation to the box:

LEMMA 2.1. *Consider an axis-parallel box* $I = I_1 \times \cdots \times I_n \subset \mathbb{R}^n$, *where* $I_1, \ldots, I_n$ *are half-open intervals, say* $I_i = [a_i, a_i + l_i)$, *of integer lengths* $l_1, \ldots, l_n \in \mathbb{Z}$. *Let* $M = (m_{ij})_{i,j}$ *be a lower-triangular unipotent matrix with inverse* $M^{-1} = (m'_{ij})_{i,j}$. *Then, the set* $\mathbb{Z}^n \cap MI$ *has size* $l_1 \cdots l_n$ *and the following algorithm selects an element* $v = (v_1, \ldots, v_n)$ *of* $\mathbb{Z}^n \cap MI$ *uniformly at random.*

Algorithm 2.1 Finding a random lattice point in a transformed box

for $i \leftarrow 1, \ldots, n$ **do**
  Pick an element δ_i of $\{0, \ldots, l_i - 1\}$ uniformly at random.
  $v_i \leftarrow \left\lceil a_i - \sum_{j<i} m'_{ij} v_j \right\rceil + \delta_i$
end for

PROOF. By definition, for any $v \in \mathbb{R}^n$, we have $v \in MI$ if and only if

$$\sum_j m'_{ij} v_j \in [a_i, a_i + l_i) \qquad \text{for all } i.$$

The inverse matrix M^{-1} is also lower-triangular and unipotent, so the sum on the left-hand side is $v_i + \sum_{j<i} m'_{ij} v_j$.

Hence, $v \in MI$ if and only if

$$a_i - \sum_{j<i} m'_{ij} v_j \le v_i < a_i - \sum_{j<i} m'_{ij} v_j + l_i \qquad \text{for all } i.$$

The integer solutions v_i to these inequalities are

$$v_i = \left\lceil a_i - \sum_{j<i} m'_{ij} v_j \right\rceil + \delta_i \qquad \text{for } \delta_i \in \{0, \ldots, l_i - 1\}. \qquad \square$$

Our algorithm will work with approximations of random real numbers and we will show that a given precision suffices with large probability to decide whether a particular polynomial inequality $f(x) > 0$ holds. The following lemma will help with this analysis. (Here, ε will be the precision to which we have computed $f(x)$.)

LEMMA 2.2. *Let* $n \ge 1$. *For any monic polynomial* $f \in \mathbb{R}[X]$ *of degree* n *and any* $\varepsilon > 0$, *the set of* $x \in \mathbb{R}$ *with* $|f(x)| \le \varepsilon$ *has measure at most* $2n\varepsilon^{1/n}$.

PROOF. Let $r_1, \dots, r_n$ be the complex roots of f, so that $f(X) = (X - r_1) \cdots (X - r_n)$. Any $x \in \mathbb{R}$ with $|f(x)| \leq \varepsilon$ must have distance at most $\varepsilon^{1/n}$ from one of the roots r_i. For any given root r_i, the set of $x \in \mathbb{R}$ with $|x - r_i| \leq \varepsilon^{1/n}$ forms an interval of length at most $2\varepsilon^{1/n}$. The claim follows by summing over all roots. □

3. Cubic rings

In this section, we describe an algorithm to construct random cubic rings.

We first remind the reader of some basic definitions. A *cubic ring* is a (commutative unitary) ring which is isomorphic to $\mathbb{Z}^3$ as a $\mathbb{Z}$-module. A *cubic integral domain* is a cubic ring which is an integral domain. They are exactly the orders in cubic number fields. The *signature* of a cubic integral domain is the number of real embeddings of its field of fractions.

3.1. Levi's parameterization. In this subsection, we recall Levi's parameterization of cubic rings by binary cubic forms.

For any ring R, let $\mathcal{V}(R)$ denote the set of binary cubic forms $f = aX^3 + bX^2Y + cXY^2 + dY^3$ with $a, b, c, d \in R$. We can naturally identify $\mathcal{V}(R)$ with R^4 by identifying a cubic form f with its coefficient vector (a, b, c, d).

Define an action of $\mathrm{GL}_2(R)$ on $\mathcal{V}(R)$ by

$$(Mf)(v) = \det(M)^{-1} \cdot f(M^T v) \qquad \text{for } M \in \mathrm{GL}_2(R) text and f \in \mathcal{V}(R) \text{ and } v \in R^2.$$

The discriminant $\mathrm{disc}(f)$ is a homogeneous degree 4 polynomial in the coefficients a, b, c, d. It is invariant under the action of the group

$$\mathrm{SL}_2^{\pm}(R) = \{g \in \mathrm{GL}_2(R) : \det(g) = \pm 1\}.$$

In [**12**], Levi described a bijection

$$\mathrm{GL}_2(\mathbb{Z}) \backslash \mathcal{V}(\mathbb{Z}) \longleftrightarrow \{\text{cubic rings}\}$$

with the following properties:

a) If an orbit $\mathrm{GL}_2(\mathbb{Z})f$ corresponds to the cubic ring S, then $\mathrm{disc}(f) = \mathrm{disc}(S)$ and there is a group isomorphism $\mathrm{Stab}_{\mathrm{GL}_2(\mathbb{Z})}(f) \cong \mathrm{Aut}(S)$, where we denote by $\mathrm{Aut}(S)$ the automorphism group of the ring S.
b) The *irreducible orbits* (orbits consisting of cubic forms f that are irreducible over $\mathbb{Q}$) correspond to the cubic integral domains. If an irreducible orbit $\mathrm{GL}_2(\mathbb{Z})f$ corresponds to the cubic integral domain S, then the signature of S is the number of roots of f in $\mathbb{P}^1(\mathbb{R})$.
c) Concretely, the cubic ring S corresponding to an orbit $\mathrm{GL}_2(\mathbb{Z})f$ with $f = aX^3 + bX^2Y + cXY^2 + dY^3$ has a $\mathbb{Z}$-basis of the form $(1, \omega_1, \omega_2)$ with

$$\begin{aligned} \omega_1 \cdot \omega_2 &= -ad \cdot 1, \\ \omega_1 \cdot \omega_1 &= -ac \cdot 1 - b \cdot \omega_1 + a \cdot \omega_2, \\ \omega_2 \cdot \omega_2 &= -bd \cdot 1 - d \cdot \omega_1 + c \cdot \omega_2. \end{aligned}$$

3.2. The Siegel set and Gauss' fundamental domain. We next describe Gauss' well-known fundamental domain for the action of $\mathrm{GL}_2(\mathbb{Z})$ on $\mathrm{SL}_2^{\pm}(\mathbb{R})$.

Any element g of $\mathrm{SL}_2^{\pm}(\mathbb{R})$ can be written uniquely as a product nak with

$$n \in N(\mathbb{R}) = \left\{ \begin{pmatrix} 1 & \\ t & 1 \end{pmatrix} : t \in \mathbb{R} \right\}, \quad a \in A(\mathbb{R}) = \left\{ \begin{pmatrix} s^{-1} & \\ & s \end{pmatrix} : s \in \mathbb{R}^\times \right\}, \quad k \in O_2(\mathbb{R}).$$

In fact, the resulting bijection $N(\mathbb{R})\times A(\mathbb{R})\times O_2(\mathbb{R}) \leftrightarrow \mathrm{SL}_2^{\pm}(\mathbb{R})$ is a diffeomorphism. Let $\mathrm{d}t, \mathrm{d}^\times s, \mathrm{d}^\times k$ be Haar measures on the groups $N(\mathbb{R}) \cong \mathbb{R}$, $A(\mathbb{R}) \cong \mathbb{R}^\times$, and $O_2(\mathbb{R})$, respectively. Then, the pushforward of the measure $\mathrm{d}t \cdot s^{-2}\mathrm{d}^\times s \cdot \mathrm{d}^\times k$ is a Haar measure $\mathrm{d}^\times g$ on $\mathrm{SL}_2^{\pm}(\mathbb{R})$.

Matrices of the form

$$n(t) := \begin{pmatrix} 1 & \\ t & 1 \end{pmatrix} \qquad \text{or} \qquad a(s) := \begin{pmatrix} s^{-1} & \\ & s \end{pmatrix}$$

act on $\mathcal{V}(\mathbb{R}) \cong \mathbb{R}^4$ as

$$\begin{pmatrix} 1 & & & \\ 3t & 1 & & \\ 3t^2 & 2t & 1 & \\ t^3 & t^2 & t & 1 \end{pmatrix} \qquad \text{or} \qquad \begin{pmatrix} s^{-3} & & & \\ & s^{-1} & & \\ & & s & \\ & & & s^3 \end{pmatrix}.$$

Let $s_{\min} := \sqrt{\sqrt{3}/2}$. The *Siegel set*

$$\mathcal{F} := \{n(t)a(s)k : |t| \leq \tfrac{1}{2},\ s \geq s_{\min},\ k \in O_2(\mathbb{R})\} \subset \mathrm{SL}_2^{\pm}(\mathbb{R}) \tag{3.1}$$

contains the famous fundamental domain

$$\mathcal{F}' := \{n(t)a(s)k : |t| \leq \tfrac{1}{2},\ s \geq \sqrt[4]{1-t^2},\ k \in O_2(\mathbb{R})\} \subset \mathrm{SL}_2^{\pm}(\mathbb{R}). \tag{3.2}$$

This set $\mathcal{F}'$ (called the *Gauss set*) is a fundamental domain for the left action of $\mathrm{GL}_2(\mathbb{Z})$ on $\mathrm{SL}_2^{\pm}(\mathbb{R})$. More precisely, we have

$$\sum_{h \in \mathrm{GL}_2(\mathbb{Z})} \alpha(hg) = 4 \text{ for all } g \in \mathrm{SL}_2^{\pm}(\mathbb{R}),$$

where $\alpha(g) = \alpha(t,s)$ for $g = n(t)a(s)k$ with

$$\begin{aligned} &\alpha(g) = 0 && \text{if } g \text{ lies outside } \mathcal{F}', && \text{i.e., } |t| > \frac{1}{2} \text{ or } s < \sqrt[4]{1-t^2}, \\ &\alpha(g) = 1 && \text{if } g \text{ lies in the interior of } \mathcal{F}', && \text{i.e., } |t| < \frac{1}{2} \text{ and } s > \sqrt[4]{1-t^2}, \\ &\alpha(g) \in [0,1] && \text{always.} \end{aligned} \tag{3.3}$$

(For simplicity, we don't give the formulas for $\alpha(g)$ on the measure 0 boundary of $\mathcal{F}'$. The sum is 4 because for any matrix g in $\mathcal{F}'$, the matrices $\left(\begin{smallmatrix} \pm 1 & \\ & \pm 1 \end{smallmatrix}\right) g$ also lie in $\mathcal{F}'$.)

3.3. Bhargava's averaging trick. We now fix a signature $r \in \{1,3\}$ and let $\mathcal{V}^r(\mathbb{R}) \subset \mathcal{V}(\mathbb{R})$ be the set of squarefree cubic forms with exactly r real roots. Let $\mathcal{W}^r(\mathbb{R}) \subset \mathcal{V}^r(\mathbb{R})$ be the subset consisting of those cubic forms f with $|\mathrm{disc}(f)| = 1$. The group $\mathrm{SL}_2^{\pm}(\mathbb{R})$ acts transitively on $\mathcal{W}^r(\mathbb{R})$. The set $\mathcal{W}^r(\mathbb{R})$ is a 3-dimensional smooth submanifold of $\mathcal{V}(\mathbb{R})$. It follows that the pushforward of $\mathrm{d}^\times g$ along the map $\mathrm{SL}_2^{\pm}(\mathbb{R}) \to \mathcal{W}^r(\mathbb{R})$, $g \mapsto g^{-1}f_0$, is independent of the choice of $f_0 \in \mathcal{W}^r(\mathbb{R})$. Denote this pushforward by $\mathrm{d}f$. Bhargava's averaging over fundamental domains / thickening of cusps trick (see for example [**4**, section 2.2] or [**7**, section 5.3]) relies on the following lemma:

LEMMA 3.1. *Fix an integrable subset $U \subseteq \mathcal{W}^r(\mathbb{R})$. We define the function η on $\mathcal{W}^r(\mathbb{R})$ to be the following average of the indicator functions of linear transforms* *of U by elements g of the fundamental domain $\mathcal{F}'$:*

$$\eta(f) := \int_{\mathcal{F}'} \chi_{gU}(f)\mathrm{d}^\times g$$

and let

$$C := \int_{\mathcal{W}^r(\mathbb{R})} \chi_U(f)\mathrm{d}f.$$

We then have for all $f \in \mathcal{W}^r(\mathbb{R})$:

$$\sum_{h\in\mathrm{GL}_2(\mathbb{Z})} \eta(hf) = 4C.$$

PROOF. We have

$$\begin{aligned}
\sum_{h\in\mathrm{GL}_2(\mathbb{Z})} \eta(hf) &= \sum_{h\in\mathrm{GL}_2(\mathbb{Z})} \int_{\mathcal{F}'} \chi_{gU}(hf)\mathrm{d}^\times g \\
&= \sum_{h\in\mathrm{GL}_2(\mathbb{Z})} \int_{\mathrm{SL}_2^\pm(\mathbb{R})} \alpha(g)\chi_{gU}(hf)\mathrm{d}^\times g \\
&= \sum_{h\in\mathrm{GL}_2(\mathbb{Z})} \int_{\mathrm{SL}_2^\pm(\mathbb{R})} \alpha(g)\chi_{h^{-1}gU}(f)\mathrm{d}^\times g \\
&= \sum_{h\in\mathrm{GL}_2(\mathbb{Z})} \int_{\mathrm{SL}_2^\pm(\mathbb{R})} \alpha(hg)\chi_{gU}(f)\mathrm{d}^\times g \\
&= 4\int_{\mathrm{SL}_2^\pm(\mathbb{R})} \chi_{gU}(f)\mathrm{d}^\times g = 4\int_{\mathrm{SL}_2^\pm(\mathbb{R})} \chi_U(g^{-1}f)\mathrm{d}^\times g \\
&= 4\int_{\mathcal{W}^r(\mathbb{R})} \chi_U(f)\mathrm{d}f = 4C.
\end{aligned}$$

□

3.4. A ball and its $\mathrm{GL}_2(\mathbb{R})$-transforms. Let $\|\cdot\|$ be the norm on $\mathcal{V}(\mathbb{R})$ associated to the following positive definite quadratic form q on $\mathcal{V}(\mathbb{R})$:

$$q(aX^3 + bX^2Y + cXY^2 + dY^3) = 5a^2 + b^2 + c^2 + 5d^2 + 2ac + 2bd.$$

LEMMA 3.2. *The quadratic form q is positive definite and invariant under the action of the orthogonal group $O_2(\mathbb{R}) \subset \mathrm{GL}_2(\mathbb{R})$.*

PROOF. One can check that for any $f \in \mathcal{V}(\mathbb{R})$,

$$\int_{\{(x,y)\in\mathbb{R}^2 : x^2+y^2\leq 1\}} f(x,y)^2\mathrm{d}(x,y) = \frac{\pi}{64}\cdot q(f).$$

The left-hand side is clearly positive definite and $O_2(\mathbb{R})$-invariant. □

LEMMA 3.3. *Consider the closed unit ball of radius 1 in $\mathcal{V}(\mathbb{R})$. The maximum values of $|a|, |b|, |c|, |d|$ on this set are $\frac{1}{2}, \frac{\sqrt{5}}{2}, \frac{\sqrt{5}}{2}, \frac{1}{2}$, respectively.*

PROOF. To achieve the maximum value of $|a|$, for instance, the derivatives of $q(f)$ with respect to b, c, d need to vanish. One easily checks that this occurs exactly at the two points $\pm\frac{1}{2}(X^3 - XY^2)$.

The maximum for $|b|$ occurs at $\pm\frac{\sqrt{5}}{2}(X^2Y - \frac{1}{5}Y^3)$. The maxima for $|c|$ and $|d|$ follow by symmetry. □

Consider the open unit ball in $\mathcal{V}(\mathbb{R})$:

$$B = \{f \in \mathcal{V}(\mathbb{R}) : \|f\| < 1\}$$

By Lemma 3.3, B is contained in the box $I := I_1 \times \cdots \times I_4$ given by

$$I_1 := \{a \in \mathbb{R} : |a| \le \tfrac{1}{2}\}, \qquad I_2 := \{b \in \mathbb{R} : |b| \le \tfrac{\sqrt{5}}{2}\},$$
$$I_3 := \{c \in \mathbb{R} : |c| \le \tfrac{\sqrt{5}}{2}\}, \qquad I_4 := \{d \in \mathbb{R} : |d| \le \tfrac{1}{2}\}.$$

Hence, for any $\lambda, s > 0$, the set $\lambda \cdot a(s)B$ is contained in the box $\lambda \cdot a(s)I$ with side lengths

$$l_1(\lambda, s) := \lambda s^{-3}, \qquad l_2(\lambda, s) := \sqrt{5}\lambda s^{-1},$$
$$l_3(\lambda, s) := \sqrt{5}\lambda s, \qquad l_4(\lambda, s) := \lambda s^3.$$

The box $\lambda \cdot a(s)I$ is in turn contained in the box $I'(\lambda, s) = I'_1(\lambda, s) \times \cdots \times I'_4(\lambda, s)$ with integer side lengths $l'_i(\lambda, s)$, where

$$I'_i(\lambda, s) := \left[-\tfrac{1}{2}l'_i(\lambda, s), \tfrac{1}{2}l'_i(\lambda, s)\right) \text{ with } l'_i(\lambda, s) := \lfloor 1 + l_i(\lambda, s) \rfloor \in \mathbb{Z} \text{ for } i = 1, \dots, 4.$$

Let $s_{\max} = \sqrt[3]{\lambda/2}$. For $s_{\min} \le s \le s_{\max}$, we can bound the side lengths $l'_i(\lambda, s)$ from above as follows:

$$\begin{aligned} l'_1(\lambda, s) &\le L'_1(\lambda) \cdot s^{-3}, & l'_2(\lambda, s) &\le L'_2(\lambda) \cdot s^{-1}, \\ l'_3(\lambda, s) &\le L'_3(\lambda) \cdot s, & l'_4(\lambda, s) &\le L'_4(\lambda) \cdot s^3, \end{aligned} \tag{3.4}$$

where

$$L'_1(\lambda) := s_{\max}^3 + \lambda, \qquad L'_2(\lambda) := s_{\max} + \sqrt{5}\lambda,$$
$$L'_3(\lambda) := s_{\min}^{-1} + \sqrt{5}\lambda, \qquad L'_4(\lambda) := s_{\min}^{-3} + \lambda.$$

3.5. The high-level algorithm. Fix a radius $R > 0$ large enough so that $R \cdot B \cap \mathcal{W}^r(\mathbb{R}) \ne \emptyset$. (For example, you can take $R = 7/4$ for $r = 1$ and $R = 5/4$ for $r = 3$.)

We now give an algorithm for computing random irreducible $\mathrm{GL}_2(\mathbb{Z})$-orbits. This first algorithm uses real arithmetic and we later explain how to deal with precision issues.

THEOREM 3.4. *Let $T \ge 1$ and $\lambda = RT^{1/4}$. Let $s_{\min} = \sqrt{\sqrt{3}/2}$ and $s_{\max} = \sqrt[3]{\lambda/2}$.*

a) *The following algorithm either fails or produces an element f of an irreducible $\mathrm{GL}_2(\mathbb{Z})$-orbit in $\mathcal{V}^r(\mathbb{Z})$ with $0 < |\mathrm{disc}(f)| \le T$.*
b) *The probability of returning an element of any given such orbit $\mathrm{GL}_2(\mathbb{Z})f$ is (for fixed r and T) proportional to $1/\#\mathrm{Stab}_{\mathrm{GL}_2(\mathbb{Z})}(f)$.*
c) *The probability of success is $> p_0$ for a constant p_0 (depending only on r and R, but not on T) if there is at least one such orbit $\mathrm{GL}_2(\mathbb{Z})f$.*

Algorithm 3.1 Finding a random orbit of cubic forms

1: Pick an element t of $(-\frac{1}{2}, \frac{1}{2})$ uniformly at random.
2: Pick an element s of $(s_{\min}, \infty)$ at random with probability measure proportional to $s^{-2}\mathrm{d}^\times s$.
3: **return** FAIL **unless** $\sqrt[4]{1-t^2} < s < s_{\max}$.
4: **return** FAIL **with probability** $1 - \frac{l'_1(\lambda,s)\cdots l'_4(\lambda,s)}{L'_1(\lambda)\cdots L'_4(\lambda)}$.

5: Pick a point $f \in \mathcal{V}(\mathbb{Z}) \cap n(t)I'(\lambda, s)$ uniformly at random using Algorithm 2.1.
6: **return** FAIL **unless** $f \in \mathcal{V}^r(\mathbb{R})$ and $0 < |\mathrm{disc}(f)| \leq T$ and f is irreducible over $\mathbb{Q}$ and $f \in R|\mathrm{disc}(f)|^{1/4} \cdot n(t)a(s)B$.
7: **return** f.

REMARK 3.5. Repeat the algorithm until it succeeds. According to c), if there is such an orbit, it will succeed after at most $1/p_0$ attempts on average.

REMARK 3.6. The smallest absolute discriminant of an irreducible orbit with signature $r = 3$ is 49. The smallest absolute discriminant of an irreducible orbit with signature $r = 1$ is 23. (See for example tables of number fields of degree 3 and small discriminant, such as the one available at [**13**].)

REMARK 3.7. The following algorithm satisfies a) and c), but every orbit has the same probability of being returned: Compute a random orbit $\mathrm{GL}_2(\mathbb{Z})f$ using Algorithm 3.1. Return this orbit $\mathrm{GL}_2(\mathbb{Z})f$ with probability $\#\,\mathrm{Stab}_{\mathrm{GL}_2(\mathbb{Z})}(f)/3$ and return FAIL otherwise. (If f corresponds to the cubic integral domain S with field of fractions K, the group $\mathrm{Stab}_{\mathrm{GL}_2(\mathbb{Z})}(f) \cong \mathrm{Aut}(S) \subseteq \mathrm{Aut}(K)$ is either trivial or cyclic of order 3.)

PROOF OF THEOREM 3.4. For $s_{\min} \leq s \leq s_{\max}$, we have

$$0 \leq \frac{l'_1(\lambda, s) \cdots l'_4(\lambda, s)}{L'_1(\lambda) \cdots L'_4(\lambda)} \leq 1,$$

according to (3.4), so line 4 of the algorithm makes sense. Claim a) is clear.

Now, consider any irreducible cubic form $f \in \mathcal{V}^r(\mathbb{Z})$ with $0 < |\mathrm{disc}(f)| \leq T$.

Since $|t| \leq \frac{1}{2}$, we have $s_{\min} = \sqrt{\sqrt{3}/2} \leq \sqrt[4]{1 - t^2}$. The algorithm returns the cubic form f with probability

$$P(f) := \frac{\int_{-1/2}^{1/2} \int_{\sqrt[4]{1-t^2}}^{s_{\max}} p(f, t, s) \cdot s^{-2} \mathrm{d}^\times s \, \mathrm{d}t}{\int_{-1/2}^{1/2} \int_{s_{\min}}^{\infty} s^{-2} \mathrm{d}^\times s \, \mathrm{d}t},$$

where

$$p(f, t, s) := \frac{l'_1(\lambda, s) \cdots l'_4(\lambda, s)}{L'_1(\lambda) \cdots L'_4(\lambda)} \cdot \frac{\chi_{n(t)I'(\lambda, s)}(f)}{\#(\mathcal{V}(\mathbb{Z}) \cap n(t)I'(\lambda, s))} \cdot \chi_{R|\mathrm{disc}(f)|^{1/4} \cdot n(t)a(s)B}(f).$$

We have $\#(\mathcal{V}(\mathbb{Z}) \cap n(t)I'(\lambda, s)) = l'_1(\lambda, s) \cdots l'_4(\lambda, s)$ by Lemma 2.1.

Since $0 < |\mathrm{disc}(f)| \leq T$ and $\lambda = RT^{1/4}$, we moreover have

$$R|\mathrm{disc}(f)|^{1/4} \cdot n(t)a(s)B \subseteq \lambda \cdot n(t)a(s)B \subseteq \lambda \cdot n(t)a(s)I \subseteq n(t)I'(\lambda, s). \tag{3.5}$$

Hence,

$$p(f, t, s) = \frac{\chi_{R|\mathrm{disc}(f)|^{1/4} \cdot n(t)a(s)B}(f)}{L'_1(\lambda) \cdots L'_4(\lambda)}.$$

Moreover, since $f = aX^3 + \cdots + dY^3$ is irreducible, we have $a \neq 0$, so $|a| \geq 1$. On the other hand, if $p(f, t, s) \neq 0$, then by (3.5), we have $f \in \lambda \cdot n(t)a(s)I$, which implies $|a| \leq \frac{1}{2}\lambda s^{-3}$. Therefore, if $p(f, t, s) \neq 0$, then $1 \leq \frac{1}{2}\lambda s^{-3}$, so $s \leq \sqrt[3]{\lambda/2} = s_{\max}$. (Bhargava calls this inequality "cutting off the cusp".) Hence, the upper bound on s in the integral in the numerator of $P(f)$ can be omitted without changing the value of the integral:

$$P(f) = \frac{\int_{-1/2}^{1/2}\int_{\sqrt[4]{1-t^2}}^{\infty} p(f,t,s)\cdot s^{-2}\mathrm{d}^\times s\,\mathrm{d}t}{\int_{-1/2}^{1/2}\int_{s_{\min}}^{\infty} s^{-2}\mathrm{d}^\times s\,\mathrm{d}t}.$$

We now multiply both the numerator and the denominator by the finite number $\int_{O_2(\mathbb{R})}\mathrm{d}^\times k$. Since B is $O_2(\mathbb{R})$-invariant and since the measure $\mathrm{d}t\cdot s^{-2}\mathrm{d}^\times s\cdot \mathrm{d}^\times k$ on $N(\mathbb{R})\times A(\mathbb{R})\times O_2(\mathbb{R})$ corresponds to the Haar measure $\mathrm{d}^\times g$ on $\mathrm{SL}_2^{\pm}(\mathbb{R})$, it follows together with the definitions of $\mathcal{F}$ and $\mathcal{F}'$ in (3.1) and (3.2) that

$$P(f) = \frac{\int_{\mathcal{F}'} p(f,g)\mathrm{d}^\times g}{\int_{\mathcal{F}}\mathrm{d}^\times g}$$

with

$$p(f,g) := \frac{\chi_{R|\mathrm{disc}(f)|^{1/4}\cdot gB}(f)}{L_1'(\lambda)\cdots L_4'(\lambda)}.$$

Note that $|\mathrm{disc}(|\mathrm{disc}(f)|^{-1/4}\cdot f)| = 1$ and therefore $|\mathrm{disc}(f)|^{-1/4}\cdot f\in\mathcal{W}^r(\mathbb{R})$ because $\mathrm{disc}(f)$ is a homogeneous polynomial of degree 4 in the coefficients of f. Setting $U := R\cdot B\cap\mathcal{W}^r(\mathbb{R})$, we can therefore rewrite $p(f,g)$ as

$$p(f,g) := \frac{\chi_{gU}(|\mathrm{disc}(f)|^{-1/4}\cdot f)}{L_1'(\lambda)\cdots L_4'(\lambda)},$$

so

$$P(f) = \frac{\int_{\mathcal{F}'}\chi_{gU}(|\mathrm{disc}(f)|^{-1/4}\cdot f)\cdot\mathrm{d}^\times g}{L_1'(\lambda)\cdots L_4'(\lambda)\cdot\int_{\mathcal{F}}\mathrm{d}^\times g}$$

By Lemma 3.1 applied to the element $|\mathrm{disc}(f)|^{-1/4}\cdot f$ of $\mathcal{W}^r(\mathbb{R})$, we obtain

$$\sum_{h\in\mathrm{GL}_2(\mathbb{Z})} P(hf) = D(\lambda) := \frac{C}{L_1'(\lambda)\cdots L_4'(\lambda)},$$

with

$$C = \frac{4\cdot\int_{\mathcal{W}^r(\mathbb{R})}\chi_U(f')\mathrm{d}f'}{\int_{\mathcal{F}}\mathrm{d}^\times g}.$$

The constant C is positive because U is a nonempty open subset of $\mathcal{W}^r(\mathbb{R})$ and $\int_{\mathcal{F}}\mathrm{d}^\times g<\infty$.

On the other hand,

$$\sum_{h\in\mathrm{GL}_2(\mathbb{Z})} P(hf) = \#\,\mathrm{Stab}_{\mathrm{GL}_2(\mathbb{Z})}(f)\cdot\sum_{f'\in\mathrm{GL}_2(\mathbb{Z})f} P(f').$$

We conclude that for any irreducible cubic form $f\in\mathcal{V}^r(\mathbb{Z})$ with $0<|\mathrm{disc}(f)|\le T$, we have the following probability of returning an element of its $\mathrm{GL}_2(\mathbb{Z})$-orbit:

$$\sum_{f'\in\mathrm{GL}_2(\mathbb{Z})f} P(f') = \frac{D(\lambda)}{\#\,\mathrm{Stab}_{\mathrm{GL}_2(\mathbb{Z})}(f)}.$$

Since $D(\lambda)$ is independent of f, this proves b).

For c), we sum over all $\mathrm{GL}_2(\mathbb{Z})$-orbits of irreducible cubic forms $f\in\mathcal{V}^r(\mathbb{Z})$ with $0<|\mathrm{disc}(f)|\le T$. The number of such orbits is $\asymp T$ for $T\to\infty$. (See [**9**].) In particular, it is $\gg T$ as long as there is at least one such orbit. Since $\mathrm{Stab}_{\mathrm{GL}_2(\mathbb{Z})}(f)$ has size at most 3 by Remark 3.7, the sum of $1/\#\,\mathrm{Stab}_{\mathrm{GL}_2(\mathbb{Z})}(f)$ over all orbits is also $\gg T$.

By definition, $L'_i(\lambda) \ll \lambda$ for all i, so $D(\lambda) \gg \lambda^{-4} \gg T^{-1}$.
It follows that the probability of success is

$$\sum_{\substack{f\in\mathcal{V}^r(\mathbb{Z})\\ \text{irreducible}\\ 0<|\mathrm{disc}(f)|\le T}} P(f) = \sum_{\substack{[f]\in\mathrm{GL}_2(\mathbb{Z})\backslash\mathcal{V}^r(\mathbb{Z})\\ \text{irreducible}\\ 0<|\mathrm{disc}(f)|\le T}} \sum_{f'\in\mathrm{GL}_2(\mathbb{Z})f} P(f')$$

$$= \sum_{\substack{[f]\in\mathrm{GL}_2(\mathbb{Z})\backslash\mathcal{V}^r(\mathbb{Z})\\ \text{irreducible}\\ 0<|\mathrm{disc}(f)|\le T}} \frac{D(\lambda)}{\#\,\mathrm{Stab}_{\mathrm{GL}_2(\mathbb{Z})}(f)} \gg T\cdot T^{-1} = 1.$$

This proves c). □

3.6. Finite precision real arithmetic. The above algorithm uses real arithmetic. In practice, we need to work with approximations of those real numbers to finite precision. Note that we need to be able to decide the inequalities appearing in the algorithm and to compute $\lfloor x\rfloor$ or $\lceil x\rceil$ for various real numbers x appearing in the algorithm. By analyzing the probability with which a given precision suffices to decide the inequalities and to compute the floor and ceiling values, we show:

THEOREM 3.8. *Algorithm* 3.1 *can be implemented on a random access machine (with a random bit generator) with expected running time* $\widetilde{\mathcal{O}}(\log T)$.

PROOF. We use the well-known method of doubling the precision until it suffices. See Algorithm 3.2 below for the concrete implementation.

We leave it to the reader to verify that Algorithm 3.2 is functionally equivalent to Algorithm 3.1. (Line 5 of Algorithm 3.2 corresponds to line 3 of Algorithm 3.1. Lines 10 and 12–14 correspond to line 5.)

We show that the probability that the algorithm doesn't finish within the first i-th iterations (due to insufficient precision) is $\mathcal{O}(T^v\varepsilon^u) = \mathcal{O}(T^v 2^{-u2^i})$ for constants $u, v > 0$ and that the running time of the i-th iteration is $\widetilde{\mathcal{O}}(\log T + p) = \widetilde{\mathcal{O}}(\log T + 2^i)$, where $p := 2^i$ and $\varepsilon := 2^{-i}$. The expected total running time is then

$$\begin{aligned}
&\ll \sum_{i\ge 1} \min(1, T^v 2^{-u2^i})\cdot \widetilde{\mathcal{O}}(\log T + 2^i)\\
&\ll \sum_{1\le i\le \lfloor \log_2(\frac{v}{u}\log_2 T)\rfloor} \widetilde{\mathcal{O}}(\log T + 2^i) + \sum_{i>\lfloor \log_2(\frac{v}{u}\log_2 T)\rfloor} T^v 2^{-u2^i}\cdot \widetilde{\mathcal{O}}(\log T + 2^i)\\
&\ll \widetilde{\mathcal{O}}(\log T) + \sum_{j\ge \lfloor \frac{v}{u}\log_2 T\rfloor} T^v 2^{-uj}\cdot \widetilde{\mathcal{O}}(\log T + j)\\
&\overset{(*)}{\ll} \widetilde{\mathcal{O}}(\log T) + \sum_{j'\ge 0} 2^{-uj'}\cdot \widetilde{\mathcal{O}}(\log T + j')\\
&\ll \widetilde{\mathcal{O}}(\log T)
\end{aligned}$$

as claimed, where the step marked by $(*)$ uses the substitution $j' := j - \lfloor \frac{v}{u}\log_2 T\rfloor$, noting that $2^{-uj'} \asymp T^v 2^{-uj}$ and $\lfloor \frac{v}{u}\log_2 T\rfloor \ll \log T$.

All real numbers computed in the algorithm are $\ll T^{\mathcal{O}(1)}$. It follows that the real numbers in the i-th iteration can be computed to absolute precision $\mathcal{O}(T^{\mathcal{O}(1)}\varepsilon)$ in time $\widetilde{\mathcal{O}}(\log T + p)$.

Now, we explain the failure probabilities and running times of the individual steps of the algorithm:

Line 5: The probability that the inequalities in line 5 cannot be decided from the given approximation is $\mathcal{O}(T^{\mathcal{O}(1)}\varepsilon)$ since both sides of the inequalities are known to an absolute precision of $\mathcal{O}(T^{\mathcal{O}(1)}\varepsilon)$ and the probability that the values are within this distance is $\mathcal{O}(T^{\mathcal{O}(1)}\varepsilon)$ as σ is uniformly distributed on $(0, 1)$.

Line 8: This works similarly.

Line 7: Being able to compute $l_i'(\lambda, s) = \lfloor 1 + l_i(\lambda, s) \rfloor$ from the approximation of $1+l_i(\lambda, s)$ is equivalent to $l_i(\lambda, s)$ not lying within a distance of $\mathcal{O}(T^{\mathcal{O}(1)}\varepsilon)$ from any integer $k > 0$. The closest integer to $l_i(\lambda, s)$ is $\ll T$, so we only need to consider $\ll T$ such integers k. We first bound the probability that $|l_4(\lambda, s)-k| = |\lambda s^3 - k| \ll T^{\mathcal{O}(1)}\varepsilon$. As $\lambda \asymp T^{\mathcal{O}(1)}$, this is equivalent to $|s^3 - \lambda^{-1}k| \ll T^{\mathcal{O}(1)}\varepsilon$. By Lemma 2.2, the set of such values $s \in \mathbb{R}$ has Lebesgue measure $\mathcal{O}(T^{\mathcal{O}(1)}\varepsilon^{1/3})$. As s is bounded from below (by $\sqrt{\sqrt{3}/2}$), the probability measure (proportional to $s^{-2}\mathrm{d}^\times s$) of this set of values s is also $\mathcal{O}(T^{\mathcal{O}(1)}\varepsilon^{1/3})$. Summing over the $\ll T$ values k, we see that the probability that the approximation is insufficient for computing $l_4(\lambda, s)$ is $\mathcal{O}(T^{\mathcal{O}(1)}\varepsilon^{1/3})$. A similar argument works for $l_3(\lambda, s)$. For $l_1(\lambda, s)$, we instead use that the inequality $|l_1(\lambda, s) - k| = |\lambda s^{-3} - k| \ll T^{\mathcal{O}(1)}\varepsilon$ is equivalent to $|s^3 - \lambda k^{-1}| \ll T^{\mathcal{O}(1)}\varepsilon$ since $1 \leq k \ll T$ and $1 \ll s \ll T^{\mathcal{O}(1)}$. We then proceed as before. The same argument works for $l_2(\lambda, s)$.

Line 9: Since $l_i'(\lambda, s) \ll T^{\mathcal{O}(1)}$ and Δ_i is chosen uniformly at random from $(0, 1)$, the probability that $\Delta_i \cdot l_i'(\lambda, s)$ lies within a distance $\mathcal{O}(T^{\mathcal{O}(1)}\varepsilon)$ from an integer is at most $\mathcal{O}(T^{\mathcal{O}(1)}\varepsilon)$.

Line 10: The number $\lceil -\frac{1}{2}l_1'(\lambda, s) \rceil$ can be computed with integer arithmetic since $l_1'(\lambda, s) \in \mathbb{Z}$.

Lines 12–14: We compute $\lceil x \rceil$ for various numbers x. Note that in each line, x is a polynomial in t of degree at most 3 whose leading coefficient has absolute value ≥ 1. By the same argument as for line 7, the probability that any of these values cannot be computed is $\mathcal{O}(T^{\mathcal{O}(1)}\varepsilon^{1/3})$.

Line 16: This involves only integer arithmetic.

Line 17: It suffices to check that $\operatorname{disc}(f) > 0$ if $r = 3$ and $\operatorname{disc}(f) < 0$ if $r = 1$.

Line 18: If

$$q(a(s)^{-1}n(-t)f) - R^2|\operatorname{disc}(f)|^{1/2} \ll T^{\mathcal{O}(1)}\varepsilon,$$

then

$$s^6 \cdot (q(a(s)^{-1}n(-t)f) - R^2|\operatorname{disc}(f)|^{1/2}) \ll T^{\mathcal{O}(1)}\varepsilon.$$

The left-hand side is a polynomial in s of degree 12 with leading coefficient $5a^2 \geq 5$. By Lemma 2.2, the Lebesgue measure of the set of values s satisfying the inequality is therefore $\mathcal{O}(T^{\mathcal{O}(1)}\varepsilon^{1/12})$. As $1 \ll s \ll T^{\mathcal{O}(1)}$, the probability measure of the corresponding set of values s is also $\mathcal{O}(T^{\mathcal{O}(1)}\varepsilon^{1/12})$.

Line 19: Note that a cubic form $f \in \mathcal{V}(\mathbb{Z})$ is reducible over $\mathbb{Q}$ if and only if it has a rational root. For $a \neq 0$, this is equivalent to the polynomial $f(X, 1) \in \mathbb{Z}[X]$ of degree 3 having a rational root. The denominator of such a root must divide $a \ll T$. Hence, it suffices to check whether

$f(X,1)$ has a root in $\frac{1}{a}\mathbb{Z}$. The real roots of $f(X,1)$ can be approximated with accuracy $\frac{1}{a}$ for example using a binary search and Sturm sequences in time $\widetilde{\mathcal{O}}(\log T)$. □

Algorithm 3.2 Finding a random orbit of cubic forms (using bit operations)

for $i \leftarrow 1, 2, \dots$ **do**

  Let $p = 2^i$ and $\varepsilon = 2^{-p}$. In this iteration, we will compute all occurring real numbers to absolute precision $\mathcal{O}(T^{\mathcal{O}(1)}\varepsilon)$. If the computed precision is insufficient to decide an occurring inequality or to compute an occurring floor or ceiling value, we immediately go to the next iteration, starting over with the next value of i.

  Pick uniformly random elements $\tau, \sigma, \pi, \Delta_1, \dots, \Delta_4$ of $(0,1)$ to absolute precision ε by picking the first p binary digits of each of the numbers, keeping any digits that were already picked in the previous iteration.

  Compute $t = \tau - \frac{1}{2}$.

  return FAIL **unless** $\frac{s_{\min}^2}{s_{\max}^2} < \sigma < \frac{s_{\min}^2}{\sqrt{1-t^2}}$.

  Compute $s = \frac{s_{\min}}{\sqrt{\sigma}}$.

  Compute $l_i'(\lambda, s) = \lfloor 1 + l_i(\lambda, s) \rfloor$ for $i = 1, \dots, 4$.

  return FAIL **if** $\pi > \frac{l_1'(\lambda,s)\cdots l_4'(\lambda,s)}{L_1'(\lambda)\cdots L_4'(\lambda)}$.

  Compute $\delta_i = \lfloor \Delta_i \cdot l_i'(\lambda, s) \rfloor$ for $i = 1, \dots, 4$.

  Compute $a = \lceil -\frac{1}{2} l_1'(\lambda, s) \rceil + \delta_1$.

  return FAIL **unless** $a \neq 0$.

  Compute $b = \lceil -\frac{1}{2} l_2'(\lambda, s) + 3ta \rceil + \delta_2$.

  Compute $c = \lceil -\frac{1}{2} l_3'(\lambda, s) - 3t^2 a + 2tb \rceil + \delta_3$.

  Compute $d = \lceil -\frac{1}{2} l_4'(\lambda, s) + t^3 a - t^2 b + tc \rceil + \delta_4$.

  Let $f = aX^3 + bX^2Y + cXY^2 + dY^3$.

  return FAIL **unless** $0 < |\mathrm{disc}(f)| \leq T$.

  return FAIL **unless** $f \in \mathcal{V}^r(\mathbb{R})$.

  return FAIL **unless** $q(a(s)^{-1} n(-t) f) < R^2 |\mathrm{disc}(f)|^{1/2}$.

  return FAIL **unless** f is irreducible over $\mathbb{Q}$.

  return f.

end for

COROLLARY 3.9 (cf. Theorem 1.1). *There are algorithms which for a given number $r \in \{1,3\}$ and T with*

$$T \geq \begin{cases} 49, & r = 3, \\ 23, & r = 1 \end{cases}$$

compute in expected time $\widetilde{\mathcal{O}}(\log T)$ a cubic integral domain S of signature r with $|\mathrm{disc}(S)| \leq T$ and such that

a) *the probability of returning any given such ring S is for fixed T proportional to $1/\#\mathrm{Aut}(S)$ or*
b) *all such rings occur with the same probability.*

(Here, cubic rings are represented by a corresponding binary cubic form.)

PROOF. Claim a) follows immediately from Theorems 3.4 and 3.8 and Remarks 3.5 and 3.6. For b), we also use Remark 3.7. □

REMARK 3.10. To compute a random cubic number field K with signature r and $|\operatorname{disc}(K)| \leq T$ (either uniformly or with probability proportional to $1/\#\operatorname{Aut}(K)$), one can compute cubic integral domains S until finding one which is the ring of integers of its field of fractions K. There are well-known algorithms for testing whether S is the ring of integers of its field of fractions, whose running time is dominated by the factorization of the integer $\operatorname{disc}(S)$. (See for example [**8**, section 6.1].)

In [**1**], Bach gave an algorithm that generates a uniformly random integer $1 \leq x \leq N$ together with its factorization. The expected running time of his algorithm is that required for $\mathcal{O}(\log N)$ primality tests of numbers $1 \leq p \leq N$. It could be an interesting problem to efficiently find a uniform random cubic integral domain together with the factorization of its discriminant.

REMARK 3.11. One of the referees raised the question whether it would be possible to generate cubic rings in narrow discriminant bands. There is indeed an algorithm similar to the one described above (but slightly more complicated) to generate a uniformly random cubic integral domain S of signature r with discriminant in $[T, T + T^{5/6+\varepsilon}]$ in expected time $\widetilde{\mathcal{O}}_\varepsilon(\log T)$. We briefly comment on the main differences:

The ball B of radius 1 centered at the origin should be replaced by a ball B' centered at a (fixed) point on $W_r(\mathbb{R})$ of radius $T^{-1/6+\varepsilon}$. In addition to $n(t)$ and $a(s)$, we need to first apply a random element k of $O_2(\mathbb{R})$ to this ball B' (as, unlike B, the ball B' is not $O_2(\mathbb{R})$-invariant). In Algorithm 3.1, we pick a random integral point in $n(t)I'(\lambda, s)$, where $I'(\lambda, s)$ is a box with integer side lengths containing $\lambda \cdot a(s)B$. We now need to instead pick a random integral point in $n(t)I''(\lambda, s, k)$, where $I''(\lambda, s, k)$ is a box with integer side lengths containing $[T^{1/4}, (T + T^{5/6+\varepsilon})^{1/4}] \cdot a(s)kB'$. The volume of this box is at most $C \cdot f(T, s)$ for some constant $C > 0$ and

$$f(T, s) := \begin{cases} T^{1/3+4\varepsilon} & \text{if } s^3 \leq T^{1/12+\varepsilon}, \\ T^{1/4+3\varepsilon} s^3 & \text{if } T^{1/12+\varepsilon} \leq s^3 \leq s_{\max}^3. \end{cases}$$

To make up for the fact that the volume of this box now very much depends on s, one needs to actually choose s at random with probability measure proportional to $f(T, s)s^{-2}\mathrm{d}^\times s$, and then change line 4 of Algorithm 3.1 to make up for the difference between this upper bound on the volume of the box and its actual volume.

(Even for ε slightly smaller than 0, the resulting algorithm would be correct, but the probability of success would go to 0 as $T \to \infty$. The lower bound 5/6 for the exponent of the width of the interval $[T, T + T^{5/6+\varepsilon}]$ corresponds to the second order term in the count of cubic integral domains [**7**].)

4. Implementation

An implementation of the algorithm described above is available at `https://github.com/fagu/random-orbits`. It makes use of the FLINT library [**15**] for large integer and polynomial arithmetic, and in particular for arbitrary-precision interval arithmetic [**11**].

Figure 1 gives the approximate average time it takes to generate one random cubic integral domain S with signature r and $|\operatorname{disc}(S)| \leq 2^t$, where the probability of obtaining a given ring S is proportional to $1/\#\operatorname{Aut}(S)$. For example, it takes roughly 10^{-3} seconds to generate a random ring with $r = 3$ and $|\operatorname{disc}(S)| \leq 2^{200}$ and roughly one second to generate a random ring with $r = 3$ and $|\mathrm{disc}(S)| \leq 2^{200\,000}$. (These computations were performed on an Intel Core i7-1355U processor.)

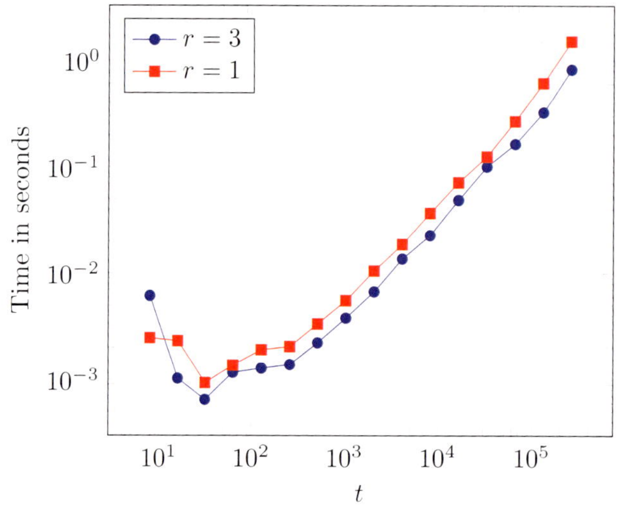

FIGURE 1. Average time to generate one random ring

5. Other parameterizations

The method described above can be adapted to other parameterizations by prehomogeneous vector spaces.

For example, it is well-known that there is a bijection between the set of $\mathrm{GL}_2(\mathbb{Z})$-orbits of primitive binary quadratic forms and the set of ideal classes of orders in quadratic number fields. One obtains an algorithm which constructs a random pair (S, I), where S is a quadratic integral domain with given signature and $|\mathrm{disc}(S)| \leq T$ and I is an invertible ideal class of S with expected running time $\widetilde{\mathcal{O}}(\log T)$. The probability the algorithm returns a given such pair (S, I) is proportional to $\frac{\mathrm{Reg}(S)}{\omega_S}$, where $\mathrm{Reg}(S)$ is the regulator of S and ω_S is the number of roots of unity in S.

The method can also be used to construct random quartic or quintic rings as in [**3**], [**5**] (with bounds derived as in [**4**] and [**6**]). Unfortunately, the constant factor in the running time crucially depends on the dimension of the prehomogeneous vector space. Quintic rings are parameterized by orbits in a 40-dimensional prehomogeneous vector space and the constant factor becomes impractically large.

References

[1] E. Bach, *How to generate factored random numbers*, SIAM J. Comput. **17** (1988), no. 2, 179–193, DOI 10.1137/0217012. Special issue on cryptography. MR935336

[2] K. Belabas, *A fast algorithm to compute cubic fields*, Math. Comp. **66** (1997), no. 219, 1213–1237, DOI 10.1090/S0025-5718-97-00846-6. MR1415795

[3] M. Bhargava, *Higher composition laws. III. The parametrization of quartic rings*, Ann. of Math. (2) **159** (2004), no. 3, 1329–1360, DOI 10.4007/annals.2004.159.1329. MR2113024

[4] M. Bhargava, *The density of discriminants of quartic rings and fields*, Ann. of Math. (2) **162** (2005), no. 2, 1031–1063, DOI 10.4007/annals.2005.162.1031. MR2183288

[5] M. Bhargava, *Higher composition laws. IV. The parametrization of quintic rings*, Ann. of Math. (2) **167** (2008), no. 1, 53–94, DOI 10.4007/annals.2008.167.53. MR2373152

[6] M. Bhargava, *The density of discriminants of quintic rings and fields*, Ann. of Math. (2) **172** (2010), no. 3, 1559–1591, DOI 10.4007/annals.2010.172.1559. MR2745272

[7] M. Bhargava, A. Shankar, and J. Tsimerman, *On the Davenport-Heilbronn theorems and second order terms*, Invent. Math. **193** (2013), no. 2, 439–499, DOI 10.1007/s00222-012-0433-0. MR3090184

[8] H. Cohen, *A course in computational algebraic number theory*, Graduate Texts in Mathematics, vol. 138, Springer-Verlag, Berlin, 1993, DOI 10.1007/978-3-662-02945-9. MR1228206

[9] H. Davenport and H. Heilbronn, *On the density of discriminants of cubic fields. II*, Proc. Roy. Soc. London Ser. A **322** (1971), no. 1551, 405–420, DOI 10.1098/rspa.1971.0075. MR491593

[10] D. S. Dummit and J. Voight, *The 2-Selmer group of a number field and heuristics for narrow class groups and signature ranks of units*, Proc. Lond. Math. Soc. (3) **117** (2018), no. 4, 682–726, DOI 10.1112/plms.12143. MR3873132

[11] F. Johansson, *Arb: efficient arbitrary-precision midpoint-radius interval arithmetic*, IEEE Trans. Comput. **66** (2017), no. 8, 1281–1292, DOI 10.1109/TC.2017.2690633. MR3681746

[12] Friedrich Levi, *Kubische Zahlkörper und binäre kubische Formenklassen*, Berichte über die Verhandlungen der Königlich Sächsischen Gesellschaft der Wissenschaften zu Leipzig, Mathematisch-Physische Klasse **66** (1914), 26–37.

[13] The LMFDB Collaboration, *The L-functions and modular forms database*, `https://www.lmfdb.org`, 2024, [Online; accessed 18 January 2024].

[14] G. Malle, *Cohen-Lenstra heuristic and roots of unity*, J. Number Theory **128** (2008), no. 10, 2823–2835, DOI 10.1016/j.jnt.2008.01.002. MR2441080

[15] The FLINT team, *FLINT: Fast Library for Number Theory*, 2024, Version 3.1.3, `https://flintlib.org`.

Universität Paderborn, Fakultät EIM, Institut für Mathematik, Warburger Str. 100, 33098 Paderborn, Germany

Email address: `fabian.gundlach@uni-paderborn.de`

Contemporary Mathematics
Volume **840**, 2026
https://doi.org/10.1090/conm/840/16805

A Shimura-Belyi map of degree 33, and number fields with Galois group 33T55 $= \Sigma\mathrm{L}_2(\mathbf{F}_{32})$

Noam D. Elkies

Abstract. Let K be the totally real quintic field $\mathbf{Q}(2\cos(\pi/11))$, and let A be the quaternion algebra over K ramified only at four of its five real places. Then $A^*/\{\pm 1\}$ contains the $(2,3,11)$ triangle group, which is the last example in Takeuchi's list (1977) of arithmetic triangle groups, and the only one associated to a quaternion algebra over a field of degree at least 5. We study the Shimura curve $C = \mathcal{X}_0(\mathfrak{p}_{32})$ of genus 2 associated to a congruence subgroup $\Gamma_0(\mathfrak{p}_{32})$. We find that C has the Weierstrass model

$$(0.1)\qquad \begin{aligned} C : y^2 &= -28x^6 - 76x^5 - 11x^4 - 30x^3 - 91x^2 + 164x - 52 \\ &= -(2x-1)(14x^5 + 45x^4 + 28x^3 + 29x^2 + 60x - 52), \end{aligned}$$

and determine the degree-33 Belyi function on C that represents the cover $\mathcal{X}_0(\mathfrak{p}_{32}) \to \mathcal{X}(1)$ of Shimura modular curves. The preimages of $\mathbf{Q}$-rational points j on $\mathcal{X}(1)$ then yield infinitely many number fields F_j with $[F_j : \mathbf{Q}] = 33$ such that $\mathrm{Gal}(F_j/\mathbf{Q}) \cong \Sigma\mathrm{L}_2(\mathbf{F}_{32})$ with quintic resolvent K. Searching for examples with small $\Delta = \mathrm{disc}(F_j/\mathbf{Q})$ finds fields with Δ as small as $2^{78}11^{30}$ (root-discriminant 45.52), as well as the field

$$(0.2)\qquad F_{-121} = \mathbf{Q}[x]/(x^{33} - 6x^{22} + 14x^{11} + 2)$$

with smaller Galois group and $\Delta = 2^{32}11^{34}$ (root-discriminant only 23.17, now the LMFDB's second-smallest for a number field of degree 33). Along the way we find that $J(C)$ is 31-isogenous with the Jacobian of the curve

$$(0.3)\qquad C' : y^2 = 8x^5 - 23x^4 + 38x^3 - 7x^2 - 16x + 8,$$

which has $(J_{C'}(K))_{\mathrm{tors}} \cong \mathbf{Z}/155\mathbf{Z}$, including a Galois orbit of five K-rational points on C', giving an exotic "Weierstrass torsion packet".

At several points in this investigation, data in the LMFDB and affiliated resources made the analysis and computations simpler or more efficient. In turn the new number fields were added to the LMFDB, which did not previously include any fields with these Galois groups.

1. Introduction

Define $a = 2\cos(\pi/11) = 1.9189859\ldots$ and $K = \mathbf{Q}(a)$, so K is the totally real quintic field with LMFDB label `5.5.14641.1`. The ring of integers O_K is $\mathbf{Z}[a]$. We choose a as the generator because it is the largest root of the "`polredabs` polynomial" used by the LMFDB;[1] namely, the minimal polynomial equation satisfied

---

2020 *Mathematics Subject Classification.* Primary 11G10, 12F12, 11Y50; Secondary 11F41.
Key words and phrases. Shimura curve, Belyi map, Galois group, Weierstrass torsion packet.
Supported in part by the Simons Foundation (Grant #550031 under the Collaboration on Arithmetic Geometry, Number Theory, and Computation).

[1] The "`polredabs` polynomial" of a number field F is a canonical defining polynomial $P_1 \in \mathbf{Z}[X]$ for F, named for the GP function that computes it. It minimizes the T_2-norm $\sum_{P(z)=0} |z|^2$ over all monic $P \in \mathbf{Z}[X]$ such that $F = \mathbf{Q}[X]/(P)$, and thus tends to have reasonably small coefficients.

by a is

$$a^5 - a^4 - 4a^3 + 3a^2 + 3a - 1 = 0. \tag{1.1}$$

Because K is the real subfield of the cyclotomic field $\mathbf{Q}(\boldsymbol{\mu}_{11})$, its Galois group $\mathrm{Gal}(K/\mathbf{Q})$ is $(\mathbf{Z}/11\mathbf{Z})^\times/\{\pm 1\} \cong \mathbf{Z}/5\mathbf{Z}$. We fix the generator σ of $\mathrm{Gal}(K/\mathbf{Q})$ corresponding to the generator ± 2 of $(\mathbf{Z}/11\mathbf{Z})^\times/\{\pm 1\}$; explicitly, $\sigma(a) = 2 - a^2 = -2\cos(2\pi/11)$.

Let A be the quaternion algebra over K ramified only at four of the five embeddings $K \hookrightarrow \mathbf{R}$, namely the embeddings taking a to $\sigma^i(a)$ for $i = 1, 2, 3, 4$ (all five choices of a set of four embeddings are equivalent under $\mathrm{Gal}(K/\mathbf{Q})$). Because the $i = 0$ embedding is unramified, it gives an isomorphism $i : \mathsf{A} \otimes_K \mathbf{R} \xrightarrow{\sim} M_2(\mathbf{R})$. Fix a maximal order $O_{\mathsf{A}} \subset \mathsf{A}$, and let Γ be the subgroup of $O_{\mathsf{A}}^\times$ consisting of elements of norm 1. Then $i(\Gamma)$ is a discrete co-compact subgroup of $\mathrm{SL}_2(\mathbf{R})$, so $\Gamma/\{\pm 1\}$ acts on the upper half-plane $\mathcal{H}$ with quotient a compact Riemann surface. We denote this quotient surface by $\mathcal{X}(1)$. For any ideal $\mathfrak{N}$ of O_K we define a congruence subgroup $\Gamma(\mathfrak{N}) \trianglelefteq \Gamma$ by $\Gamma(\mathfrak{N}) = \{g \in \Gamma : g \equiv 1 \bmod \mathfrak{N}\}$, and thus a Shimura curve $\mathcal{X}(\mathfrak{N})$ covering $\mathcal{X}(1)$. There is an isomorphism $O_{\mathsf{A}}/\mathfrak{N}O_{\mathsf{A}} \xrightarrow{\sim} M_2(O_K/\mathfrak{N})$ of O_K-algebras, unique up to conjugation in Γ, which we use to define the usual congruence subgroups $\Gamma_0(\mathfrak{N}), \Gamma_1(\mathfrak{N})$ with $\Gamma(N) \trianglelefteq \Gamma_1(\mathfrak{N}) \trianglelefteq \Gamma_0(\mathfrak{N}) \subseteq \Gamma_0(\mathfrak{N})$, and the corresponding Shimura curves $\mathcal{X}_0(\mathfrak{N}), \mathcal{X}_1(\mathfrak{N})$ and maps $\mathcal{X}(\mathfrak{N}) \to \mathcal{X}_1(\mathfrak{N}) \to \mathcal{X}_0(\mathfrak{N}) \to \mathcal{X}(1)$.

These Shimura curves and maps for A are of particular interest to us for the following reasons. According to the list of arithmetic triangle groups in [**11**], $\Gamma/\{\pm 1\}$ is the $(2, 3, 11)$ triangle group in $\mathrm{Aut}(\mathcal{H})$, and is the only arithmetic triangle group coming from a quaternion algebra over a number field of degree 5 or more. Since Γ is a triangle group, its covers by $\mathcal{X}_0(\mathfrak{N}), \mathcal{X}_1(\mathfrak{N}), \mathcal{X}(\mathfrak{N})$ are Belyi maps (finite covers of $\mathbf{P}^1$ ramified above only three points). When $\mathfrak{N}$ is invariant under $\mathrm{Gal}(K/\mathbf{Q})$, the curves $\mathcal{X}_0(\mathfrak{N}), \mathcal{X}_1(\mathfrak{N}), \mathcal{X}(\mathfrak{N})$ are defined over $\mathbf{Q}$, though in general the action of $\mathrm{SL}_2(O_K/\mathfrak{N})$ on $\mathcal{X}(N)$ is defined only over some cyclotomic extension of K. Other than the case $\mathfrak{N} = O_K$ of $\mathcal{X}(1)$ itself, this first happens for $\mathfrak{N} = \mathfrak{p}_{11} := (2 + a)O_K$, which is the prime of K over the totally ramified rational prime 11, and for $\mathfrak{N} = \mathfrak{p}_{32} := 2O_K$, which is the prime of K over the inert rational prime 2. The corresponding Shimura curves $\mathcal{X}_0(\mathfrak{N})$ have genus 1 for $\mathfrak{N} = \mathfrak{p}_{11}$ and 2 for $\mathfrak{N} = \mathfrak{p}_{32}$, see for instance [**12**, Table 4.5] (with $d_F = \mathrm{disc}(K/\mathbf{Q}) = 11^4 = 14641$).

For $\mathfrak{N} = \mathfrak{p}_{11}$ the group $\mathrm{SL}_2(O_K/\mathfrak{N})$ is $\mathrm{SL}_2(\mathbf{F}_{11})$, and the curve $\mathcal{X}(\mathfrak{p}_{11})$ with its action of $\mathrm{SL}_2(\mathbf{F}_{11})$ is the same as the classical modular curve $\mathrm{X}(11)$, because the curve $\mathcal{X}(\mathfrak{p}_{11})$ and its Shimura-Belyi map to $\mathcal{X}(1)$ have the same ramification as $\mathrm{X}(11) \to \mathrm{X}(1)$ (the elliptic point of order 11 corresponds to the classical cusp), and the monodromy generators determine the map. We choose the rational coordinate $j = j_\wedge$ on $\mathcal{X}(1)$ that takes values $1728, 0, \infty$ on the elliptic points, so that we can use the same formulas for both maps.

For $\mathfrak{N} = \mathfrak{p}_{32}$ there is no such shortcut. We compute equations for the genus-2 curve[2] $\mathcal{X}_0(\mathfrak{p}_{32})$ and the degree-33 map $\mathcal{X}_0(\mathfrak{p}_{32}) \to \mathcal{X}(1)$. The genus and degree are barely small enough that it would have been feasible, albeit arduous, to find the curve and map using known techniques such as those of [**4**, **5**]. Fortunately the computation was considerably streamlined and simplified by the use of data already in the LMFDB and in affiliated resources (namely [**3**]).

[2]We did not attempt to give explicit equations for the curve $\mathcal{X}_1(\mathfrak{p}_{32})$, which is already of genus 32, let alone the genus-1241 curve $\mathcal{X}(\mathfrak{p}_{32})$.

The degree-33 extension of function fields $\mathbf{Q}(\mathcal{X}_0(\mathfrak{p}_{32}))/\mathbf{Q}(\mathcal{X}(1))$ has Galois group $\Sigma L_2(\mathbf{F}_{32}) = SL_2(\mathbf{F}_{32}) \rtimes Gal(\mathbf{F}_{32}/\mathbf{F}_2)$ acting on the 33 points of $\mathbf{P}^1(\mathbf{F}_{32})$, with quintic resolvent K. (The quintic resolvent is the subfield of the Galois closure fixed by the index-5 normal subgroup $SL_2(\mathbf{F}_{32})$. Here no further cyclotomic extension is needed because $\mathbf{F}_{32}$ has characteristic 2 and K already contains $\boldsymbol{\mu}_2$.) By Hilbert irreducibility, specialization to rational points of $\mathcal{X}(1)$ yields infinitely many distinct degree-33 extensions of $\mathbf{Q}$ with the same Galois group and quintic resolvent. The group $\Sigma L_2(\mathbf{F}_{32})$, in its transitive action on $\mathbf{P}^1(\mathbf{F}_{32})$, is called "33T55" in the LMFDB. The LMFDB did not include any number fields with Galois group 33T55, and apparently none was known. Now that we have the explicit map $\mathcal{X}_0(\mathfrak{p}_{32}) \to \mathcal{X}(1)$, we can generate infinitely many such fields F_j by specializing to rational $j \in \mathcal{X}(1)$. We searched for specializations with small discriminant Δ. By far the smallest Δ that we found was $2^{78}11^{30}$, for $j_{\mathsf{A}} = -121$; but this is a CM point on $\mathcal{X}(1)$ (though not on X(1)), so we expect a smaller Galois group, and indeed the `polredabs` polynomial $x^{33} - 6x^{22} + 14x^{11} + 2$ for F_{-121} shows its Galois group is solvable (it turns out to be "33T9 = $C_{33} : C_{10}$"). This field was not in the LMFDB either; it is now the number field of second-smallest discriminant among the LMFDB's degree 33 fields. The smallest discriminant we found for a 33T55 field is $2^{78}11^{30}$, for the field with $j_{\mathsf{A}} = 287496 = 66^3$ (which is a CM point on X(1) but not on $\mathcal{X}(1)$), with `polredabs` polynomial

```
x^33 - 7*x^32 + 20*x^31 + 4*x^30 - 132*x^29 + 132*x^28 + 1056*x^27
 - 2464*x^26 - 1276*x^25 + 14608*x^24 - 17512*x^23 - 12100*x^22
+ 90024*x^21 - 102608*x^20 - 104060*x^19 + 482064*x^18 - 177386*x^17
    - 703978*x^16 + 1731268*x^15 + 196988*x^14 - 4514752*x^13
  + 3205048*x^12 + 7176752*x^11 - 9604496*x^10 - 5377240*x^9
   + 16445836*x^8 + 468512*x^7 - 17393508*x^6 + 995588*x^5
 + 9545932*x^4 - 292820*x^3 - 2225432*x^2 + 102487*x + 219615 .
```

This is now the number field of sixth-smallest discriminant among the LMFDB's degree 33 fields, following the $SL_2(\mathbf{F}_{32})$ field of discriminant $2^{62}157^{16}$ computed by Bosman [**2**].

The rest of this paper is organized as follows. We start (§2) by reviewing the degree-12 Belyi map $X_0(11) \to X(1)$ of classical modular curves, then show that it is the same as the Shimura-Belyi map $\mathcal{X}_0(\mathfrak{p}_{11}) \to \mathcal{X}(1)$, and describe the roles of each of this curve's five $\mathbf{Q}$-rational points as points on both $X_0(11)$ and $\mathcal{X}_0(\mathfrak{p}_{11})$. In the next section we close in on the degree-33 Belyi map $\mathcal{X}_0(\mathfrak{p}_{32}) \to \mathcal{X}(1)$ by combining information from the geometry of the map (§3.1) and from Hilbert modular forms for K (§3.2) to guess the model (0.1) for $\mathcal{X}_0(\mathfrak{p}_{32})$ (§3.3); then (§3.4) we find a degree-33 Belyi map on this curve, and prove (§3.5) that it has geometric Galois group $SL_2(\mathbf{F}_{32})$ and is thus the desired Shimura-Belyi map. Along the way we find another genus-2 curve C' whose Jacobian has 155-torsion over K, generated by divisors $(Q) - (P)$ where P is a rational Weierstrass point and Q is in a $Gal(K/\mathbf{Q})$ orbit of K-rational points that give an exotic "Weierstrass torsion packet" in the sense of [**9, 10**]; we describe this situation briefly in §3.6. In the final section (§4), we describe the search for fields F_j of small discriminant, and list the best examples we found (1). An Appendix exhibits GP code that computes a polynomial $P_{33}(X)$ with coefficients in $\mathbf{Z}[j]$ that generates the extension $\mathbf{Q}(\mathcal{X}_0(p_{32}))/\mathbf{Q}(\mathcal{X}(1))$.

2. The Shimura-Belyi map $\mathcal{X}_0(\mathfrak{p}_{11}) \to \mathcal{X}(1)$, and the Q-points on $\mathcal{X}_0(\mathfrak{p}_{11})$

We claim that as a Belyi map, the degree-12 cover $\mathcal{X}_0(\mathfrak{p}_{11}) \to \mathcal{X}(1)$ is identical with the cover $\mathrm{X}_0(11) \to \mathrm{X}(1)$ of classical modular curves. Indeed the two covers have the same geometric Galois group $\mathrm{PSL}_2(\mathbf{F}_{11})$, and the monodromy generators g_2, g_3, g_{11} of $\mathcal{X}_0(\mathfrak{p}_{11}) \to \mathcal{X}(1)$ above the elliptic points of index $2, 3, 11$ have the same cycle structures $2^6, 3^4, (1, 11)$ as those of $\mathrm{X}_0(11) \to \mathrm{X}(1)$ above the elliptic points of index $2, 3$ and the cusp respectively. These cycle structures, together with the condition $g_2 g_3 g_{11} = 1$, determine g_2, g_3, g_{11} up to conjugation in $\mathrm{PGL}_2(\mathbf{F}_{11})$; so the covers must be isomorphic as claimed.

We choose the projective coordinate j_{A} on $\mathcal{X}(1)$ that takes the same values $1728, 0, \infty$ at the elliptic points of index $2, 3, 11$ as the classical j-invariant on the elliptic points of index $2, 3$ and the cusp. We can then re-use the known equations for $\mathrm{X}_0(11)$ and the cover $\mathrm{X}_0(11) \to \mathrm{X}(1)$ in the Shimura setting. The curve has Weierstrass equation $y^2 + y = x^3 - x^2 - 10x - 20$, and

$$\begin{aligned} j = \frac{1}{x-16}\Big(&-11x^6 + 148x^5 + 643x^4 - 2704x^3 - 6780x^2 + 1781x + 3308 \\ &- (x^5 + 23x^4 - 697x^3 + 1031x^2 + 2170x + 353)y\Big). \end{aligned} \tag{2.1}$$

eliminating y yields a relation $(x-16)j^2 + A(x)j + B(x) = 0$ where A, B are determined by

$$\begin{aligned} B(x) &= -(x^4 - 20x^3 + 62x^2 + 116x + 97)^3, \\ 1728^2 - 1728A(x) + B(x) &= -(x^6 - 30x^5 + 243x^4 - 256x^3 \\ &\qquad - 1053x^2 + 654x + 7793)^2. \end{aligned} \tag{2.2}$$

It is well-known that the elliptic curve $X_0(11)$ has Mordell-Weil group $\mathbf{Z}/5\mathbf{Z}$. We choose the generator $T = (16, -61)$, and list for each $n = 0, 1, 2, 3, 4$ the (x, y) coordinates of nT, and the value of j, computed from (2.1) or (for $n = 4$) from (2.2):

n	$(x(nT), y(nT))$	$j(nT) = j_{\mathsf{A}}(nT)$	$\mathrm{X}_0(11)$	$\mathcal{X}_0(\mathfrak{p}_{11})$
0	(∞, ∞)	∞	cusp $i\infty$	$\mathrm{CM}(-(2+a)^3)$
1	$(16, -61)$	∞	cusp 0	$e_{11} = \mathrm{CM}(-(2+a))$
2	$(5, -6)$	$-121 = -11^2$	121a1, c2	$\mathrm{CM}(-(2+a)^3)$
3	$(5, 5)$	$-32768 = -2^{15}$	121b1, b2 $= \mathrm{CM}(-11)$	non-CM
4	$(16, 60)$	$-24729001 = -131^3 11$	121a2, c1	non-CM

It remains to explain the table's columns labeled $\mathrm{X}_0(11)$ and $\mathcal{X}_0(\mathfrak{p}_{11})$. The Fricke involution w_{11} of $\mathrm{X}_0(11)$ switches the two cusps, so it must be the map $P \mapsto T - P$. In particular $3T$ is the fixed point. We recognize $j(3T) = -2^{15}$ as the j-invariant of elliptic curves such as 121b1 and 121b2 with complex multiplication (CM) by the quadratic order of discriminant -11; such an elliptic curve is 11-isogenous with its quadratic twist by $\mathbf{Q}(\sqrt{-11})$ (and indeed 121b1 and 121b2 are related by such a quadratic twist). The remaining points $2T, 4T$ yields j-invariants -11^2 and $-131^3 11$ of non-CM elliptic curves that are related by an 11-isogeny.

The Shimura curve $\mathcal{X}_0(\mathfrak{p}_{11})$ also has an involution w, coming from the normalizer of $\Gamma_0(\mathfrak{p}_{11})$ in $\mathrm{SL}_2(\mathbf{R})$. But this involution is not the same as the Fricke involution $P \mapsto T - P$ of $\mathrm{X}_0(11)$. Instead this involution must take T to itself, because T is the simple pole of j_{A} and is thus the unique elliptic point of $\mathcal{X}_0(\mathfrak{p}_{11})$, related to itself by a "$\mathfrak{p}_{11}$-isogeny". Since $j_{\mathsf{A}} = \infty$ is an elliptic point of order 11, it is a CM point corresponding to the quadratic extension $\mathbf{Q}(\boldsymbol{\mu}_{11})$ of K generated by $\sqrt{-(2+a)}$. The pole of j_{A} of multiplicity 11 at $(x, y) = (\infty, \infty)$ then gives a $\mathfrak{p}_{11}$-isogeny between $j_{\mathsf{A}} = \infty$ and $j_{\mathsf{A}}(w(\infty, \infty)) = j_{\mathsf{A}}(2T) = -121$. Thus these are also CM points on $\mathcal{X}_0(\mathfrak{p}_{11})$, but of discriminant $-(2+a)^3$. Here it is the points $3T, 4T$ that yield non-CM points $j = -2^{15}$ and $j = -131^3 11$ related by a $\mathfrak{p}_{11}$-isogeny. We shall encounter both this isogenous pair and the CM point $j_{\mathsf{A}} = -121$ in the final section of this paper.

3. The Shimura-Belyi map $\mathcal{X}_0(\mathfrak{p}_{32}) \to \mathcal{X}(1)$

3.1. Geometry. We start by using the geometry of the Belyi map $C = \mathcal{X}_0(\mathfrak{p}_{32}) \to \mathcal{X}(1)$ and the involution w of $\mathcal{X}_0(\mathfrak{p}_{32})$ to constrain the preimages on C of the elliptic points on $\mathcal{X}(1)$. We then use the quotient map $\mathcal{X}_1(\mathfrak{p}_{32}) \to \mathcal{X}_0(\mathfrak{p}_{32}) = C$ to further constrain the arithmetic of J_C.

Since $\mathcal{X}(1)$ has genus zero, we identify a map $C \to \mathcal{X}(1)$ with a rational function on C by choosing a rational coordinate on $\mathcal{X}(1)$. We retain the coordinate of the previous section, which takes the values $1728, 0, \infty$ on the elliptic points of indices $2, 3, 11$ respectively. Again we denote the resulting function on C by j_{A}. The monodromy generators above the elliptic points are elements of order $2, 3, 11$ in $\mathrm{SL}_2(\mathbf{F}_{32})$; thus their cycle structures are respectively $(1, 2^{16}), 3^{11}, 11^3$. (Check: the Euler characteristic of C is then $33\chi(\mathbf{P}^1) - 16 - 2\cdot 11 - 10\cdot 3 = 66 - 16 - 22 - 30 = -2$, consistent with $g(C) = 2$.) We denote by P the unique multiplicity-1 preimage of 1728. This is the unique elliptic point on $\mathcal{X}_0(\mathfrak{p}_{32})$, and is thus fixed by the involution w of $\mathcal{X}_0(\mathfrak{p}_{32})$ coming from the normalizer of $\Gamma_0(\mathfrak{p}_{32})$ in $\mathrm{SL}_2(\mathbf{R})$. It turns out that w is the hyperelliptic involution of $\mathcal{X}_0(\mathfrak{p}_{32})$, making P a Weierstrass point. This could be checked directly by counting fixed points of w: the hyperelliptic involution fixes the six Weierstrass points, while a non-hyperelliptic involution of a genus-2 curve fixes only four points. Instead of attempting to do this calculation directly, we give an alternative argument in §3.2 using the decomposition of the relevant space of Hilbert modular forms. Alternatively, we can add the statement that w is hyperelliptic to several other guesses that we shall make on the way to computing a Belyi map, statements that we justify in §3.5 by proving that our map has the correct Galois group.

The ramification behavior of our map means that the polar divisor of our rational function $j_{\mathsf{A}} \in \mathbf{Q}(C)$, and the zero divisors of j_{A} and $j_{\mathsf{A}} - 1728$, are given by

$$(j_{\mathsf{A}})_\infty = 11D_3, \quad (j_{\mathsf{A}})_0 = 3D_{11}, \quad (j_{\mathsf{A}})_{1728} = 2D_{16} + (P) \tag{3.1}$$

for some divisors D_3, D_{11}, D_{16} of degrees $3, 11, 16$ on C. Because the map is ramified only above $\{1728, 0, \infty\}$, the differential $\mathrm{d}j_{\mathsf{A}}$ has divisor

$$(\mathrm{d}j_{\mathsf{A}}) = (\mathrm{d}j_{\mathsf{A}})_0 - (\mathrm{d}j_{\mathsf{A}})_\infty = 2D_{11} + D_{16} - 12D_3. \tag{3.2}$$

Now the three divisors in (3.1) are linearly equivalent to each other, and the divisor of a differential of a nonconstant function is linearly equivalent to the canonical divisor K on C. This gives relations among $D_3, D_{11}, D_{16}, (P), K$ in $\mathrm{Pic}(C)$:

$$11D_3 \sim 3D_{11} \sim 2D_{16} + (P), \quad 2D_{11} + D_{16} - 12D_3 \sim K. \tag{3.3}$$

The first equation gives

$$D_3 \sim 3D_1, \quad D_{11} \sim 11D_1 \tag{3.4}$$

where D_1 is the degree-1 divisor $4D_3 - D_{11}$. Then

$$D_{16} \sim K - 2D_{11} + 12D_3 \sim K - 22D_1 + 36D_1 = K + 14D_1, \tag{3.5}$$

whereupon the remaining condition $3D_{11} \sim 2D_{16} + (P)$ of (3.3) gives $33D_1 \sim 2K + 28D_1 + (P)$ and finally

$$5D_1 \sim 2K + (P). \tag{3.6}$$

Since P is a Weierstrass point, $K \sim 2(P)$, so (3.6) gives $5D_1 \sim 4(P) + (P) = 5(P)$, whence $5(D_1 - (P)) \sim 0$. (The factor of 5 can be identified with the numerator of the factor

$$\frac{5}{66} = 1 - \frac{1}{2} - \frac{1}{3} - \frac{1}{11} \tag{3.7}$$

in the formula for the hyperbolic area of $\mathcal{X}(1)$.) It remains to show that the class $[D_1 - (P)] \in \mathrm{Pic}^0(C) = J_C$ is actually 5-torsion, not zero. Suppose it were zero; that is, suppose $D_1 \sim (P)$. Then $D_3 \sim 3D_1 \sim 3(P)$ would imply that P is one of the points in D_3: by Riemann-Roch $h^0(3P) = 2 = h^0(K) = h^0(2P)$. But the divisors $(j_{\mathsf{A}})_\infty = 11D_3$ and $(j_{\mathsf{A}})_{1728} = 2D_{16} + (P)$ are disjoint, so D_3 cannot contain P. □

The curve $C = \mathcal{X}_0(\mathfrak{p}_{32})$ also admits a cyclic cover $\mathcal{X}_1(\mathfrak{p}_{32}) \to \mathcal{X}_1(\mathfrak{p}_{32})/\mathbf{F}_{32}^* \cong \mathcal{X}_0(\mathfrak{p}_{32})$ over K. The map is unramified because it is cyclic of degree 31, and the only elliptic point of $\mathcal{X}_0(\mathfrak{p}_{32})$ has index 2 which is coprime with 31. By Kummer theory this unramified cyclic cover gives a subgroup of J_C that isomorphic over K with $\boldsymbol{\mu}_{31}$. We shall use this to choose C from a pair of candidate curves whose Jacobians are related by a 31-isogeny.

3.2. Modular forms. We next use modular forms to surmise that J_C is in the isogeny class of the simple factor of the Jacobian of $\mathrm{X}_0(242)$ corresponding to the modular form 242.2.a.f.

The LMFDB does not at present include Shimura modular forms (other than those associated to congruence subgroups of $\mathrm{SL}_2(\mathbf{Z})$), but we can get at J_C indirectly via Hilbert cusp forms of parallel weight 2 with base field K. We find that at level $\mathfrak{p}_{32}$ there is a unique Hecke orbit of such forms, namely the orbit labeled 5.5.14641.1-32.1-a, with dimension 2 (as expected because C has genus 2) and eigenvalue field $\mathbf{Q}(\sqrt{5})$. This also gives us a proof that the involution w of $\mathcal{X}_0(\mathfrak{p}_{32})$ must be hyperelliptic: otherwise this space of cusp forms would split into $+1$ and -1 subspaces.

Since the level is invariant under $\mathrm{Gal}(K/\mathbf{Q})$, so are the Hecke eigenvalues, and in particular the Hecke eigenvalue at a prime $\mathfrak{p}$ depends only on its norm $\mathbf{N}\mathfrak{p}$. Consulting the LMFDB's "home page" for 5.5.14641.1-32.1-a, we tabulate Hecke eigenvalues for $\mathbf{N}\mathfrak{p} \leq 131$, and also for $\mathfrak{p} = 3O_K$, which has norm $3^5 = 243$ and is the first $\mathfrak{p}$ above a prime inert in K that does not divide the level:

(3.8)

$\mathbf{N}\mathfrak{p}$	11	23	32	43	67	89	109	131	$\cdots$	243	$\cdots$
$a_\mathfrak{p}$	e	$2e$	1	$-9e-3$	$5e+8$	$-5e-5$	$8e+14$	$11e$	$\cdots$	$20e+4$	$\cdots$

Here e is a solution of $e^2 + e - 1 = 0$. In particular the coefficient field is $\mathbf{Q}(e) = \mathbf{Q}(\sqrt{5})$, and J_C has K-endomorphisms by $\mathbf{Z}[e]$. But C is defined over $\mathbf{Q}$, and

$\mathrm{Gal}(K/\mathbf{Q})$ has no nontrivial action on $\mathbf{Z}[e]$, so the endomorphisms of $\mathbf{Z}[e]$ on J_C must be defined over $\mathbf{Q}$.

Since J_C is a RM surface over $\mathbf{Q}$, it is a simple factor up to isogeny of some modular Jacobian $J_0(N) = J_{X_0(N)}$, and thus corresponds to a $\mathrm{Gal}(\bar{\mathbf{Q}}/\mathbf{Q})$-orbit of modular eigenforms ϕ of weight 2 for $\Gamma_0(N)$. Since C has good reduction away from $\{2, 11\}$, the level N must be of the form $2^{e_2}11^{e_{11}}$ for some integers e_2, e_{11}. We search the LMFDB and find that the first spaces of such forms with coefficients in $\mathbf{Q}(\sqrt{5})$ occur for $N = 242$ (with $(e_2, e_{11}) = (1, 2)$), namely 242.2.a.d and 242.2.a.f, which are related by quadratic twist by $\mathbf{Q}(\sqrt{-11})$. For a rational prime $p \equiv \pm 1 \bmod 11$, the q^p coefficient of ϕ should equal the $T_\mathfrak{p}$ eigenvalue of 5.5.14641.1-32.1-a where $\mathfrak{p}$ is any of the primes of K with $\mathbf{N}\mathfrak{p} = (p)$. We find that this condition is already satisfied by 242.2.a.f, at least for the primes $p \leq 131$ tabulated in (3.8). (Note that the LMFDB page for 242.2.a.f gives the eigenvales in terms of $\beta = -e$. For classical modular forms the LMFDB gives only 100 coefficients, but this is enough information to extend the q-expansion further in Magma.) Moreover, the q^3 coefficient a_3 of 242.2.a.f is $-e + 1$, and we calculate that the T_{3O_K} eigenvalue $20e + 4$ of 5.5.14641.1-32.1-a equals $\lambda^5 + \bar{\lambda}^5$ where $\lambda, \bar{\lambda}$ are the roots of the characteristic equation $F^2 - a_3 F + 3$.

At this point we are almost certain that 242.2.a.f is our desired modular form ϕ. We cannot claim to have proven this, because for all we know ϕ might have level $N = 2^{e_2}11^{e_{11}}$ beyond the LMFDB's range. If we knew in advance that $N = 242$ then we would already have a proof.[3] Once we surmise an explicit Weierstrass model for the curve, and find a degree-33 Belyi map on that curve, we shall be able to identify that map with the Shimura-Belyi map $\mathcal{X}_0(\mathfrak{p}_{32}) \to \mathcal{X}(1)$ by proving that it has the correct Galois group.

3.3. The curves C' and C. We next combine the information from §§3.1,3.2 to surmise an equation for $\mathcal{X}_0(\mathfrak{p}_{32})$.

Here we could not simply look up the curve in the LMFDB. The LMFDB's 66158 genus-2 curves over $\mathbf{Q}$ include 99 whose Jacobians' $\mathbf{Q}$-endomorphism algebra are orders in real quadratic fields, in 92 isogeny classes, each linked to a classical modular eigenform ψ of weight 2 with coefficients in the same quadratic field. But this is a small fraction of the LMFDB's 80387 weight-2 eigenforms with quadratic coefficients. Moreover, even when ψ does link to an isogeny class of curves, the LMFDB might not contain every curve in the class. This already happens for the first two such ψ, of levels $N = 23$ and $N = 29$, where the LMFDB contains a genus-2 curve of discriminant N^2 whose Jacobian also has discriminant N^2, but does not contain the modular curve $X_0(N)$ because its discriminant (23^6 or 29^5) falls outside the LMFDB's range $|\Delta| \leq 10^6$. The homepage of our form 242.2.a.f does not link to any genus-2 curves.

Given an eigenform ψ for $\Gamma_0(N)$ whose coefficients generate a quadratic field, and which thus corresponds to an isogeny class of abelian surfaces, one can try to find a genus-2 Jacobian in the isogeny class by numerically integrating ψ over cycles of $X_0(N)$ to compute a period lattice to enough precision that its Igusa invariants

---

[3] We did guess that $N = 242$ by analogy with the known case of $\mathcal{X}_0(\mathfrak{p}_{27})$ for the $(2, 3, 7)$ triangle group. The $(2, 3, 7)$ triangle group is $\Gamma(1)$ for the quaternion algebra over $\mathbf{Q}(2\cos(\pi/7))$ ramified only at two of this field's three real embeddings, and we found in [4] that the Shimura curve $\mathcal{X}_0(\mathfrak{p}_{27})$, with $\mathfrak{p}_{27} = 3\mathbf{Z}[2\cos(\pi/7)]$, is an elliptic curve of conductor $147 = 3^1 7^2$, suggesting $N = 2^1 11^2$ for the presesnt curve $\mathcal{X}_0(\mathfrak{p}_{32})$.

can be recognized as rational numbers. In [**3**] we report on a computation that tried this for each of the 16929 such ψ with $N \le 10^4$. We succeeded for most of them, including all cases where ψ has coefficients in a field of narrow class number 1 such as $\mathbf{Q}(\sqrt{5})$ and $\mathbf{Q}(\sqrt{2})$ with the $\bar{\mathbf{Q}}$-endomorphism algebra contained in the coefficient field, in which case the isogeny class always contains a surface with a principal polarization defined over $\mathbf{Q}$. In particular we have already found a curve C' whose Jacobian is in the isogeny class corresponding to 242.2.a.f. In the github repository (see the end of the bibliographic entry for [**3**]) this curve is reported as [[35,79,77,43,14,2],[0,-1,-1]], which is to say

$$y^2 - (x^2 + x)y = 2x^5 + 14x^4 + 43x^3 + 77x^2 + 79x + 35; \tag{3.9}$$

completing the square and substituting $x - 2$ for x gives a reduced model in narrow Weierstrass form,

$$C' : y^2 = 8x^5 - 23x^4 + 38x^3 - 7x^2 - 16x + 8. \tag{3.10}$$

This looks promising, because C' has a Weierstrass point (at infinity), and $J_{C'}$ has a 5-torsion point over $\mathbf{Q}$, represented by the divisor $x^2 - x - 1 = 0$, $y = 4x + 1$ with Weil function

$$4x^5 - x^4 + 14x^3 + 17x^2 - 2x - 4 - (3x^2 + 5x + 2)y. \tag{3.11}$$

Moreover, $\mathcal{X}_0(\mathfrak{p}_{32})$ must have good reduction outside $\{2, 11\}$, and C' satisfies this condition. (This does not follow automatically from the good reduction of $J_{C'}$ outside $\{2, 11\}$; see (3.14). Over K, the curve C' also has good reduction at $\mathfrak{p}_{11}$: the right-hand side factors as $8(x + 3)^5 \bmod 11$, and translating x to $x - 3$ yields

$$y^2 = 8x^5 - 13 \cdot 11x^4 + 94 \cdot 11x^3 - 31 \cdot 11^2x^2 + 56 \cdot 11^2x + 40 \cdot 11^2; \tag{3.12}$$

so we can choose a local uniformizer π at $\mathfrak{p}_{11}$, scale (x, y) to $(\pi^2 x, \pi^5 y)$, and divide through by π^{10} to get a model that remains smooth mod $\mathfrak{p}_{11}$. But we saw that the Jacobian of $\mathcal{X}_0(\mathfrak{p}_{32})$ must contain a subgroup $\boldsymbol{\mu}_{31}$ over K; it turns out that $J_{C'}$ instead contains a subgroup $\mathbf{Z}/31\mathbf{Z}$, which makes C' a curve of independent interest (see §3.6 below), but means C' cannot be our desired curve $\mathcal{X}_0(\mathfrak{p}_{32})$.

This suggests that $\mathcal{X}_0(\mathfrak{p}_{32})$ might be a curve whose Jacobian is the quotient of $J_{C'}$ by its 31-torsion group. We use period matrices to compute a putative equation for such a curve:

$$\begin{aligned} y^2 &= -28x^6 - 76x^5 - 11x^4 - 30x^3 - 91x^2 + 164x - 52 \\ &= -(2x - 1)(14x^5 + 45x^4 + 28x^3 + 29x^2 + 60x - 52). \end{aligned} \tag{3.13}$$

This curve, too, has a rational Weierstrass point, good reduction outside $\{2, 11\}$, good reduction at $\mathfrak{p}_{11}$ over K, and a Jacobian with a $\mathbf{Q}$-rational 5-torsion point. Our isogeny class contains the Jacobians of at least two further curves, related by 5-isogenies with those we already found, namely

$$\begin{aligned} y^2 &= -7(256x^6 - 16x^5 + 40x^4 - 1249x^3 + 56x^2 - 784x), \\ y^2 &= 8x^5 - 1199x^4 - 7546x^3 + 10769x^2 + 67760x - 100672. \end{aligned} \tag{3.14}$$

But neither of them can be $\mathcal{X}_0(\mathfrak{p}_{32})$. Indeed each curve in (3.14) fails two criteria: neither curve's Jacobian has a rational 5-torsion point, and each curve has bad reduction also at 7 (though their Jacobians still have good reduction outside $\{2, 11\}$).

We found no further Jacobians in this isogeny class. We thus proceed on the assumption that $\mathcal{X}_0(\mathfrak{p}_{32}) = C$.

3.4. The Shimura-Belyi map. We now compute the Shimura-Belyi map using the known technique of mod-p search followed by p-adic lift. We choose $p = 5$, the smallest prime at which the elliptic points $j = 0, 1728$ are distinct. If we had not surmised C in advance, we would have to try all pairs (C, D) of a genus-2 curve C mod 5 with a 5-torsion class $D \in J_C$, which would be more complicated to implement and would take about 5^3 times longer to run. Starting from C the search took little more than an hour on two processors, each trying a different choice of D (since there are two choices up to the hyperelliptic involution).

It is convenient to put the Weierstrass point P at infinity, even though this makes the coefficients of C larger. We thus work with the model

$$y^2 = 64x^5 - 1095x^4 - 1248x^3 - 1224x^2 - 640x - 112. \tag{3.15}$$

The divisors that differ from K by 5-torsion are

$$D: 8x^2 + 3x + 2 = 0,\ y = (77x - 66)/8; \quad D': x^2 + x + 2 = 0,\ y = 37x - 2; \tag{3.16}$$

and their images under the hyperelliptic involution $\iota : (x, y) \mapsto (x, -y)$. (Here and later we do not reduce mod 5 because we shall soon need the same formulas over the 5-adic numbers.) We find the functions

$$f_1 = 32x^3 + 29x^2 + 24x - (y + 4), \quad f_2 = 16x^3 + 21x^2 + y + 12 \tag{3.17}$$

with divisors $(f_1) = 2D + \iota^* D' - 6P$, $(f_2) = D + 2D' - 6P$.

We search for sections s_3, s_{11} of $5P - D$ and $13P - D'$, which we normalize by setting the "leading coefficient" (coefficient of xy and $x^5 y$ respectively) equal 1. Thus

$$s_3 = y - \frac{77x - 66}{8} + c_0(8x^2 + 3x + 2) \tag{3.18}$$

for some $c_0 \in \mathbf{F}_5$, and

$$s_{11} = A(x)(y - (37x - 2)) + B(x)(x^2 + x + 2) \tag{3.19}$$

for some monic polynomial A of degree 4 and some polynomial B of degree at most 4. Thus we have 5^{10} candidates, which is just under 10^7. For each s_3 and s_{11} we compute the sections

$$\frac{s_3^{11} \iota^*(f_1^4 f_2^2)}{(8x^2 + 3x + 2)^{10}(x^2 + x + 2)^4}, \quad \frac{s_{11}^3 f_1 \iota^* f_2}{(8x^2 + 3x + 2)(x^2 + x + 2)^3} \tag{3.20}$$

of $35P - D$, find the linear combination that cancels the leading $x^{15}y$ coefficient and is thus a section of $34P - D$, and test whether its norm from C to $\mathbf{P}^1$ (which is a degree-32 polynomial in x) is a square. This finds two solutions mod 5, one of which is spurious (the linear combination is actually a section of $32P - D$); the other search, with D, D' replaced by $D', \iota^* D$, finds only a spurious solution.

We now take the vector of coefficients of our s_3 and s_{11}, lift it to $\mathbf{Z}_5$, and regard it as an approximation to the actual 5-adic solution. Four iterations of Newton's algorithm improve the error from $O(5)$ to $O(5^{16})$, which is enough to recognize c_0

as $-17/8$ and the quartics $A(t)$ and $B(t)$ as

$$A(x) = x^4 + \frac{115x^3 + 10x^2 + 356x - 248}{3},$$

$$B(x) = \frac{545x^4 + 2991x^3 + 3298x^2 - 12620x - 1464}{3}. \tag{3.21}$$

We finally obtain our Shimura-Belyi function as the ratio of the sections (3.20) of $35P - D$, composed with the fractional linear transformation of $\mathbf{P}^1$ that puts the ramified points at $1728, 0, \infty$. See the Appendix for GP code that computes j_{A} as a function of x and y, and eliminates y to find the relation $P_{33}(j_{\mathsf{A}}, x) = 0$ between j_{A} and x.

3.5. Conclusion of the proof. At this point we have found a Belyi map whose monodromy generators $g_2, g_3, g_{11} \in S_{33}$ have the same cycle structures as those of the Shimura-Belyi map $\mathcal{X}_0(\mathfrak{p}_{32}) \to \mathcal{X}(1)$. For the degree-12 map $\mathcal{X}_0(\mathfrak{p}_{11}) \to \mathcal{X}(1)$ the cycle structures suffice to determine the map, because the cycle structures $(2^6), (3^4), (1, 11)$ together with the condition $g_2 g_3 g_{11} = 1$ determine g_2, g_3, g_{11} uniquely up to conjugation in S_{12}. This is no longer true for the cycle structures $(1, 2^{16}), (3^{11}), (11^3)$ in S_{33}.[4] We next show that the Belyi map constructed in §3.4 has geometric Galois group $\mathrm{SL}_2(\mathbf{F}_{32})$. This will suffice, because in $\mathrm{SL}_2(\mathbf{F}_{32})$ the equation $g_2 g_3 g_{11} = 1$ does have a solution in elements of order 2, 3, 11 that is unique up to conjugation in $\Sigma\mathrm{L}_2(\mathbf{F}_{32}) \subset S_{33}$.

Denote the geometric Galois group by G_0, and the Galois group over $\mathbf{Q}$ by G. We soon guess that G_0 and G must be $\mathrm{SL}_2(\mathbf{F}_{32})$ and $\Sigma\mathrm{L}_2(\mathbf{F}_{32})$ respectively by specializing j_{A} to "random" rational numbers and computing the Galois group of the resulting degree-33 extension of $\mathbf{Q}$, or by factoring specializations modulo many primes. But such calculations cannot prove that G_0 and G are the expected groups. We do, however, learn that G_0 and G must contain $\mathrm{SL}_2(\mathbf{F}_{32})$ and $\Sigma\mathrm{L}_2(\mathbf{F}_{32})$ respectively. Indeed it is already enough to take $j_{\mathsf{A}} = -2$ and factor the resulting polynomial mod 23: there are two linear factors and an irreducible factor of degree 31, so G and G_0 must contain elements of order 31, and the only transitive subgroups of S_{33} with such elements are $\mathrm{SL}_2(\mathbf{F}_{32}), \Sigma\mathrm{L}_2(\mathbf{F}_{32}), A_{33}, S_{33}$. Moreover G_0 cannot be $\Sigma\mathrm{L}_2(\mathbf{F}_{32})$ or S_{33}, because it is generated by even permutatios of orders 2, 3, 11.

But it takes more work to exclude the possibility that $G_0 = A_{33}$. We use the method that we introduced in [**5**, §3] (and was later used also in [**1**]). Here we show that G_0 does not act transitively on the $\binom{33}{4}$ 4-element subsets of the 33 roots. If it did, then we would have a degree $\binom{33}{4}$ cover C_4 of $\mathbf{P}^1$ by a connected curve of

[4]We construct a counterexample as follows. Let $\alpha = \bar{a} + 1 \in \mathbf{F}_{32}$, and let $g_2, g_3 \in S_{33}$ be the permutations $z \mapsto z + \alpha$ and $z \mapsto (z+1)/z$ of $\mathbf{P}^1(\mathbf{F}_{32})$, correspondinging to the matrices $\left[\begin{smallmatrix} 1 & \alpha \\ 0 & 1 \end{smallmatrix}\right], \left[\begin{smallmatrix} 1 & 1 \\ 1 & 0 \end{smallmatrix}\right] \in \mathrm{SL}_2(\mathbf{F}_{32})$. Then $g_2 g_3(z) = (\bar{a}z + 1)/z$ corresponds to the matrix $\left[\begin{smallmatrix} \bar{a} & 1 \\ 1 & 0 \end{smallmatrix}\right]$ of trace $\bar{a}$, so $g_2 g_3$ has order 11. Now let $g_2' = \tau g_2 \tau$ where τ is the simple transposition $(0\ 1)$, and let $g_{11}' = (g_2' g_3)^{-1}$. Then g_2', g_3, g_{11}' have the same cycle structures as g_2, g_3, g_{11} and satisfy the same equation $g_2' g_3 g_{11}' = 1$. But $g_2' g_{11}'^3$ has order 70, so the subgroup generated by g_2', g_3, g_{11}' cannot be isomorphic with $\mathrm{SL}_2(\mathbf{F}_{32})$. Indeed that subgroup is the full alternating group A_{33}, because $g_2' g_{11}'^2$ has order 31 and A_{33} has no proper transitive subgroup whose order is a multiple of $31 \cdot 70$.

genus at most

$$1 + \frac{1}{2}\Big(1 - \frac{1}{2} - \frac{1}{3} - \frac{1}{11}\Big)\binom{33}{4} = 1551, \tag{3.22}$$

so over a q-element field $\mathbf{F}_q$ the curve C_4 would have at most $q + 1 + 2 \cdot 1551\sqrt{q}$ points by the Weil bound. Since we expect that in fact $G_0 = \mathrm{SL}_2(\mathbf{F}_{32})$, the action should have 5 orbits, and so C_4 will be a union of 5 curves and should have about $5q$ points if $\mathbf{F}_q$ contains a quotient field of O_K. So the assumption $G_0 = A_{33}$ will yield a contradiction if q is large enough, and we expect to succeed if q is comfortably larger than $(2 \cdot 1551/(5-1))^2$. We chose $q = 10^6 + 33$, the smallest prime above 10^6 that splits in K. Using the Lemma in [**5**, §3] we counted most of the points of $\mathbf{F}_q(C_4)$ by factoring q polynomials over $\mathbf{F}_q$ and discarding the handful that had repeated factors. This took under 10 minutes on a single processor. We found that $|C_4(\mathbf{F}_q)| \geq 4779840$, while the Weil bound for a curve of genus at most 1551 is $|C_4(\mathbf{F}_q)| \leq 4102085$. This contradiction shows that $G_0 \neq A_{33}$ and completes the proof.

3.6. Scenic detour: 155-torsion in $J_{C'}(K)$ and an exotic Weierstrass torsion packet on C'. Using the table (3.8) of Hecke eigenvalues or otherwise, we soon see that $\#J_C(K)_{\mathrm{tors}} \mid 155$; for example, $\#J_C(O_K/\mathfrak{p}_{11}) = 155$, and there can be no 11-torsion because $\#J_C(O_K/\mathfrak{p}) = 4 \cdot 155$ for $\mathbf{N}\mathfrak{p} = 23$. The same bound thus holds for any other abelian surface K-isogenous with J_C. Computing $\#J_C(O_K/\mathfrak{p})$ for further primes $\mathfrak{p}$ suggests that $J_C(O_K/\mathfrak{p})$ has 155-torsion for every prime $\mathfrak{p}$ of O_K other than $\mathfrak{p}_{32}$, and thus that the isogeny class may contain a surface with 155-torsion over K.

We find that in fact $(J_{C'}(K))_{\mathrm{tors}} \cong \mathbf{Z}/155\mathbf{Z}$. We already know a 5-torsion point of $J_{C'}(\mathbf{Q})$, so we need only find 31-torsion. A search for points of low height on $C'(K)$ other than the Weierstrass point soon finds the $\mathrm{Gal}(K/\mathbf{Q})$ orbit of

$$Q : (x, y) = (a^4 - 2a^3 + a, 11(13a^4 - 35a^3 + 7a^2 + 28a - 8)). \tag{3.23}$$

We find that the function

$$f = y + (a^3 - a^2 - a)(4x^2 - 20x + 14) + (7x^2 - 13x + 8) \tag{3.24}$$

on C' has divisor $2Q + \iota(\sigma Q) + D - 5(P)$, where ι is the hyperelliptic involution, P is the Weierstrass point at infinity, σ is our generator of $\mathrm{Gal}(K/\mathbf{Q})$ taking a to $2 - a^2$, and D is the divisor $2x^2 - 10x + 7 = 0$, $y = -22x + (33/2)$ such that $5D \sim 5K$. Since $\iota(\sigma Q) + \sigma Q \sim K \sim 2(P)$, we deduce

$$(\sigma Q) - (P) \sim 2((Q) - (P)) + D - K. \tag{3.25}$$

Applying σ, and noting that P, D, K are σ-invariant, gives

$$(\sigma^2 Q) - (P) \sim 2((\sigma Q) - (P)) + D - K \sim 4((Q) - (P)) + 3(D - K), \tag{3.26}$$

and then

$$(\sigma^3 Q) - (P) \sim 8((Q) - (P)) + 7(D - K), \tag{3.27}$$

$$(\sigma^4 Q) - (P) \sim 16((Q) - (P)) + 15(D - K), \tag{3.28}$$

and finally

$$(Q) - (P) = (\sigma^5 Q) - (P) \sim 32((Q) - (P)) + 31(D - K). \tag{3.29}$$

Hence $31((Q)-(P))$ is equivalent to the 5-torsion divisor $31(D-K)$, whence $(Q)-(P)$ is a 155-torsion divisor.

Thus $C'(\bar{\mathbf{Q}})$ contains at least 10 non-Weierstrass points P' such that the divisor $(P')-(P)$ is torsion, namely the $\mathrm{Gal}(K/\mathbf{Q})$ orbits of Q and ιQ. This is an "unlikely intersection" for a curve of genus 2 with no automorphisms other than ι and the identity: such curves vary over a 3-dimensional moduli space, so we might expect at most three pairs of non-Weierstrass points in the "Weierstrass torsion packet". I thank Bjorn Poonen for using his gp program to check that there are no further such P' on our curve. Poonen notes that already in [**7**, p.293] there is an example of a curve C_{27} with six pairs, all in $C_{27}(\mathbf{Q})$ and differing by 27-torsion divisors; but our C' seems to have the largest torsion order among the known examples of such unlikely intersections of a genus-2 curve C with $J_C(\bar{\mathbf{Q}})_{\mathrm{tors}}$.

4. Specializations to $\Sigma\mathrm{L}_2(\mathbf{F}_{32})$ extensions of Q with small discriminant

We now have a family of degree-33 fields F_j with generic Galois group $\Sigma\mathrm{L}_2(\mathbf{F}_{32})$. For any rational $j = j_{\mathsf{A}} \neq 0, 1728$, we can compute the discriminant Δ of F_j; this does not require factoring a large polynomial discriminant because we know that all prime factors of Δ other than 2 and 11 appear in the denominator of j or the numerator of j or $j - 1728$. We first tried all $j \in \mathbf{Z}$ with $|j| \leq 10 \cdot 1728$, and also all $j = 1728m/n$ for small m, n. This revealed the outlier F_{-121} with $\Delta = 2^{32}11^{34}$, which is the only example we found for which $\mathrm{Gal}(F_j/\mathbf{Q})$ is smaller than $\Sigma\mathrm{L}_2(\mathbf{F}_{32})$.

Listing j's in order of increasing Δ, we recognized several of the best j values as j-invariants of elliptic curves with small and smooth conductor. We thus asked the LMFDB for the list of all elliptic curves in the database whose conductors are $\{2, 3, 5, 7, 11\}$-smooth, and tried all of their j-invariants. There were 137612 curves, with 6086 distinct j-invariants other than 0, 1728. This found several new small values of Δ, corresponding to j-values outside the range of our first search.

In Table 1 (next page) we list the 17 values of j_{A} that yield the fields F_j of smallest Δ that we found. In each case we list also: the LMFDB labels of the elliptic curves $E_{\min}$ of minimal conductor that have j-invariant j_{A}; the factorization of Δ; and the root-discriminant rd $= \Delta^{1/33}$ rounded to the nearest .01 (which is the LMFDB's convention), as a measure of the size of Δ.

Two of those fields arise more than once. The triple coincidence includes $j_{\mathsf{A}} = -2^{15}$ and $j_{\mathsf{A}} = -131^3 11$, which we already know are related by a $\mathfrak{p}_{11}$-isogeny and thus give rise to isomorphic Galois representations mod $\mathfrak{p}_{32}$, but we have no explanation for the third value of j_{A} that yields the same field. Nor can we explain the double appearance of the field of discriminant $2^{36}5^{30}11^{24}$. Curiously $j_{\mathsf{A}} = -15^3$ and $j_{\mathsf{A}} = 255^3$ yield different fields of the same discriminant $2^{32}7^{16}11^{34}$; they are j-invariants of 2-isogenous elliptic curves with complex multiplication in discriminant -7 and -28, but again we have no explanation (except that we might expect the occasional collision of Δ's when both fields are ramified above the same primes), and it might be a mere curiosity.

TABLE 1. degree-33 fields F_j with small discriminant Δ

j_{A}	$E_{\min}$	Δ	rd
$-121 = -11^2$	121a1, 121c2	$2^{32}11^{34}$	23.17
$287496 = 66^3$	32a1, 32a2	$2^{78}11^{30}$	45.52
$10976 = 2^5 7^3$	128a1, b1, c1, d1	$2^{88}11^{28}$	48.57
$-14641/80 = -11^4/2^4 5$; $-616295051/64000 = -(23 \cdot 37/40)^3$	1210c1, 1210j1; 1210d1, 1210i1	$2^{36}5^{30}11^{24}$	52.63
$-937024/9 = -2^6 11^4/3^2$	11616f1, g1, v1, w1	$2^{62}3^{30}11^{24}$	57.11
$2048 = 2^{11}$	200b2, 200d2	$2^{38}5^{16}11^{34}$	57.34
$-3375 = -15^3$	49a2, 49a4	$2^{32}7^{16}11^{34}$	59.51
$16581375 = (3 \cdot 5 \cdot 17)^3$	49a1, 49a3	$2^{32}7^{16}11^{34}$	59.51
$54000 = 2^4 3^3 5^3$	36a1, 36a2	$2^{52}3^{16}11^{34}$	60.07
$8000 = 20^3$	256a1, a2, d1, d2	$2^{78}11^{34}$	60.88
$-121945/32 = -29^3 5/2^5$	50a2, 50b3	$2^{48}5^{22}11^{28}$	61.30
$-3072 = -2^{12}3$	216a1, 216d1	$2^{40}3^{38}11^{28}$	62.79
$43307231/82944 = 11^2 71^3/(2^{10}3^4)$	726a1, 726f1	$2^{32}3^{30}11^{34}$	62.89
$-32768 = -2^{15}$; $-24729001 = -131^3 11$; $-128667913/4096 = -227^3 11/2^{12}$	121b1, 121b2; 121a2, 121c1; 242a1, 242b1	$2^{32}11^{48}$	64.07

Six of the seventeen j_{A} values are among the thirteen rational j-invariants of CM curves, for discriminants $-7, -8, -11, -12, -16, -28$. Of the remaining seven such discriminants, -3 and -4 correspond to ramified points $j_{\mathsf{A}} = 0, 1728$ of the cover, and the discriminant -27 value $-160^3 3$ (curves 27a1, 27a2) was already on our list of candidates; we added the remaining four (discriminants $-19, -43, -67, -163$), but found no further small Δ. Three of the j_{A} values are j-invariants of non-CM elliptic curves involved in sporadic isogenies, of degrees 11 and 15 (conductors 121 and 50); we tried the remaining ones (from the 17-isogenous curves of conductor 14450 — the 37-isogenous pair in conductor 1225 was already in our list), but again the resulting number fields had considerably larger discriminants than those in Table 1.

Acknowledgements

This work was supported in part by the Simons Foundation. The calculations were done with Maxima, (Pari-)GP, and Magma. I also thank Bjorn Poonen for information about Weierstrass torsion packets, Drew Sutherland for observing that the model of C' in [**3**] is not quite reduced, and John Jones for quickly adding the new fields F_j to the LMFDB. Finally I am grateful to the referees for their careful reading of the initial manuscript and for their detailed corrections and suggestions.

References

[1] Dominik Barth and Andreas Wenz, *On Elkies' method for bounding the transitivity degree of Galois groups*, J. Symbolic Comput. **108** (2022), 17–22, DOI 10.1016/j.jsc.2021.05.001. MR4265710

[2] Johan Bosman, *Modular forms applied to the computational inverse Galois problem.* preprint, 2011 (`arXiv:1109.6879v1`).

[3] Edgar Costa, Noam D. Elkies, Sachi Hashimoto, Aashraya Jha, Kimball Martin, Bjorn Poonen, and John Voight, *A database of curves with modular Jacobians* (2025), in preparation; data at `https://github.com/edgarcosta/ModularAbelianSurfaces/blob/master/olddata/label_to_curve_QQ.txt`

[4] Noam D. Elkies, *Shimura curves for level-3 subgroups of the* $(2,3,7)$ *triangle group, and some other examples*, Algorithmic number theory, Lecture Notes in Comput. Sci., vol. 4076, Springer, Berlin, 2006, pp. 302–316, DOI 10.1007/11792086_22. MR2282932

[5] Noam D. Elkies, *The complex polynomials* $P(x)$ *with* $\mathrm{Gal}(P(x)-t)\cong M_{23}$, ANTS X—Proceedings of the Tenth Algorithmic Number Theory Symposium, Open Book Ser., vol. 1, Math. Sci. Publ., Berkeley, CA, 2013, pp. 359–367, DOI 10.2140/obs.2013.1.359. MR3207422

[6] Hang Fu and Michael Stoll, *Elliptic curves with common torsion x-coordinates and hyperelliptic torsion packets*, Proc. Amer. Math. Soc. **150** (2022), no. 12, 5137–5149, DOI 10.1090/proc/16102. MR4494592

[7] Franck Leprévost, *Jacobiennes de certaines courbes de genre 2: torsion et simplicité* (French), J. Théor. Nombres Bordeaux **7** (1995), no. 1, 283–306, DOI 10.5802/jtnb.144. Les Dix-huitièmes Journées Arithmétiques (Bordeaux, 1993). MR1413580

[8] The LMFDB Collaboration, *The L-functions and modular forms database*, `http://www.lmfdb.org` and `http://beta.lmfdb.org`, 2024 [Online; accessed December 2024 and January 2025].

[9] Bjorn Poonen, *Genus-two curves with 22 torsion points* (English, with English and French summaries), C. R. Acad. Sci. Paris Sér. I Math. **330** (2000), no. 7, 573–576, DOI 10.1016/S0764-4442(00)00222-6. MR1760441

[10] Bjorn Poonen, *Computing torsion points on curves*, Experiment. Math. **10** (2001), no. 3, 449–465. MR1917430

[11] Kisao Takeuchi, *Commensurability classes of arithmetic triangle groups*, J. Fac. Sci. Univ. Tokyo Sect. IA Math. **24** (1977), no. 1, 201–212. MR463116

[12] John Voight, *Shimura curves of genus at most two*, Math. Comp. **78** (2009), no. 266, 1155–1172, DOI 10.1090/S0025-5718-08-02163-7. MR2476577

[13] John Voight, *Quaternion algebras*, Graduate Texts in Mathematics, vol. 288, Springer, Cham, [2021] ©2021, DOI 10.1007/978-3-030-56694-4. MR4279905

Appendix A. GP code

We give GP code that uses the formulas in §3.4 to compute a polynomial $P_{33}(j,X)$, of degree 33 in X, such that $P_{33}(j_{\mathsf{A}},x)=0$. Then $F_j=\mathbf{Q}[x]/(P_{33}(j,x))$ for any $j\in\mathbf{Q}$ such that the degree-33 polynomial $P_{33}(j,\cdot)$ is irreducible.

```
Y; /* to make Y the "main variable" */

C0 = -(2*X-1) * (14*X^5 + 45*X^4 + 28*X^3 + 29*X^2 + 60*X - 52);
C = 4*X^6 * subst(C0, X, (X+2)/(2*X));
/* C = 64*X^5 - 1095*X^4 - 1248*X^3 - 1224*X^2 - 640*X - 112 */

f1 = 32*X^3 + 29*X^2 + 24*X - (Y+4);
f2 = 16*X^3 + 21*X^2 + (Y+12);

c0 = -17/8;
```

```
A = X^4 + (115*X^3 + 10*X^2 + 356*X - 248)/3;
B = (545*X^4 + 2991*X^3 + 3298*X^2 - 12620*X - 1464) / 3;

s3 = Y - (77*X-66)/8 + c0*(8*X^2+3*X+2);
s11 = A * (Y - (37*X-2))  +  B * (X^2+X+2);

DENOM = s3^11 * Mod(subst(f1^4*f2^2,Y,-Y), Y^2 - C);
NUM = 2^16 * 12^3 * s11^3 * f1 * Mod(subst(f2,Y,-Y), Y^2 - C);

/* now j = ((X^2 + X + 2) * (8*X^2 + 3*X + 2)^9 * NUM) / DENOM */

P33 = norm((X^2 + X + 2) * (8*X^2 + 3*X + 2)^9 * NUM - j*DENOM);
P33 /= -2^59 * (X^2 + X + 2)^8 * (8*X^2 + 3*X + 2)^21

/*
P33 = (j-1728)^2 * X^33 + ... - 2^33 * (j^2 + 10726*j + 2^12*71^3)
check:  polredabs(subst(P33,j,-121)) = X^33 - 6*X^22 + 14*X^11 + 2
*/
```

Department of Mathematics, Harvard University, Cambridge, MA 02138
Email address: elkies@math.harvard.edu

Contemporary Mathematics
Volume **840**, 2026
https://doi.org/10.1090/conm/840/16806

Fine structure in some landscapes of L-functions

David W. Farmer, Sally Koutsoliotas, Stefan Lemurell,
and David P. Roberts

Abstract. We present three examples in an emerging theory of fine structure in landscapes of L-functions. These examples illustrate how the classical principles of zero repulsion and zero rigidity underlie the novel phenomena of ditch avoidance and congruence bias.

In our LuCaNT 2023 paper [**3**] we organized collections of L-functions by representing each L-function as an "L-point" in a Euclidean region, thereby obtaining what we call landscapes. The last paragraph in [**3**, §3.1] asked for an explanation of a striation phenomenon very visible in the landscape presented as [**3**, Figure 3.1]. The report [**8**] continued the study of this particular landscape, emphasizing certain ***fine structure*** that underlies the striation phenomenon. It also sketched how we expect such fine structure to be present in general landscapes. This theory is still mainly in an experimental stage, and it is important to have data from as many landscapes as possible so as to, for example, understand how the strength of fine structure varies in different regimes.

Sections 1, 2, and 3 draw increasingly complex landscapes and explain with reference to these examples what we mean by fine structure. Section 4 uses zeros of the L-functions in question to heuristically explain striations in terms of ***ditch avoidance***. Section 5 explains how the nature of the ditches causes a ***congruence bias*** that we did not even suspect when we wrote [**3**]. Section 6 was added in the revision stage of this paper; it uses the newly found [**6**] as an essential starting point, and takes next steps towards making the theory rigorous. Because of the examples-first structure of this paper, some readers may want to take an early look at the later more general sections. For example, understanding trapped zeros as illustrated by Figure 4 facilitates the reading of the first three sections.

We follow the general conventions of [**3**]. Our analysis uses only the functional equation, which we write in the form

$$\Lambda_L(s) := G_L(s)L(s) = \varepsilon_L\overline{\Lambda}_L(1-s)$$

where $\overline{f}(z) := \overline{f(\overline{z})}$. We refer to $G_L(s)$ as "the Γ-factors," but it also contains a factor $N^{s/2}$ where N is the conductor of the L-function. We refer to ε_L as "the sign," and it is a complex number on the unit circle. Central to our theory is that the Γ-factors and the sign of a given L-function give a "best first guess" at the locations of the critical zeros. These best first guesses involve the Riemann-Siegel theta function and close analogs, and are slightly too complicated to describe here.

The degree 2 data we use come from numerical computations of [**9**], with initial segments of the data being made completely rigorous in the LMFDB [**7**]. Our degree 3 data were generated using "L-functions out of nothing" [**2**, **5**] and are thus heuristic; data are available at [**4**].

1. Fixed sign, one spectral parameter

The two simplest nontrivial landscapes come from Maass forms on $\mathrm{SL}(2,\mathbb{Z})$. The corresponding degree 2 L-functions have $G_L(s) = \Gamma_\mathbb{R}(s+\delta+i\lambda)\Gamma_\mathbb{R}(s+\delta-i\lambda)$, where $(\delta, \varepsilon_L) = (0,1)$ for even Maass forms, and $(1,-1)$ for odd Maass forms. The landscapes are drawn in Figure 1, with ditches represented by vertical segments. They occur where the best first guess places a zero at $\frac{1}{2}+\lambda i$, in which case it also places a zero at $\frac{1}{2}-\lambda i$. The ditches occur at the local minima of $w_\pm(t) = \cos\left(2t\log\left(\frac{t}{e\pi}\right) \mp \frac{\pi}{4}\right)$, where $+$ and $-$ on the left indicate the even and odd cases.

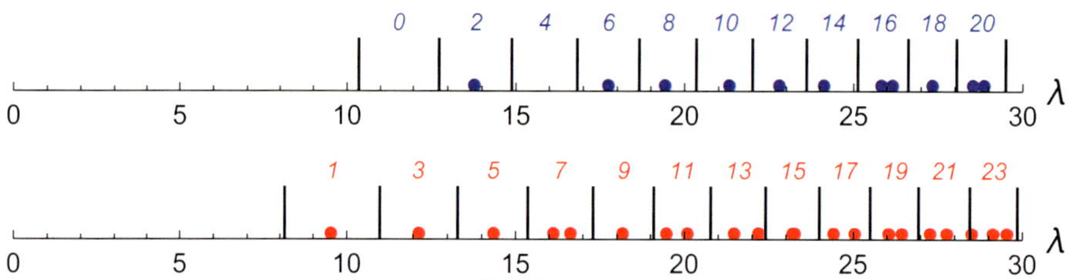

FIGURE 1. L-points of even (top) and odd (bottom) Maass forms on $\mathrm{SL}(2,\mathbb{Z})$.

A key role in our theory in this instance is the number z_1 of ***trapped zeros*** of a given L-function. A zero is trapped when its imaginary part is in $(-\lambda,\lambda)$. An interval between adjacent ditches is labeled by the expected number of trapped zeros, based on the best first guess for their locations. A key point is that this number, as well as its analogs in later sections, is computable, as the ditches are, from $G_L(s)$ and ε_L only.

One can see that there is a tendency of L-points to avoid ditches, and that this tendency is stronger in the even case. Not evident from the picture is that, for all the drawn L-points, the number of trapped zeros agrees with the expected number of trapped zeros. The ditch avoidance, and the matching of actual and expected numbers of trapped zeros, are aspects of fine structure.

A more analytic way to capture ditch avoidance is to consider the sum $f_\varepsilon(T) = \sum_{\lambda<T}^{\varepsilon} w_\varepsilon(\lambda)$, with the superscript on the sum indicating that only λ of parity ε are included. If there were no ditch avoidance phenomenon, since w_ε is 0 on average one would expect $f_\varepsilon(T)$ to remain very small as the cutoff T increases, and moreover, to exhibit chaotic oscillations. However, the observed behavior is very different as follows. In the interval $[0,1400]$ the data from [**9**] says that there are 77926 and 80044 different λ coming from even and odd Maass forms, respectively. One has $f_+(1400) \approx 6250.1$ and $f_-(1400) \approx 236.1$. The numerical fits $f_+(T) \approx 0.111\,T^{1.509}$ and $f_-(T) \approx 3.433\,T^{0.582}$ are quite accurate over the entire interval $[0,1400]$. Thus

ditch avoidance persists well past the window of Figure 1, being strong in the even case and much weaker in the odd case. This purely experimental finding on $f_+(T)$ will be improved in Section 6, while the experimental finding on $f_-(T)$ is too subtle for current theory.

2. Varying sign, one spectral parameter

Our next example concerns L-functions of degree 3 and conductor 4. If the local representation at 2 is Steinberg, then the sign is a cube root of 1, and otherwise the sign is not restricted. In Figure 2 we combine both cases, with the argument of the sign on the vertical axis, and the spectral parameter λ on the horizontal axis. We could draw the ditches sharply, like we did in Figure 1 of [8]. However we present them instead as they were first encountered in our numerical explorations.

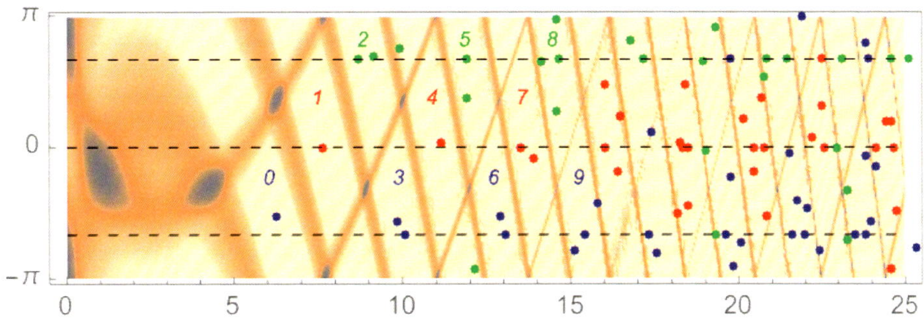

FIGURE 2. L-points with Γ-factor $G_L = 4^{s/2}\Gamma_{\mathbb{R}}(s+i\lambda)\Gamma_{\mathbb{C}}(s+\frac{3}{2}-i\lambda/2)$.

An upward-sloping ditch occurs when the best first guess places a zero at $\frac{1}{2}-\lambda i$ while a downward-sloping ditch occurs when it places a zero at $\frac{1}{2}+\frac{\lambda}{2}i$. The L-points have a strong tendency to avoid the former and a weaker tendency to avoid the latter. Here a zero is trapped if its imaginary part is in $(-\lambda, \frac{\lambda}{2})$. As before, the numbers indicate the expected number of trapped zeros. In the above figure, for each L-point the actual number z_1 of trapped zeros is within one of the expected number. To make the actual number readable on the picture, we color points blue, red, or green according to whether z_1 is 0, 1, or 2 modulo 3. In this figure most of the L-points have the same color as their region; the first deviation being the green L-point in the blue region 3 actually has only 2 trapped zeros.

3. Two spectral parameters

Our final example revisits Figure 3.2 of [**3**], but now with superimposed ditches and L-point coloring. Striations were clearly visible before, but are now explained by ditches which go between them.

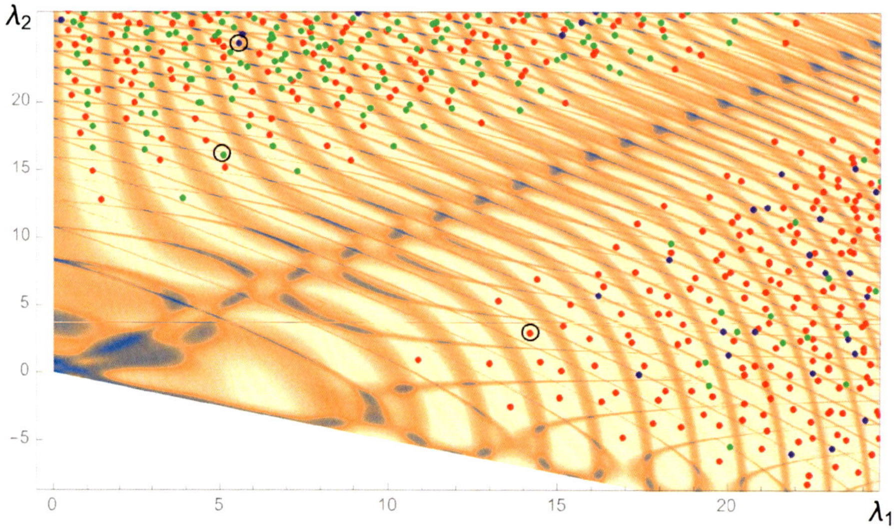

FIGURE 3. L-points with $G_L(s)=\Gamma_{\mathbb{R}}(s+i\lambda_1)\Gamma_{\mathbb{R}}(s+1+i\lambda_2)\Gamma_{\mathbb{R}}(s+1+i\lambda_3)$

As illustrated in Figure 4 below, each L-function has an associated vector (z_1, z_2), where z_j is the number of critical zeros with imaginary part in $(\lambda_j, \lambda_{j+1})$. Each region in Figure 3 also has such an integral vector: the expected vector of trapped zeros for L-points in that region. From the general recipe from [**8**], we focus on the difference $c = z_2 - z_1$ in $\mathbb{Z}/3$. L-points are colored blue, red, or green, according to their class c. About 80% of the L-points are in their correct region, so that one can infer the color of regions from the figure. In the upper regime, area is divided evenly between red = 1 and green = 2, with no area allocated to blue = 0. In the lower regime, each c occurs, with red = 1 occurring in hexagons, and the remaining classes occurring only in small triangles, which are six times smaller.

4. Zero repulsion and zero rigidity force ditch avoidance

Here we explain the section title via our third example, the other two examples being similar but easier. The three plots in Figure 4 show the trivial and critical zeros of the circled L-points in Figure 3, in order of increasing horizontal coordinate. Note that there is an evident ***zero repulsion*** between the trivial zeros and critical zeros, and the effect is greater when the trivial zeros are closer to the critical line.

The best guess at zeros is indicated by horizontal segments. They come from the first term of the Riemann-Siegel formula, which depends only on $G_L(s)$ and ε_L. The fact that the actual zeros do not stray too far from their predicted locations is ***zero rigidity***. One can see rigidity in the case of the ζ-function by executing `Plot[{RiemannSiegelZ[t], 2 Cos[RiemannSiegelTheta[t]]}, {t, 0, 100}]` in *Mathematica* or on `wolframalpha.com` .

Recall that the ditches occur where a predicted zero agrees with a row of trivial zeros. Zero repulsion says that there should not be an actual zero nearby. Zero rigidity says that there should be an actual zero nearby. We resolve this contradiction by a ***ditch avoidance*** principle that says that L-points are less likely to occur near ditches.

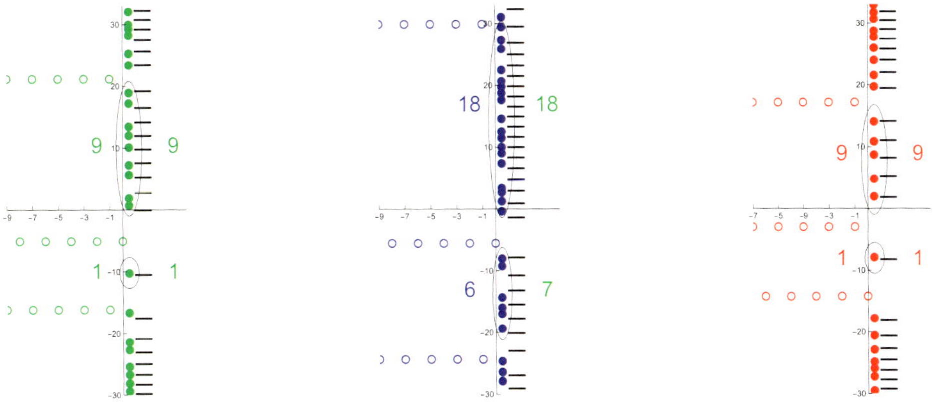

FIGURE 4. The zeros of the three circled L-functions in Figure 3

Zero repulsion and zero rigidity together make the actual number of zeros likely to be close to the predicted number of zeros. Deviations from prediction can be expected to be greater as conductors or spectral parameters increase. In Figure 4, the prediction is correct for the outer two L-functions. However, the blue L-point indexing the inner L-function is in a triangle in Figure 3 corresponding to the color green. For L-points in this triangle, the prediction is $(7, 18)$. This particular L-function with its actual $(z_1, z_2) = (6, 18)$ is off by one.

5. Region sizes cause congruence bias

The outline in [**8**] says that for degree d L-functions, both regions and L-points have a class in $\mathbb{Z}/d\mathbb{Z}$. For fixed ε, the probability measure on $\mathbb{Z}/d\mathbb{Z}$ induced by the region sizes can be calculated and it differs markedly from the uniform distribution. This deviance was already discussed in the last paragraph of Section 3, where classes 0, 1, 2 have an area split of $(0, 1/2, 1/2)$ and $(1/8, 3/4, 1/8)$ in the upper and lower regions, respectively. To the extent that L-points tend to be in their correct region, their class coming from the configuration of trapped zeros would be biased too.

In our three examples, congruence bias looks as follows. For Figure 1, on each line only one parity occurs for z_1. One can say that the extreme congruence bias is enforced by the parity of the L-function. In Figure 2, each dashed line is 50% in its dominant color and 25% in the other colors. But the L-points are almost entirely in the dominant color. In Figure 3, the L-point bias is moderately close to the region bias, in both the upper and lower regimes.

6. Towards a rigorous theory of fine structure

While revising this paper, we found a paper by C. J. Mozzochi [**6**]. Mozzochi works with $w_q(t) = \cos(2t\log\left(\frac{qt}{e\pi}\right) - \frac{\pi}{4})$, where $q > 0$ is a parameter. Summing over λ of both parities, he remarks that, translating to our style of cutoffs, the "trivial bound" for $\sum_{\lambda<T} w_q(\lambda)$ from the Selberg trace formula is $O(T^{3/2})$. His Theorem 1 then says that for q the reciprocal of an integer, the exponent $3/2$ cannot be lowered. Thus, taking $q = 1$, [**6**] is proving the indefinite continuation of ditch avoidance in the classical context of $SL_2(\mathbb{Z})$, already in 1999! Moreover, it proves that the λ also avoid other collections of ditches, one for each reciprocal integer.

We are in the process of generalizing and sharpening Mozzochi's theorem to

$$\sum_{\lambda<T}^{\varepsilon} w_q(\lambda) = \frac{a_q^\varepsilon T^{3/2}}{3\pi^{3/2}} + o(T^{3/2}),$$

with the expectation that we can obtain a power-saving error term. Thus we are refining by fixing the parity ε and giving a specific growth constant. The factor a_q^ε is 0 if q is irrational and is given by an expression involving Kloosterman sums if q is rational. For $q = 1, 2, 3, 4$ our formula gives $a_q^+ = 2, 4/3, 3/8, 0$ and $a_q^- = 0, 0, -9/8, -4/3$. As two examples, least square fits as in §1 except with the exponent fixed at 3/2 give $\sum_{\lambda<T}^+ w_1(\lambda) \approx 2 \cdot 0.05966\, T^{3/2}$ and $\sum_{\lambda<T}^- w_4(\lambda) \approx -4/3 \cdot 0.05984\, T^{3/2}$. These numerics compare well with $1/(3\pi^{3/2}) \approx 0.05986$.

We expect this approach will give similar results for ditch avoidance of λ coming from classical Maass forms of any level and character. A promising future direction is to see what the Selberg trace formula gives for larger groups, such as congruence subgroups of $SL_3(\mathbb{Z})$ as in §2 and §3. Another promising direction is to use the Kuznetsov trace formula instead. Note that the explanation of ditch avoidance via these trace formulae does not make direct reference to L-functions or their zeros! Thus one is still in search of an appropriate tool for rigorously understanding congruence bias.

References

[1] Farmer, David W., Currently there are no reasons to doubt the Riemann Hypothesis: the zeta function beyond the realm of computation. preprint. arXiv:2211.11671.

[2] David W. Farmer, Sally Koutsoliotas, and Stefan Lemurell, *Maass forms on* GL(3) *and* GL(4), Int. Math. Res. Not. IMRN **22** (2014), 6276–6301, DOI 10.1093/imrn/rnt145. MR3283005

[3] David W. Farmer, Sally Koutsoliotas, Stefan Lemurell, and David P. Roberts, *The landscape of L-functions: degree 3 and conductor 1*, LuCaNT: LMFDB, computation, and number theory, Contemp. Math., vol. 796, Amer. Math. Soc., Providence, RI, 2024, pp. 313–338, DOI 10.1090/conm/796/16007. MR4732693

[4] https://github.com/davidfarmer/LandscapeData

[5] https://github.com/lemurell/Euler/

[6] C. J. Mozzochi, *On the distribution of the eigenvalues of the hyperbolic Laplacian for* PSL(2, **Z**). *II*, J. Number Theory **86** (2001), no. 2, 284–301, DOI 10.1006/jnth.2000.2558. MR1813114

[7] The LMFDB Collaboration, The L-functions and modular forms database, https://www.lmfdb.org/ModularForm/GL2/Q/Maass/, 2025.

[8] Roberts, David P., Landscapes of L-functions, Oberwolfach Workshop 2436, report 40/2024, https://publications.mfo.de/handle/mfo/4254, p. 26-28.

[9] Holger Then, *Maass cusp forms for large eigenvalues*, Math. Comp. **74** (2005), no. 249, 363–381, DOI 10.1090/S0025-5718-04-01658-8. MR2085897

American Institute of Mathematics, Caltech 8-32, 1200 E California Blvd., Pasadena CA 91125
Email address: farmer@aimath.org

Department of Physics and Astronomy, Bucknell University, Lewisburg, PA 17837
Email address: s.koutsoliotas@bucknell.edu

Department of Mathematical Sciences, Chalmers University of Technology, 41296 Gothenburg, Sweden
Email address: sj@chalmers.se

Division of Science and Mathematics, University of Minnesota Morris, Morris, MN 56267
Email address: roberts@morris.umn.edu

Contemporary Mathematics
Volume **840**, 2026
https://doi.org/10.1090/conm/840/16807

Computing a database of rigorous Maass forms

David Lowry-Duda

ABSTRACT. We announce a database of rigorously computed Maass forms with trivial character on congruence subgroups $\Gamma_0(N)$ for squarefree level N. We also briefly describe the methods of computation and future plans.

1. Computation of Maass Forms

Let $\Gamma_0(N)$ be a congruence subgroup in $\mathrm{SL}(2,\mathbb{Z})$. Each matrix $\gamma = \left(\begin{smallmatrix} a & b \\ c & d \end{smallmatrix}\right) \in \mathrm{SL}(2,\mathbb{R})$ acts on the upper halfplane $\mathcal{H}$ in the standard way, given by $\gamma z = \frac{az+b}{cz+d}$. The quotient space $\Gamma_0(N)\backslash\mathcal{H}$ is a noncompact Riemann surface. In the hyperbolic metric on $\mathcal{H}$, the weight k Laplace-Beltrami operator, Δ_k, takes the form

$$\Delta_k = y^2\Big(\frac{\partial^2}{\partial x^2} + \frac{\partial^2}{\partial y^2}\Big) - iky\frac{\partial}{\partial x}.$$

Let χ denote a Dirichlet character of modulus N. Then Maass cuspforms of weight k and character χ on $\Gamma_0(N)$ are eigenfunctions f of Δ_k that satisfy

- $\Delta_k f + \lambda f = 0$,
- $f(\gamma z) = \chi(\gamma)(cz+d)^k f(z)$ for all $\gamma = \left(\begin{smallmatrix} a & b \\ c & d \end{smallmatrix}\right) \in \Gamma_0(N)$, and
- $f \in L^2(\Gamma_0(N)\backslash\mathcal{H})$.

Maass cuspforms constitute the discrete component of the spectral resolution of Δ_k and therefore form building blocks for all $L^2(\Gamma_0(N)\backslash\mathcal{H})$. The problem of giving numerical examples of Maass cuspforms has been widely considered since the 1970s. Despite this, it is difficult to give any nontrivial example of a Maass cuspform.

Each Maass form of weight k and character χ on $\Gamma_0(N)$ has a Fourier expansion of the form

$$f(z) = \sum_{n\neq 0} a(n)\sqrt{y}K_{ir}(2\pi|n|y)e^{2\pi i n x} \tag{1}$$

where $z = x + iy$, $K_\nu(y)$ is the modified Bessel function of the second kind, and the eigenvalue λ of f is given by $\lambda = \frac{1}{4} + r^2$ for $r \in \mathbb{R}$. This r is called the spectral parameter of f. To describe a Maass form, it's necessary to describe the eigenvalue λ (or equivalently the spectral parameter r) and the coefficients $a(n)$. In practice, most data associated to a generic Maass form is conjectured to be

---

2020 *Mathematics Subject Classification.* Primary 11F30, 11F03; Secondary 11Y99, 11-11.

This work was supported by the Simons Collaboration in Arithmetic Geometry, Number Theory, and Computation via the Simons Foundation grant 546235.

transcendental and algebraically independent of standard constants; hence one gives approximations to the eigenvalue and coefficients to specify a Maass form.

In [**The05**], H. Then gives an extensive list of references to earlier computations of Maass forms. Until 2006, methods to generate numerical examples were fundamentally *heuristic*, meaning that there is no guarantee that a claimed approximation is actually close to a true Maass form.

Booker, Strömbergsson and Venkatesh [**BSV06**] gave the first method of computing *rigorous* Maass forms on $\mathrm{SL}(2,\mathbb{Z})$. Stated more precisely, they gave a method that gives explicit intervals containing the eigenvalue and coefficients, and hence bounds for how close a putative Maass form is to a true Maass form. There are now three algorithms to generate rigorous Maass cuspforms:

(1) Quasimode construction, a generalization of [**BSV06**] to general level and character by Child [**Chi22**].
(2) A rigorous implementation of the Selberg trace formula due to Seymour-Howell [**SH22**],
(3) A rigorous version of Hejhal's algorithm due to Seymour-Howell and Lowry-Duda, to be described in forthcoming work [**LDSH**].

It is notable that Booker (and to a lesser extent, Strömbergsson) helped develop each of these algorithms.

In this note, we describe how an initial database of 35416 rigorous Maass cuspforms (all of weight 0, on congruence subgroups $\Gamma_0(N)$ with N squarefree) was computed using these algorithms and inserted into the LMFDB [**LMF19**]. We briefly describe the three algorithms and how they interact in §2. In §3, we detail the data now available. And in §4, we give additional comments on the current and future status of this database.

2. Algorithms for Rigorously Computing Maass Forms

We say that a Maass cuspform f_j is computed rigorously if there is an explicit interval $I_{j,\lambda}$ for the eigenvalue and explicit intervals $I_{j,n}$ for the coefficients such that there is exactly one true Maass form

$$f_j(z) = \sum_{n\neq 0} a_j(n)\sqrt{y}K_{ir_j}(2\pi|n|y)e^{2\pi inx}$$

with $\lambda_j = \frac{1}{4} + r_j^2 \in I_{j,\lambda}$ and $a_j(n) \in I_{j,n}$ for all n. In this note, we impose the additional restriction that there can be no missing eigenvalues, or rather that every eigenvalue up to λ_j must also have been produced.

We now briefly describe each of the three methods of computation.

Remark 1. A folklore conjecture states that the vector space of Maass forms on $\mathrm{SL}(2,\mathbb{Z})$ with a given eigenvalue is one-dimensional. This would imply that a sufficiently tight interval $I_{j,\lambda}$ would be sufficient to uniquely specify a Maass form on $\mathrm{SL}(2,\mathbb{Z})$. This is true for all of the Maass forms in the initial database. In general, however, there are spaces with nontrivial character in which multiple Maass forms have the same eigenvalue. In this case, both $I_{j,\lambda}$ and a finite number of $I_{j,n}$ are necessary to uniquely identify the form f_j.

2.1. Rigorous Trace Formula. With extreme simplification, the Selberg trace formula in [**SH22**, Theorem 3.1] describes sums over coefficients weighted by "nice" analytic test functions F, taking the form

$$\sum_{j>0} F(r_j)a_j(n) = (\text{complicated but explicit}).$$

This sum is over all Maass forms f_j on $\Gamma_0(N)$. The omitted right hand side is in terms of data associated to conjugacy classes of the group $\Gamma_0(N)$ and the Fourier transform $\widehat{F}$.

One significant difficulty in this approach comes from the hyperbolic contribution. These terms include special values $L(1, (d/\cdot))$ for d of the form $(t^2 - 4n)$ for $t \in \mathbb{Z}$, and hence involve computing class numbers and regulators of quadratic fields by the class number formula. Choosing test functions F with the property that $\widehat{F}$ has compact support allows one to compute data for only finitely many quadratic fields. By carefully combining different choices of these test functions, it's possible to isolate intervals for individual eigenvalues and coefficients.

REMARK 2. This relationship between class numbers and Maass forms can also be used in reverse. With ample Maass form data, one can compute *more* class numbers. See [**BBD+24**] for more in this direction.

This algorithm is distinguished in that it can *guarantee that all Maass forms have been found* in an eigenvalue range. Every eigenvalue in the database was isolated and certified with the trace formula. In so doing, low-quality approximations to each Maass form were also obtained for the other methods to refine.

2.2. Quasimode construction. In [**Chi22**], Child extends the quasimode construction technique in [**BSV06**] to general level. If $\widetilde{f}_j$ is a putative Maass form and $(\Delta - \lambda)\widetilde{f}_j$ has small L^2 norm, then a spectral resolution shows that $\widetilde{f}_j$ is close to a true eigenfunction. Child (and BSV) show that it is sufficient to obtain strong bounds along the boundary of the fundamental domain. This method can be used to certify heuristic approximations to Maass forms.

Quasimode construction requires *extremely precise* approximations (i.e. hundreds or thousands of digits of precision). This works particularly well in level one, where heuristic approximations are the easiest to generate. All level 1 forms in the database were computed using quasimode construction (and the trace formula guaranteed that no forms were omitted).

2.3. Rigorous Hejhal. Hejhal's algorithm [**Hej99**] is one of the better-known algorithms for producing heuristic approximations to Maass forms. Strömberg described how to adapt Hejhal's heuristic algorithm for general level multiplier systems in [**Str05**].

The fundamental idea is to use automorphy to construct linear systems of equations for the coefficients. Rapid decay from the Bessel functions in (1) shows that the truncation $\widetilde{f}_j$ to the first M coefficients is close to f_j. Taking an appropriate linear combination of $\widetilde{f}_j$ at points $z_j = x_j + iY$ along a fixed horocycle, one obtains equations of the form

$$a_j(n)\sqrt{Y}K_{ir_j}(2\pi|n|Y) = \frac{1}{2Q}\sum_{j=1-Q}^{Q} \widetilde{f}(z_j)e(-nx_j) + (\text{truncation error}).$$

If the horocycle points are chosen so that each z_j is outside the fundamental domain, then we compute the pullbacks z_j^* to the fundamental domain and insert those instead. This introduces nonlinear mixing into the system, heuristically improving the condition number of the system of equations. Solving this system would give an approximation to a Maass form.

The challenge is that the spectral parameter r_j is also unknown. Instead, one guesses values of r_j and obtains heuristic coefficients for a form that would have eigenvalue $\frac{1}{4} + r_j^2$. Based on the behavior of these heuristic guesses, one can try to adjust the values r_j to try to find eigenvalues yielding coefficients with behaviors that behave more like actual coefficients.

The rigorous implementation of Hejhal's algorithm makes the truncation and other errors explicit and tracks how an initial approximation (e.g. coming from the trace formula), guaranteed to some initial precision, behaves under iteration of Hejhal's algorithm. With a sufficiently accurate initial approximation, iteration will yield provably stronger approximations. We applied this algorithm to the low-quality approximations coming from the trace formula for levels $N \geq 2$; unfortunately, the initial approximations were sometimes too weak and the algorithm failed to improve the precision.

3. Data Computed

We rigorously computed 35416 Maass cuspforms on $\Gamma_0(N)$ across squarefree N from 1 to 105. For each Maass form, we compute the eigenvalue, the first 1000 coefficients, and a portrait (constructed using [**LD25**], building on similar methods in [**LD22**]). As an example, we consider the Maass form with the smallest eigenvalue on $\Gamma_0(15)$ This has the following data:

Level	15
Weight	0
Character	15.1
Symmetry	odd
Fricke Sign	-1
Spectral Parameter	$1.51842933602416588036746726626 \pm 6 \cdot 10^{-12}$
Spectral Index	1

In addition, the first 1000 coefficients are stored as (center, error) pairs.

The level, weight, character, and spectral parameter are defined as in §1. Here, the character is specified by its Conrey label. The Spectral Index gives the index of the eigenvalue among the sorted list of all eigenvalues of Maass forms of that level, weight, and character, starting at 1.

Each Maass form is also an eigenfunction under the reflection operator $J(z) = -\overline{z}$ and either has "even" symmetry (when $f_j(-\overline{z}) = f_j(z)$) or "odd" symmetry (when $f_j(-\overline{z}) = -f_j(z)$). Further, each Maass form is an eigenfunction under the Fricke involution. The Symmetry and Fricke Sign pieces of data give the eigenvalues under these operators, respectively.

REMARK 3. When available, we give additional heuristic digits of precision for spectral parameters. For example, the spectral parameter in this example is specified to more than 12 digits. We expect several of these digits to be accurate, but cannot yet guarantee this.

REMARK 4. In principle the symmetry type and Fricke sign can be computed from the rest of the Maass form data. But for 15423 of the 35416 Maass forms, the coefficients aren't computed with enough precision to identify the Fricke sign.

These pieces of data form the label, uniquely identifying Maass forms with a fixed eigenvalue. The full label takes the form

$$\mathsf{Label} = \mathsf{Level.Weight.ConreyIndex.SpecIndex.Disambig.}$$

The final term in the label is designed to disambiguate between the finitely many forms of the same level, weight, character, and eigenvalue: it gives the lexicographical index of the sequence $\{(\mathrm{Re}(a_j(n)), \mathrm{Im}(a_j(n)))\}_{n\geq 1}$ among those Maass forms. The form above has the label 15.0.1.1.1.

In the initial database, the weight is 0, the character is trivial, and the disambiguation index is always 1. Thus forms can also be specified by Level.SpecIndex, and the form above has short label 15.1. Both links work in the LMFDB.

In total, the database consists of approximately 4.954 GB of Maass form data.

It is easiest to compute level 1 forms and the database includes 2202 level 1 forms, each verified using the trace formula and quasimode construction algorithm. For each level $2 \leq n \leq 105$, we computed as many Maass forms as we could while guaranteeing that no eigenvalues were omitted.

4. Comments on Database Construction

Finally, we conclude with several small comments.

(1) For 15423 forms, the precision from the trace formula isn't high enough for the current rigorous implementation of Hejhal's algorithm to refine. For these forms, the database includes reasonable estimates on the eigenvalue but very poor estimates for the coefficients. Different methods to handle these problems are in development. The plots for the forms with low-quality coefficients are thus informal.

(2) By restricting to squarefree level, we omit Maass forms coming from induced representations of Hecke characters. These are Maass cuspforms that are *explicitly computable* (see e.g. [**Maa49**]). Algorithms to compute these efficiently have been recently implemented in PARI/GP by Molin and Page [**MP22**]. We also miss more complicated (and potentially interesting) Artin representations.

Forthcoming work of Booker, Bober, Knightly, Krishnamurthy, Lee, Lowry-Duda, and Seymour-Howell seeks to work out an explicit, computable trace formula for general level and weight.

(3) The database currently omits all L-functions of Maass forms. In principle, these can be constructed from current data by adjusting existing numerical algorithms. But current precision often doesn't allow accurate computation of the L-functions. This is left as a future project, after more accurate approximations are available.

Acknowledgements

We thank Andrew Booker, Min Lee, Brendan Hassett, Andrei Seymour-Howell, and Drew Sutherland for guidance and support throughout the computation. We also thank John Voight for several suggestions on how to improve the initial form of the database, as well as David Roe, who helped improve the database following these suggestions. This task was easier thanks to the work of Edgar Costa and David Roe, and the entire process they helped create around contributing to the LMFDB [**CR21**].

References

[BBD+24] Ce Bian, Andrew R. Booker, Austin Docherty, Michael J. Jacobson Jr., and Andrei Seymour-Howell, *Unconditional computation of the class groups of real quadratic fields*, LuCaNT: LMFDB, computation, and number theory, Contemp. Math., vol. 796, Amer. Math. Soc., Providence, RI, 2024, pp. 29–53, DOI 10.1090/conm/796/15996. MR4732682

[BSV06] Andrew R. Booker, Andreas Strömbergsson, and Akshay Venkatesh, *Effective computation of Maass cusp forms*, Int. Math. Res. Not., posted on 2006, Art. ID 71281, 34, DOI 10.1155/IMRN/2006/71281. MR2249995

[Chi22] Kieran Child, *Twist-minimal trace formula for holomorphic cusp forms*, Res. Number Theory **8** (2022), no. 1, Paper No. 11, 27, DOI 10.1007/s40993-021-00302-9. MR4356511

[CR21] Edgar Costa and David Roe, *Zen and the art of database maintenance*, Arithmetic geometry, number theory, and computation, Simons Symp., Springer, Cham, 2021, pp. 277–292, DOI 10.1007/978-3-030-80914-0_8. MR4427966

[Hej99] Dennis A. Hejhal, *On eigenfunctions of the Laplacian for Hecke triangle groups*, Emerging applications of number theory (Minneapolis, MN, 1996), IMA Vol. Math. Appl., vol. 109, Springer, New York, 1999, pp. 291–315, DOI 10.1007/978-1-4612-1544-8_11. MR1691537

[LD22] David Lowry-Duda, *Visualizing modular forms*, Arithmetic geometry, number theory, and computation, Simons Symp., Springer, Cham, 2021, pp. 537–557, DOI 10.1007/978-3-030-80914-0_19. MR4427977

[LD25] David Lowry-Duda, *Portrait code for Maass forms*, 2025, `https://github.com/davidlowryduda/maass_portraits`.

[LDSH] David Lowry-Duda and Andrei Seymour-Howell, *A rigorous implementation of Hejhal's algorithm*, Forthcoming.

[LMF19] The LMFDB Collaboration, *The L-functions and modular forms database*, `http://www.lmfdb.org`, 2019, [Online; accessed 30 October 2019].

[Maa49] Hans Maass, *Automorphe Funktionen und indefinite quadratische Formen* (German), S.-B. Heidelberger Akad. Wiss. Math.-Nat. Kl. **1949** (1949), no. 1, 42. MR31946

[MP22] Pascal Molin and Aurel Page, *Computing groups of Hecke characters*, Res. Number Theory **8** (2022), no. 4, Paper No. 91, 26, DOI 10.1007/s40993-022-00396-9. MR4496695

[SH22] Andrei Seymour-Howell, *Rigorous computation of Maass cusp forms of squarefree level*, Res. Number Theory **8** (2022), no. 4, Paper No. 83, 15, DOI 10.1007/s40993-022-00393-y. MR4493409

[Str05] Fredrik Strömberg, *Computational aspects of Maass waveforms*, Ph.D. thesis, Uppsala Universitet, 2005.

[The05] Holger Then, *Maass cusp forms for large eigenvalues*, Math. Comp. **74** (2005), no. 249, 363–381, DOI 10.1090/S0025-5718-04-01658-8. MR2085897

ICERM, 121 South Main Street, Box E, 11th Floor, Providence, RI, 02903
Email address: `david@lowryduda.com`

Contemporary Mathematics
Volume **840**, 2026
https://doi.org/10.1090/conm/840/16808

The spine of a supersingular ℓ-isogeny graph

Taha Hedayat, Sarah Arpin, and Renate Scheidler

ABSTRACT. Supersingular elliptic curve ℓ-isogeny graphs over finite fields offer a setting for a number of quantum-resistant cryptographic protocols. The security analysis of these schemes typically assumes that these graphs behave randomly. Motivated by this debatable assertion, we explore structural properties of these graphs. We detail the behavior, governed by congruence conditions on p, of the ℓ-isogeny graph over $\mathbb{F}_p$ when passing to the spine, i.e. the subgraph induced by the $\mathbb{F}_p$-vertices in the full ℓ-isogeny graph. We describe the diameter of the spine and offer numerical data on the number of vertices, over both $\mathbb{F}_p$ and $\overline{\mathbb{F}}_p$, in the center of the ℓ-isogeny graph. Our plots of these counts exhibit a wave-shaped pattern which supports the assertion that centers of supersingular ℓ-isogeny graphs exhibit the same behavior as those of random $(\ell+1)$-regular graphs.

1. Introduction

Supersingular elliptic curve isogeny graphs have undergone a surge of research activity in recent years, in part due to their suitability as a mathematical foundation for quantum-safe cryptographic applications. In particular, the path-finding problem in these graphs seems to be intractable even on a quantum computer. For a prime p, the supersingular isogeny graph $\mathcal{G}_\ell(\overline{\mathbb{F}}_p)$ has as its vertex set the $\overline{\mathbb{F}}_p$-isomorphism classes of supersingular elliptic curves, labeled by their j-invariants in $\mathbb{F}_{p^2}$. The directed edges of $\mathcal{G}_\ell(\overline{\mathbb{F}}_p)$ are the ℓ-isogenies between elliptic curves representing vertices, where ℓ is a (usually small) prime. Given two vertices, represented by two elliptic curves E, E' over $\overline{\mathbb{F}}_p$, the path-finding problem in $\mathcal{G}_\ell(\overline{\mathbb{F}}_p)$ asks to find a path from E to E' comprised of ℓ-isogenies. The presumed intractability of this problem provides security for a number of cryptographic protocols, including [**CLG09**, **DFKL+20**, **FFK+23**] and their variants.

It is well-known that supersingular ℓ-isogeny graphs are optimal expander graphs, and are in fact Ramanujan graphs when $p \equiv 1 \pmod{12}$. The security analysis of supersingular isogeny based cryptographic schemes typically assumes that $\mathcal{G}_\ell(\overline{\mathbb{F}}_p)$ behaves like a "random" Ramanujan graph, a supposition that has

2020 *Mathematics Subject Classification.* 14H52, 11G20, 11-04, 11-11, 05C40.

Key words and phrases. Supersingular elliptic curve ℓ-isogeny graph, spine, graph diameter, graph center.

The first and third author acknowledge the support of the *Natural Sciences and Engineering Research Council of Canada* (NSERC).

The second author is a faculty fellow of the Commonwealth Cyber Initiative and is supported by an AMS-Simons Travel Grant.

since been called into question. For example, the p-power Frobenius acting on $\mathcal{G}_\ell(\overline{\mathbb{F}}_p)$ pairs up paths with their Frobenius conjugate paths. It also fixes vertices in $\mathbb{F}_p$ and those ℓ-isogenies between them that are defined over $\mathbb{F}_p$. Moreover, path-finding becomes substantially easier when the start and end vertices belong to $\mathbb{F}_p$ [**DG16, CJS14**]. Such special structural features exhibited by the subgraph of $\mathcal{G}_\ell(\overline{\mathbb{F}}_p)$ induced by the $\mathbb{F}_p$-vertices, referred to as the *spine* of $\mathcal{G}_\ell(\overline{\mathbb{F}}_p)$, may make it possible to distinguish a supersingular elliptic curve ℓ-isogeny graph from a random optimal expander or Ramanujan graph.

These questions prompted the authors of [**ACNL+23**] to launch a thorough investigation into the spine $\mathcal{S}_\ell^p$ of $\mathcal{G}_\ell(\overline{\mathbb{F}}_p)$. To that end, they considered the supersingular isogeny $\mathbb{F}_p$-graph $\mathcal{G}_\ell(\mathbb{F}_p)$, where the vertices are now $\mathbb{F}_p$-isomorphism classes of supersingular elliptic curves and the edges are $\mathbb{F}_p$-rational ℓ-isogenies. The structure of this graph is well understood and was described in detail in [**DG16**]. There is a natural two-step process of passing from $\mathcal{G}_\ell(\mathbb{F}_p)$ to the spine $\mathcal{S}_\ell^p \subset \mathcal{G}_\ell(\overline{\mathbb{F}}_p)$: vertices in $\mathcal{G}_\ell(\mathbb{F}_p)$ corresponding to twists of curves are identified in $\mathcal{S}_\ell^p$ (with any $\mathbb{F}_p$-isogenies between them turning into loop edges), and edges arising from ℓ-isogenies not defined over $\mathbb{F}_p$ are then added. In [**ACNL+23**], the possible behaviors exhibited by the connected components of $\mathcal{G}_\ell(\mathbb{F}_p)$ under this process were analyzed in detail. Here, we expand on this exploration, refining the results of [**ACNL+23**] and offering new findings.

Our contribution herein is two-fold. In Section 4, we describe all the ways, characterized by explicit congruence conditions on p, in which components of $\mathcal{G}_\ell(\mathbb{F}_p)$ can behave when passing to $\mathcal{S}_\ell^p$ for the arguably most interesting case of $\ell = 2$. In Section 5, we do the same for $\ell = 3$ and give a road map of how to extend our approach to isogeny degrees $\ell \geq 5$.

Leveraging our spine structure results, we describe the diameter (the largest possible directed distance between any pair of vertices) in any component of $\mathcal{S}_\ell^p$ in Section 6. This is followed by an extensive numerical investigation of the center of $\mathcal{G}_\ell(\overline{\mathbb{F}}_p)$, i.e. the set of vertices for which the largest distance to any other vertex is minimal. Center vertices can be thought of as having increased connectivity to the rest of the graph compared to the vertices outside the center. The fact that Frobenius is a graph automorphism on $\mathcal{G}_\ell(\overline{\mathbb{F}}_p)$ that fixes precisely the $\mathbb{F}_p$-vertices might suggest that spine vertices are more prominently represented in the center of $\mathcal{G}_\ell(\overline{\mathbb{F}}_p)$ than vertices not defined over $\mathbb{F}_p$. Our numerical experiments for $\ell = 2, 3$ and the first 2260 primes $p \neq 2, 3$ (i.e. $5 \leq p < 20000$) demonstrate that this is in fact not the case for this range of parameters, thereby providing evidence against this claim. Plots of counts of $\mathbb{F}_p$-vertices belonging to the center of $\mathcal{G}_\ell(\overline{\mathbb{F}}_p)$ exhibit a wave pattern, where wave peaks become higher and are spaced increasingly far apart for larger values of p. Similar wave shapes appear for $\ell = 3$, and this behavior is more pronounced when plotting the size of the entire center of $\mathcal{G}_2(\overline{\mathbb{F}}_p)$ rather than just the number of its spine vertices. Although visually striking, this pattern is in fact evidence that the center size of $\mathcal{G}_\ell(\overline{\mathbb{F}}_p)$ behaves like that of a random $(\ell + 1)$-regular graph.

1.1. Accompanying data and code. The data used to create the figures in this work, as well as the SageMath [**S+25**] code used to generate that data, can be found at the GitHub repository [**Hed25**]. Consult the README.md file for a list of the included files and their functionalies. This code is modified from its original version, which was created by the first author in the summer of 2023. Data collected in `.csv` files was generated using SageMath 10.4 [**S+25**] on a MacBook Pro, Apple M3, with 16 GB of memory, running macOS Sonoma 14.6.1.

1.2. Acknowledgments. We thank Jonathan Love, Jonathan Komada Eriksen and Thomas Decru for their observations and comments in Section 6.3. We are also grateful to our anonymous referees for their comprehensive reviews and their constructive and helpful comments.

2. Supersingular elliptic curve isogeny graphs

Let p be a prime, $\mathbb{F}_p$ the finite field of p elements and $\overline{\mathbb{F}}_p$ a fixed algebraic closure of $\mathbb{F}_p$. We consider supersingular elliptic curves over $\overline{\mathbb{F}}_p$ and recall that every $\overline{\mathbb{F}}_p$-isomorphism class of such curves contains a representative that is defined over $\mathbb{F}_{p^2}$. Isomorphism classes of supersingular elliptic curves are classified by their j-invariant, which is thus an element of $\mathbb{F}_{p^2}$. The j-invariants of curves with extra automorphisms may or may not be supersingular; specifically, $j = 1728$ is supersingular if and only if $p \equiv 3 \pmod 4$, and $j = 0$ is supersingular if and only if $p \equiv 2 \pmod 3$.

While the endomorphism ring of a supersingular elliptic curve defined over $\overline{\mathbb{F}}_p$ is isomorphic to a maximal order in a quaternion algebra, the ring of endomorphisms defined over $\mathbb{F}_p$ of any elliptic curve over $\mathbb{F}_p$ is isomorphic to an imaginary quadratic order. Specifically, when $p \equiv 1 \pmod 4$, all supersingular elliptic curves defined over $\mathbb{F}_p$ have $\mathbb{F}_p$-endomorphism ring isomorphic to the maximal order $\mathbb{Z}[\sqrt{-p}]$ of discriminant $-4p$. When $p \equiv 3 \pmod 4$, all such curves have $\mathbb{F}_p$-endomorphism ring isomorphic to either the maximal order $\mathbb{Z}[(1+\sqrt{-p})/2]$ of discriminant $-p$ or its index 2 suborder $\mathbb{Z}[\sqrt{-p}]$ of discriminant $-4p$. For any quadratic discriminant Δ, denote by $h(\Delta)$ the class number (i.e. the size of the class group) of the quadratic order of discriminant Δ.

PROPOSITION 2.1 (Class number parity, $p \equiv 3 \pmod 4$). *If $p \equiv 3 \pmod 4$, then $h(-p)$ is odd.*

PROOF. For any fundamental discriminant $\Delta < 0$, the 2-rank of the class group of the quadratic order $\mathcal{O}_\Delta$ of discriminant Δ, i.e. its number of 2-Sylow factors, is one less than the number of prime factors of Δ by genus theory (see [**JW09**, p. 170] for example). Hence $h(-p)$ is odd when $p \equiv 3 \pmod 4$. □

When $p \equiv 1 \pmod 4$, the 2-rank of the class group of $\mathbb{Q}(\sqrt{-p})$ is 1, so $h(-4p)$ is even. But this does not determine the 2-adic valuation of $h(-4p)$.

Now fix a prime ℓ. We associate two directed graphs to the set of supersingular elliptic curves and their ℓ-isogenies. For both graphs, the set of vertices does not depend on ℓ, but the set of edges does. Two isogenies are said to be $\mathbb{F}_p$-*equivalent* (resp., $\overline{\mathbb{F}}_p$-*equivalent*) if they are equal up to post-composition with an $\mathbb{F}_p$-automorphism (resp., an $\overline{\mathbb{F}}_p$-automorphism).

DEFINITION 2.2 (Supersingular elliptic curve isogeny graphs). Let p and ℓ be primes. Define the following two graphs.

- The *supersingular elliptic curve ℓ-isogeny graph over* $\mathbb{F}_p$, denoted $\mathcal{G}_\ell(\mathbb{F}_p)$, is the directed graph whose vertices are $\mathbb{F}_p$-isomorphism classes of supersingular elliptic curves and whose edges are ℓ-isogenies of these curves up to $\mathbb{F}_p$-equivalence.

- The *supersingular elliptic curve ℓ-isogeny graph over $\overline{\mathbb{F}}_p$*, denoted $\mathcal{G}_\ell(\overline{\mathbb{F}}_p)$, is the directed graph whose vertices are $\overline{\mathbb{F}}_p$-isomorphism classes of supersingular elliptic curves and whose edges are ℓ-isogenies of these curves up to $\overline{\mathbb{F}}_p$-equivalence.

By identifying ℓ-isogenies with their duals, the two graphs become undirected. Note that due to post-composition by the extra automorphisms at j-invariants 0 and 1728, this process may identify two or three directed edges in $\mathcal{G}_\ell(\overline{\mathbb{F}}_p)$ with the same edge in the other direction; see [**ACNL+23**, Rem. 2.3]. We will work with the undirected variants when it is convenient and edge direction does not matter.

It is well-known that the undirected variant of $\mathcal{G}_\ell(\overline{\mathbb{F}}_p)$ is an optimal expander graph, and is in fact an $(\ell+1)$-regular Ramanujan graph when $p \equiv 1 \pmod{12}$. We recall three key results governing the structure of $\mathcal{G}_\ell(\overline{\mathbb{F}}_p)$:

- For any j-invariant j, the neighbors of j in $\mathcal{G}_\ell(\overline{\mathbb{F}}_p)$ are precisely the roots of $\Phi_\ell(j, Y) \pmod p$. The multiplicity of an edge joining j to j' is the multiplicity of the root j' of $\Phi_\ell(j, Y) \pmod p$. An explicit list of modular polynomials of can be found at [**Sut**].
- Two (not necessary distinct) j-invariants j, j' are joined by a multi-edge in $\mathcal{G}_\ell(\overline{\mathbb{F}}_p)$ if both are roots of the polynomial $\mathrm{Res}_\ell(X) \pmod p$, where

$$\mathrm{Res}_\ell(X) = \mathrm{Res}\left(\Phi_\ell(X,Y), \frac{\partial}{\partial Y}\Phi_\ell(X,Y);\ Y\right), \tag{2.1}$$

  the resultant of the level ℓ modular polynomial $\Phi_\ell(X,Y)$ and its partial derivative with respect to Y when both are considered as polynomials in Y with coefficients in $\mathbb{Z}[X]$.
- The roots of the Hilbert class polynomial $H_\Delta(X)$ for the imaginary quadratic order $\mathcal{O}_\Delta$ of discriminant Δ are precisely the j-invariants with endomorphism ring isomorphic to $\mathcal{O}_\Delta$. Such a j-invariant is supersingular if and only if p does not split in $\mathcal{O}_\Delta$ (see [**Lan87**, Ch. 13, Thm. 12].

The structure of $\mathcal{G}_\ell(\mathbb{F}_p)$ for $p \geq 5$ was first described in [**DG16**]. Following volcano terminology first introduced in [**FM02**], vertices in $\mathcal{G}_\ell(\mathbb{F}_p)$ corresponding to elliptic curves with $\mathbb{F}_p$-endomorphism ring $\mathbb{Z}[(1+\sqrt{-p})/2]$ are said to lie on the *surface*; those with $\mathbb{F}_p$-endomorphism rings $\mathbb{Z}[\sqrt{-p}]$ lie on the *floor*. When $p \equiv 1 \pmod 4$, all vertices are on the floor. We restate the main structure theorem of [**DG16**] here. Recall that two vertices of a graph are *adjacent* if they are joined by an edge, in which case each vertex is said to be *incident* with the edge. Adjacent vertices are also referred to as *neighbors*.

THEOREM 2.3. [**DG16**, Thm. 2.7], *Structure of $\mathcal{G}_\ell(\mathbb{F}_p)$ Let $p \geq 5$ be a prime.*

(1) *If $\ell = 2$ and $p \equiv 1 \pmod 4$, then $\mathcal{G}_2(\mathbb{F}_p)$ consists of $h(-4p)$ vertices joined in adjacent pairs.*

(2) *If $\ell = 2$ and $p \equiv 3 \pmod 8$, then $\mathcal{G}_2(\mathbb{F}_p)$ consists of $4h(-p)$ vertices organized into $h(-p)$* tripod *formations. Each tripod consists of a single vertex on the surface that is adjacent to three distinct vertices on the floor.*

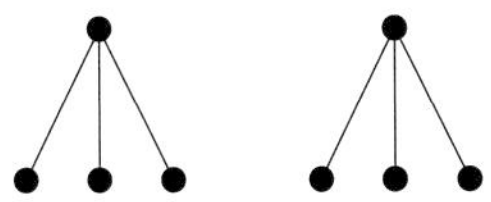

(3) *If* $\ell = 2$ *and* $p \equiv 7 \pmod 8$, *then* $\mathcal{G}_2(\mathbb{F}_p)$ *consists of* $2h(-p)$ *vertices, organized into* volcanoes. *Each volcano contains a (possibly degenerate) cycle consisting of vertices on the surface, each of which is adjacent to a unique vertex on the floor.*

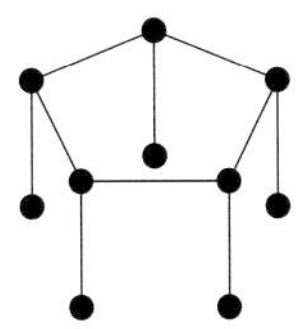

(4) *If* $\ell > 2$, *with* $\ell \neq p$, *then* $\mathcal{G}_\ell(\mathbb{F}_p)$ *is a disoint union of (possibly degenerate) cycles. If* $p \equiv 3 \pmod 4$, *then each cycle contains either surface vertices or floor vertices, but not both.*

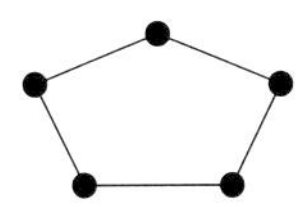

In cases (3) and (4), the length of each cycle is the order of the class generated by a prime ideal above ℓ *in the class group of the corresponding quadratic order. When* $p \equiv 3 \pmod 4$, *vertices on the surface may be incident with loop edges.*

Note that cycles may be degenerate, i.e. consist of one vertex only. In particular, for $\ell > 2$, $\mathcal{G}_\ell(\mathbb{F}_p)$ may consist of isolated vertices (possibly with loops); for example, when the Legendre symbol $(\frac{-p}{\ell}) = -1$ or $\mathbb{Q}(\sqrt{-p})$ has class number 1.

We characterize loops and multi-edges in $\mathcal{G}_\ell(\mathbb{F}_p)$ when $p > \ell$.

PROPOSITION 2.4 (Loops in $\mathcal{G}_\ell(\mathbb{F}_p)$). *Suppose* $p > \ell$.

(1) *If* $p \equiv 3 \pmod 4$, *then a vertex on the surface of* $\mathcal{G}_\ell(\mathbb{F}_p)$ *is incident with a loop if and only if* $4\ell - p$ *is a perfect square. In this case, every vertex on the surface is incident with two distinct loops, corresponding to dual degree* ℓ *endomorphisms.*

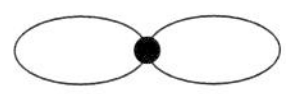

(2) *No vertex of* $\mathcal{G}_\ell(\mathbb{F}_p)$ *on the floor is incident with a loop.*

PROOF. Loops in $\mathcal{G}_\ell(\mathbb{F}_p)$ correspond to degree ℓ endomorphisms over $\mathbb{F}_p$, which in turn correspond to elements of norm ℓ in the appropriate quadratic order.

(1) For brevity, put $\omega = (1 + \sqrt{-p})/2$ and let $\alpha = a + b\omega \in \mathbb{Z}[\omega]$ be the element of norm ℓ corresponding to loop incident with a vertex on the surface of $\mathcal{G}_\ell(\mathbb{F}_p)$. Since $p > \ell$, we obtain

$$4p > 4\ell = 4N(\alpha) = (2a + b)^2 + b^2 p \geq b^2 p.$$

so $|b| \le 1$. If $b = 0$, then $\ell = N(\alpha) = a^2$ which is impossible since ℓ is prime. This forces $b = \pm 1$, so $4\ell - p = (2a \pm 1)^2$ is a perfect square. Conversely, if $4\ell - p = m^2$ with $m \in \mathbb{Z}$, then m must be odd. Thus, $\alpha = a + \omega \in \mathbb{Z}[\omega]$ with $a = (m-1)/2$ is an element of norm ℓ. Note that $\alpha = (m + \sqrt{-p})/2$, so the two loops at j correspond to the conjugate elements $(m \pm \sqrt{-p})/2 \in \mathbb{Z}[\omega]$ of norm ℓ.

(2) Now let $\alpha = a + b\sqrt{-p} \in \mathbb{Z}[\sqrt{-p}]$. Then $N(\alpha) = a^2 + b^2 p$. If $b \neq 0$, then $N(\alpha) \ge p > \ell$, whereas if $b = 0$, then $N(\alpha) = a^2$ which is not prime and hence also distinct from ℓ. Hence $\mathbb{Z}[\sqrt{-p}]$ contains no elements of norm ℓ. □

In particular, $\mathcal{G}_\ell(\mathbb{F}_p)$ can only contain loops if $p \equiv 3 \pmod 4$ and $4\ell - p$ is a perfect square. Applying this to $\ell = 2, 3$ immediately yields the following.

COROLLARY 2.5 (Loops in $\mathcal{G}_\ell(\mathbb{F}_p)$ for $\ell = 2, 3$). *For $\ell = 2, 3$ and $p > \ell$, the graph $\mathcal{G}_\ell(\mathbb{F}_p)$ contains no loops except when $(\ell, p) \in \{(2, 7), (3, 11)\}$, where $\mathcal{G}_\ell(\mathbb{F}_p)$ has one vertex that is incident with two pairs of loops.*

We now turn to multi-edges in $\mathcal{G}_\ell(\mathbb{F}_p)$ that are not loops.

PROPOSITION 2.6 (Multi-edges in $\mathcal{G}_\ell(\mathbb{F}_p)$). *Suppose $p > \ell$.*

(1) *If $p \equiv 3 \pmod 4$, then $\mathcal{G}_\ell(\mathbb{F}_p)$ contains no directed multi-edges except possibly loops unless $(\ell, p) = (2, 3)$. Here, $\mathcal{G}_2(\mathbb{F}_3)$ has one directed triple edge.*

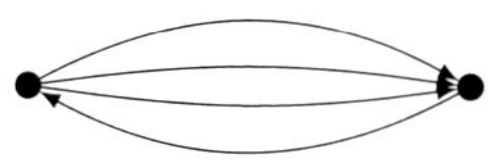

(2) *If $p \equiv 1 \pmod 4$, then $\mathcal{G}_\ell(\mathbb{F}_p)$ contains directed multi-edges if and only if it has at least two vertices and $2\ell - p$ is a perfect square. In this case, the vertices in $\mathcal{G}_\ell(\mathbb{F}_p)$ are joined pairwise by pairs of double edges in opposite directions.*

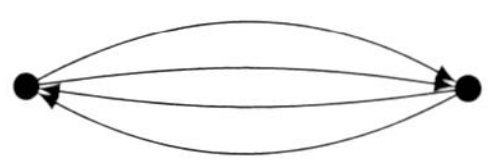

PROOF. A directed multi-edge in $\mathcal{G}_\ell(\mathbb{F}_p)$ that is not a loop corresponds to multiple $\mathbb{F}_p$-non-equivalent ℓ-isogenies $\phi, \psi : E \to E'$ where E and E' are not isomorphic over $\mathbb{F}_p$. Then $\hat{\phi}\psi$ is an endomorphism on E of degree ℓ^2, and we have a multi-edge if and only if $\hat{\phi}\psi$ is not the multiplication by ℓ map on E. Thus, multi-edges arise from elements $\alpha \neq \pm\ell$ of norm ℓ^2 in the corresponding quadratic order which are not squares up to sign.

(1) Assume $p \equiv 3 \pmod 4$. Note that ℓ does not ramify in $\mathbb{Z}[\omega]$ as $\ell \neq p$. Suppose first that ℓ splits in $\mathbb{Z}[\omega]$, and write $(\ell) = \mathfrak{l}\bar{\mathfrak{l}}$ with prime ideals $\mathfrak{l}, \bar{\mathfrak{l}}$ of $\mathbb{Z}[\omega]$. Note that this precludes the case $p = 3$, as $\ell < p$ implies $\ell = 2$ in this case, but 2 is inert in $\mathbb{Q}(\sqrt{-3})$. Let $\alpha \in \mathbb{Z}[\omega]$ with $\alpha \neq \pm\ell$ and $N(\alpha) = \ell^2$. Unique prime ideal factorization, together with $(\alpha) \neq (\ell)$, forces $(\alpha) = \mathfrak{l}^2$ or $(\alpha) = \bar{\mathfrak{l}}^2$. Assume the former; the case $(\alpha) = \bar{\mathfrak{l}}^2$ is entirely analogous. Then $\mathfrak{l}^2$ is principal. Since $h(-p)$ is odd by Proposition 2.1, $\mathfrak{l}$ must be

principal. This means that α is a square in $\mathbb{Z}[\omega]$ up to sign, which we precluded.

Now assume that ℓ is inert in $\mathbb{Z}[\omega]$. Since (ℓ) is the only prime ideal of norm ℓ^2, unique prime ideal factorization implies $(\alpha) = (\ell)$. The assumption $\alpha \neq \ell$ now forces $p = 3$ (the only case where $\mathbb{Z}[\omega]$ has non-trivial units) and hence $\ell = 2$ since $\ell < p$. Thus, $\alpha = \pm 2\zeta^k$ where $k = 1, 2$ and ζ is a primitive cube root of unity, so $\pm\alpha \in \{1+\sqrt{-3}, 1-\sqrt{-3}\}$. Indeed, among the four $\mathbb{F}_3$-isomorphism classes of supersingular elliptic curves over $\mathbb{F}_3$, exactly two, represented by the curves $E_\pm : y^2 = x^3 \pm x$, are 2-isogenous over $\mathbb{F}_3$. The curve E_+ is on the floor, whereas E_- is on the surface and has, up to sign, three $\mathbb{F}_3$-rational automorphisms $(x, y) \mapsto (x + i, y)$ for $i = 0, 1, 2$. So there are three $\mathbb{F}_3$-inequivalent 2-isogenies from E_- to E_+, producing three directed edges. Their duals differ only by post-composition by an automorphism on E_- and are hence equivalent, yielding one edge in the opposite direction.

(2) Now suppose $p \equiv 1 \pmod 4$. Let $\alpha \in \mathbb{Z}[\sqrt{-p}]$ with $N(\alpha) = \ell^2$, $\alpha \neq \pm\ell$ and $\pm\alpha$ not a square in $\mathbb{Z}[\sqrt{-p}]$. Write $\alpha = a + b\sqrt{-p}$ with $a, b \in \mathbb{Z}$. Then $a^2 + b^2 p = \ell^2$ and $b \neq 0$, so $|a| < \ell$. We have

$$b^2 p = \ell^2 - a^2 = (\ell - |a|)(\ell + |a|), \tag{2.2}$$

so p divides $\ell - |a|$ or $\ell + |a|$. Since $1 \leq \ell - |a| \leq \ell < p$, the first of these possibilities cannot happen; hence $p \mid \ell + |a|$. Write $\ell + |a| = kp$ for some $k \in \mathbb{Z}$. Then $k \geq 1$ and $k\ell < kp = \ell + |a| < 2\ell$, forcing $k = 1$ and hence $p = \ell + |a|$. By (2.2), we have $b^2 = \ell - |a| = 2\ell - p$, so $2\ell - p$ is a perfect square. Conversely, if $2\ell - p = b^2$ for some $b \in \mathbb{Z}$, then the 4 elements $\alpha = \pm(p - \ell) \pm b\sqrt{-p}$ all have norm ℓ^2, are distinct from $\pm\ell$, and are not squares in $\mathbb{Z}[\sqrt{-p}]$.

Note that as in the case $p \equiv 3 \pmod 4$, we have $(\alpha) = \mathfrak{l}^2$ or $(\alpha) = \bar{\mathfrak{l}}^2$, where $\mathfrak{l}$ and $\bar{\mathfrak{l}}$ are the two prime ideals above (ℓ) in $\mathbb{Z}[\sqrt{-p}]$. So the ideal class of $\mathfrak{l}$ has order 2 in the class group of $\mathbb{Q}(\sqrt{-p})$. This means that if $\mathcal{G}_\ell(\mathbb{F}_p)$ has more than one vertex and $2\ell - p$ is a perfect square (which rules out $\ell = 2$ as $p \geq 5$), all the cycles in $\mathcal{G}_\ell(\mathbb{F}_p)$ as described in part (4) of Theorem 2.3 have length 2. They correspond to two pairs of directed edges and their respective duals. □

Proposition 2.6 shows that if $p > \ell > 2$, then $\mathcal{G}_\ell(\mathbb{F}_p)$ can only contain directed double edges when $p \equiv 1 \pmod 4$ and $p \leq 2\ell - 1$. The cases $\ell = 2, 3$ can now once again be easily deduced.

COROLLARY 2.7 (Multi-edges in $\mathcal{G}_\ell(\mathbb{F}_p)$ for $\ell = 2, 3$). *For $\ell = 2, 3$ and $p > \ell$, apart from the loops of Corollary 2.5, $\mathcal{G}_\ell(\mathbb{F}_p)$ contains no directed multi-edges except when $(\ell, p) \in \{(2, 3), (3, 5)\}$.*

A list of prime pairs (ℓ, p) with $2 \leq \ell < 100$ and $p > \ell$ for which $\mathcal{G}_\ell(\mathbb{F}_p)$ has directed multi-edges, including loops, can be found in the file called `loops & multi-edges in G_l(Fp).pdf` at the GitHub repository [**Hed25**]. A notebook with code to generate visual depictions of $\mathcal{G}_\ell(\mathbb{F}_p)$ and $\mathcal{G}_\ell(\overline{\mathbb{F}}_p)$, under the name `Graph_Viz.ipynb`, is also available at [**Hed25**].

3. The spine of $\mathcal{G}_\ell(\overline{\mathbb{F}}_p)$

In this section, we review the relationship between $\mathcal{G}_\ell(\mathbb{F}_p)$ and $\mathcal{G}_\ell(\overline{\mathbb{F}}_p)$; specifically the process of moving from $\mathbb{F}_p$-isomorphism classes of elliptic curves to $\overline{\mathbb{F}}_p$-isomorphism classes, and from ℓ-isogenies defined over $\mathbb{F}_p$ to those defined over $\overline{\mathbb{F}}_p$. This material is a summary of [**ACNL+23**].

Definition 3.1 (Spine). The *spine* $\mathcal{S}_\ell^p$ is the subgraph of $\mathcal{G}_\ell(\overline{\mathbb{F}}_p)$ induced by the vertices in $\mathbb{F}_p$. Specifically, the vertices of $\mathcal{S}_\ell^p$ are the $\overline{\mathbb{F}}_p$-isomorphism classes of supersingular elliptic curves defined over $\mathbb{F}_p$, and its edges are all the ℓ-isogenies joining these vertices.

The terminology was first introduced in [**ACNL+23**] and stems from the fact that $\mathcal{G}_\ell(\overline{\mathbb{F}}_p)$ can be obtained from $\mathcal{S}_\ell^p$ by adding pairs of conjugate j-invariants and the corresponding conjugate ℓ-isogenies for any j-invariant in $\mathbb{F}_p$, conjuring an image of a ribcage anchored at a spine. There are natural maps between the graphs $\mathcal{G}_\ell(\mathbb{F}_p)$, $\mathcal{G}_\ell(\overline{\mathbb{F}}_p)$, and $\mathcal{S}_\ell^p$:

Definition 3.2. Define $\Gamma : \mathcal{G}_\ell(\mathbb{F}_p) \to \mathcal{G}_\ell(\overline{\mathbb{F}}_p)$ to take vertices of $\mathcal{G}_\ell(\mathbb{F}_p)$ to their $\overline{\mathbb{F}}_p$-isomorphism classes and edges to their $\overline{\mathbb{F}}_p$-equivalence classes.

Next, define $\Theta : \mathrm{Im}(\Gamma) \to \mathcal{G}_\ell(\overline{\mathbb{F}}_p)$ to add edges between vertices in $\mathrm{Im}(\Gamma)$ that correspond to isogenies defined over $\overline{\mathbb{F}}_p$ and not defined over $\mathbb{F}_p$. In particular, $\Theta(\mathrm{Im}(\Gamma)) = \mathcal{S}_\ell^p$.

Lastly, define $\Omega = \Theta \circ \Gamma : \mathcal{G}_\ell(\mathbb{F}_p) \to \mathcal{S}_\ell^p \subseteq \mathcal{G}_\ell(\overline{\mathbb{F}}_p)$.

The following graph structural changes are possible under Ω:

Definition 3.3.

- **Stacking:** Two connected components of $\mathcal{G}_\ell(\mathbb{F}_p)$ *stack* under the map Γ of Definition 3.2 if they are the same graph when labeled by j-invariants.

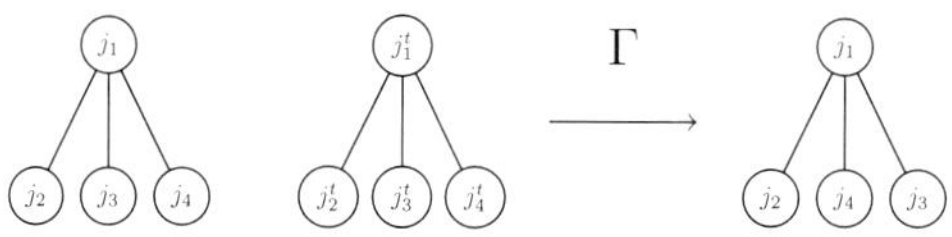

- **Folding:** A connected component of $\mathcal{G}_\ell(\mathbb{F}_p)$ *folds* under the map Γ if it only contains vertices corresponding to both quadratic twists for every j-invariant appearing as a vertex in the component.

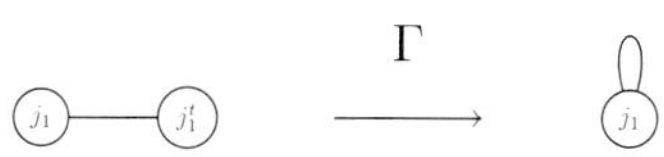

- **Attachment at a vertex:** Two components of $\mathcal{G}_\ell(\mathbb{F}_p)$ have a *vertex attachment* under the map Γ if they both contain a vertex with the same j-invariant but the neighbors of that shared j-invariant are different in the two components.

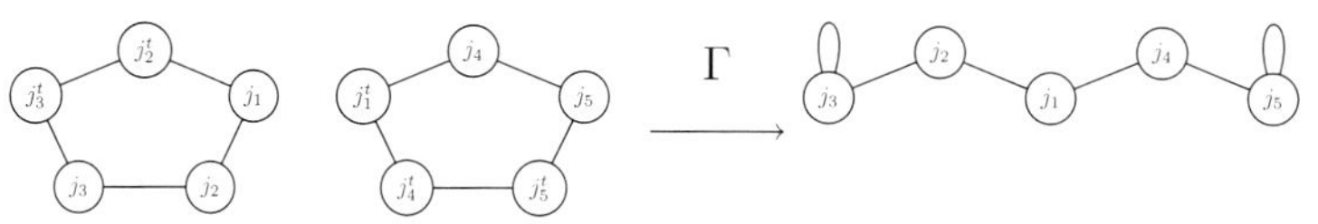

- **Attachment by a new edge:** Two connected components of $\mathcal{G}_\ell(\mathbb{F}_p)$ have an *edge attachment* if a new edge appears under the map Θ of Definition 3.2 which connects these two components.

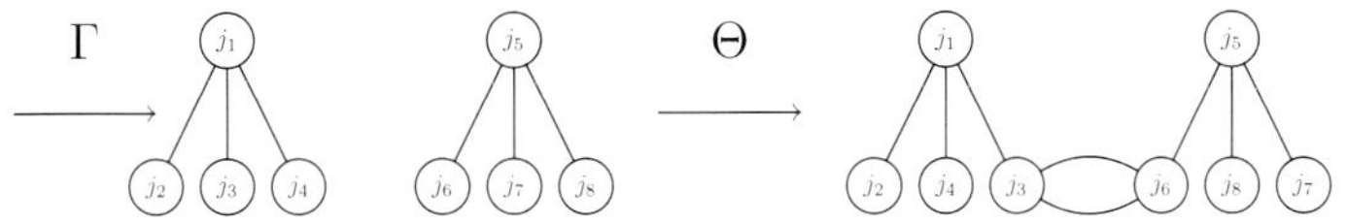

The map Γ is always 2-to-1 on vertices, and also 2-to-1 on edges outside of the neighborhoods of $j = 0, 1728$. Definition 3.3 immediately implies the following.

LEMMA 3.4. *Folding and stacking are mutually exclusive.*

PROOF. If two components stack, then one component's vertices correspond to the quadratic twists of the other component's vertices. If a component folds, then the quadratic twists of its vertices belong to that same component. $\square$

As shown in [**ACNL+23**, Prop. 3.16 and Cor. 3.24], vertex attachment is only possible at the j-invariant 1728 and only for $\ell > 2$. Attachment by a new edge implies a double edge in $\mathcal{G}_\ell(\overline{\mathbb{F}}_p)$ by [**ACNL+23**, Cor. 3.15]. We recall the precise theorems from [**ACNL+23**] describing the changes to $\mathcal{G}_\ell(\mathbb{F}_p)$ under the map Ω:

THEOREM 3.5 ([**ACNL+23**, Thm. 3.29]). *Let $\ell = 2$. Under the map Γ : $\mathcal{G}_2(\mathbb{F}_p) \to \mathcal{G}_2(\overline{\mathbb{F}}_p)$ of Definition* 3.2*, only stacking and folding are possible. Under the map $\Theta : \mathrm{Im}(\Gamma) \to \mathcal{G}_2(\overline{\mathbb{F}}_p)$, at most one attachment by a new edge is possible. In particular, attachment by a vertex is not possible.*

THEOREM 3.6 ([**ACNL+23**, Thm. 3.18]). *Let p and $\ell > 2$ be distinct primes such that the order of a prime ideal $\mathfrak{l}$ above ℓ in the class group of $\mathbb{Q}(\sqrt{-p})$ is odd. Under the map $\Gamma : \mathcal{G}_\ell(\mathbb{F}_p) \to \mathcal{G}_\ell(\overline{\mathbb{F}}_p)$:*

- *the two components containing vertices corresponding to $j = 1728$ fold and attach at the vertex $j = 1728$;*
- *all other components stack.*

Under the map Θ : $\mathrm{Im}(\Gamma) \to \mathcal{G}_\ell(\overline{\mathbb{F}}_p)$, the number of new edges is bounded by the degree of $Res_\ell(X) \pmod p$, with $Res_\ell(X)$ given in (2.1).

In the next two sections, building on the results of [**ACNL+23**, Sec. 3], we explicitly describe the structure of $\mathcal{S}_2^p$ and and $\mathcal{S}_3^p$ in terms of specific congruence conditions on p. We also provide a road map for extending this approach in principle to any ℓ and outline obstacles one might encounter. Small primes are treated separately elsewhere: a detailed description of the graphs $\mathcal{G}_\ell(\mathbb{F}_p)$, $\mathcal{S}_\ell^p$ and $\mathcal{G}_\ell(\overline{\mathbb{F}}_p)$ for $2 \leq p \leq 31$ can be found under the file name:
`SmallCharacteristicGraphDescription.pdf`.

This information can also be generated with the notebook `Small_Prime_Information.ipynb`, and the notebook `Graph_Viz.ipynp` generates images of all three graphs. All these sources are available at [**Hed25**].

4. Structure of the spine $\mathcal{S}_2^p$

In this section, we provide congruence conditions on p that govern the structure of $\mathcal{S}_2^p$, the spine for $\ell = 2$. This case carries the most interest, especially when $p \equiv 3 \pmod 4$, due to the volcano structure of $\mathcal{G}_2(\mathbb{F}_p)$. For much of this section, we only consider primes $p \geq 17$; for details on the primes $2 \leq p \leq 13$, consult the sources cited at the end of Section 3.

As expected, our investigation makes extensive use of the modular polynomial $\Phi_2(X, X)$ and the polynomial $\mathrm{Res}_2(X)$ defined in (2.1). Explicitly:

$$\Phi_2(X, X) = -(X - 1728)(X - 8000)(X + 3375)^2, \tag{4.1}$$

$$\mathrm{Res}_2(X) = -2^2 X^2 (X - 1728)(X + 3375)^2 (X^2 + 191025X - 121287375)^2. \tag{4.2}$$

By Corollary 2.5, $\mathcal{G}_2(\mathbb{F}_p)$ contains loops only for $p = 7$. We recall a well-known result about loops in $\mathcal{G}_2(\overline{\mathbb{F}}_p)$, cf. [**Gha24**, Ex. 3].

LEMMA 4.1 (Loops in $\mathcal{G}_2(\overline{\mathbb{F}}_p)$). *Let $p \neq 2, 7$. Loops occur in $\mathcal{G}_2(\overline{\mathbb{F}}_p)$ at vertices corresponding to precisely the following j-invariants, all belonging to $\mathbb{F}_p$:*

$$\begin{aligned} 1728 &\text{ if } p \equiv 3 \pmod 4, \\ 8000 &\text{ if } p \equiv 5, 7 \pmod 8, \\ -3375 &\text{ if } p \equiv 3, 5, 6 \pmod 7. \end{aligned}$$

PROOF. The loops in $\mathcal{G}_2(\overline{\mathbb{F}}_p)$ are precisely the roots of the polynomial $\Phi_2(X, X)$ of (4.1). Its linear factors are the Hilbert class polynomials $H_{-4}(X)$, $H_{-8}(X)$ and $H_{-7}(X)$. The congruence conditions on p characterize when p is inert in the corresponding imaginary quadratic fields. □

The j-invariants listed in Lemma 4.1 need not be distinct for primes $p \leq 5$; these small primes are handled explicitly in the document: `SmallCharacteristicGraphDescription.pdf` at [**Hed25**].

LEMMA 4.2. [**ACNL+23**, Prop. 3.22] *Let $j \in \mathbb{F}_p$ be a supersingular j-invariant, and denote by v_j, w_j the two distinct vertices of $\mathcal{G}_2(\mathbb{F}_p)$ representing isomorphism classes with j-invariant j. If $j \neq 1728$, then v_j and w_j have neighbors in $\mathcal{G}_2(\mathbb{F}_p)$ that correspond to identical j-invariants. Specifically:*

1. *If $p \equiv 1 \pmod 4$, then the unique neighbor of v_j and the unique neighbor of w_j are vertices representing the same j-invariant j'.*
2. *If $p \equiv 3 \pmod 4$ and the vertices v_j, w_j are on the floor, then the unique neighbor of v_j and the unique neighbor of w_j are vertices with the same j-invariant j'.*
3. *If $p \equiv 3 \pmod 4$ and the vertices v_j, w_j are on the surface, then v_j has three neighbors with distinct j-invariants j_1, j_2, j_3 and w_j has three neighbors with the same j-invariants j_1, j_2, j_3.*

The following proposition corrects [**ACNL+23**, Cor. 3.28].

PROPOSITION 4.3 (Folding for $\ell = 2$). *Let $p > 7$ be prime. The only connected components of $\mathcal{G}_2(\mathbb{F}_p)$ which fold are those containing vertices with $j = 8000$ (if $p \equiv 5, 7 \pmod 8$) and/or $j = 1728$ (if $p \equiv 3 \pmod 4$).*

PROOF. A component folds if and only if it contains only vertices corresponding to both $\mathbb{F}_p$-twists of a single supersingular elliptic curve j-invariant when $\ell = 2$, see [**ACNL+23**, Cor. 3.28]. Such a component necessarily contains two adjacent vertices corresponding to a pair of twists which are 2-isogenous over $\mathbb{F}_p$. The roots of $\Phi_2(X, X)$ are precisely $8000, 1728, -3375$; see (4.1).

Since -3375 is a double root of $\Phi_2(X, X)$, there are two distinct 2-isogenies between elliptic curves with j-invariant $j = -3375$. There are no loops or multi-edges in $\mathcal{G}_2(\mathbb{F}_p)$ for $p > 7$ by Corollaries 2.5 and 2.7, so these isogenies do not correspond to edges in $\mathcal{G}_2(\mathbb{F}_p)$.

In [**ACNL+23**], the authors mistakenly declare that $j = 8000$ is only supersingular for $p \equiv 5 \pmod 8$, when in fact $j = 8000$ is supersingular for $p \equiv 5, 7 \pmod 8$. The two models that the authors list for the curve over $\mathbb{Z}$:

$$E_{8000} : y^2 = x^3 - 4320x + 96768, \quad E^t_{8000} : y^2 = x^3 - 17280x - 774144$$

are twists by $\sqrt{-2}$ (not $\sqrt{2}$, as stated by the authors). Since -2 is not a square modulo p for $p \equiv 5, 7 \pmod 8$, E and E' reduce to supersingular quadratic twists over $\mathbb{F}_p$ for such primes p. Moreover, E and E' have a $\mathbb{Z}$-rational 2-isogeny between them, so they represent adjacent vertices on a connected component of $\mathcal{G}_2(\mathbb{F}_p)$.

Likewise, $\mathbb{Z}$-models for $j = 1728$ show that the $\mathbb{F}_p$-isomorphism classes are 2-isogenous over $\mathbb{F}_p$. □

Since edge attachment forces a double edge, we need to identify double edges in $\mathcal{G}_2(\overline{\mathbb{F}}_p)$ that join two vertices in $\mathbb{F}_p$. This amounts to ascertaining when the roots of the polynomial $\mathrm{Res}_2(X)$ in (4.2) are supersingular and belong to $\mathbb{F}_p$. For $j = 0$, this requires $p \equiv 2 \pmod 3$, and for the roots 1728 and -3375 of $\mathrm{Res}_2(X)$, this was addressed in Lemma 4.1. So we need only consider the quadratic factor of $\mathrm{Res}_2(X)$, which is in fact the Hilbert class polynomial $H_{-15}(X)$.

LEMMA 4.4 (Multi-edges in $\mathcal{G}_2(\overline{\mathbb{F}}_p)$). *Let $p \geq 7$. In addition to the loops in Lemma 4.1, $\mathcal{G}_2(\overline{\mathbb{F}}_p)$ has multi-edges when $p = 7, 13$ or $p \equiv 11, 14 \pmod{15}$.*

PROOF. By (4.2), additional double edges correspond to the roots modulo p of $H_{-15}(X) = X^2 + 191025X - 121287375$, which are $(-191025 \pm 85995\sqrt{5})/2$. For p odd, their reductions modulo p belong to $\mathbb{F}_p$ if and only if $p \mid 85995 = 3^3 \cdot 5 \cdot 7^2 \cdot 13$ or 5 is a quadratic residue modulo p. So assuming $p \geq 7$, $H_{-15} \pmod p$ has roots in $\mathbb{F}_p$ if and only if $p = 7, 13$ or $(\frac{5}{p}) = 1$. The latter condition holds if and only if $p \equiv \pm 1 \pmod 5$.

For $p \geq 7$, the roots of $H_{-15}(X) \pmod p$ are j-invariants of supersingular elliptic curves if and only if $(\frac{-15}{p}) = -1$. Note that this holds for $p = 7$ and $p = 13$, in which case $H_{-15}(X)$ has the double root $-191025/2 \in \mathbb{F}_p$; else it has two distinct roots in $\mathbb{F}_p$. Assuming $(\frac{5}{p}) = 1$, the condition $(\frac{-15}{p}) = -1$ reduces to $(\frac{-3}{p}) = -1$, or equivalently, $p \equiv 2 \pmod 3$. Finally, $p \equiv \pm 1 \pmod 5$ and $p \equiv 2 \pmod 3$ if and only if $p \equiv 11$ or $14 \pmod{15}$. □

PROPOSITION 4.5 (New edges and edge attachment for $\ell = 2$). *Suppose $p \geq 17$.*

(1) *If $p \not\equiv 7 \pmod 8$, then new edges appear at the supersingular j-invariants which are roots of $H_{-15}(X)$. The new edge joining the two distinct roots of $H_{-15}(X)$ is attaching. Thus, edge attachment happens when $p \equiv 11$, 29, 41, 59, 89, 101 $\pmod{120}$.*

(2) *If $p \equiv 7 \pmod 8$, then attachment by an edge can only happen between vertices distinct from -3375, 1728 and 0 whose j-invariants are roots of $H_{-15}(X)$.*

PROOF. Part (1) is an extension of [**ACNL+23**, Cor. 3.30], where the authors proved the result under the assumption that $p > 101$. We contribute the computations for $p \leq 101$ to complete the result.

New edges correspond to distinct roots of $H_{-15}(x)$ that are supersingular j-invariants in $\mathbb{F}_p$. Since $p \neq 7, 13$, the roots of $H_{-15}(x)$ are distinct modulo p and are supersingular for $p \equiv 11$ or $14 \pmod{15}$ by Lemma 4.4. Combining these congruence conditions with $p \not\equiv 7 \pmod 8$ results in the set of congruence classes listed in part (1). It thus suffices to verify this for $p = 29, 41, 59, 71, 89$, and 101. For $p = 29, 41$ and 59, edge attachment holds by the connectivity of $\mathcal{G}_2(\overline{\mathbb{F}}_p)$ and the fact that all the supersingular j-invariants belong to $\mathbb{F}_p$. For $p = 89$ and 101, direct computation of $\mathcal{S}_2^p$ confirms the edge attachment.

The case $p \equiv 7 \pmod 8$ is covered in [**ACNL+23**, Prop. 3.25]. □

For $p \equiv 11, 14 \pmod{15}$ with $p \equiv 7 \pmod 8$, or equivalently, $p \equiv 71$ or $119 \pmod{120}$, the new edges of Proposition 4.5 may or may not be attaching.

Triple edges in $\mathcal{G}_\ell(\overline{\mathbb{F}}_p)$ also play a role in ascertaining edge attachment. The following theorem characterizes the occurrence of triple edges in $\mathcal{G}_2(\overline{\mathbb{F}}_p)$.

PROPOSITION 4.6 (Triple edges in $\mathcal{G}_2(\overline{\mathbb{F}}_p)$). *For brevity, let $\mathcal{P}_2 = \{2, 3, 5, 7, 13\}$.*

(1) *If $p \in \mathcal{P}_2$, then $\mathcal{G}_2(\overline{\mathbb{F}}_p)$ is identical to the spine $\mathcal{S}_2^p$ and consists of a single vertex with a triple loop.*
(2) *If $p \notin \mathcal{P}_2$, and $p \equiv 2 \pmod 3$, then there is a triple edge from the vertex corresponding to $j = 0$ to the vertex corresponding to $j = 54000$.*
(3) *For all other primes p, $\mathcal{G}_2(\overline{\mathbb{F}}_p)$ does not contain triple edges.*

PROOF. A proof for the primes $p \in \mathcal{P}_2$ can be found in the document `SmallCharacteristicGraphDescription.pdf` at [**Hed25**], so suppose $p \notin \mathcal{P}_2$. The primes p for which $\mathcal{S}_2^p$ contains a triple edge are those for which the polynomials $\mathrm{Res}_2(X)$ of (4.2) and the polynomial

$$\begin{aligned}\mathrm{Res}_2^{(2)}(X) &= \mathrm{Res}\left(\Phi_2(X,Y), \frac{\partial^2}{\partial Y^2}\Phi_2(X,Y);\ Y\right)\\ &= -2^2 \cdot 3X(X-405)(X^2 - 2571X + 1492425)\end{aligned}$$

share a common root in $\mathbb{F}_p$. The respective sets of roots of these two polynomials:

$$\left\{0, 1728, -3375, \frac{-191025 \pm 85995\sqrt{5}}{2}\right\}, \quad \left\{0, 405, \frac{2571 \pm 39\sqrt{421}}{2}\right\}.$$

The j-invariant 0 is a common root of both polynomials, so when $j = 0$ is supersingular (i.e. $p \equiv 2 \pmod 3$), the graph $\mathcal{S}_2^p$ has a triple edge from $j = 0$. Solving $\Phi_2(0, Y) = 0$ for Y over $\mathbb{Z}$, we see that this triple edge is to the vertex $j = 54000$ and is hence not a loop (as $54000 \not\equiv 0 \pmod p$ for $p \geq 7$). We show that $\mathrm{Res}_2(X)$ and $\mathrm{Res}_2^{(2)}(X)$ have no other shared roots when $p \notin \mathcal{P}_2$, thus ruling out any other triple edges in $\mathcal{S}_2^p$.

For brevity, let $f(X) = X^2 - 2571X + 1492425$ denote the quadratic factor of $\mathrm{Res}_2^{(2)}(X)$. We observe that the root $j = 1728$ of $\mathrm{Res}_2(X)$ is not a root of $\mathrm{Res}_2^{(2)}(X)$. For $p \notin \mathcal{P}_2$, note that $1728 \not\equiv 0, 405 \pmod p$ and $f(1728) = 3^6 \cdot 7^2 \not\equiv 0 \pmod p$. Similarly, the root $j = -3375$ of $\mathrm{Res}_2(X)$ is not a root of $\mathrm{Res}_2^{(2)}(X)$, as $-3375 \not\equiv 0$ or 405 modulo p and $f(-3375) = 3^6 \cdot 5^2 \cdot 7 \cdot 13^2 \not\equiv 0 \pmod p$ when $p \notin \mathcal{P}_2$.

Finally, we establish that $H_{-15}(X)$ and $f(X)$ have no shared root. Equating the roots of these two polynomials modulo p yields the following sequence of implications:

$$\begin{aligned}
\frac{-191025 \pm 85995\sqrt{5}}{2} &\equiv \frac{2571 \pm 39\sqrt{421}}{2} \pmod{p},\\
(85995)^2 \cdot 5 &\equiv \left(193596 \pm 39\sqrt{421}\right)^2 \pmod{p},\\
-504351432 &\equiv \pm 15100488\sqrt{421} \pmod{p},\\
(-504351432)^2 &\equiv (15100488)^2 \cdot 421 \pmod{p},\\
158371952330592000 &\equiv 0 \pmod{p}.
\end{aligned}$$

The only prime $p \notin \mathcal{P}_2$ that divides 158371952330592000 is $p = 11$. The constant coefficients of $H_{-15}(X)$ and $f(X)$ are both multiples of 11, and it is now easy to verify that the only common root of these two polynomials modulo 11 is 0. □

With these ingredients, we are ready to explicitly describe the graph structure of $\mathcal{S}_2^p$. As in Theorem 2.3, we consider three cases according to the structure of $\mathcal{G}_2(\mathbb{F}_p)$, in Theorems 4.7, 4.8, and 4.9, respectively.

We refer to `SmallCharacteristicGraphDescription.pdf` at [**Hed25**] for primes p with $2 \leq p \leq 13$.

THEOREM 4.7 (Spine structure, $p \equiv 1 \pmod 4$ and $\ell = 2$). *Let $p \geq 17$ with $p \equiv 1 \pmod 4$. When mapping $\mathcal{G}_2(\mathbb{F}_p)$ into the spine $\mathcal{S}_2^p$, there may be stacking, folding, or attachment by new edges. The following congruence conditions on p determine precisely which of these occur and how often.*

(1) $p = 29$*: $\mathcal{G}_2(\mathbb{F}_{29})$ has three components: the component containing the two vertices with $j = 8000$ folds, and there is an edge attachment connecting this folded component to the other component. The spine $\mathcal{S}_2^p$ is the entire 2-isogeny graph $\mathcal{G}_2(\overline{\mathbb{F}}_p)$.*
(2) $p \equiv 29, 101 \pmod{120}$, $p \neq 29$*: the component containing the two vertices with $j = 8000$ folds, all other components stack, and there is an edge attachment connecting two stacked components. The spine $\mathcal{S}_2^p$ consists of one isolated vertex with $j = 8000$ with a loop, one component with four vertices, and the remaining $h(-4p)/2 - 5$ vertices are joined in pairs.*
(3) $p \equiv 41, 89 \pmod{120}$*: all components stack, and there is an edge attachment. The spine $\mathcal{S}_2^p$ consists of one connected component with four vertices, and the remaining $h(-4p)/2 - 4$ vertices are joined in pairs.*
(4) $p \equiv 13, 37, 53, 61, 77, 109 \pmod{120}$*: the component containing the two vertices with $j = 8000$ folds, all other components stack, and there are no edge attachments. The spine $\mathcal{S}_2^p$ consists of one isolated vertex with $j = 8000$ with a loop, and the remaining $h(-4p)/2 - 1$ vertices connect in pairs.*
(5) $p \equiv 1, 17, 49, 73, 97, 113 \pmod{120}$*: all components stack and there are no edge attachments. The spine $\mathcal{S}_2^p$ is $h(-4p)/2$ vertices joined in pairs.*

The five cases are summarized in Table 4.1.

	Edge attachment		*No edge attachment*
No fold	$p \equiv 41, 89 \pmod{120}$		$p \equiv 1, 17, 49, 73, 97, 113 \pmod{120}$
One fold	*w/ folded comp.*	*not w/ folded comp.*	$p \equiv 13, 37, 53, 61, 77, 109 \pmod{120}$
	$p = 29$	$p \equiv 29, 101 \pmod{120}$, $p \neq 29$	

TABLE 4.1. Spine structure for $p \equiv 1 \pmod 4$

PROOF. The case $p = 29$ can be verified by directly computing $\mathcal{S}_2^{29}$.

We consider the cases for stacking and folding first, followed by those of edge attachment. The individual congruence conditions can be combined by the Chinese Remainder Theorem to provide the statements in this theorem.

Folding and stacking: By Proposition 4.3, the only connected components which could possibly fold are those containing $j = 8000$ and $j = 1728$. Since $p \equiv 1 \pmod 4$, $j = 1728$ is not a supersingular j-invariant. The j-invariant 8000 is supersingular over $\mathbb{F}_p$ whenever $p \equiv 5, 7 \pmod 8$. Combining this with our assumption that $p \equiv 1 \pmod 4$, the connected component with vertices having j-invariant 8000 will fold whenever $p \equiv 5 \pmod 8$. The rest of the components necessarily stack.

Edge attachment: Every edge that appears in $\mathcal{G}_2(\overline{\mathbb{F}}_p)$ but does not already belong to $\mathcal{G}_2(\mathbb{F}_p)$ results in a double edge by [**ACNL+23**, Lem. 3.14]. If a new edge is not attaching, this would result in a triple edge in the case of $p \equiv 1 \pmod 4$ since the connected components of $\mathcal{G}_2(\mathbb{F}_p)$ are precisely edges, see Theorem 2.3. By Proposition 4.6, a triple edge only occurs when a double edge is added between vertices with an existing $\mathbb{F}_p$-edge between them, so there are no attaching triple edges for $p > 13$, $p \equiv 1 \pmod 4$.

By Proposition 4.5, attachment by a new edge happens when $p \equiv 11, 29, 41, 59, 89, 101 \pmod{120}$. Combining this with our assumption that $p \equiv 1 \pmod 4$ yields the congruence conditions $p \equiv 29, 41 \pmod{60}$.

To ascertain whether or not this edge attaches to the folding component (corresponding to $j = 8000$), we observe that if $p \equiv 1 \pmod 4$, $j = 8000$ is supersingular if and only if $p \equiv 5 \pmod 8$. Since $p > 13$, $j = 8000$ is a root of $H_{-15}(X)$ only for $p = 29$. This explains the second column of Table 4.1. The third column is obtained by sorting the remaining congruence classes modulo 120 by their congruence classes modulo 8. □

THEOREM 4.8 (Spine structure, $p \equiv 3 \pmod 8$ and $\ell = 2$). *Let $p \geq 17$ with $p \equiv 3 \pmod 8$. When mapping $\mathcal{G}_2(\mathbb{F}_p)$ into the spine $\mathcal{S}_2^p$, there may be stacking, folding, or attachment by new edges. The connected component of $\mathcal{G}_2(\mathbb{F}_p)$ containing the two vertices with $j = 1728$ always folds. The following congruence conditions on p determine precisely which of these occur and how often.*

(1) $p = 59$*: the folded component gets edge attached to another component by an edge between two vertices on the floor.*
(2) $p \equiv 11, 59 \pmod{120}$ *and* $p \neq 11, 59$*: an edge attachment takes place between two stacked components with the attaching edge incident to two vertices on the floor.*
(3) $p \equiv 19, 43, 67, 83, 91, 107 \pmod{120}$*: no edge attachment takes place.*

The three cases are summarized in Table 4.2.

No edge attachment	$p \equiv 19, 43, 67, 83, 91, 107 \pmod{120}$
EA w/ folded comp.	$p = 59$
EA not w/ folded comp.	$p \equiv 11, 59 \pmod{120}$ *and* $p \neq 11, 59$

TABLE 4.2. Spine structure for $p \equiv 3 \pmod 8$

PROOF. The case $p = 59$ can be again be observed directly by computing $\mathcal{S}_2^{59}$. As before, we consider stacking and folding first, followed by edge attachment. Chinese remaindering again produces the specified congruence conditions on p.

Folding and stacking: By Proposition 4.3, the only connected components which could possibly fold are those containing $j = 8000$ or $j = 1728$. Since $p \equiv 3 \pmod 8$, $j = 8000$ is not a supersingular j-invariant, but $j = 1728$ is supersingular, and folding happens for the component containing the two vertices with $j = 1728$.

Edge attachment: By Proposition 4.5, attachment by a new edge happens when $p \equiv 11, 29, 41, 59, 89, 101 \pmod{120}$. Combining this with $p \equiv 3 \pmod 8$ yields $p \equiv 11, 59 \pmod{120}$.

We can determine when the new edge attaches to the folded component. The folded component is a tripod whose surface vertex corresponds to $j = 1728$ and whose floor vertices have j-invariants 1728, 287496, and 287496, by Theorem 2.3, Proposition 4.3, and the factorization $\Phi_2(1728, X) = (X - 1728)(X - 287496)^2$. For $p > 13$, the j-invariant $j = 1728$ is not a root of $H_{-15}(X)$, and $j = 287496$ is a root of $H_{-15}(X)$ precisely when $p = 59$. In the remaining cases, the edge attachment occurs between stacking components. This produces Table 4.2. □

THEOREM 4.9 (Spine structure, $p \equiv 7 \pmod 8$ and $\ell = 2$). *Let $p \geq 17$ with $p \equiv 7 \pmod 8$. When mapping $\mathcal{G}_2(\mathbb{F}_p)$ into the spine $\mathcal{S}_2^p$, there may be stacking, folding, or attachment by new edges. Only the unique connected component of $\mathcal{G}_2(\mathbb{F}_p)$ containing $j = 1728$ and $j = 8000$ folds. The following congruence conditions on p determine the presence of new edges.*

(1) $p \equiv 71, 119 \pmod{120}$: *there is a new double-edge in $\mathcal{S}_2^p$ which may or may not be an attachment.*
(2) $p \equiv 7, 23, 31, 47, 79, 103 \pmod{120}$: *edge attachment does not occur.*

PROOF. We proceed as in the proofs of the previous two theorems. By Proposition 4.3, the only components of $\mathcal{G}_2(\mathbb{F}_p)$ which could possibly fold are those containing $j = 8000$ and $j = 1728$. For $p \equiv 7 \pmod 8$, both these are supersingular, and are distinct for $p > 13$. By [**ACNL+23**, Ex. 3.8], exactly one of the occurrences of 1728 in $\mathcal{G}_2(\mathbb{F}_p)$ lies on the surface of a volcano. Since $h(-p)$ is odd by Proposition 2.1 and an ideal above 2 is necessarily non-principal, any volcano surface contains an odd number of vertices greater than 1, and the number of connected components is a divisor of $h(-p)$ and hence odd. Since stacking occurs in pairs, an odd number of connected components must fold. We show that precisely one connected component folds, and that this is the unique component containing $j = 1728, 8000$ and their twists. A connected component of the volcano shape (Theorem 2.3 (3)) with an odd number of vertices which folds must satisfy two conditions: First, there must be one j-invariant which is associated to one vertex on the surface and one vertex on the floor, and these vertices must be adjacent. This occurs precisely for $j = 1728$. This condition alone implies that the component folds, and as a result there must be precisely one j-invariant which is 2-isogenous to its quadratic twist across the rim of the volcano. This occurs precisely for $j = 8000$ by Lemma 4.1. Hence the unique component containing $j = 1728, 8000$ folds, and no other component folds.

A new edge is added when $p \equiv 11, 14 \pmod{15}$ by Proposition 4.5, but this may or may not be an edge attachment. Other than this edge, there are no new non-loop edges, so no further edge attachment can occur. Combining $p \equiv 11, 14 \pmod{15}$ with $p \equiv 7 \pmod 8$ gives $p \equiv 71, 119 \pmod{120}$. □

REMARK 4.10. Theorem 4.9 shows that precisely one connected component of $\mathcal{G}_2(\mathbb{F}_p)$ folds for $p \equiv 7 \pmod 8$. The structure of this folding component tells us that the smallest positive integer k such that there exists an $\mathbb{F}_p$-rational isogeny of degree 2^k between an elliptic curve with $j = 8000$ and an elliptic curve with $j = 1728$ is $k = (r-1)/2$, where r is the (necessarily odd) order of the class of a prime ideal above 2 in the class group of $\mathbb{Q}(\sqrt{-p})$.

REMARK 4.11. Whether or not the new edge of Theorem 4.9 (1) produces an edge attachment depends entirely on whether the roots of $H_{-15}(X)$ belong to the same connected component of $\mathcal{G}_2(\mathbb{F}_p)$. When $p \equiv 7 \pmod 8$, the volcano structure of this graph leaves too many possibilities. There is no single simple condition to establish the existence of an endomorphism, so we are not able to determine if the new (double) edge is attaching or not. We provide examples for each case.

EXAMPLE 4.12 ($p = 71$, no edge attachment). For $p = 71$, edge attachment is in principle possible by Theorem 4.9; however, it does not occur. The graph $\mathcal{G}_2(\mathbb{F}_{71})$ consists of a single component. It has 7 vertices on the surface joined to 7 vertices on the floor. This component folds and there is a new edge, but clearly this cannot be an edge attachment: any time $\mathcal{G}_2(\mathbb{F}_p)$ consists of a single connected component, edge attachment is not possible. See Figure 4.1.

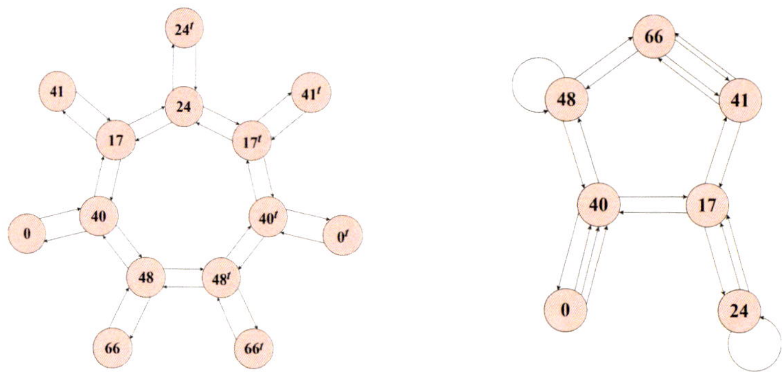

(A) (B)

FIGURE 4.1. (A): The graph $\mathcal{G}_2(\mathbb{F}_{71})$. Vertex labels j, j^t represent the j-invariant of a curve and its quadratic twist. (B): The spine graph $\mathcal{S}_2^{71}$.

EXAMPLE 4.13 ($p = 1319$, edge attachment). For $p = 1319 \equiv 119 \pmod{120}$, edge attachment is possible by Theorem 4.9, and in fact it occurs. The graph $\mathcal{G}_2(\mathbb{F}_{1319})$ has five connected components. Each component is a volcano with 9 vertices on the surface and 9 vertices on the floor. Two pairs of connected components

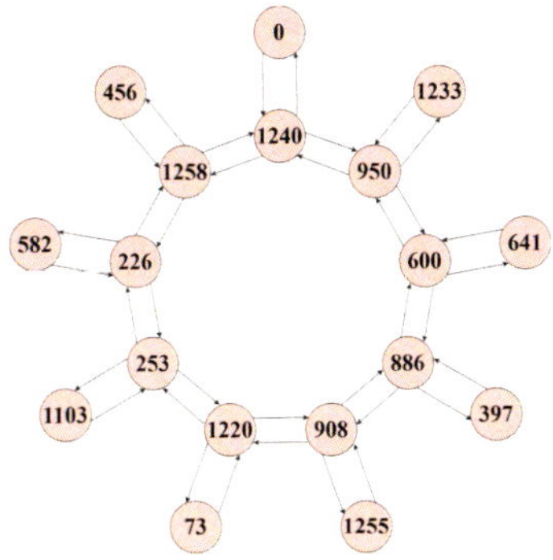

(A)

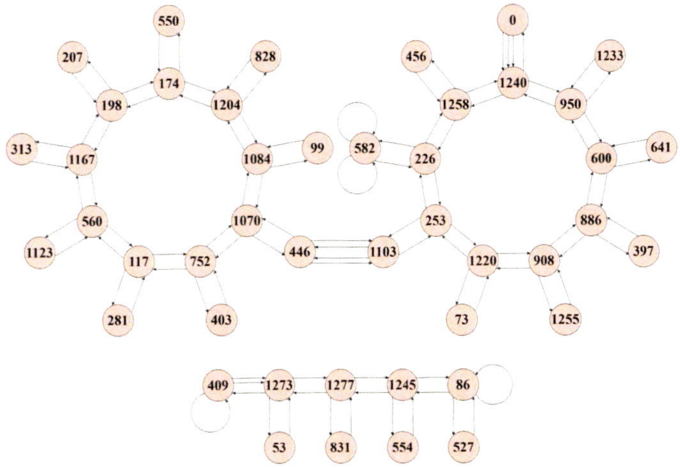

(B)

FIGURE 4.2. (A): One (of five) connected components of $\mathcal{G}_2(\mathbb{F}_{1319})$. (B): The spine graph $\mathcal{S}_2^{1319}$.

stack and the remaining connected component folds. The two stacked components attach at a new edge at the vertices with j-invariants 446 and 1103. See Figure 4.2.

5. Structure of $\mathcal{S}_\ell^p$ for $\ell \geq 3$

The process of moving from $\mathcal{G}_\ell(\mathbb{F}_p)$ to $\mathcal{S}_\ell^p$ for $\ell \geq 3$ is less involved than the analogous procedure for $\ell = 2$ due to the substantially simpler structure of $\mathcal{G}_\ell(\mathbb{F}_p)$ as a collection of disjoint cycles. Note that if $(\frac{-p}{\ell}) = -1$, then $\mathcal{G}_\ell(\mathbb{F}_p)$ has no edges, including loops, by Theorem 2.3 and Corollary 2.5. In this case, all components (i.e. isolated vertices) stack, none fold, there is no vertex attachment, and only new edges are introduced. We provide an overview of the general method for computing spine structures, with particular focus on the case $\ell = 3$, where we describe the structure of $\mathcal{S}_3^p$ in a manner similar to Theorems 4.7–4.9.

STEP 1 (Loops). Determine the vertices in $\mathcal{G}_\ell(\overline{\mathbb{F}}_p)$ belonging to $\mathbb{F}_p$ that are incident with loops by factoring $\Phi_\ell(X, X)$ over $\mathbb{F}_p$ and determining the congruence conditions on p such that these vertices correspond to supersingular j-invariants in $\mathbb{F}_p$. For edge attachment (investigated in step 3), also ascertain which of these loops belong to $\mathcal{G}_\ell(\mathbb{F}_p)$ via Proposition 2.4.

By Corollary 2.5, $\mathcal{G}_3(\mathbb{F}_p)$ contains loops only for $p = 11$. The loops in $\mathcal{G}_3(\overline{\mathbb{F}}_p)$ are given as follows.

LEMMA 5.1 (Loops in $\mathcal{G}_3(\overline{\mathbb{F}}_p)$). *Let $p \neq 2, 3, 11$. Loops occur in $\mathcal{G}_3(\overline{\mathbb{F}}_p)$ at vertices corresponding to precisely the following j-invariants, all belonging to $\mathbb{F}_p$:*

$$\begin{aligned} 0 \text{ and } 54000 &\text{ if } p \equiv 2 \pmod 3, \\ 8000 &\text{ if } p \equiv 5, 7 \pmod 8, \\ -32768 &\text{ if } p \equiv 2, 3, 6, 7, 8, 10 \pmod{11}. \end{aligned}$$

PROOF. We have

$$\Phi_3(X, X) = -X(X-54000)(X-8000)^2(X+32768)^2 = H_{-3}H_{-12}H_{-8}^2(X)H_{-11}(X)^2.$$

The congruence conditions on p come from the inertness of p in the corresponding quadratic fields. □

The j-invariants of Lemma 5.1 are not all distinct for $p \leq 5$.

STEP 2 (Folding and vertex attachment). If $p \equiv 3 \pmod 4$, then there are two different components containing 1728 by Theorem 3.6. The two components fold and get attached at vertex $j = 1728$, and all other components stack.

If $p \equiv 1 \pmod 4$, then the components that can fold are those containing two adjacent vertices in $\mathcal{G}_\ell(\mathbb{F}_p)$ with the same j-invariant, where $\mathcal{G}_\ell(\mathbb{F}_p)$ does not have a loop by Corollary 2.5 (but of course $\mathcal{G}_\ell(\overline{\mathbb{F}}_p)$ does). These vertices, along with loops (see step 1), correspond to vertices $j \in \mathbb{F}_p$ such that $\Phi_\ell(j, j) = 0$.

PROPOSITION 5.2 (Folding and vertex attachment for $\ell = 3$). *If $p \equiv 11 \pmod{12}$, then only the two components containing $j = 1728$ fold and attach at 1728. If $p \equiv 5 \pmod{12}$, then only the component containing $j = 0$ folds and there is no vertex attachment. Else there is neither folding nor vertex attachment.*

PROOF. For $p \equiv 11 \pmod{12}$, this is Theorem 3.6. Suppose $p \equiv 5 \pmod{12}$. Then $j = 0$ and $j = 54000$ are supersingular (they are the roots of $H_{-3}(X)$ and $H_{-12}(X)$, respectively), whereas 1728 is not. Two elliptic curves defined over $\mathbb{F}_p$ with $j = 0$ are $E_0 : y^2 = x^3 + 1$ and $E_0^t : y^2 = x^3 - 3$, its twist by a cube root of $\sqrt{-3}$ in $\mathbb{F}_{p^2}$. Since -3 is not a square in $\mathbb{F}_p$, they are non-isomorphic over $\mathbb{F}_p$. There is a 3-isogeny from E_0 to E_0^t with kernel $\langle(0, 1)\rangle$ and hence defined over $\mathbb{F}_p$, i.e. corresponding to an edge in $\mathcal{G}_3(\mathbb{F}_p)$.

By [**ACNL+23**, Ex. 3.20], the only elliptic curves over $\mathbb{F}_p$ with an $\mathbb{F}_p$ rational 3-isogeny to their twist are those with j-invariants 0 and 54000. Using the same reasoning as in the proof of [**ACNL+23**, Thm. 3.18], we see that all their occurrences in $\mathcal{G}_3(\mathbb{F}_p)$ belong to the same component, and this is the only component that folds.

For all other primes p, we have $(\frac{-p}{3}) = -1$, so $\mathcal{G}_3(\mathbb{F}_p)$ contains no edges. □

STEP 3 (New edges and edge attachment). From the factorization of the polynomial $\mathrm{Res}_\ell(X)$ into Hilbert class polynomials, determine the new edges; it may not be possible to ascertain whether or not they attach. All new edges are multi-edges by [**ACNL+23**, Cor. 3.15], but not all multi-edges are new edges. Loops cannot be attaching edges. We sketch the idea for characterizing multi-edges and loops for $\ell = 3$. Consider the resultant

$$\begin{aligned} \mathrm{Res}_3(X) = -3^3 X^2 (X-8000)^2 (X-1728)^2 (X+32768)^2 \\ \times H_{-20}(X) H_{-32}(X) H_{-35}(X). \end{aligned} \tag{5.1}$$

The roots of (5.1) are j-invariants represented by vertices incident with multi-edges in $\mathcal{G}_3(\overline{\mathbb{F}}_p)$. We find the roots and generate congruence conditions on p under which these roots are supersingular and in $\mathbb{F}_p$. Furthermore, we determine values of p for which distinct roots in $\mathbb{Z}$ coincide modulo p and place them into a list of excluded p-values, denoted $\mathcal{P}_3$.

To determine which multi-edges are loops, we note that loops occur precisely at the j-invariants listed in Lemma 5.1. Substituting these j-values into the polynomial (5.1), we find the values of p for which multi-edges are loops and add these p-values to the list of excluded primes $\mathcal{P}_3$.

We obtain three spine structure theorems, differentiated by number of folding components, for all primes p outside the following set of excluded small values:

$$\mathcal{P}_3 = \{5, 7, 11, 13, 17, 19, 23, 29, 31, 41, 47, 59, 61, 71, 79, 89, 101, 139, 151, 199, 271\}.$$

The spines for primes p with $2 \leq p \leq 31$, along with the graphs $\mathcal{G}_\ell(\mathbb{F}_p)$ and $\mathcal{G}_\ell(\overline{\mathbb{F}}_p)$, are explicitly described in the document:
`SmallCharacteristicGraphDescription.pdf` at [**Hed25**].

For the remaining primes in $\mathcal{P}_3$, the corresponding information can be produced with the notebook `Small_Prime_Information.ipynb`. The notebook `Graph_Viz.ipynp` generates images of all three graphs. These sources are all available at [**Hed25**]. For the remainder of this section, we only consider primes $p \notin \mathcal{P}_3$.

THEOREM 5.3 (Spine structure, $\ell = 3$, no folding). *Suppose $p \notin \mathcal{P}_3$. In the following cases, no connected component of $\mathcal{G}_3(\mathbb{F}_p)$ folds and no vertex attachment takes place. In addition to new loops, new edges are added as follows.*

(1) *None when*

$$\begin{aligned} p \equiv\ & 1, 13, 37, 43, 67, 73, 97, 109, 121, 157, 163, 169, 187, 193, \\ & 253, 277, 283, 289, 307, 313, 337, 361, 373, 397, 403, 421, \\ & 433, 457, 493, 517, 523, 529, 541, 547, 577, 589, 613, 643, \\ & 667, 673, 697, 709, 733, 757, 781, 787, 793, 817 \pmod{840} \end{aligned}$$

(2) *One when*

$$\begin{aligned} p \equiv\ & 61, 103, 127, 181, 211, 223, 229, 241, 247, 331, 349, 367, \\ & 379, 409, 463, 481, 487, 499, 571, 583, 601, 607, 649, 661, \\ & 703, 727, 739, 769, 823, 829 \pmod{840}. \end{aligned}$$

(3) *Two that do no share any vertices when*

$$p \equiv 19, 79, 139, 151, 319, 451, 619, 631, 691, 751, 799, 811 \pmod{840}$$

(4) *Three that do no share any vertices when*

$$p \equiv 31, 199, 271, 391, 439, 559 \pmod{840}.$$

THEOREM 5.4 (Spine structure, $\ell = 3$, one component folds). *Suppose $p \notin \mathcal{P}_3$. In the following cases, the connected component of $\mathcal{G}_3(\mathbb{F}_p)$ containing $j = 0$ folds and no vertex attachment takes place. In addition to new loops, new edges are added as follows.*

(1) *None when*

$$p \equiv 17, 29, 53, 113, 137, 149, 173, 197, 221, 233, 257, 281, 293, 317, 353, 377, 389, 401, 437, 449, 473, 533, 557, 569, 593, 617, 641, 653, 677, 701, 713, 737, 773, 797, 809, 821 \pmod{840}.$$

(2) *One when*

$$p \equiv 41, 89, 101, 209, 269, 341, 461, 509, 521, 629, 689, 761 \pmod{840}$$

THEOREM 5.5 (Spine structure, $\ell = 3$, two components folds). *Suppose $p \notin \mathcal{P}_3$. In the following cases, the two connected components of $\mathcal{G}_3(\mathbb{F}_p)$ containing $j = 1728$ fold and get attached at $j = 1728$. In addition to new loops, new edges are added as follows.*

(1) *None when*

$$p \equiv 83, 107, 227, 323, 347, 443, 467, 563, 587, 683, 803, 827 \pmod{840}.$$

(2) *One when*

$$p \equiv 11, 23, 47, 143, 167, 179, 263, 383, 407, 491, 503, 527, 611, 647, 659, 743, 767, 779 \pmod{840}.$$

(3) *Two that do no share any vertices when*

$$p \equiv 59, 71, 131, 191, 239, 251, 299, 359, 419, 431, 599, 731 \pmod{840}.$$

(4) *Three that do no share any vertices when*

$$p \equiv 311, 479, 551, 671, 719, 839 \pmod{840}.$$

Note that Theorem 5.3 covers exactly the setting when $\mathcal{G}_\ell(\mathbb{F}_p)$ has no edges. Theorem 5.4 deals with the case when $j = 0$ is supersingular and $j = 1728$ is not, while in Theorem 5.5, both $j = 0$ and $j = 1728$ are supersingular.

One obstacle to obtaining general explicit structure results about $\mathcal{S}_\ell^p$ for $p \geq 5$, stated only in terms of congruence conditions on p, is the fact that degrees of Hilbert class polynomials grow as the corresponding discriminant increases in absolute value. Already for $\ell = 5$, this becomes a problem: the polynomial $\mathrm{Res}_5(X)$ has degree 54 and decomposes over $\mathbb{Z}$ into linear, quadratic and quartic irreducible factors. While formulas exist for the roots of degree 4 polynomials, the conditions on p become increasingly complicated, and for irreducible factors of degree 5 and higher, such root formulas may no longer exist.

6. Experiments on the Structural Properties of Isogeny Graphs

As mentioned in the introduction, $\mathcal{G}_\ell(\overline{\mathbb{F}}_p)$ is an optimal expander graph and in fact a Ramanujan graph for $p \equiv 1 \pmod{12}$ (when neither 0 nor 1728 is supersingular). The fact that we can partition the vertices of $\mathcal{G}_\ell(\overline{\mathbb{F}}_p)$ into two categories, namely the $\mathbb{F}_p$-vertices and the $(\mathbb{F}_{p^2} \setminus \mathbb{F}_p)$-vertices, casts doubt upon the assumption that $\mathcal{G}_\ell(\overline{\mathbb{F}}_p)$ behaves like a random graph. To shed further light on this question, we gathered a substantial amount of data on graph-theoretic invariants of spines $\mathcal{S}_\ell^p$ for $\ell \leq 5$ and many primes p. We provide an in-depth study of a selection of these invariants in this section.

6.1. Relevant notions and definitions. Let G be a directed graph with vertex set $V(G)$ and edge set $E(G)$. Recall that G is said to be *strongly connected* if it contains a directed path between any two vertices. The *distance* $d(v, w)$ from a vertex v to a vertex w is the length (i.e. the number of edges) in a shortest path from v to w in G. If no such path exists, we set $d(v, w) = \infty$. Also, $d(v, v) = 0$.

DEFINITION 6.1 (Eccentricity, diameter, radius, center).

(1) The *(out-)eccentricity* of a vertex $v \in V(G)$ is the quantity
$$ecc^+(v) = \max\{d(v, w) : w \in V(G) \text{ and } d(v, w) \neq \infty\}.$$
(2) If every component of G is strongly connected, then the *diameter* and radius of G are the respective quantities
$$\begin{aligned} diam(G) &= \max\{ecc^+(v) : v \in V(G)\}, \\ rad(G) &= \min\{ecc^+(v) : v \in V(G)\}. \end{aligned}$$
Otherwise, the diameter and radius of G are infinite, i.e. $diam(G) = rad(G) = \infty$.
(3) If the radius of G is finite then the *center* of G is the set
$$cen(G) = \{v \in V(G) : ecc^+(v) = rad(G)\}.$$

The notion of out-eccentricity captures the furthest distance required to travel from a vertex v to any other vertex of G, and the diameter is the largest of any such distance. Similarly, the radius is the smallest distance between any two vertices of G, and the center of G is comprised of the set of vertices for which the distance to any other vertex is minimal (i.e. takes on the value of the radius). As the name suggests, center vertices can be thought of as "central" in the sense that they are better connected to the entire graph compared to vertices outside the center.

6.2. Diameter of the spine $\mathcal{S}_2^p$. In light of the structure theorems from Section 4, the diameters of the connected components of $\mathcal{S}_2^p$ can be explicitly computed in almost all cases. Again, for this entire section assume $p > 13$.

REMARK 6.2 (Diameters for $\ell > 2$). One could use the structure theorems from Section 5 to make similar statements about the diameter of $\mathcal{S}_3^p$ (or even $\mathcal{S}_\ell^p$ for $p \geq 5$). However, compared to the case $\ell = 2$, less is known about the order r of an ideal class of a prime ideal above an odd prime ℓ in the appropriate class group, and it is harder to obtain concrete statements. Indeed, the $\ell > 2$ case is similar to the $p \equiv 7 \pmod 8$ case for $\ell = 2$ (Theorem 6.5), where the lengths of the cycles in $\mathcal{G}_\ell(\mathbb{F}_p)$ depend on the quantity r.

Following the structure theorems in Section 4, we determine the diameters of the connected components of $\mathcal{S}_2^p$ in the cases $p \equiv 1 \pmod 4$ and $p \equiv 3 \pmod 8$. Examples illustrating each case can be found in `SpineDiameter_examples.ipynb` at [**Hed25**]. For $p \equiv 7 \pmod 8$, we cannot determine edge attachments, which prevents us from classifying the diameter of $\mathcal{S}_2^p$ completely in this case.

Recall that (the undirected version of) $\mathcal{G}_2(\mathbb{F}_p)$ consists of pairs of vertices joined by a single edge when $p \equiv 1 \pmod 4$. Mapping into the spine $\mathcal{S}_2^p$, the generic behavior for these components is to stack. The diameters of the connected components of $\mathcal{S}_2^p$ thus depend on the number of folds and edge attachments.

THEOREM 6.3 (Spine Diameters, $p \equiv 1 \pmod 4$ and $\ell = 2$). *Let $p \geq 17$ with $p \equiv 1 \pmod 4$. There are $h(-4p)/2$ vertices in $\mathcal{S}_2^p$. The following congruence conditions on p completely determine the diameters of the components of $\mathcal{S}_2^p$:*

(1) *If $p = 29$, then $\mathcal{S}_2^p = \mathcal{G}_2(\overline{\mathbb{F}}_p)$, so it contains three vertices (diameter 2).*
(2) *If $p \equiv 29, 101 \pmod{120}$, then one connected component of $\mathcal{S}_2^p$ has a single vertex with a loop, one connected component has four vertices (diameter 3), and the remaining $h(-4p)/2 - 5$ vertices are joined in pairs (diameter 1).*
(3) *If $p \equiv 41, 89 \pmod{120}$, then one connected component of $\mathcal{S}_2^p$ has four vertices (diameter 3), and the remaining $h(-4p)/2 - 4$ vertices are joined in pairs (diameter 1).*
(4) *If $p \equiv 13, 37, 53, 61, 77, 109 \pmod{120}$, then one connected component of $\mathcal{S}_2^p$ is a single vertex with a loop, and the remaining $h(-4p)/2 - 1$ vertices are joined in pairs (diameter 1).*
(5) *If $p \equiv 1, 17, 49, 73, 97, 113 \pmod{120}$, then $\mathcal{S}_2^p$ consists of $h(-4p)/2$ vertices joined in pairs, so each connected component of $\mathcal{S}_2^p$ has diameter 1.*

PROOF. This is a direct result of applying Theorem 4.7 to the possible graph structure of $\mathcal{G}_2(\mathbb{F}_p)$ given in Theorem 2.3. □

THEOREM 6.4 (Spine Diameters, $p \equiv 3 \pmod 8$ and $\ell = 2$). *Let $p \geq 17$ with $p \equiv 3 \pmod 8$. There are $2h(-p)$ vertices in $\mathcal{S}_2^p$. The following congruence conditions on p completely determine the diameters of the components of $\mathcal{S}_2^p$:*

(1) *If $p = 59$, then $\mathcal{S}_2^p$ is a single connected component formed by a tripod edge joined by an edge to a folded tripod component (diameter 4).*
(2) *If $p \equiv 11, 59 \pmod{120}$ ($p \neq 59$), then one connected component of $\mathcal{S}_2^p$ consists of two adjacent vertices (diameter 1), one connected component is 8 vertices in two tripod shapes joined by a double edge (diameter 5), and the remaining $2h(-p) - 9$ vertices are adjacent in groups of four in tripod formation.*
(3) *If $p \equiv 19, 43, 67, 83, 91, 107 \pmod{120}$, then one connected component of $\mathcal{S}_2^p$ consists of two adjacent vertices (diameter 1) and the remaining $2h(-p) - 2$ vertices are joined in groups of four in tripods from case (2) of Theorem 2.3 (diameter 2).*

PROOF. Follows directly from applying Theorem 4.8 to Theorem 2.3. □

THEOREM 6.5 (Spine Diameters, $p \equiv 7 \pmod 8$ and $\ell = 2$). *Let $p \geq 17$ with $p \equiv 7 \pmod 8$. There are $h(-p)$ vertices in $\mathcal{S}_2^p$. Let r denote the order of the ideal class generated by either of the prime ideals above 2 in the class group of $\mathbb{Q}(\sqrt{-p})$. If $p \equiv 7, 23, 31, 47, 79, 103 \pmod{120}$, then the diameter of the spine is $(r+3)/2$.*

PROOF. Follows directly from applying Theorem 4.9 to Theorem 2.3. □

If $p \equiv 71, 119 \pmod{120}$, then the diameter of $\mathcal{S}_2^p$ is uncertain. Attaching edges will approximately double the diameter of the resulting connected component, but there are no clear congruence conditions for when this occurs. See Figure 6.1 for a plot visualizing the mean diameters of the components of $\mathcal{S}_2^p$ for a range of primes $p \equiv 7 \pmod 8$. The notebook used to collect the data in this figure can be found in the file `SpineDiameter.ipynb` at [**Hed25**].

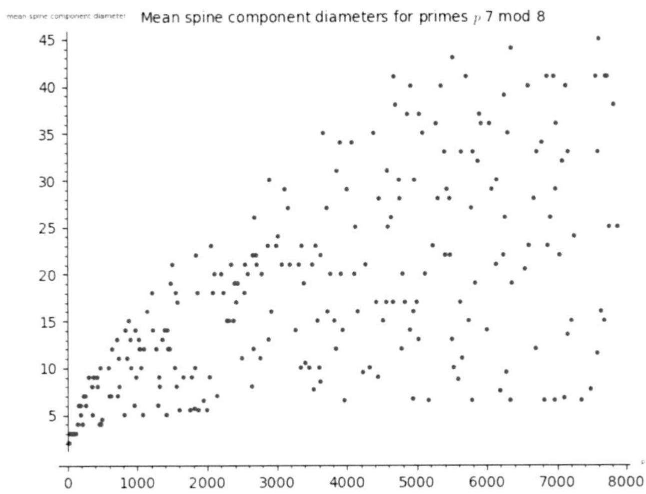

FIGURE 6.1. Mean spine component diameters in $\mathcal{G}_2(\overline{\mathbb{F}}_p)$, for 250 primes $p \equiv 7 \pmod 8$ with $23 \leq p \leq 7879$.

6.3. Center of $\mathcal{G}_\ell(\overline{\mathbb{F}}_p)$. In this work, we computed the centers of supersingular elliptic curve 2- and 3-isogeny graphs and counted the number of center vertices belonging to their respective spines. The accompanying data are listed in the file `center012925.csv`. They were generated with the notebook `Center_DataGeneration.ipynb` and plotted using `Center_DataProcessing.ipynb`. All these sources can be found at [**Hed25**].

Recall that the center of a graph (Definition 6.1) is the set of vertices with minimal out-eccentricity. These vertices are well-connected to every other vertex in the graph. Considering that the p-power Frobenius map is a graph automorphism on $\mathcal{G}_\ell(\overline{\mathbb{F}}_p)$ that fixes the vertices of $\mathcal{S}_\ell^p$, one might expect the spine vertices to be over-represented in the center of $\mathcal{G}_\ell(\overline{\mathbb{F}}_p)$. To see this, we note that for any vertex $v \in \mathbb{F}_p$, the set of distances from v displays a symmetry whereby distances can be paired up. Specifically, if w is any vertex of $\mathcal{G}_\ell(\overline{\mathbb{F}}_p)$ and w^p its Frobenius conjugate, then $d(v, w) = d(v, w^p)$. This property does not hold for vertices $v \in \mathbb{F}_{p^2} \setminus \mathbb{F}_p$. So the set of distances from vertices in $\mathbb{F}_p$ only supports "half the randomness" of the set of distances from vertices outside $\mathbb{F}_p$.

In our first experiment, we counted the number of $\mathbb{F}_p$-vertices in the center of $\mathcal{G}_2(\overline{\mathbb{F}}_p)$ as p ranges through the 2260 primes from 5 to 19997. Immediately, a striking wave-like pattern emerged. In order to ascertain whether this behavior was particular to just the $\mathbb{F}_p$-vertices in the center, we repeated the experiment for the entire center of $\mathcal{G}_2(\overline{\mathbb{F}}_p)$ using the same range of primes. This produced a very similar pattern where the wave shapes are even more pronounced. The results are plotted in blue in Figure 6.2. Analogous results for $\ell = 3$ look similar.

We are indebted to Jonathan Love for the following explanation. The observed wave pattern in Figure 6.2 closely matches the expected behavior of the minimum value of an integer-valued distribution with slow growing mean – in this case, the out-eccentricity whose minimum value (which is the radius) grows approximately as $\log(p/12)$ – and very small standard deviation. Each wave corresponds to primes p

for which $\mathcal{G}_2(\overline{\mathbb{F}}_p)$ has a fixed radius r. As p grows, so does $p/12$ (the number of vertices in $\mathcal{G}_2(\overline{\mathbb{F}}_p)$), allowing fewer vertices with out-eccentricity r until no more such vertices exist. At this point, the radius of $\mathcal{G}_2(\overline{\mathbb{F}}_p)$ jumps to $r+1$, an out-eccentricity value taken on by many more vertices, which starts the next wave. Our experiments confirm that the radius of $\mathcal{G}_2(\overline{\mathbb{F}}_p)$ is generally only slightly larger than $\log_2(p/12)$ in the range of primes under investigation. Thus, if a wave corresponding to a given radius r peaks at a prime p, then the peak of the next wave, corresponding to radius $r+1$, should be located close to $2p$; in other words, the distance between consecutive wave crests doubles each time. Our data bears this out as well.

The green 'Discrete Gaussian' points in Figure 6.2 were obtained via a Discrete Gaussian sampler as follows. For any prime p with $p \equiv 1 \pmod{12}$, simulate a 3-regular graph G with $(p-1)/12$ vertices. Assign an out-eccentricity to any vertex of G by sampling from a normal distribution with mean $1.8\log(p)$ and standard deviation 0.38 and plot the floor function of the sampled values. Although out-eccentricities of adjacent vertices in $\mathcal{G}_2(\overline{\mathbb{F}}_p)$ are in actuality not independent, it is evident that this model matches our observed center size data quite closely.

Additionally, we thank Thomas Decru and Jonathan Komada Eriksen for the observation that the likelihood of a vertex v of $\mathcal{G}_\ell(\overline{\mathbb{F}}_p)$ to belong to the center can be estimated by the discrepancy between the theoretically possible and the actual number of ways in which it can achieve $ecc^+(v) = r$, where r is the radius of $\mathcal{G}_\ell(\overline{\mathbb{F}}_p)$. Let T_ℓ be the tree rooted at v where v has $\ell+1$ children, all other interior vertices have ℓ children, and all leaf nodes are located at level r. Then T_ℓ models an idealized version of the possible paths of length at most r from v in $\mathcal{G}_\ell(\overline{\mathbb{F}}_p)$, assuming no cycles are encountered. The difference $\epsilon = |V(T_p)| - |(\mathcal{G}_\ell(\overline{\mathbb{F}}_p))|$ is a measure of the likelihood for a vertex of $\mathcal{G}_\ell(\overline{\mathbb{F}}_p)$ to lie in the center. For $\ell = 2$, the number of vertices in T_2 is $|V(T_2)| = 1 + 3(2^r - 1)$. The quantity $\epsilon = |V(T_2)| - |V(\mathcal{G}_2(\overline{\mathbb{F}}_p))|$, scaled by $1/12$ for best fit, is plotted in red in Figure 6.2.

To see if elliptic curves with extra automorphisms had an effect on the size of the center of $\mathcal{G}_2(\mathbb{F}_2)$, we separated our data into congruence classes of $p \pmod 4$ and $p \pmod 3$, and the plots are found in Figure 6.3. No definitive pattern emerges for the congruence classes of $p \pmod 3$. While the data points for both congruences classes of p modulo 4 are spread out over the entire data range, higher center size counts (i.e. more data points in the wave peaks) appear for $p \equiv 3 \pmod 4$. This is to be expected because the radius r of $\mathcal{G}_\ell(\overline{\mathbb{F}}_p)$ tends to be larger in this case and hence easier to attain. In particular, the vertex associated to $j = 1728$ has only one neighbor distinct from itself, whereas a generic vertex of $\mathcal{G}_\ell(\overline{\mathbb{F}}_p)$ is expected to have three neighbors. A larger radius r makes it easier for a random vertex to achieve out-eccentricity at most r and thus belong to the center of $\mathcal{G}_\ell(\overline{\mathbb{F}}_p)$. We also investigated the likelihood of 1728 belonging to the center of $\mathcal{G}_2(\overline{\mathbb{F}}_p)$. Overwhelmingly, this is not the case: out of the 1135 primes $p \equiv 3 \pmod 4$ with $5 \leq p < 20000$, the only primes p for which 1728 lies in the center of $\mathcal{G}_2(\overline{\mathbb{F}}_p)$ are $p = 7, 11, 19$.

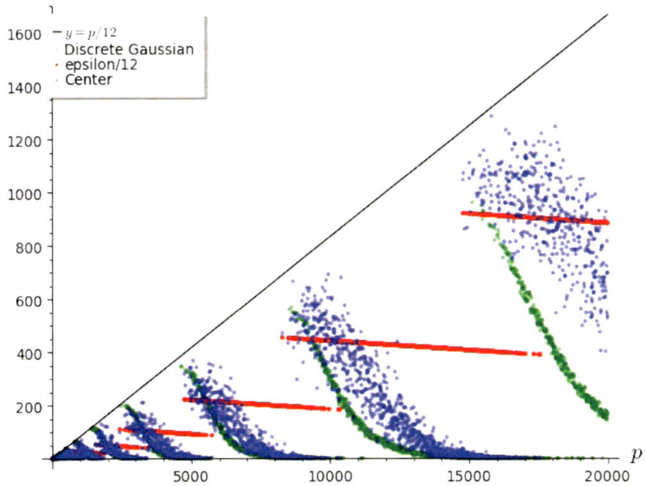

FIGURE 6.2. Size of the center of $\mathcal{G}_2(\overline{\mathbb{F}}_p)$ (in blue) and discrete Gauss sampling with $\mu = 1.8\log(p)$, $\sigma = 0.38$, (in green). The red segments represent the estimate $\epsilon/12$ for a vertex of $\mathcal{G}(\overline{\mathbb{F}}_p)$ to belong to the center, and the line $y = p/12$ (in black) approximates the total number of vertices in $\mathcal{G}_2(\overline{\mathbb{F}}_p)$. COLOR ONLINE.

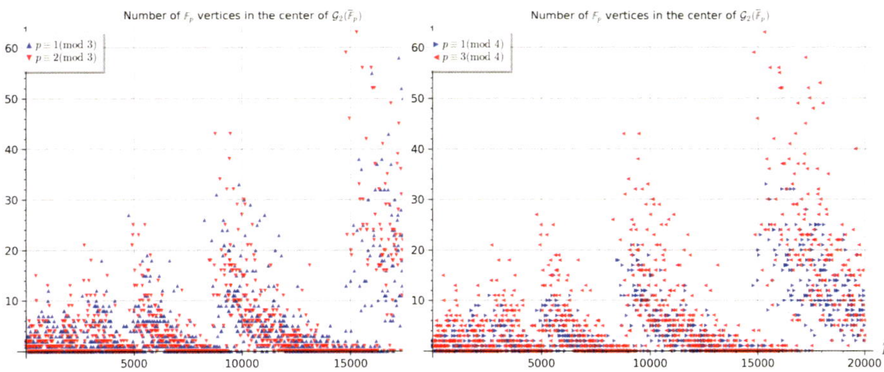

(A) $p \pmod 3$ (B) $p \pmod 4$

FIGURE 6.3. Number of $\mathbb{F}_p$-vertices in the center. Left: dark triangles pointing up (in blue), light triangles pointing down (in red) represent $p \equiv 1 \pmod 3$, $p \equiv 2 \pmod 3$ (resp.). Right: dark triangles pointing right (in blue), light triangles pointing left (in red) represent $p \equiv 1 \pmod 4$, $p \equiv 3 \pmod 4$ (resp.). COLOR ONLINE.

7. Conclusions and future work

The spine $\mathcal{S}_\ell^p$ of the supersingular ℓ-isogeny graph $\mathcal{G}_\ell(\overline{\mathbb{F}}_p)$, our main protagonist, is obtained by mapping the supersingular ℓ-isogeny graph $\mathcal{G}_\ell(\mathbb{F}_p)$ into $\mathcal{G}_\ell(\overline{\mathbb{F}}_p)$ via a natural two-step process. When passing from $\mathbb{F}_p$-isomorphism classes of curves to $\overline{\mathbb{F}}_p$-isomorphism classes, vertices of $\mathcal{G}_\ell(\mathbb{F}_p)$ representing j-invariants of twists are identified, leading to either stacking or folding of connected components of $\mathcal{G}_\ell(\mathbb{F}_p)$. Components may be joined via attachment at a vertex (for $\ell > 2$ only, and only for j-invariant 1728) or an edge. Passing from ℓ-isogenies over $\mathbb{F}_p$ to those over $\overline{\mathbb{F}}_p$ subsequently introduces new edges.

The authors of [**ACNL+23**] provided the first major insight into this process. Our structure theorems in Sections 4 and 5 offer a refinement of their work by characterizing this behavior almost completely in the cases $\ell = 2, 3$ via congruence conditions on p, and outlining a general road map for determining $\mathcal{S}_\ell^p$ for larger primes ℓ. For any pair (ℓ, p), the graphs $\mathcal{G}_\ell(\mathbb{F}_p)$, $\mathcal{S}_\ell^p$ and $\mathcal{G}_\ell(\overline{\mathbb{F}}_p)$ can be explicitly generated using our code at [**Hed25**]; for small primes p and $\ell = 2, 3$, they are described explicitly in a separate document there and cited throughout this paper.

Our structure theorems make it possible to determine the diameter of $\mathcal{S}_2^p$, i.e. the largest distance between any two vertices of $\mathcal{S}_2^p$. This explicitly shows that the diameter tends to be very small when $p \not\equiv 7 \pmod 8$; in the case $p \equiv 7 \pmod 8$, the diameter is determined by the order of the ideal class represented by a prime ideal above 2 in the class group of $\mathbb{Q}(\sqrt{-p})$.

It is natural to ask how the spine is situated inside the full ℓ-isogeny graph. To that end, we considered the centers of both $\mathcal{S}_2^p$ and $\mathcal{G}_2(\mathbb{F}_p)$, i.e. the collection of vertices whose furthest distance to any other vertex is minimal. The count of center vertices in $\mathcal{G}_\ell(\overline{\mathbb{F}}_p)$ defined over $\mathbb{F}_p$, as well as the size of the full center of $\mathcal{G}(\overline{\mathbb{F}}_p)$, follow a remarkable wave-like pattern as p grows. A similar pattern was observed for $\ell = 3$. The community-sourced explanation shows that this pattern is evidence of supersingular elliptic curve isogeny graphs behaving as random graphs.

Our continuing exploration of graph theoretic features of $\mathcal{S}_\ell^p$ gives us new insight into the cryptographically relevant heuristic assumptions we make about $\mathcal{G}_\ell(\overline{\mathbb{F}}_p)$. Beyond the findings reported herein, we conducted extensive numerical experiments generating a substantial volume of data on both internal and external connectivity properties of $\mathcal{S}_\ell^p$, as well as counts and proportions of vertices in the periphery of $\mathcal{S}_\ell^p$ and $\mathcal{G}_\ell(\overline{\mathbb{F}}_p)$. Our findings raise a number intriguing questions (and answers, thanks to our community); analyzing and understanding the results of our experiments is very much a work in progress as we strive to shed further light on the structural features and patters found in supersingular elliptic curve ℓ-isogeny graphs.

References

[ACNL+23] Sarah Arpin, Catalina Camacho-Navarro, Kristin Lauter, Joelle Lim, Kristina Nelson, Travis Scholl, and Jana Sotáková, *Adventures in supersingularland*, Exp. Math. **32** (2023), no. 2, 241–268, DOI 10.1080/10586458.2021.1926009. MR4592945

[CJS14] Andrew Childs, David Jao, and Vladimir Soukharev, *Constructing elliptic curve isogenies in quantum subexponential time*, J. Math. Cryptol. **8** (2014), no. 1, 1–29, DOI 10.1515/jmc-2012-0016. MR3163097

[CLG09] Denis X. Charles, Kristin E. Lauter, and Eyal Z. Goren, *Cryptographic hash functions from expander graphs*, J. Cryptology **22** (2009), no. 1, 93–113, DOI 10.1007/s00145-007-9002-x. MR2496385

[DFKL+20] Luca De Feo, David Kohel, Antonin Leroux, Christophe Petit, and Benjamin Wesolowski, *SQISign: compact post-quantum signatures from quaternions and isogenies*, Advances in cryptology—ASIACRYPT 2020. Part I, Lecture Notes in Comput. Sci., vol. 12491, Springer, Cham, 2020, pp. 64–93. MR4210303

[DG16] Christina Delfs and Steven D. Galbraith, *Computing isogenies between supersingular elliptic curves over* $\mathbb{F}_p$, Des. Codes Cryptogr. **78** (2016), no. 2, 425–440, DOI 10.1007/s10623-014-0010-1. MR3451433

[FFK+23] Luca De Feo, Tako Boris Fouotsa, Péter Kutas, Antonin Leroux, Simon-Philipp Merz, Lorenz Panny, and Benjamin Wesolowski, *SCALLOP: scaling the CSI-FiSh*, Public-key cryptography—PKC 2023. Part I, Lecture Notes in Comput. Sci., vol. 13940, Springer, Cham, 2023, pp. 345–375, DOI 10.1007/978-3-031-31368-4_13. MR4591147

[FM02] Mireille Fouquet and François Morain, *Isogeny volcanoes and the SEA algorithm*, Algorithmic number theory (Sydney, 2002), Lecture Notes in Comput. Sci., vol. 2369, Springer, Berlin, 2002, pp. 276–291, DOI 10.1007/3-540-45455-1_23. MR2041091

[Gha24] Wissam Ghantous, *Loops, multi-edges and collisions in supersingular isogeny graphs*, Adv. Math. Commun. **18** (2024), no. 4, 935–955, DOI 10.3934/amc.2022038. MR4762477

[Hed25] Taha Hedayat, *Lucant-2025-supersingular-ell-isogeny-spine*, `https://github.com/TahaHedayat/LUCANT-2025-Supersingular-Ell-Isogeny-Spine`, 2025.

[JW09] Michael J. Jacobson Jr. and Hugh C. Williams, *Solving the Pell equation*, CMS Books in Mathematics/Ouvrages de Mathématiques de la SMC, Springer, New York, 2009. MR2466979

[Lan87] Serge Lang, *Elliptic functions*, 2nd ed., Graduate Texts in Mathematics, vol. 112, Springer-Verlag, New York, 1987. With an appendix by J. Tate, DOI 10.1007/978-1-4612-4752-4. MR890960

[S+25] William A. Stein et al., *Sage Mathematics Software (Version 10.4)*, The Sage Development Team, 2025, `http://www.sagemath.org`.

[Sut] Andrew V. Sutherland, *Modular polynomials*, `https://math.mit.edu/~drew/ClassicalModPolys.html`, Accessed: 2025-01-17.

Department of Mathematics and Statistics, University of Calgary, 2500 University Drive NW, Calgary, AB T2N 1N4, Canada
Email address: `taha.hedayat@ucalgary.ca`

Virginia Tech Department of Mathematics, 225 Stanger Street, Blacksburg, VA 24061, USA
Email address: `sarpin@vt.edu`

Department of Mathematics and Statistics, University of Calgary, 2500 University Drive NW, Calgary, AB T2N 1N4, Canada
Email address: `rscheidl@ucalgary.ca`

Contemporary Mathematics
Volume **840**, 2026
https://doi.org/10.1090/conm/840/16809

Explicit modularity of K3 surfaces with complex multiplication of large degree

Edgar Costa, Andreas-Stephan Elsenhans, Jörg Jahnel, and John Voight

Abstract. We consider the transcendental motives of three K3 surfaces X conjectured to have complex multiplication (CM). Under this assumption, we match these to explicit algebraic Hecke quasi-characters ψ_X, and CM abelian threefolds A. This provides substantial evidence that a power of A corresponds to X under the Kuga–Satake correspondence.

1. Introduction

K3 surfaces provide a rich class of objects to study in number theory and the Langlands program, testing conjectures that connect arithmetic geometry and automorphic forms through Galois representations and L-functions.

The case where the Picard number ρ achieves its maximum $\rho = 20$ has been well-studied. Potential modularity was established through their association with algebraic Hecke quasi-characters (also called Hecke Grossencharacters or just Hecke characters) by Shioda–Inose [**SI77**, §6, Theorem 6] (see also Livné [**Liv95**]): the transcendental cohomology has complex multiplication (CM) by an imaginary quadratic field. Over $\mathbb{Q}$, an explicit correspondence with classical modular forms of weight 3 was worked out by Elkies–Schütt [**ES13**].

Recent efforts towards incorporating K3 surfaces into the *L-functions and Modular Forms Database (LMFDB)* [**LMFDB**] has renewed questions of explicit modularity for K3 surfaces, but less is known about modularity for K3 surfaces of lower Picard number. In general, a complex K3 surface X with $\rho(X) \leq 16$ does not admit a Shioda–Inose structure. Piatetski-Shapiro–Shafarevich [**PŠŠ73**] expressed the L-function of a K3 surfaces with complex multiplication as a product of Hecke L-functions over some finite extension via the Kuga–Satake correspondence and applying the corresponding statement for abelian varieties. The theory of complex multiplication for K3 surfaces was further developed by Rizov [**Riz05**]. Building on this work, Valloni [**Val23**] considers K3 surfaces with CM by the full ring of integers and studied their fields of definition; and more recently, Ito [**Ito25**] is more explicit about the properties of the Hecke quasi-characters that appear in the equality of L-series. Livné–Schütt–Yui [**LSY10**] established modularity for the finitely many K3 Delsarte surfaces (up to twist): they are (CM) quotients of Fermat surfaces.

This paper advances these efforts in a new direction, through computation. We remain focused on explicit examples of K3 surfaces with apparent CM of large degree. Indeed, there has been recently renewed interest [**BGS24**] in moduli of K3 surfaces with extra Hodge endomorphisms. Our main result matches the transcendental cohomology of certain K3 surfaces with algebraic Hecke quasi-characters, as follows. For a complex surface X, we let $\mathrm{T}(X)_{\mathbb{Q}} \subseteq \mathrm{H}^2(X, \mathbb{Q})$ be the transcendental subspace (see section 2). If moreover X is defined over a number field $F \subset \mathbb{C}$, then for ℓ prime, we have via comparison $\mathrm{T}(X) \otimes \mathbb{Q}_\ell \hookrightarrow \mathrm{H}^2_{\text{ét}}(X, \mathbb{Q}_\ell)$ and we let $\rho_{\mathrm{T}(X),\ell}\colon \mathrm{Gal}_F \circlearrowright (\mathrm{T}(X) \otimes \mathbb{Q}_\ell)$ be the associated Galois representation.

Let $X = X_i$ for $i = 1, 2, 3$ be one of the three K3 surfaces obtained from the following affine models:

$$\begin{aligned} X_1&\colon w^2 = xyz(x^3 - 3xy^2 + y^3 - 3x^2z - 3xyz + 9y^2z + 6yz^2 + z^3) \\ X_2&\colon w^2 = xyz(7x^3 - 7x^2y + y^3 + 49x^2z - 21xyz - 7y^2z + 98xz^2 + 49z^3) \\ X_3&\colon w^2 = xyz\begin{pmatrix} 49x^3 - 304x^2y + 361xy^2 + 361y^3 + 570x^2z - 2793xyz \\ +2888y^2z + 2033xz^2 - 5415yz^2 + 2299z^3 \end{pmatrix}. \end{aligned} \tag{1.1}$$

More precisely, we take X_i to be the smooth projective surface obtained from the taking branched double cover of $\mathbb{P}^2$ defined by (1.1) and blowing up the $15 = \binom{6}{2}$ double points in the branch locus of 6 lines. Then $\dim_{\mathbb{Q}_\ell} \mathrm{T}(X_i) = 22 - 16 = 6$, and there is substantial numerical evidence that in each case, $\mathrm{T}(X_i)_{\mathbb{Q}}$ has CM by K_i, where $K_i = F_i(\sqrt{-1})$ is the cyclic sextic field defined in Table 1 by their LMFDB label. For this evidence, see Elsenhans–Jahnel [**EJ16**, §5] and the end of section 2.

THEOREM 1.2. *For $i = 1, 2, 3$ and $X = X_i$, the following statements hold.*

(a) *Suppose $\mathrm{T}(X)_{\mathbb{Q}}$ has CM by K. Then for all primes ℓ,*

$$\rho_{\mathrm{T}(X),\ell} \simeq \mathrm{Ind}_{\mathrm{Gal}_K}^{\mathrm{Gal}_{\mathbb{Q}}} \psi_{X,\ell} \tag{1.3}$$

where ψ_X is of ∞-type $\{(0,2),(1,1),(1,1)\}$ defined in Table 2. *In particular, we have*

$$L(\mathrm{T}(X), s) = L(s, \psi_X). \tag{1.4}$$

(b) *Let $A = A_i = \mathrm{Jac}(C_i)$ be the Jacobian defined in* Table 1. *Then*

$$\begin{aligned} \rho_{\mathrm{H}^1(A),\ell} &\simeq \mathrm{Ind}_{\mathrm{Gal}_K}^{\mathrm{Gal}_{\mathbb{Q}}} \psi_{A,\ell} \\ L(\mathrm{H}^1(A), s) &= L(s, \psi_A) \end{aligned} \tag{1.5}$$

where ψ_A is of ∞-type $\{(0,1),(0,1),(0,1)\}$ defined in Table 3.

(c) *We have*

$$\begin{aligned} \rho_{\mathrm{H}^2(A),\ell} &\simeq \mathrm{Ind}_{\mathrm{Gal}_F}^{\mathrm{Gal}_{\mathbb{Q}}} \mathbb{Q}_\ell(1) \oplus \mathrm{Ind}_{\mathrm{Gal}_K}^{\mathrm{Gal}_{\mathbb{Q}}} (\psi_X \oplus \psi')_\ell \\ L(\mathrm{H}^2(A), s) &= \zeta_F(s+1) L(s, \psi_X) L(s, \psi'), \end{aligned} \tag{1.6}$$

where $\mathbb{Q}_\ell(1)$ is the Tate twist, $F \subseteq K$ is the unique cubic subfield, and ψ' is of ∞-type $\{(0,2),(0,2),(1,1)\}$ defined in Table 4.

This provides substantial evidence that a power of A corresponds to X under the Kuga–Satake correspondence [**KS67**]: for more, see Remark 3.2. Our computations are performed in Magma [**BCP97**]; the code is available at `https://github.com/edgarcosta/K3withCM/`. There is a natural Galois action on algebraic Hecke quasi-characters by $\psi^\sigma = \psi \circ \sigma$ for $\sigma \in \mathrm{Gal}(K \mid \mathbb{Q})$, with $L(s, \psi^\sigma) = L(s, \psi)$. Up to this Galois action, the characters ψ_X, ψ_A, and ψ' in Theorem 1.2 are unique.

TABLE 1. Polynomials defining CM fields and genus 3 curves.

i	K_i	F_i	Defining equation for C_i
1	6.0.419904.1	3.3.81.1	$y^2 = x^7 + 6x^5 + 9x^3 + x$
2	6.0.153664.1	3.3.49.1	$y^2 = x^7 + 7x^5 + 14x^3 + 7x$
3	6.0.8340544.1	3.3.361.1	$y^2 = x^7 + 1786x^5 + 44441x^3 + 278179x$
4	6.0.59105344.1	3.3.961.1	$y^2 = x^7 + 961x^5 - 3694084x^3 + 1832265664x$

TABLE 2. Uniquely defining properties of ψ_X, up to $\mathrm{Gal}(K \mid \mathbb{Q})$.

i	$\mathrm{cond}(\psi_{X_i})$	p	$L_p(\psi_{X_i}, T)$
1	64.1	17	$1 - 6T + 15pT^2 + 12p^2T^3 + 15p^3T^4 - 6p^4T^5 + p^6T^6$
2	3136.1	13	$1 - 2T + 19pT^2 + 4p^2T^3 + 19p^3T^4 - 2p^4T^5 + p^6T^6$
3	23104.1	37	$1 + 14T - 5pT^2 - 28p^2T^3 - 5p^3T^4 + 14p^4T^5 + p^6T^6$
4	61504.13	29	$1 - 38T - 9pT^2 + 52p^2T^3 - 9p^3T^4 - 38p^4T^5 + p^6T^6$

TABLE 3. Uniquely defining properties of ψ_A, up to $\mathrm{Gal}(K \mid \mathbb{Q})$.

i	$\mathrm{cond}(\psi_{A_i})$	p	$L_p(\psi_{A_i}, T)$
1	4096.1	17	$1 - 6T + 15T^2 - 52T^3 + 15pT^4 - 6p^2T^5 + p^3T^6$
2	25088.1	13	$1 + 4T + 7T^2 + 40T^3 + 7pT^4 + 4p^2T^5 + p^3T^6$
3	184832.1	37	$1 + 4T + 15T^2 - 152T^3 + 15pT^4 + 4p^2T^5 + p^3T^6$
4	3936256.41	29	$1 + 4T + 51T^2 + 216T^3 + 51pT^4 + 4p^2T^5 + p^3T^6$

TABLE 4. Uniquely defining properties of ψ', up to $\mathrm{Gal}(K \mid \mathbb{Q})$.

i	$\mathrm{cond}(\psi'_i)$	p	$L_p(\psi'_i, T)$
1	1.1	17	$1 + 42T + 1023T^2 + 1132pT^3 + 1023p^2T^4 + 42p^4T^5 + p^6T^6$
2	1.1	13	$1 + 34T + 631T^2 + 652pT^3 + 631p^2T^4 + 34p^4T^5 + p^6T^6$
3	1.1	37	$1 + 82T + 4423T^2 + 5452pT^3 + 4423p^2T^4 + 82p^4T^5 + p^6T^6$
4	1.1	29	$1 + 74T + 3067T^2 + 3268pT^3 + 3067p^2T^4 + 74p^4T^5 + p^6T^6$

REMARK 1.7. To each Hecke quasi-character ψ for a CM extension $K \supseteq F$, by restriction of the automorphic representation we can also associate a Hilbert modular form f over F with matching Galois representation and L-function. As such forms f have nontrivial character (and in parts (a) and (c), weights $(3, 3, 1)$ and $(3, 1, 1)$, respectively), they currently fall outside the database of Hilbert modular forms in the LMFDB. We hope to see them in a future expansion of the database.

Table 1 comes from Weng [**Wen01**, §6] and is certified correct [**CMSV19**]. For the fourth row ($i = 4$), we were not able to find a matching K3 surface (among double covers of $\mathbb{P}^2$ branched along 6 lines, possibly due to the nontrivial class group), but part (b) still holds; it would be interesting to produce a K3 surface in this case (not necessarily a degree 2 model).

2. Setup

Let X be a polarized K3 surface over a number field F. We denote by X^{al} its base change to F^{al}, an algebraic closure of F. We are interested in studying the Galois representations that arise from $\mathrm{H}^2_{\text{ét}}(X^{\mathrm{al}}, \mathbb{Q}_\ell)$, for a prime ℓ. Let $\mathrm{NS}(X)$ denote the Néron–Severi group of X. Under the canonical isomorphism $\mathrm{NS}(X) \cong \mathrm{Pic}(X)$, we may identify $\mathrm{NS}(X^{\mathrm{al}}) \cong \mathrm{H}^2(X_{\mathbb{C}}, \mathbb{Z}) \cap \mathrm{H}^{1,1}(X, \mathbb{C}) \subsetneq \mathrm{H}^2(X_{\mathbb{C}}, \mathbb{Z})$. Let $\rho(X) := \mathrm{rk}\,\mathrm{NS}(X)$ be the Picard number.

Let $\mathrm{T}(X)$ be the transcendental lattice of X, the orthogonal complement of $\mathrm{NS}(X_{\mathbb{C}})$ in $\mathrm{H}^2(X, \mathbb{Z})$. The space $\mathrm{T}(X)$ is a sub-Hodge structure of $\mathrm{H}^2(X_{\mathbb{C}}, \mathbb{Z})$ with Hodge numbers $(1, 20 - \rho(X), 1)$. Let $E = E(X)$ be the algebra of endomorphisms of $\mathrm{T}(X)$ that respect the Hodge structure. Zarhin [**Zar83**, Theorems 1.5.1, 1.6] has shown that E is either a totally real field or a CM field.

The Galois representation $\rho_{\mathrm{H}^2,\ell}\colon \mathrm{Gal}(F^{\mathrm{al}} \,|\, F) \to \mathrm{GO}(\mathrm{H}^2(X_{\mathbb{C}}, \mathbb{Z}) \otimes \mathbb{Q}_\ell)$ decomposes as $\rho_{\mathrm{H}^2,\ell} = \rho_{\mathrm{NS},\ell} \oplus \rho_{\mathrm{T},\ell}$, and we focus on

$$\rho_{\mathrm{T},\ell}\colon \mathrm{Gal}(F^{\mathrm{al}} \,|\, F) \to \mathrm{GO}(\mathrm{T}(X) \otimes \mathbb{Q}_\ell).$$

and its associated [**Ser70**] L-function $L(\mathrm{T}(X), s)$. In the case that $\dim_E \mathrm{T}(X) = 1$, in fact E is necessarily a CM field, and by class field theory we have $L(\mathrm{T}(X), s) = L(s, \psi)$ for some algebraic Hecke quasi-character ψ over E.

We consider K3 surfaces $X \to \mathbb{P}^2$ as (resolutions of) branched over 6 lines in general (and in particular in good) position. In this case, $\rho(X) \geq 16$, with equality when the lines are in very general position. We restrict to the K3 surfaces identified in (1.1). As mentioned in the introduction, there is strong numerical evidence [**EJ16**, §5] that these K3 surfaces have complex multiplication (CM). We further computed 100 digit approximations to the period lattices using the method of Elsenhans–Jahnel [**EJ24**, §6]. In fact, this CM is apparently by the maximal order $\mathbb{Z}_{K_i}$ in each case. More precisely, for each surface we found numerical approximations of six period integrals $\tau_1, \ldots, \tau_6$ that form a basis of the period lattice such that the ratios τ_i/τ_1 for $i = 1, \ldots, 6$ coincide with a $\mathbb{Z}$-basis of the maximal order of the conjectural endomorphism field. For the surface X_1 and for chosen cycles,

$$(\tau_1, \ldots, \tau_6) \approx (2.6402, 11.6474, 7.60232, -7.6023i, -4.96206i, -6.68537i);$$

with respect to the eigenvalues 0.467911 and i, the period vector is an eigenvector of $\begin{pmatrix} 2 & -1 & 1 & & & \\ -1 & 2 & -2 & & & \\ 0 & -1 & 2 & & & \\ & & & 2 & -1 & -1 \\ & & & -1 & 2 & 0 \\ & & & -2 & 1 & 2 \end{pmatrix}$, $\begin{pmatrix} & & & -1 & 1 & 0 \\ & & & 0 & -1 & -1 \\ & & & -1 & 0 & 0 \\ 0 & 0 & 1 & & & \\ -1 & 0 & 1 & & & \\ 1 & 1 & -1 & & & \end{pmatrix}$ and the cup form is $\begin{pmatrix} -1 & 0 & 1 & & & \\ 0 & 0 & 1 & & & \\ 1 & 1 & -1 & & & \\ & & & 0 & 1 & 1 \\ & & & 1 & -1 & 0 \\ & & & 1 & 0 & 0 \end{pmatrix}$. The data for the other examples are very similar. Computing periods to a precision of 100 decimal places took about half an hour on a standard desktop.

3. Proof of main result

Under the assumption that the L-function matches an algebraic Hecke quasi-character, to find the correct one we need to bound its conductor. It seems difficult in general to obtain such a bound by computing the conductor of the L-function of the K3 surface. We can however produce a finite list of possibilities as follows. We start with the list of bad primes of the K3 surface and the primes above them in K. To bound the exponents of these primes, recall that (by the $\mathfrak{p}$-adic logarithm) the unit groups $(\mathbb{Z}_{K,\mathfrak{p}}/\mathfrak{p}^e)^\times$ as $e \to \infty$ have a bounded number of invariant factors. So to show that the exponent is bounded, we just need to show that the order of the finite part of the Hecke quasi-character is bounded, using the following lemma.

LEMMA 3.1. *Let ψ be an algebraic Hecke quasi-character over K of modulus $\mathfrak{N}$ and let $M \subset \mathbb{C}$ be the field generated by the values of ψ. Let $\chi\colon (\mathbb{Z}_K/\mathfrak{N})^\times \to \mathbb{C}^\times$ be the Dirichlet character defined by $\chi(a) = \psi(a\mathbb{Z}_K)\psi_\infty(a)$. Then $\mathbb{Q}(\chi) \subseteq M$.*

PROOF. By definition, an algebraic Hecke quasi-character takes values in a number field. From the idelic formulation, we conclude that the subfield generated by the restriction of ψ to the infinite places is contained in M, hence also $\mathbb{Q}(\chi)$. □

PROOF OF THEOREM 1.2. We first prove (a). We compute the bad primes for X by checking if the reduction no longer leads to 15 distinct intersection points of the 6 lines. We bound the exponents of the primes using Lemma 3.1. Following Watkins [**Wat11**, § 5.2], using Magma we compute the full list of algebraic Hecke quasi-characters ψ with the required ∞-type, conductor bounded as above, and $\mathbb{Q}(\psi) \subseteq K_i$. More precisely, we start with the principal character ψ_0 of the chosen ∞-type and its associated Dirichlet character χ_0 (see Lemma 3.1). Next, we enumerate the Dirichlet characters χ whose lifts to Hecke characters twist ψ_0 to give a primitive character ψ with $\mathbb{Q}(\psi) \subseteq K_i$. Concretely, we require that $\chi' := \chi/\chi_0$ be primitive, trivial on units, and satisfy $\mathbb{Q}(\chi) \subseteq K_i$. Because all these conditions can be phrased on the abstract character group, we apply the filters there—-iterating over every element would be impractical for large levels. For example, for X_3 we consider characters of conductor $\mathfrak{N} = \mathfrak{p}_2^7 \cdot 7 \cdot 11 \cdot \mathfrak{p}_{19}$, where $\mathfrak{p}_p$ is the unique prime dividing p. The Dirichlet character group modulo $\mathfrak{N}$ is isomorphic to

$$(\mathbb{Z}/4\mathbb{Z})^5 \oplus (\mathbb{Z}/8\mathbb{Z})^2 \oplus (\mathbb{Z}/24\mathbb{Z})^2 \oplus \mathbb{Z}/48\mathbb{Z} \oplus (\mathbb{Z}/240\mathbb{Z})^3 \oplus \mathbb{Z}/5040\mathbb{Z}$$

which contains over 10^{20} elements; of these, only 279 936 satisfy our requirements.

We then compute $L_p,(\mathrm{T}(X), T)$ using a method of Elsenhans–Jahnel [**EJ16**] based on a trace formula involving a matrix expansion. In the style of Sherlock Holmes, we eliminate all but one (up to the action of $\mathrm{Gal}(K \mid \mathbb{Q})$) by finding primes p uniquely identifying $L_p(\mathrm{T}(X), T) = L_p(\psi, T)$. It was enough to consider good primes $p < 250$ totally split in K_i to obtain a unique match for each example in a single pass. After a match was found, we identified a prime p which, together with the conductor, uniquely identifies the character (up to the Galois action).

Part (b) is proven in the same way as part (a). For part (c), we note that $\mathrm{H}^2(A) \simeq \bigwedge^2 \mathrm{H}^1(A)$, so applying (b) and identifying characters we find that $\mathrm{H}^2(A) \simeq V_1 \oplus V_2 \oplus V_3$ as representations of $\mathrm{Gal}_\mathbb{Q}$, where $V_1 \simeq \mathrm{Ind}_{\mathrm{Gal}_F}^{\mathrm{Gal}_\mathbb{Q}} \mathbb{Q}_\ell(1)$ and $\dim_K V_2 = \dim_K V_3 = 1$. The relationship between the two characters ψ_A and ψ_X can be further encoded by the equality $\psi_X(\mathfrak{p}) = \psi_A(\sigma_1(\mathfrak{p}))\psi_A(\sigma_2(\mathfrak{p}))$ for all unramified primes $\mathfrak{p}$ of degree 1 and where $\sigma_1, \sigma_2 \in \mathrm{Gal}(K \mid \mathbb{Q})$ are the two elements of order 3. We finish as in (a), checking on distinguishing primes. □

REMARK 3.2. The Kuga–Satake construction [**KS67**] (see also van Geemen [**Geo08**, §5]) attaches to a complex polarized K3 surface X a complex abelian variety such that there is an embedding $\mathrm{T}(X)(1) \hookrightarrow H^1(A) \otimes H^1(A)$ of Hodge structures. In our case, this relationship is made explicit in Theorem 1.2(c) via comparison, in the sense that X and A have associated to Hecke characters ψ_X and ψ_A, with ψ_X appearing as a symmetric product of ψ_A. This strongly suggests that

X and A are connected via the Kuga–Satake construction, at least up to isogeny and powers.

Acknowledgements

We thank Eran Assaf and David Roe for helpful conversations and Eva Bayer-Fluckiger and the anonymous referees for corrections and suggestions. Costa (SFI-MPS-Infrastructure-00008651, AS) and Voight (SFI-MPS-Infrastructure-00008650, JV) were supported by Simons Foundation grants.

References

[BGS24] Eva Bayer-Fluckiger, Bert van Geemen, and Matthias Schütt. K3 surfaces with real or complex multiplication, 2024.

[BCP97] Wieb Bosma, John Cannon, and Catherine Playoust, *The Magma algebra system. I. The user language*, J. Symbolic Comput. **24** (1997), no. 3-4, 235–265, DOI 10.1006/jsco.1996.0125. Computational algebra and number theory (London, 1993). MR1484478

[CMSV19] Edgar Costa, Nicolas Mascot, Jeroen Sijsling, and John Voight, *Rigorous computation of the endomorphism ring of a Jacobian*, Math. Comp. **88** (2019), no. 317, 1303–1339, DOI 10.1090/mcom/3373. MR3904148

[ES13] Noam D. Elkies and Matthias Schütt, *Modular forms and K3 surfaces*, Adv. Math. **240** (2013), 106–131, DOI 10.1016/j.aim.2013.03.008. MR3046305

[EJ16] Andreas-Stephan Elsenhans and Jörg Jahnel, *Point counting on K3 surfaces and an application concerning real and complex multiplication*, LMS J. Comput. Math. **19** (2016), no. suppl. A, 12–28, DOI 10.1112/S1461157016000176. MR3540943

[EJ24] Andreas-Stephan Elsenhans and Jörg Jahnel, *Real and complex multiplication on K3 surfaces via period integration*, Exp. Math. **33** (2024), no. 2, 193–224, DOI 10.1080/10586458.2022.2061649. MR4755231

[Geo08] Bert van Geemen, *Real multiplication on K3 surfaces and Kuga-Satake varieties*, Michigan Math. J. **56** (2008), no. 2, 375–399, DOI 10.1307/mmj/1224783519. MR2492400

[Ito25] Rikuto Ito. Motivic modularity of CM K3 surfaces, 2025. arXiv:2406.16325

[KS67] Michio Kuga and Ichirô Satake, *Abelian varieties attached to polarized K_3-surfaces*, Math. Ann. **169** (1967), 239–242, DOI 10.1007/BF01399540. MR210717

[Liv95] Ron Livné, *Motivic orthogonal two-dimensional representations of* Gal($\overline{\mathbf{Q}}/\mathbf{Q}$), Israel J. Math. **92** (1995), no. 1-3, 149–156, DOI 10.1007/BF02762074. MR1357749

[LSY10] Ron Livné, Matthias Schütt, and Noriko Yui, *The modularity of K3 surfaces with non-symplectic group actions*, Math. Ann. **348** (2010), no. 2, 333–355, DOI 10.1007/s00208-009-0475-9. MR2672305

[LMFDB] The LMFDB Collaboration. The L-functions and modular forms database. https://www.lmfdb.org, 2024. [Online; accessed 30 December 2024].

[PŠŠ73] I. I. Pjateckiĭ-Šapiro and I. R. Šafarevič, *The arithmetic of surfaces of type* K3 (Russian), Trudy Mat. Inst. Steklov. **132** (1973), 44–54, 264. MR335521

[Riz05] Jordan Rizov. Complex multiplication for K3 surfaces, 2005. arXiv:math/0508018

[Ser70] Jean Pierre Serre, *Facteurs locaux des fonctions zêta des varietés algébriques (définitions et conjectures)* (French), Séminaire Delange-Pisot-Poitou. 11e année: 1969/70. Théorie des nombres. Fasc. 1: Exposés 1 à 15; Fasc. 2: Exposés 16 à 24, Secrétariat Math., Paris, 1970, pp. 15. MR3618526

[SI77] T. Shioda and H. Inose, *On singular K3 surfaces*, Complex analysis and algebraic geometry, Iwanami Shoten Publishers, Tokyo, 1977, pp. 119–136. MR441982

[Val23] Domenico Valloni, *Fields of definition of K3 surfaces with complex multiplication*, J. Number Theory **242** (2023), 436–470, DOI 10.1016/j.jnt.2022.04.013. MR4490456

[Wat11] Mark Watkins, *Computing with Hecke Grössencharacters* (English, with English and French summaries), Actes de la Conférence "Théorie des Nombres et Applications", Publ. Math. Besançon Algèbre Théorie Nr., vol. 2011, Presses Univ. Franche-Comté, Besançon, 2011, pp. 119–135. MR2894271

[Wen01] Annegret Weng, *A class of hyperelliptic CM-curves of genus three*, J. Ramanujan Math. Soc. **16** (2001), no. 4, 339–372. MR1877806

[Zar83] Yu. G. Zarhin, *Hodge groups of $K3$ surfaces*, J. Reine Angew. Math. **341** (1983), 193–220, DOI 10.1515/crll.1983.341.193. MR697317

Department of Mathematics, Massachusetts Institute of Technology, 77 Massachusetts Avenue, MA 02139, USA
Email address: edgarc@mit.edu
URL: https://edgarcosta.org

School of Mathematics and Statistics, Carslaw Building (F07), University of Sydney, NSW 2006, Australia

Institut für Mathematik, Emil-Fischer Strasse 30, D-97074 Würzburg, Germany
Email address: Stephan.Elsenhans@uni-wuerzburg.de
URL: https://www.mathematik.uni-wuerzburg.de/computeralgebra/team/elsenhans-stephan-prof-dr/

Department Mathematik, Univ. Siegen, Walter-Flex-Str. 3, D-57068 Siegen, Germany
Email address: jahnel@mathematik.uni-siegen.de
URL: http://www.uni-math.gwdg.de/jahnel

Department of Mathematics, Dartmouth College, 6188 Kemeny Hall, Hanover, NH 03755, USA

School of Mathematics and Statistics, Carslaw Building (F07), University of Sydney, NSW 2006, Australia
Email address: jvoight@gmail.com
URL: https://jvoight.github.io/

Contemporary Mathematics
Volume **840**, 2026
https://doi.org/10.1090/conm/840/16810

Experimental investigations on Lehmer's conjecture for elliptic curves

Sven Cats, John Michael Clark, Charlotte Dombrowsky, Mar Curcó Iranzo, Krystal Maughan, and Eli Orvis

Abstract. In this short note, we give a method for computing a non-torsion point of smallest canonical height on a given elliptic curve $E/\mathbb{Q}$ over all number fields of a fixed degree. We then describe data collected using this method, and investigate related conjectures of Lehmer and Lang using these data.

1. Introduction

Let E be an elliptic curve over a number field K. We denote by $\overline{K}$ a fixed algebraic closure of K and by $\hat{h}$ the canonical height function on $E(\overline{K})$. Recall that $\hat{h}(P) = 0$ if and only if P is a torsion point. There is much interest in studying the canonical heights of non-torsion points. In particular, we have the following conjecture, which is known as Lehmer's conjecture because of its analogy with a conjecture of D.H. Lehmer from 1933 [**7**]. It describes how the smallest possible height of a non-torsion point $P \in E(\overline{K})$ varies with the (minimal) field $K(P)$ over which P is defined.

Conjecture 1.1 (Lehmer). *Let*

$$C_E := \inf\left\{\hat{h}(P) \cdot [K(P) : K]\right\},$$

where the infimum ranges over the non-torsion points $P \in E(\overline{K}) - E(\overline{K})_{tors}$. Then the constant C_E satisfies $C_E > 0$.

The other primary conjecture describes how the smallest possible height of a non-torsion point $P \in E(\overline{K})$ defined over an extension of a given degree varies with the curve E. Denote by j_E, Δ_E the j-invariant and minimal discriminant of E/K. We write $N_{K/\mathbb{Q}} : K \to \mathbb{Q}$ for the norm map and see Definition 2.1 for the height function $h : \mathbb{P}^1(\overline{K}) \to \mathbb{R}_{\geq 0}$.
Consider the quantity $M_E = \max\{h(j_E), \log |N_{K/\mathbb{Q}}\Delta_E|, 1\}$.

Conjecture 1.2 (Lang). *Let*

$$C_{K,d} := \inf\left\{\frac{\hat{h}(P)}{M_{E'}}\right\},$$

2020 *Mathematics Subject Classification.* Primary 11G05; Secondary 14G40.

where the infimum ranges over all elliptic curves E'/K and the non-torsion points $P \in E'(\overline{K}) - E'(\overline{K})_{tors}$ for which $K(P)$ is contained in a degree d extension of K. Then the constant $C_{K,d}$ satisfies $C_{K,d} > 0$.

Although there is theoretical progress on these conjectures and their generalisations to abelian varieties over number fields, very little experimental work has been done investigating the values of C_E and $C_{K,d}$. In this short paper, we describe a database of quadratic points of small height on 17,834 elliptic curves over the rationals $K = \mathbb{Q}$. In 728 of the cases, the point in the database is provably the point of smallest height on the given elliptic curve over any quadratic field. The computations to collect our data required just over 800 hours of CPU time. We use these data to investigate the constants in Conjectures 1.1 and 1.2.

We proceed first with a brief background on heights, followed by a description of the theoretical results underlying the algorithm used to build our database. We then discuss some preliminary observations about the resulting data, and possible future work.

2. Computing minimal heights over field extensions

Let K be a number field with fixed algebraic closure $\overline{K}$, and let E be an elliptic curve over K, given by an affine Weierstrass equation with coefficients in K.

DEFINITION 2.1. Let $x : E(\overline{K}) \to \mathbb{P}^1(\overline{K})$ denote the map taking the x-coordinate and $h : \mathbb{P}^1(\overline{K}) \to \mathbb{R}_{\geq 0}$ the (absolute logarithmic) Weil height on $\mathbb{P}^1(\overline{K})$, as defined in [**6**], Section B.2. By a standard abuse of notation, we also denote by $h : E(\overline{K}) \to \mathbb{R}_{\geq 0}$ the map defined by $P \mapsto h(x(P))$.

We denote the canonical height on E/K by

$$\hat{h} : E(\overline{K}) \to \mathbb{R}_{\geq 0}, \quad P \mapsto \lim_{n\to\infty} \frac{1}{4^n} h(2^n P).$$

Recall that the canonical height is the unique quadratic form $E(\overline{K}) \to \mathbb{R}_{\geq 0}$ with the property that the function $P \mapsto |h(P) - \hat{h}(P)|$ is bounded.

Let E be an elliptic curve over a number field K and let $\mathscr{F}$ be a set of finite field extensions of K with the following properties:

- If $F \in \mathscr{F}$ and $F' \subset F$, then $F' \in \mathscr{F}$.
- The set of degrees $\{[F : K] : F \in \mathscr{F}\}$ is finite.

Consider the infimum

$$C_{E,\mathscr{F}} := \inf_{F\in\mathscr{F}, P\in E(F)-E(F)_{\text{tors}}} \left\{\hat{h}(P) \cdot [F : K]\right\}.$$

REMARK 2.2. The first property ensures that whenever $F \in \mathscr{F}$ and $P \in E(F)$, the set $\mathscr{F}$ also contains the minimal field of definition $K(P)$ of P. The second property ensures that the subset of number fields in $\mathscr{F}$ of discriminant bounded by a given value is finite. In turn, using (for example) Lemma 2.3 below, this implies that a Northcott property holds for all fields in $\mathscr{F}$: There are finitely many points of bounded height on E over fields in $\mathscr{F}$. Thus the minimum height of such points exists and it follows that $C_{E,\mathscr{F}} > 0$. The fact that $C_{E,\mathscr{F}} > 0$ also follows directly from Theorem 2.4 and it is predicted by Conjecture 1.1 since $C_{E,\mathscr{F}} \geq C_E$.

In this section we explain how to explicitly compute $C_{E,\mathscr{F}}$ using a lower bound on the Weil height $h(P)$ of the x-coordinate, and an upper bound on the difference $\left|h(P) - \hat{h}(P)\right|$ with the canonical height.

We proceed in two steps: First we determine a finite set $\mathscr{F}' \subset \mathscr{F}$ such that $C_{E,\mathscr{F}'} = C_{E,\mathscr{F}}$. Then we explain how to solve the finite problem of determining $C_{E,\mathscr{F}'}$. Computational challenges arise when $\mathscr{F}'$ is large; we discuss these in the next section, where we consider the case $K = \mathbb{Q}$ and $\mathscr{F} = \{F/\mathbb{Q} : [F : \mathbb{Q}] \leq 2\}$.

As noted under Definition 2.1, we can fix $B_E \in \mathbb{R}_{>0}$ such that

$$\left|h(P) - \hat{h}(P)\right| \leq B_E \tag{1}$$

for all $P \in \cup_{F\in\mathscr{F}} E(F)$, see for example [**10**] for an explicit value of B_E. For now, any B_E satisfying (1) will do, but for our explicit computations it is useful to have B_E as small as possible. We will use a modified version of the bound given in [**3**], which we describe in Section 2.1.

LEMMA 2.3. *Let $D \in \mathbb{R}_{\geq 0}$, $F \in \mathscr{F}$, and $d = [F : K]$. Let δ_K be the number of Archimedean places of K. Define $\Delta(D, E, F) \in \mathbb{R}_{>0}$ by*

$$\Delta(D, E, F) := \exp\left(d\delta_K \log d + d(2d - 2)B_E + (2d - 2)D\right).$$

If the discriminant Δ_F of F satisfies $|\Delta_F| \geq \Delta(D, E, F)$, then $\hat{h}(P) \geq \frac{D}{d}$ for all $P \in E(F) - E(F)_{tors}$ satisfying $K(P) = F$. Further, if $[F : K] = [F' : K]$, then $\Delta(D, E, F) = \Delta(D, E, F')$.

PROOF. By Theorem 2 in [**9**] we have $h(P) \geq \frac{1}{2d-2}\left(\frac{1}{d}\log|\Delta_F| - \delta_K \log d\right)$. The first part of the lemma follows by combining $|\Delta_F| \geq \Delta(D, E, F)$ and Equation (1). The second part of the lemma is clear from the definition of $\Delta(D, E, F)$. □

We can now reduce $\mathscr{F}$ to a finite set.

THEOREM 2.4. *Let $D' \in \mathbb{R}_{\geq 0}$ be such that $C_{E,\mathscr{F}} \leq D'$ and*

$$\mathscr{F}' = \{F \in \mathscr{F} : |\Delta_F| \leq \Delta(D', E, F)\}.$$

Then, $\mathscr{F}'$ is finite and $C_{E,\mathscr{F}'} = C_{E,\mathscr{F}}$.

PROOF. By our initial assumptions on $\mathscr{F}$, the set $\{[F : K] : F \in \mathscr{F}\}$ is finite. Therefore, it follows from the second part of Lemma 2.3 that the maximum $\Delta = \max\{\Delta(D', E, F) : F \in \mathscr{F}\}$ exists. The set $\mathscr{F}_\Delta = \{F \in \mathscr{F} : |\Delta_F| \leq \Delta\}$ is finite by the Hermite–Minkowski Theorem, and hence its subset $\mathscr{F}' \subset \mathscr{F}_\Delta$ is also finite. The first part of Lemma 2.3 implies that $C_{E,\mathscr{F}'} = C_{E,\mathscr{F}}$. □

In principle, we can therefore compute $C_{E,\mathscr{F}}$ as follows: Do an initial search to find $F' \in \mathscr{F}, P' \in E(F')$ with $K(P') = F'$ such that $D' = \hat{h}(P')[F' : K]$ is small. Then $C_{E,\mathscr{F}} \leq D'$ and we write $\mathscr{F}' \subset \mathscr{F}$ for the associated finite set of fields from Theorem 2.4. In theory any F', P' work, but in practice it is worth spending more time in the initial search, as a smaller D' decreases the number of fields in $\mathscr{F}'$ to be considered later. For each $F \in \mathscr{F}'$ do a finite search to find the points $P \in E(F)$ such that

$$h(P) \leq \frac{D'}{[F : K]} + B_E. \tag{2}$$

If $\mathscr{F}'$ is not too large[1] and we can list it explicitly, we obtain in this way the finite list of F, P satisfying (2), among which is a number field $F_E \in \mathscr{F}$ and a $P_E \in E(F_E) - E(F_E)_{\text{tors}}$ such that $C_{E,\mathscr{F}} = \hat{h}(P_E) \cdot [F_E : K]$.

2.1. A modified CPS height bound. Let E be an elliptic curve over a number field K and $\mathscr{F}$ a set of extensions of K such that the set of degrees $\{[F : K] : F \in \mathscr{F}\}$ is finite. In this subsection we compute a bound $B_E \in \mathbb{R}_{\geq 0}$ such that $h(P) - \hat{h}(P) \leq B_E$ for all $P \in E(F)$ as F ranges over $\mathscr{F}$. As mentioned, there is previous work (for example Silverman [**10**] and Bruin [**2**]) computes a bound on $h(P) - \hat{h}(P)$ for all $P \in E(\overline{K})$, but in practice a smaller bound is desirable. For a given $F \in \mathscr{F}$, Cremona, Prickett and Siksek describe a bound $B_{E,F}$ for $h(P) - \hat{h}(P)$ for all $P \in E(F)$ in Theorem 1 of [**3**], which is small enough for our purposes, and we now explain how to modify it to work for all $F \in \mathscr{F}$ at once.

Indeed, for $F \in \mathscr{F}$, the bound $B_{E,F}$ is of the form

$$B_{E,F} = \frac{1}{[F : K]} \sum_v M_v,$$

where the sum ranges over the set of archimedean places of F and the set of primes of F for which E/F_v has bad reduction. The M_v are certain local invariants associated to E/F_v. Since the set of degrees $\{[F : K] : F \in \mathscr{F}\}$ is finite, and since there are only finitely many extensions of $\mathbb{R}$ and $\mathbb{Q}_p$ of any given degree, the set $\{B_{E,F} : F \in \mathscr{F}\}$ attains its maximum and we can set $B_E = \max\{B_{E,F} : F \in \mathscr{F}\}$.

REMARK 2.5. We have implemented the above procedure for computing B_E in the case $K = \mathbb{Q}$ and $\mathscr{F} = \{F/\mathbb{Q} : [F : \mathbb{Q}] \leq 2\}$.

3. Computational Results

We implemented the strategy in Section 2 in the case of quadratic fields using MAGMA version 21.2-2[**1**] and Sagemath version 10.6 [**12**]. In particular, for every elliptic curve in the Cremona database [**8**] of conductor at most 3,000, we conducted an initial search to find points of small height. We then calculated a bound $\Delta = \Delta(D', E, F)$ as in Lemma 2.3, using the height bound in Section 2.1 as our B_E, as well as a bound B on the logarithmic height of the points that needed to be searched as in Equation (2). For curves for which $\Delta < 10^5$ and $B < 50$, we searched all possible x-coordinates to obtain provably the smallest point, using the Sagemath implementation of [**4**]. In order to keep the computations feasible, for curves where either bound exceeded the numbers described above, we searched only over quadratic fields with $|\Delta_K| \leq 1,000$. We believe this choice is sufficient as in the provable cases the point of smallest height was usually found lying in a field with small discriminant. The resulting datasets and the code used to produce them are available at `https://github.com/EliOrvis/LehmersConjectureForECs`.

[1] What this means exactly depends on the efficiency of the used algorithms and the available memory and computing power. See also Section 4.

The datasets contain the following fields:

- the Cremona label for the curve;
- the discriminant of the quadratic field over which the point of smallest height over all quadratic fields is defined;
- the coordinates of the point of smallest height over all quadratic fields;
- the height of this point.

REMARK 3.1. In view of the abundance of data in the LMFDB on generators of the Mordell-Weil group of the elliptic curves E in our database and of their quadratic twists, it is not necessary to conduct an initial point search to compute the first bound Δ as these can be found in the LMFDB. Similarly, one could use the LMFDB precomputed rational points defined on E itself or one of its quadratic twists to perform the search for points over quadratic fields. Implementing these changes could improve our algorithm. We thank an anonymous referee for this suggestion.

3.1. Description of data. We ran an initial search on 17,834 elliptic curves, which required just over 800 hours of CPU time running on a server operating Red Hat Enterprise Linux 8.10. We were able to then verify that we found the point of smallest height over all quadratic fields for 86 of these curves. For 542 curves, the initial search failed to find a point, and so there is no point for these curves in our dataset. Among the remaining curves, the first curves in our list (ordered by conductor) for which the discriminant bound obtained by the initial search was too big were the curves with Cremona label 11a1 and 11a2, of conductor 11.

Among all curves in our dataset, the smallest height we found was the point

$$(27, -119, 1) \quad \text{on the elliptic curve} \quad y^2 + xy + y = x^3 + x^2 - 2990x + 71147, \tag{3}$$

which has Cremona label 1470l1, and height 0.0099641079999....

We note that the point in (3) was also found by Elkies [**5**], although his normalization of the height makes his value half of ours. At the same time, Taylor found points of much smaller height on elliptic curves defined over quadratic fields [**11**] in unpublished work. Our methods, however, differ from both of these previous computations, in that we search broadly over elliptic curves by conductor, whereas these prior computations were targeted searches in families of elliptic curves likely to contain points of small height.

We also make some observations about the quadratic fields over which points of smallest height are defined. In our dataset, the point of smallest height that we found was defined over $\mathbb{Q}$ for 2,199 of the elliptic curves. The next most common fields were the two cyclotomic quadratic fields: $\mathbb{Q}(\sqrt{-3})$ with 1,610 elliptic curves and $\mathbb{Q}(\sqrt{-4})$ with 1,191 elliptic curves. This remained consistent when restricting to curves where our point is provably the smallest over any quadratic field: in this case, the most common field was $\mathbb{Q}(\sqrt{-3})$ with 20 curves, followed by $\mathbb{Q}(\sqrt{-4})$ with 14, and then $\mathbb{Q}$ with 11. Finally, we note that among curves where we have provably found the point of smallest height over all quadratic fields, this point always agrees with the point found in our initial search. Thus, we suspect that for many of the remaining curves, the point in our dataset is in fact the smallest over all quadratic fields.

3.2. Remarks on Conjectures 1.1 and 1.2. We conducted some preliminary investigations into Conjectures 1.1 and 1.2 using the data we collected. The resulting charts can be found at
`https://github.com/EliOrvis/LehmersConjectureForECs`.
Unfortunately, we did not find a discernible relationship between the smallest height point in our dataset and either the conductor or the discriminant of the elliptic curve.

Acknowledgements

We would like to thank our mentors, Nicole Looper and Shiva Chidambaram, for their guidance throughout the project. We would also like to thank Joseph Silverman, for his mentorship, and Andrew Sutherland, for his computational insights. We thank John Voight for referring us to the article [**4**]. We would like to thank the organizers Serin Hong, Hang Xue, Alina Bucur, Renee Bell, Brandon Levin, Anthony Várilly-Alvarado, Isabel Vogt and David Zureick-Brown of the 2024 Arizona Winter School on Abelian Varieties. Finally, we would like to thank the National Science Foundation and the Clay Mathematics Institute, and the anonymous referees for their helpful feedback.

References

[1] Wieb Bosma, John Cannon, and Catherine Playoust, *The Magma algebra system. I. The user language*, J. Symbolic Comput. **24** (1997), no. 3-4, 235–265, DOI 10.1006/jsco.1996.0125. Computational algebra and number theory (London, 1993). MR1484478

[2] Peter Bruin, *Bornes optimales pour la différence entre la hauteur de Weil et la hauteur de Néron-Tate sur les courbes elliptiques sur* $\overline{\mathbb{Q}}$ (French), Acta Arith. **160** (2013), no. 4, 385–397, DOI 10.4064/aa160-4-5. MR3119786

[3] J. E. Cremona, M. Prickett, and Samir Siksek, *Height difference bounds for elliptic curves over number fields*, J. Number Theory **116** (2006), no. 1, 42–68, DOI 10.1016/j.jnt.2005.03.001. MR2197860

[4] John R. Doyle and David Krumm, *Computing algebraic numbers of bounded height*, Math. Comp. **84** (2015), no. 296, 2867–2891, DOI 10.1090/mcom/2954. MR3378851

[5] N. Elkies. Nontorsion points of low height on elliptic curves over $\mathbb{Q}$. Available online at: `https://people.math.harvard.edu/~elkies/low_height.html`, last accessed on 27.05.2025.

[6] Marc Hindry and Joseph H. Silverman, *Diophantine geometry: An introduction*, Graduate Texts in Mathematics, vol. 201, Springer-Verlag, New York, 2000, DOI 10.1007/978-1-4612-1210-2. MR1745599

[7] D. H. Lehmer, *Factorization of certain cyclotomic functions*, Ann. of Math. (2) **34** (1933), no. 3, 461–479, DOI 10.2307/1968172. MR1503118

[8] The LMFDB Collaboration. The L-functions and modular forms database. `https://www.lmfdb.org`, 2025. [Online; accessed January 2025].

[9] Joseph H. Silverman, *Lower bounds for height functions*, Duke Math. J. **51** (1984), no. 2, 395–403, DOI 10.1215/S0012-7094-84-05118-4. MR747871

[10] Joseph H. Silverman, *The difference between the Weil height and the canonical height on elliptic curves*, Math. Comp. **55** (1990), no. 192, 723–743, DOI 10.2307/2008444. MR1035944

[11] G. Taylor. Nontorsion points of low height on elliptic curves over number fields. Available online at: `https://maths.straylight.co.uk/low_height`, last accessed on 10.06.2025.

[12] The Sage Developers. SageMath, the Sage Mathematics Software System (Version 10.6), 2025. `https://www.sagemath.org`.

Centre for Mathematical Sciences, University of Cambridge, Cambridge, UK
Email address: sc2173@cam.ac.uk

Department of Mathematics, University of Texas at Austin, Austin, Texas, USA
Email address: john.m.clark@utexas.edu

Mathematical Institute, Leiden University, Leiden, The Netherlands
Email address: c.k.l.dombrowsky@math.leidenuniv.nl

Mathematical Institute, Utrecht University, Utrecht, The Netherlands
Email address: m.curcoiranzo@uu.nl

Department of Computer Science, University of Vermont, Burlington, Vermont, USA
Email address: Krystal.Maughan@uvm.edu

Department of Mathematics, University of Colorado at Boulder, Boulder, Colorado, USA
Email address: eli.orvis@colorado.edu

Contemporary Mathematics
Volume **840**, 2026
https://doi.org/10.1090/conm/840/16811

Creating a dynamic database of finite groups

Lewis Combes, John Jones, Jennifer Paulhus, David Roe, Manami Roy, and Sam Schiavone

ABSTRACT. A database of abstract groups has been added to the *L-functions and Modular Forms Database* (LMFDB), available at `https://www.lmfdb.org/Groups/Abstract/`. We discuss the functionality of the database and what makes it distinct from other available databases of abstract groups. We describe solutions to mathematical problems we encountered while creating the database, as well as connections between the abstract groups database and other collections of objects in the LMFDB.

1. Introduction

Finite groups have played a profound role in mathematics for close to two centuries and, almost since their inception, mathematicians have asked classification questions about them. The quest to classify finite simple groups took up most of the 20th century. In the last 30 years, as computational power became prominent, various databases of groups have been developed. The *ATLAS of Finite Groups* which contains information on interesting groups, notably character tables, began as a book [**CCN**$^+$**85**] and is now available online [**WWT**$^+$]. The computer algebra programs GAP [**GAP21**, **BEO02**] and Magma [**BCP97**] both have complete databases of groups of order up to 2000 (except for order 1024, for which there are almost 49.5 billion distinct isomorphism classes). These computer algebra programs also include other databases, such as almost simple and perfect groups (up to orders $1.6 \cdot 10^7$ and $2 \cdot 10^6$, respectively) and transitive groups up to degree 48 [**Hul**], [**BCH**$^+$]. The *GroupNames* project [**Dok**] includes groups up to order 500 (skipping orders 256 and 384), as well as additional information such as lattice of subgroups and character tables for the groups.

We have created an online database of over 540,000 finite groups, 275 million of their subgroups, and almost 40 million irreducible complex characters. The database can be found at: `https://www.lmfdb.org/Groups/Abstract/`.

2020 *Mathematics Subject Classification.* 20-04, 20-08, 20Dxx.

Roe and Schiavone were supported by the Simons Collaboration in Arithmetic Geometry, Number Theory, and Computation via Simons Foundation grant 550033. Paulhus was partially supported by a Frank and Roberta Furbush Scholarship from Grinnell College. Combes was supported by the Engineering and Physical Sciences Research Council grant EP/R513313/1.

1.1. Highlights of the database. As we note above, there are several other databases of groups already available, so why do we need yet another one? Some of the useful features of our database are the following.

- *The database is dynamically searchable.* We have precomputed and stored many invariants and properties of the groups in the collection, and created a search interface which allows users to search on these stored values. For instance, we can quickly determine the 8 groups G of order 1440 which satisfy the short exact sequence

$$1 \to D_{10} \to G \to F_9 \to 1,$$

  with D_n the dihedral group of order $2n$ and F_q the group of linear affine transformations of the finite field $\mathbb{F}_q$; we can immediately determine that the smallest group to have an irreducible character of degree 7 is a group of order 56; in a matter of seconds we can find all 129,069 subgroups of groups of order between 1 and 255 which are non-normal, non-abelian, non-maximal, and non-trivial proper. The list of all such subgroups (a 23 megabyte file) can be downloaded in a few seconds. Each of these searches would be time-consuming and computationally intensive in computer algebra systems like Magma or GAP.
- *The database is free and easy to use.* Use of our database requires no prior knowledge of any programming language or computer algebra package, making it accessible to a broad audience, from students learning the basics of group theory to experts using groups in their research. While most of our computations were done in Magma, which is closed-source and requires purchase of a license, the data we have computed is now publicly available on the internet through the *L-functions and Modular Forms Database* (LMFDB).
- *The database fosters connections between other collections of objects in the LMFDB.* The LMFDB is a huge international collaboration to collect, curate, and connect computational mathematical work in number theory, particularly as it relates to the Langlands program. Groups are attached to many mathematical objects throughout the LMFDB, from Sato-Tate groups to automorphism groups of curves, and our database of groups facilitates more connections between these various collections. See Figure 3 for a diagram illustrating some of these connections.
- *The database aggregates groups drawn from a variety of sources.* The database is flexible in that it can accommodate groups of various types, such as polycyclic, permutation, and matrix groups, drawn from multiple sources. See Section 2 for a list of sources used to compute the data.
- *The database supports searches on subgroups.* For each group in the database, we have computed and stored information on its subgroups. Users can search for all groups containing a subgroup with a given property; for instance, one can easily find the 12 groups of order 96 that contain a normal subgroup isomorphic to A_4; see Section 3.3 for more details on the subgroup information stored.

1.2. Structure of the paper. We begin in Section 2 with information about sources for the groups currently included in the database, and how the data about them is generated. In Section 3 we describe the web interface and demonstrate a number of features of the database. In the process of designing the database and computing the corresponding data, we encountered a series of challenges which required mathematical solutions. The next two sections discuss these challenges: Section 4 describes the method by which we computed presentations for solvable groups and Section 5 details the procedures we used to deterministically label subgroups, conjugacy classes, and characters. In Section 6 we give examples of connections between the abstract groups database and other collections of objects in the LMFDB.

2. Data overview

2.1. Sources of data. The database currently contains roughly 540,000 groups, 275 million subgroups and 40 million of their irreducible complex characters.

The code used to generate the data is written in Magma [**BCP97**] and Python [**VRD09**], and it can be found in the GitHub repository `https://github.com/roed314/FiniteGroups`. The database, as with the rest of the LMFDB, uses PostgreSQL as its database management system. Computations were carried out via a combination of an AMD Epyc 7713 server with 256 2.0GHz cores and 2TB of RAM, and a distributed computation on Google Compute Engine, with a substantial use of GNU parallel [**Tan11**] in both cases.

TABLE 1. Number of groups by original source. See Section 2 for more information.

Source	Total	Solvable	Perm.	Matrix	OptimizedPC	MinPerm
Small Groups	257936	257500	257746	68042	257500	257746
Transitive Groups	235919	211279	235919	218	14499	161656
Intransitive Groups	5444	2739	5444	16	2330	5378
Classical Lie Type	2201	2	1509	2201	0	1509
CARAT	189	185	186	189	174	186
$\mathrm{GL}_n(\mathbb{F}_q)$ Subgroups	3018	2456	3000	3018	2397	3000
$\mathrm{GL}_2(\mathbb{Z}/N)$ Subgroups	29771	28819	25319	29771	24323	25254
Perfect	123	0	123	1	0	123
Chevalley	13	0	7	13	0	7
Sporadic	9	0	9	7	0	9
Small Group Auto.	283	283	283	0	73	111
Transitive Group Auto.	498	438	498	0	51	201
Auto. Groups of Curves	530	527	530	0	526	530

We compute and store data on groups from various sources. Counts of groups currently in the database from these different sources are given in Table 1. The "Total" column represents the total number of groups in the database from that source, excluding those already shown on a previous line. The "Solvable", "Perm.", and "Matrix" columns give the number of groups from each source that are solvable, the number for which we store a permutation representation, and the number for which we store a matrix representation (over $\mathbb{Z}$, $\mathbb{F}_p$, $\mathbb{F}_q$, or $\mathbb{Z}/N$), respectively. The "OptimizedPC" column counts how many groups from a given source include an optimized polycyclic presentation (see Section 4), and "MinPerm" gives the number of groups for which we store a minimal degree permutation representation.

Approximately half of the data originates from the Small Groups library in Magma [**BE99a, BE99b, BEO01, O'B90, BE01, O'B91, NOVL04, OVL05, DE05,**

DE12, DEP22]. All groups of order up to 2000 are included, except those whose order is larger than 500 and divisible by 128. In the rest of the paper, these are the groups we will call "small groups". Most of the remaining groups come from the transitive groups database [**Hul05, CH08, HR20**], from which we include all groups of degree up to 47, except those of degree 32, with orders between 512 and 40 billion. Note that our notion of equivalence in this database is abstract isomorphism rather than conjugacy within S_n; the work to divide the transitive groups into isomorphism classes is briefly described in Section 2.5. In addition to the databases of small groups and transitive groups, we use the following sources:

- classical Lie groups up to particular bounds, and Chevalley groups that don't already show up as classical Lie groups [**Tay87, RT98, HRT01**];
- additional intransitive groups which are subgroups of S_{15} not currently in the database, computed using the Magma `Subgroup` command;
- all integer matrix groups of dimension up to 6 from CARAT [**OPS08**];
- subgroups of $\mathrm{GL}_n(\mathbb{F}_q)$ computed via Magma for $n = 2$ ($q < 1000$), $n = 3$ ($q < 16$), $n = 4$ ($q < 7$) and $n = 5$ ($q = 2$);
- subgroups of $\mathrm{GL}_2(\mathbb{Z}/N\mathbb{Z})$ for N up to 125 (skipping 80, 96, 104, 112 and 120);
- all perfect groups of order up to 50,000 from the corresponding database in Magma;
- additional sporadic groups J_1, J_2, HS, J_3, McL, He, Ru, Co_3, and Co_2 with permutation and matrix representations taken from the Atlas of Sporadic Groups [**WWT$^+$**];
- automorphism groups of curves up to genus 48 [**Bre00**], and "large" automorphism groups (of order $> 4(g-1)$) of curves up to genus 101 [**Con10**].

Since the automorphism groups of small groups can be much larger than the groups themselves, we also include all automorphism groups of groups of order up to 255 and of transitive groups of degree up to 23.

2.2. Basic attributes computed for every group. For each group, we compute as many attributes as possible. Some use commands directly from Magma, such as determining if a group is abelian, while others require special functions we wrote (for example, determining if a group is metacyclic). We compute a "reasonable" presentation for the group (see Section 4) and a minimal degree permutation representation. We usually compute the lattice of subgroups up to automorphisms of the group, and up to conjugacy when possible (Sections 2.4 and 3.3), and conjugacy classes and character tables when feasible. We also determine special subgroups such as the center, commutator, Frattini, and Fitting subgroups, as well as various series for the group.

2.3. Challenging attributes computed for many groups. Many of our quantities of interest become more difficult to compute as the size or degree of the group increases. Our general approach is to try to compute everything for all groups, setting appropriate timeouts and omitting data if the computation does not finish or if we encounter errors in Magma. We have developed additional code for some problems where Magma's built-in methods were insufficient for our purposes; notable examples include computing quotients of permutation groups and isomorphism testing (discussed in Section 2.5).

2.4. Computing subgroup lattices. The subgroup lattice is an example of a particularly challenging attribute to compute. For each group, we attempt to compute the full lattice of subgroups, but there are many groups in our database where it is infeasible to compute subgroups up to conjugacy. For example, C_2^{10} has 229,755,605 subgroups, none of which are conjugate. However, it only has 11 classes of subgroups up to automorphism. We implemented the computation of the subgroup lattice up to automorphism and improved on Magma's built-in subgroup lattice methods up to conjugacy. In some cases, such as S_{47}, even working up to automorphism does not sufficiently reduce the quantity of subgroups, so we restrict our attention to normal subgroups and their complements, maximal subgroups, Sylow subgroups, and subgroups with small index and/or trivial core due to their importance in permutation representations.

Magma has built-in methods for computing both the subgroup lattice and the list of subgroups up to conjugacy (without inclusions). We found that the second was much faster than the first, and that we could recover the inclusions more quickly by computing a vector counting the intersections with each conjugacy class: if we have a group G and subgroups $H_1 \subseteq H_2$, then the vector of counts for H_2 dominates that for H_1, dramatically decreasing the number of calls to `IsConjugateSubgroup`. As noted above, when there are many subgroups up to conjugacy (e.g. abelian p-groups), we instead compute subgroups up to automorphism. The first algorithm for doing so uses the holomorph of G, i.e., the semidirect product $G \rtimes \mathrm{Aut}(G)$. Conjugacy within the holomorph translates to automorphism in the group itself, and a code snippet kindly provided by Derek Holt, using lifting through an elementary abelian series, allowed for the computation of representatives of subgroups up to automorphism in the solvable case without computing the list up to conjugacy.

However, the holomorph is implemented in Magma as a permutation group of degree equal to the order of G, and as the size of G increases, computations using the holomorph get bogged down. For non-solvable G and G with order at least 5000, we switch to a graph theoretic algorithm. First we compute a list of subgroups up to conjugacy, together with a list of automorphisms that generate the outer automorphism group of G. Taking subgroups up to conjugacy as vertices, we add an edge between H_1 and H_2 if there is an outer generator σ with $\sigma(H_1)$ conjugate to H_2. The components of the resulting graph give the subgroups up to automorphism,.

2.5. The isomorphism problem. When adding transitive permutation groups to our abstract groups database, we had to ensure that groups with multiple transitive permutation representations were only added once each. For small groups, Magma will compute its small group identification number. In a larger range, the LMFDB's transitive group section already had the needed information. However, there were multiple cases where neither approach sufficed.

Running isomorphism tests can be time consuming, so we used the following strategy. We took multiple isomorphism invariants of a group which are quick to compute and combined them into a hash. This could be precomputed for large numbers of groups. Groups with different hashes were clearly non-isomorphic. If groups produced identical hashes, we then had Magma perform a slower, but conclusive test for isomorphism. Our hash was highly effective in distinguishing non-isomorphic groups. Full details will be given in a forthcoming article by Roe [**Roe**].

3. Features of the database

The database is viewable through the LMFDB website at `https://www.lmfdb.org/Groups/Abstract/`. In this section we describe how to navigate the web interface, and highlight some particular features.

3.1. Features of the search interface. Groups can be searched through numerous query types, for example: the isomorphism class of a group's automorphism group, commutator, center, abelianization, Frattini subgroup, etc.; the order of any of these and the group itself; many boolean properties such as nilpotency, simplicity, and solubility. With the search interface, it is easy to quickly answer questions such as *"What are the nilpotent groups of order 36?"* and *"Are there any groups of order 256 with abelianization $C_2 \times C_2$?"* In this way, the database provides a useful service to researchers looking for groups with particular properties, as well as students coming to grips with the basics of group theory. It is also possible to search for groups whose orders factor in a particular way. For example, groups whose order is of the form p^3q^2r for primes p, q, r can be queried with the string `[3,2,1]`.

Groups are assigned a unique label in the database (see Section 5 for more details). These labels can be used for searching for specific groups, and match those in the GAP Small Groups database when the ranges overlap. If the label is not known beforehand, several options exist for searching for a specific group. Many families of groups can be found using strings. For example, `GL(n,q)` gives the page for $\mathrm{GL}_n(\mathbb{F}_q)$, and the strings `D10` and `F5` give the two groups in the example from Section 1.1—a full list of supported "transliterations" can be found alongside the database in the `group.families` knowl. Groups defined in Magma can also be searched for using the output from the `GroupName` function.

There is also a curated list of some interesting groups that one can quickly access, and one can also view the page of a group picked randomly from the database.

3.2. Features of a group page. On individual group pages, information is organized roughly by topic. Constructions of the group via a presentation (see Section 4), and as direct, semidirect, and non-split products are given when possible, as well as representations as a matrix or permutation group. Homological information like the Schur multiplier and commutator length is provided. Special subgroups like the center, commutator, and socle are displayed. For many groups, we provide several subgroup diagrams; one can toggle whether subgroups are shown up to conjugacy or up to automorphism (see Section 3.3), and independently whether to show all such classes of subgroups or just normal subgroups. We also display the derived series, the chief series, and the upper and lower central series of the group. Groups for which the given group is a maximal subgroup or a maximal quotient are listed under the Supergroups section.

Groups are also represented pictorially, in line with other sections of the LMFDB. Each disc in a group picture represents a conjugacy class of elements, arranged in an annulus around the middle class (of the identity), with distance according to the number of prime factors in the class' order. The size of a disc is proportional to the size of the conjugacy class, and the color is chosen based on its order. Decisions on how to display a group in picture form are essentially a matter of taste; our conventions lead to some group pictures showing striking symmetries within the conjugacy classes, such as that of `480.60` in Figure 1.

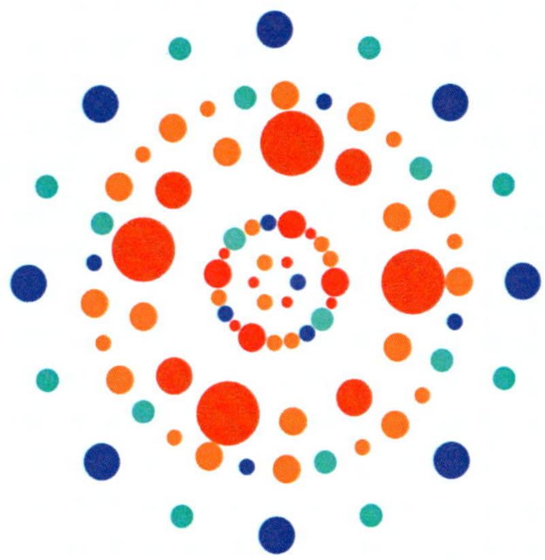

FIGURE 1. Group picture of 480.60

For many groups, we provide character tables for both complex and rational characters. Rows correspond to irreducible characters, and columns correspond to either conjugacy classes (for the complex character table) or divisions (for the rational character table); for more details on conjugacy classes and divisions, see Section 5.1. For ease of terminology, in this discussion we refer to columns as corresponding to "classes" to cover both cases simultaneously.

The headers for classes expand to show basic information such as the class's size and centralizer, as well as a representative for the class. As is standard, the row(s) above the table contain power map information on classes. For each prime p dividing the order of the group and each class represented by an element g, the entry above that column in the row labeled "pP" gives the label for the class of g^p. For each complex character, in the 2nd column of the table, we designate whether the representation is real, complex, or quaternionic.

REMARK 3.1. For some groups which lie outside the range of the database, we still generate webpages that display some basic information about the group by using on-the-fly calculations in GAP and SageMath. Two types of groups that are handled this way are groups in GAP's Small Groups database that are outside the LMFDB collection, such as 512.402873, and large abelian groups, such as $C_{12}^3 \times C_6$.

REMARK 3.2. Statistics about the abstract groups database can be found at https://www.lmfdb.org/Groups/Abstract/stats. We have computed many statistics based on the order factorization type of a group, i.e., whether the order is a prime, a product of two distinct primes, a power of one prime times another prime, etc. For instance, one can see the distribution of the nilpotency class of groups with order p^k, where p is a prime and $k \in \{3, \ldots, 6\}$.

3.3. Subgroups. One of the main features of our work is a companion database of subgroups. Subgroups have their own search functionality and labeling system (see Section 5.3). Given a group H, it is possible to search the database for all groups G containing H as a subgroup, subject to numerical constraints such as the index of H in G and the order of G. Cyclic subgroups, normal subgroups, and Sylow and Hall subgroups of a group can all be searched for. Each subgroup also has its own page. These pages give information about the ambient group and the quotient group structure (when normal). Related subgroups such as the centralizer, normalizer, normal closure and core of the subgroup are given.

The subgroups database allows us to display the subgroup lattice for many groups. This lattice is algorithmically very useful, allowing the computation of generating data about the group, such as the size of a smallest generating set (called the rank)[1] and how many such sets a group has, as well as expressions for the group as a semidirect product, its linear and permutation degrees, etc. The subgroup lattice can be viewed on a group page when it has fewer than 100 subgroup classes; see Figure 2 for an example. Inclusion is indicated with lines, and clicking a particular subgroup reveals information about it, e.g., whether it is maximal or solvable, its normalizer, its core. Subgroups are separated into levels vertically based on their orders or by the number of prime factors in their order (the user can toggle between these two options), and can be rearranged by clicking and dragging. When the diagram is too large to conveniently display on the page, the full diagram can often be viewed on a separate linked page.

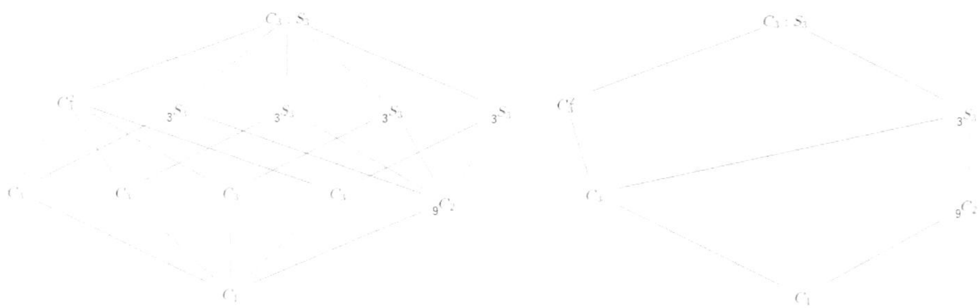

FIGURE 2. Subgroup diagrams of `18.4` up to conjugacy and autjugacy, as they appear on the LMFDB

In addition, we provide profiles of the subgroups up to automorphism and up to conjugation. For each order of subgroup, the profile lists the isomorphism types and multiplicities of classes of subgroups with that type. For example, for the elementary abelian group C_2^5, the profile for subgroups up to conjugation would list 155 conjugacy classes of subgroups of order 8 (all isomorphic to C_2^3), and one class of subgroups of the same order up to automorphism.

4. Computing presentations for solvable groups

As mentioned in Section 3.2, each group page contains various constructions of the group. In the case of solvable groups we give particularly nice presentations, exploiting the fact that solvable groups are polycyclic. A group G is *polycyclic* if it has a descending chain of subgroups

$$G = G_1 \geq G_2 \geq \cdots \geq G_{n+1} = 1$$

such that each G_{i+1} is normal in G_i, and the quotient G_i/G_{i+1} is cyclic. A polycyclic presentation is a special type of presentation for the group G that takes

[1] Although this is now more easily computable with Magma's `SmallestGeneratingSet` command

advantage of this descending chain of subgroups; for details, see [**HEO05**, Definition 8.7].

However, the polycyclic presentations in Magma use one generator for each prime dividing the order of the group (counted with multiplicity). So, for example, the cyclic group of order $2^2 \cdot 3 \cdot 5$ would have a presentation with 4 generators instead of the simpler one using a single generator. Here we describe a process for finding a more human-readable polycyclic presentation for a solvable group by using chains of subgroups. One can pass between polycyclic presentations and subnormal filtrations with cyclic quotients:

(1) given a polycyclic presentation with generators $\{g_1, \ldots, g_m\}$, the chain
$$1 \leq H_1 \leq \cdots \leq H_m = G$$
will be a filtration of G with cyclic quotients, where $H_i = \langle g_1, \ldots, g_i \rangle$;
(2) given a filtration of the form above, any choice of elements g_i generating H_i/H_{i-1} will give generators for a polycyclic presentation.

Define the *relative order* of g_i to be the order of the quotient H_i/H_{i-1}. We approach the search for a presentation by first finding a filtration and then arbitrarily choosing generators.

We first construct all minimal length chains of subgroups, each normal in the next, with cyclic relative quotients. We do this via a "top-down" approach, building them down from the top in layers until reaching the trivial subgroup. This process guarantees a polycyclic presentation with a minimal number of generators; we then compare all such minimal polycyclic presentations according to the following criteria.

(1) Maximize the number of generators with order equal to their relative order.
(2) Maximize the number of generators that commute with each other.
(3) Aim for relative orders that are non-increasing.
(4) Conjugacy relations should be "deeper" (within smaller groups in the filtration).

This process cannot feasibly be made canonical: at some points we make arbitrary choices, both for the filtration and for the choice of generators.

There are some groups for which the process of constructing and comparing all minimal length chains is too time-consuming. In these cases we instead use two other algorithms, both of which run much faster, but may not give as good a presentation. The first proceeds by greedily building the chains by starting with $H = G$ and iteratively choosing a random normal subgroup $K < H$ with H/K cyclic. The second algorithm simply refines the derived series by taking an abelian basis at each step in order to fill in the series to one with cyclic quotients.

It is important to pick the presentation at the beginning of our computations for each solvable group, as this presentation impacts certain attributes of the group, such as how we represent its elements.

5. Labels

A foundational principle of the LMFDB is to display every object with a unique label, an identifier that carries mathematically relevant information which can ideally be computed in a deterministic way. The primary role of the label is to serve as a permanent identifier for the object (so that it can be referenced in the mathematical literature, for example). Labels generally encode a sequence of invariants of an object, and provide a natural ordering. Conversely, a sequential ordering of all objects of a given type trivially allows one to label them, by their position in the sequence. We will talk of having labels and an ordering interchangeably. When sorting ordered tuples, we always mean lexicographically, and treat complex numbers as ordered pairs of real numbers for this purpose.

We label small groups as `N.i` corresponding to the GAP ID encoded as a string, where `N` is the order of the group and `i` distinguishes groups of the same order (as determined in GAP). If a group is not in the GAP Small Groups database, we replace `i` with an incrementing letter code, assigning labels to groups as they are added to our database.

The existence of automorphisms precludes having a definitive labeling *de novo*. Consequently, the labels for conjugacy classes, characters, and subgroups depend on fixing an ordered list of generators, which we do, and a specific realization of the group, i.e., as a permutation group, a polycyclic group, or as a matrix group. The realization is important since, in each case, one can easily construct an injective set map from the group to $\mathbb{Z}$, which then orders elements of the group. We then need to have a reproducible method for generating elements of a group. In the applications below, we found that picking pseudo-random group elements worked efficiently; for details on how we generate such elements, see `https://www.lmfdb.org/knowledge/show/group.pseudo_random_elements`.

5.1. Conjugacy classes and divisions. If G is a group and $g_1, g_2 \in G$, we say that g_1 and g_2 are in the same *division* if there exists $h \in G$ such that $h^{-1} g_1 h = g_2^i$ for some $i \in \mathbb{Z}$ such that $\gcd(i, |g_1|) = 1$. Divisions are unions of conjugacy classes, and rational-valued characters are constant on divisions. In fact, the irreducible rational-valued characters form a basis for the space of functions that are constant on divisions, as complex characters do for class functions. We say that a division is *maximal* if a representative generates a maximal cyclic subgroup of G.

Full details of how we label conjugacy classes and divisions are given at `https://www.lmfdb.org/knowledge/show/group.division_computing_labels`; here we describe some properties of the labeling.

Divisions are labeled with `nA` where `n` is a positive integer giving the order of a representative element and `A` is a string made up of capital letters. They are ordered first by the size of a conjugacy class within the division, and then by the number of conjugacy classes within a division.

Finding representatives for each division can be computationally expensive, so we focus first on maximal divisions. From there, we can compute labels and representatives for other classes as follows. Working through the maximal divisions in order, we use its chosen representative g and compute the classes of elements g^i with $0 < i < |g|$ and $\gcd(i, |g|) > 1$. The order of the non-maximal divisions is simply the order in which they appear from this sequence (looping over divisions, and within that, powers of an element).

The label for a conjugacy class starts with the label of the division containing it. If there is only one class within the division, the two labels are the same. When there is more than one class in a division `nA`, the conjugacy classes are labeled `nAj` where `j` is an integer. The first class encountered will then be `nA1`. Let g be an element representing this class. We then consider the conjugacy classes $[g^{-1}]$, $[g^2]$, $[g^{-2}]$, ... These correspond to conjugacy class labels `nA-1`, `nA2`, `nA-2`, ... The label for the class is the first time it appears in this sequence. Conjugacy classes which land in the same division have the same prefix, and will be grouped together in the complex character table for the group.

5.2. Characters. When feasible, we give information on irreducible rational-valued and complex-valued characters. The group $\mathrm{Gal}(\overline{\mathbb{Q}}/\mathbb{Q})$ acts on the complex characters, and the Galois orbits correspond to the rational characters, with each rational character being simply the sum of the complex characters in an orbit. We note that a rational-valued character may not arise from a rational-valued representation. The Schur index gives the smallest multiplier for a rational character so that the resulting character arises from a rational-valued representation.

Permutation representations of a group G can be realized using permutation matrices, and are thus sums of irreducible representations. Transitive permutation representations are classified by their degree n and *T-number* t [**CHM98**]. For a given irreducible representation, we can consider the "first" transitive permutation containing it to be the smallest pair (n, t).

The labels for rational characters are of the form `G.na` where `G` is the label for the group, `n` is the degree, and `a` is a lower-case letter which acts as a counter. The label for a complex character takes the form `G.nak` where `G.na` is the label for the corresponding rational character and `k` is positive integer which serves as a counter. In the character tables, characters are sorted by their labels viewed as a tuple (n, a) or (n, a, k). To completely determine the labels, we just need to know the ordering among characters with the same degree.

For rational characters, we sort by (d, m, n, t) where d is the degree (which is explicitly given in the label), m is the size of the Galois orbit of complex characters giving rise to it, and n and t refer to the smallest containing permutation representation. Any remaining ties are broken by sorting on the vector of character values using our ordering of conjugacy classes. Complex irreducible characters are given in the same order as their corresponding rational characters, with characters within a Galois orbit ordered by their vectors of values.

5.3. Subgroups. Labels for subgroups take one of four forms, depending on whether they are computed up to conjugacy or automorphism, and whether or not the deterministic process described below succeeds. (If it fails or we do not have all subgroups of a given order, we use a fallback nondeterministic label.) The deterministic labels take the form `N.i.m.a.c` (for subgroups up to conjugacy) or `N.i.m.a` (for subgroups up to automorphism). The piece `N.i` is the label of the ambient group G, `m` is the index of H in G, and `a` and `c` are alpha-numeric identifiers that distinguish H up to automorphism and conjugacy, respectively. To determine `a` and `c` we use the idea of Gassmann equivalence classes to order subgroups of the same index.

DEFINITION 5.1. Two subgroups H_1 and H_2 of G are *Gassmann equivalent* if they intersect each G-conjugacy class with the same cardinality. Equivalently, H_1 and H_2 have the same index m and the permutation representations $\pi_{H_1} : G \to S_m$ and $\pi_{H_2} : G \to S_m$ have the same character (which counts the number of cosets fixed by each conjugacy class).

For more on Gassmann equivalence and related notions see [**Sut21**], and for full details on the subgroup labeling scheme see `https://lmfdb.org/knowledge/show/group.computing_subgroup_labels`; here we describe some aspects of the labeling.

In order to label Gassmann classes, we first fix an ordering of the conjugacy classes of G as in Section 5.1. To each subgroup H we associate a vector of positive integers whose entries are the sizes of the intersections of H with each conjugacy class. We then sort the subgroups using these vectors. An analogous process works for Gassmann vectors up to automorphism, where we collect together G-conjugacy classes that are related by an automorphism of G.

If there are multiple subgroups up to conjugacy or automorphism within a given Gassmann class, we use the subgroup lattice to further order those remaining subgroups. We sort subgroups H in the same Gassmann class by considering the collection of supergroups of H; if this does not suffice to distinguish Gassmann equivalent subgroups, we resort to computing permutation representation and their associated characters.

Once we have an ordering inside the Gassmann classes, we assign a letter label to the Gassmann class and concatenate a number from the ordering of subgroups inside the Gassmann class to create an alpha-numeric value which we assign to `a`, and similarly, we assign an alpha=numeric value to `c` when we are able to work up to conjugacy. For example, the label of A_5 inside S_5 (up to conjugacy) is `120.34.2.a1.a1`. As another example, there are two classes of subgroups (up to automorphism) of $\mathrm{SL}(3,4)$ isomorphic to $\mathrm{GL}(2,4)$ with labels `60480.a.336.a1` and `60480.a.336.a2`—these are Gassmann-equivalent but not related by an automorphism—and four classes of subgroups (up to automorphism) of $\mathrm{SL}(3,4)$ of order 8, each contained in a different Gassmann class: `60480.a.7560.a1` (Q_8), `60480.a.7560.b1` ($C_2 \times C_4$), `60480.a.7560.c1` (D_4), and `60480.a.7560.d1` (C_2^3).

In some cases, we store only a few subgroups of a given order (we may store the center of a group, for example, but not other subgroups of that order) or the code for computing the labels did not finish. In these cases, we append an arbitrary letter code to the `N.i.m` part of the label, with capital letters indicating that subgroups are stored up to conjugacy and lower case indicating subgroups up to automorphism.

6. Connections and examples

We anticipate that the group home pages will be a valuable tool for everyone from undergraduate students first learning abstract algebra to experts in the field. The layout of each page, the search features, and the knowls allow for exploration of many group-theoretic concepts. Students can easily find examples of groups with common features like simple or solvable groups, they can explore Sylow subgroups (and more generally subgroup lattices), and they can learn about advanced topics such as series or character theory. Students can also explore relationships between different attributes of groups. A quick search demonstrates, for example, the existence of non-trivial groups that are perfect but not simple or A-groups[2] that are not solvable.

For researchers, the abstract groups database is intimately connected with many other mathematical objects in the LMFDB. We highlight some of those connections below.

6.1. Connections with other objects. One of the integral aims of the LMFDB is demonstrating links between different mathematical objects. The LMFDB universe gives the big picture of various connections that are related to the *Langlands program.* There are many related objects, and links between them are displayed in the database. In this section we list other databases in the LMFDB which are linked with the database of Abstract Groups. We describe the various connections to other LMFDB pages with the group pages in Figure 3. Note that, at present, the boxes representing the LMFDB sections Hypergeometric motives over $\mathbb{Q}$, Lattices, and Modular curves in Figure 3 are currently only available on the beta version of the LMFDB (`https://beta.lmfdb.org/`).

[2]A group G is an *A-group* if all its Sylow subgroups are abelian.

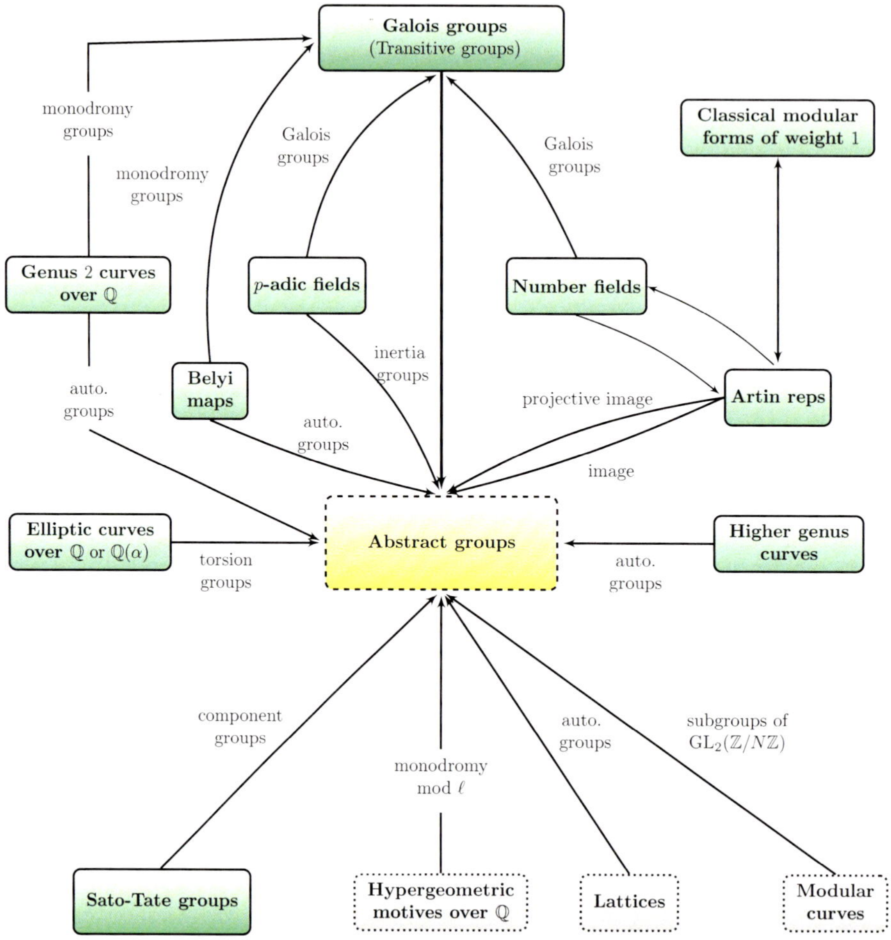

FIGURE 3. Diagram illustrating the connections between abstract group pages and other collections of objects in the LMFDB. Nodes in solid (green) and dotted (grey) rectangles are on the production and beta version of the LMFDB, respectively.

Since many objects in the LMFDB link to groups, having a built-in database of groups makes it easy for the user to find more information about the groups in question. Some groups may be familiar to the user, such as those which arise as automorphism groups of genus 2 curves; others are more complicated.

For instance, Artin representations connect to abstract groups via both their image and their projective image. One can also go from Artin representations to Galois groups (i.e., transitive permutation groups) by taking a stem field for the field cut out by the image of the representation to get a number field, and then taking the Galois group of its Galois closure. For a number field, one has the Artin representations of the Galois group of the normal closure of the field. The following example illustrates this connection further.

EXAMPLE 6.1. Consider the Artin representation with LMFDB label `4.5744.8t39.d.a`, which has the smallest conductor of all 4-dimensional irreducible

Artin representations with group `https://www.lmfdb.org/Groups/Abstract/192.1493` [**JR17**]. This group is displayed as $C_2^3 : S_4$, but since the semidirect product notation may not uniquely identify a group, the connection to a specific group page lets the user obtain lots of information about this group. Similarly, the projective image of this group is $C_2^2 : S_4$, via which the user is referred to the page `https://www.lmfdb.org/Groups/Abstract/96.227` for more details.

6.2. Applications via Galois theory. Perhaps the most celebrated application of groups is their usage in studying field extensions as a part of Galois theory. Below we give examples of how our database of finite groups, together with the Galois correspondence, can be used to study objects in other collections in the LMFDB.

EXAMPLE 6.2. Let $K = \mathbb{Q}(\alpha)$ be the number field with LMFDB label `5.1.35152.1`, where α has minimal polynomial $f := x^5 - x^4 + 2x^3 - 4x^2 + x - 1$. Then K has Galois closure $K^{\text{gal}} = \mathbb{Q}(\beta)$ with LMFDB label`20.0.3354518684571451850752.1`, where β has minimal polynomial

$$x^{20} - 2x^{19} + 10x^{17} - 15x^{16} + 40x^{14} - 64x^{13} + 46x^{12} + 8x^{11} - 32x^{10} + 8x^9 + 46x^8 - 64x^7 + 40x^6 - 15x^4 + 10x^3 - 2x + 1\,,$$

and $G := \text{Gal}(K^{\text{gal}}/\mathbb{Q}) \cong F_5$, the Frobenius group of order 20 (with LMFDB label `20.3`). Examining the lattice of subgroups up to conjugacy given on the `20.3` homepage, we see that F_5 is solvable, so the roots of f can be expressed by radicals. Computing fixed fields, we obtain the subfield lattice shown in Figure 4,

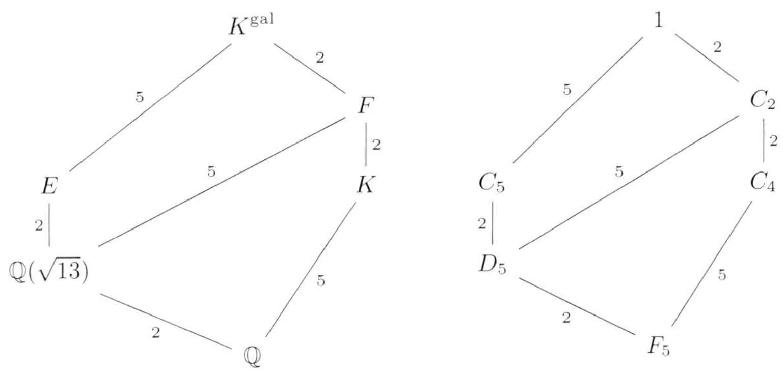

FIGURE 4. The subfield lattice of the number field with LMFDB label `20.0.3354518684571451850752.1` and the subgroup lattice of its Galois group F_5.

where F and E are the number fields given by

$$x^{10} - x^9 - 3x^8 + 5x^6 + x^5 - 5x^4 + 3x^2 - x - 1$$

and

$$x^4 + x^3 + 2x^2 - 4x + 3\,,$$

respectively. We observe that the commutator subgroup of G is C_5, the subgroup corresponding to E. Thus E is the largest abelian subfield of K^{gal}, with Galois group $\text{Gal}(E/\mathbb{Q}) \cong G^{\text{ab}} \cong C_4$.

EXAMPLE 6.3. Let $K = \mathbb{Q}_2(\alpha)$ be the p-adic field with LMFDB label `2.4.6.7`, where α has minimal polynomial $f := x^4+2x^3+2x^2+2$. Note that K is not Galois over $\mathbb{Q}_2$ and K has Galois closure $K^{\text{gal}} = \mathbb{Q}_2(\beta)$ with LMFDB label `2.12.18.59`, where β has minimal polynomial

$$x^{12} - 2x^{11} + 6x^{10} + 4x^9 + 6x^8 + 12x^7 - 4x^6 - 8x^3 + 16x^2 - 8$$

and $G := \text{Gal}(K^{\text{gal}}/\mathbb{Q}_2) \cong A_4$, the alternating group of order 12 (with LMFDB label `12.3`). The inertia group $I := I(K^{\text{gal}}/\mathbb{Q}_2)$ of K^{gal} is the abelian group C_2^2 of order 4 (with LMFDB label `4.2`). The wild inertia group of K^{gal} (i.e., the unique 2-Sylow subgroup of I) in this case is equal to I.

Examining the lattice of subgroups up to conjugacy given on the `12.3` homepage, we obtain the subfield lattice of $K^{\text{gal}}/\mathbb{Q}_2$, shown in Figure 5. It is clear from the subgroup lattice of `12.3` that $\text{Gal}(K^{\text{gal}}/K) \cong C_3$ and the extension $K/\mathbb{Q}_2$ is primitive. Also note that A_4 is solvable of length 2, so we find two intermediate fields E and F of K^{gal} such that $\text{Gal}(K^{\text{gal}}/E) \cong C_2^2$ and $\text{Gal}(K^{\text{gal}}/F) \cong C_2$. Computing fixed fields, we obtain the subfield lattice shown in Figure 5, where E and F are the p-adic fields given by $x^3 - x + 1$ and $x^6 + x^2 - 1$, respectively. The LMFDB labels for E and F are `2.3.0.1` and `2.6.6.1`, respectively. We observe that the commutator subgroup of G is C_2^2, the subgroup corresponding to E. Thus E is the largest abelian subfield of K^{gal}, with Galois group $\text{Gal}(E/\mathbb{Q}) \cong G^{\text{ab}} \cong C_3$.

This is an interesting example since K is the only degree 4 extension of $\mathbb{Q}_p$, for any p, which has Galois group $\text{Gal}(K^{\text{gal}}/\mathbb{Q}_2) \cong A_4$.

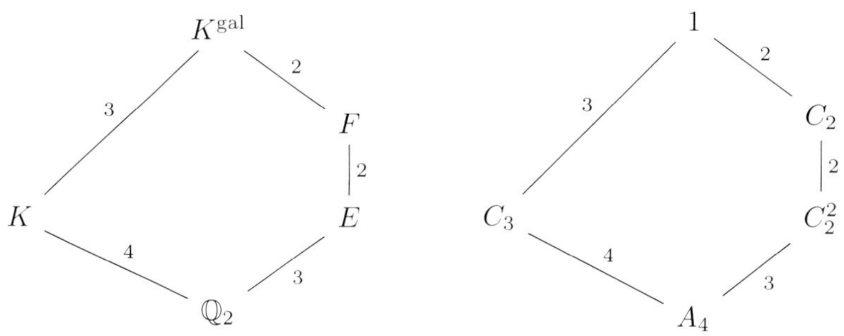

FIGURE 5. The subfield lattice of the Galois closure of the p-adic field with LMFDB label `2.4.6.7` and the subgroup lattice of its Galois group A_4.

EXAMPLE 6.4. Let G be the Sato-Tate group $J(O)$ with LMFDB label `1.4.F.48.48a`. Let G^0 be the connected component of the identity and G/G^0 be the component group of G. As noted on its homepage, $J(O)$ has the largest component group ($C_2 \times S_4$, LMFDB label `48.48`, with order 48) among Sato-Tate groups of abelian surfaces over number fields.

Let A be an abelian variety of dimension $g \leq 3$ defined over a number field k. Let K be the minimal extension of k over which all endomorphisms of A are defined, i.e., such that $\text{End}(A_K) = \text{End}(A_{\overline{k}})$. By [**FKRS12**, Proposition 2.17], then $\text{ST}_A / \text{ST}_A^0 \cong \text{Gal}(K/k)$, where ST_A is the Sato-Tate group of A.

Let $C : y^2 = x^6 - 5x^4 + 10x^3 - 5x^2 + 2x - 1$ considered over $\mathbb{Q}$, and let $A = \mathrm{Jac}(C)$. As shown in [**FKRS12**], the variety A has Sato-Tate group $J(O)$, realizing this group over $\mathbb{Q}$, and the endomorphisms of A are defined over the number field $K = \mathbb{Q}(\sqrt{-2}, \sqrt{-11}, a, b)$ where

$$a^3 - 7a + 7 = 0 \qquad \text{and} \qquad b^4 + 4b^2 + 8b + 8 = 0\,.$$

Applying the above proposition to this example, we have

$$C_2 \times S_4 \cong G/G^0 \cong \mathrm{Gal}(K/k)\,.$$

However, as shown in [**FKRS12**, Table 4], the curve C can also be used to realize 24 other Sato-Tate groups by varying the base field. In other words, by taking a number field L with $k \subseteq L \subseteq K$ and considering the base change A_L, we can obtain other Sato-Tate groups. For instance, examining [**FKRS12**, Table 8] we see that $J(T)$ (LMFDB label `1.4.F.24.13a`) is the unique Sato-Tate group occurring in genus 2 with component group $C_2 \times A_4$. Computing the fixed field L of $C_2 \times A_4 \leq C_2 \times S_4$, we find that $L = \mathbb{Q}(\sqrt{-11})$. Thus the base change A_L has Sato-Tate group $J(T)$, realizing another of the 52 possible Sato-Tate groups.

With further calculation, one can use other base changes of A to realize other Sato-Tate groups whose component groups are subgroups of $C_2 \times S_4$ by considering the lattice of subgroups given on the homepage for $C_2 \times S_4$ and computing fixed fields.

Galois theory can also be applied in a geometric context. The equivalence of categories between function field extensions in one variable and non-singular projective curves (see [**Sta18**, Tag 0BXX], for instance) allows us to study non-constant morphisms of curves by examining the associated extension of function fields. We consider an example coming from the families of higher genus curves with automorphisms collection, available at `https://www.lmfdb.org/HigherGenus/C/Aut/`.

EXAMPLE 6.5. Consider the refined passport with label `3.168-42.0.2-3-7`. It corresponds to a unique topological equivalence class of morphisms $X \to X/\mathrm{Aut}(X) \cong \mathbb{P}^1$, where X is the Klein quartic curve, a genus 3 curve with the largest possible number of automorphisms for its genus. (Such curves are known as *Hurwitz curves.*) It has automorphism group $\mathrm{Aut}(X) \cong \mathrm{PSL}_2(\mathbb{F}_7)$ (which has LMFDB label `168.42`), and all automorphisms are defined over the field $K = \mathbb{Q}(\zeta_7)$. By examining the lattice of subgroups of $\mathrm{PSL}_2(\mathbb{F}_7)$ given on its homepage and applying the Galois correspondence, we can find all intermediate covers Y with $X \to Y \to X/\mathrm{Aut}(X) \cong \mathbb{P}^1$.

Computationally, this can be accomplished using `Magma`'s `CurveQuotient` command [**BCP97**]. Calling this on each subgroup of $\mathrm{PSL}_2(\mathbb{F}_7)$ in turn, we find three intermediate covers Y of genus 1, corresponding to the subgroups of $\mathrm{PSL}_2(\mathbb{F}_7)$ isomorphic to C_2, C_3, and C_4. (All other quotients by non-trivial subgroups of $\mathrm{Aut}(X)$ result in genus 0 curves.) Each of these curves can be equipped with the structure of an elliptic curve by taking as the origin of the group law the image of $(1 : 0 : 0)$ under the appropriate quotient map. With some further computation (see [**Elk99**, Equation 2.10]), one can show that each of these elliptic curves is isomorphic to the curve $E : y^2 = 4x^3 + 21x^2 + 28x$ over K. From this we observe that $\mathrm{Jac}(X)$ decomposes as E^3 up to isogeny.

6.3. Modular curves and subgroups of $\mathrm{GL}_2(\mathbb{Z}/N\mathbb{Z})$**.** The LMFDB currently has a database of modular curves, available at `https://www.lmfdb.org/ModularCurve/Q/`. We conclude this section by describing how properties of modular curves can be deduced from their corresponding subgroups of $\mathrm{GL}_2(\mathbb{Z}/N\mathbb{Z})$. For a more comprehensive exposition of modular curves and subgroups of $\mathrm{GL}_2(\widehat{\mathbb{Z}})$, see [**RSZB22**] or [**Zyw22**].

Let E be an elliptic curve over $\mathbb{Q}$ and $E[N]$ be its N-torsion subgroup for each $N \in \mathbb{Z}_{\geq 1}$. The absolute Galois group $G_{\mathbb{Q}} := \mathrm{Gal}(\overline{\mathbb{Q}}/\mathbb{Q})$ acts on $E[N]$, and since $E[N] \cong (\mathbb{Z}/N\mathbb{Z})^2$ as abelian groups, we obtain a representation

$$\rho_{E,N} : G_{\mathbb{Q}} \to \mathrm{Aut}(E[N]) \cong \mathrm{GL}_2(\mathbb{Z}/N\mathbb{Z})\,.$$

By choosing compatible bases, we can take the inverse limit and package these together as a single representation

$$\rho_E : G_{\mathbb{Q}} \to \varprojlim_N \mathrm{GL}_2(\mathbb{Z}/N\mathbb{Z}) = \mathrm{GL}_2(\widehat{\mathbb{Z}})\,.$$

If E does not have complex multiplication, then Serre's Open Image Theorem [**Ser72**] implies that the image of ρ_E is an open subgroup of $\mathrm{GL}_2(\widehat{\mathbb{Z}})$, hence has finite index. Given an open subgroup H of $\mathrm{GL}_2(\widehat{\mathbb{Z}})$, one can define the modular curve X_H whose K-points parametrize elliptic curves E/K such that $\mathrm{img}(\rho_E) \subseteq H$ (up to conjugation). (For a precise definition of X_H, see [**RSZB22**, Section 2.3] or [**Zyw22**, Section 3].)

For each $N \in \mathbb{Z}_{\geq 1}$, let $\pi_N : \mathrm{GL}_2(\widehat{\mathbb{Z}}) \to \mathrm{GL}_2(\mathbb{Z}/N\mathbb{Z})$ be the projection map. Every open subgroup $H \leq \mathrm{GL}_2(\widehat{\mathbb{Z}})$ contains $\ker(\pi_N)$ for some N, and the smallest such $N \in \mathbb{Z}_{\geq 1}$ is called the *level* of H. If H contains $\ker(\pi_N)$, then H can be recovered from $\pi_N(H)$ as $H = \pi_N^{-1}(\pi_N(H))$. Thus to store H on a computer, one can simply store generators for $\pi_N(H)$ where N is the level of H.

Although this is a database of modular curves, much of the geometric data is computed group theoretically. Let $H \leq \mathrm{GL}_2(\widehat{\mathbb{Z}})$ be an open subgroup of level N. Letting $\Gamma_H := \pm\pi_N(H) \cap \mathrm{SL}_2(\mathbb{Z}/N\mathbb{Z})$, then one can determine the number of elliptic points and cusps of X_H by studying the action of the matrices

$$\begin{pmatrix} 0 & 1 \\ -1 & 0 \end{pmatrix}, \quad \begin{pmatrix} 0 & 1 \\ -1 & -1 \end{pmatrix}, \quad \begin{pmatrix} 1 & 1 \\ 0 & 1 \end{pmatrix}$$

on the coset space $\Gamma_H \backslash \mathrm{SL}_2(\mathbb{Z}/N\mathbb{Z})$. This data can in turn be used to compute the genus of X_H. We further illustrate this connection in the example below.

EXAMPLE 6.6. Consider the modular curve $X_0(6)$ of level 6 with LMFDB label `6.12.0.a.1`. As a containment $H \subseteq H'$ of open subgroups of $\mathrm{GL}_2(\widehat{\mathbb{Z}})$ induces a morphism $X_H \to X_{H'}$ of modular curves, we have morphisms $X_0(6) \to X_0(2)$ and $X_0(6) \to X_0(3)$. On the homepage for $X_0(6)$, it is further claimed that $X_0(6)$ is the fiber product of $X_0(2)$ and $X_0(3)$ over $X(1)$—we explain how this can be determined group theoretically.

Let H_2, H_3, H_6 be the subgroups of $\mathrm{GL}_2(\mathbb{Z}/2\mathbb{Z})$, $\mathrm{GL}_2(\mathbb{Z}/3\mathbb{Z})$, $\mathrm{GL}_2(\mathbb{Z}/6\mathbb{Z})$ corresponding to $X_0(2)$ (with label `2.3.0.a.1`), $X_0(3)$ (with label `3.4.0.a.1`), and $X_0(6)$, respectively. Note that $\mathrm{GL}_2(\mathbb{Z}/6\mathbb{Z})$ has label `288.851` and the subgroup $H_6 \cong C_2^2 \times S_3$ has subgroup label `288.851.12.c1.a1`. Taking the inverse image of H_2 and H_3 under the projection maps $\pi_{6,2} : \mathrm{GL}_2(\mathbb{Z}/6\mathbb{Z}) \to \mathrm{GL}_2(\mathbb{Z}/2\mathbb{Z})$, $\pi_{6,3}$:

$\mathrm{GL}_2(\mathbb{Z}/6\mathbb{Z}) \to \mathrm{GL}_2(\mathbb{Z}/3\mathbb{Z})$, we obtain subgroups $\widetilde{H_2} := \pi_{6,2}^{-1}(H_2)$ with subgroup label `288.851.3.a1.a1` and $\widetilde{H_3} := \pi_{6,3}^{-1}(H_3)$ with subgroup label `288.851.4.b1.a1`.

Either by examining the (very large) subgroup lattice diagram of $\mathrm{GL}_2(\mathbb{Z}/6\mathbb{Z})$ or by computing directly, we find that $\widetilde{H_2} \cap \widetilde{H_3} = H_6$ and $\langle \widetilde{H_2}, \widetilde{H_3} \rangle = \mathrm{GL}_2(\mathbb{Z}/6\mathbb{Z})$. This shows that H_6 is the fiber product of $\widetilde{H_2}$ and $\widetilde{H_3}$ over $\mathrm{GL}_2(\mathbb{Z}/6\mathbb{Z})$, and hence $X_0(6)$ is the fiber product of $X_0(2)$ and $X_0(3)$ over $X(1)$, as depicted in Figure 6.

$$
\begin{array}{ccc}
X_0(6) & \longrightarrow & X_0(2) \\
\downarrow & \ulcorner & \downarrow \\
X_0(3) & \longrightarrow & X(1)
\end{array}
\qquad\qquad
\begin{array}{ccc}
H_6 & \longrightarrow & \widetilde{H_2} \\
\downarrow & \ulcorner & \downarrow \\
\widetilde{H_3} & \longrightarrow & \mathrm{GL}_2(\mathbb{Z}/6\mathbb{Z})
\end{array}
$$

FIGURE 6. The fiber product diagram for the modular curve $X_0(6)$, and the fiber product diagram of the corresponding subgroups of $\mathrm{GL}_2(\mathbb{Z}/6\mathbb{Z})$.

Acknowledgments

We owe a large debt of gratitude to Tim Dokchitser, whose *GroupNames* website was the initial inspiration for our database and who generously shared code and database expertise with us (as well as giving us permission to re-post many of the definitions on his page as knowls[3] on the LMFDB). We are also appreciative of the advice and many helpful suggestions of Andrew Sutherland. We are grateful for various conversations and code from Michael Bush, Derek Holt, Alexander Hulpke, Bjorn Poonen, and David Roberts. We thank the American Institute of Mathematics for hosting a mini-workshop where much of the initial planning happened, and the Institute for Computational and Experimental Research in Mathematics for a workshop where the project took off. We are grateful to the referees for their comments, which have improved the paper.

References

[BCH+] Greg Butler, John Cannon, Derek Holt, Alexander Hulpke, John McKay, and Gordon Royle. Database of transitive groups. `https://magma.maths.usyd.edu.au/magma/handbook/text/781`.

[BCP97] Wieb Bosma, John Cannon, and Catherine Playoust, *The Magma algebra system. I. The user language*, J. Symbolic Comput. **24** (1997), no. 3-4, 235–265, DOI 10.1006/jsco.1996.0125. Computational algebra and number theory (London, 1993). MR1484478

[BE99a] Hans Ulrich Besche and Bettina Eick, *Construction of finite groups*, J. Symbolic Comput. **27** (1999), no. 4, 387–404, DOI 10.1006/jsco.1998.0258. MR1681346

[BE99b] Hans Ulrich Besche and Bettina Eick, *The groups of order at most 1000 except 512 and 768*, J. Symbolic Comput. **27** (1999), no. 4, 405–413, DOI 10.1006/jsco.1998.0259. MR1681347

[BE01] Hans Ulrich Besche and Bettina Eick, *The groups of order $q^n \cdot p$*, Comm. Algebra **29** (2001), no. 4, 1759–1772, DOI 10.1081/AGB-100002131. MR1853124

[3] A *knowl* is a feature of the LMFDB, where definitions and results are stored within webpages and can be revealed by clicking. The term was coined to reflect that each knowl contains a little bit of knowlege.

[BEO01] Hans Ulrich Besche, Bettina Eick, and E. A. O'Brien, *The groups of order at most 2000*, Electron. Res. Announc. Amer. Math. Soc. **7** (2001), 1–4, DOI 10.1090/S1079-6762-01-00087-7. MR1826989

[BEO02] Hans Ulrich Besche, Bettina Eick, and E. A. O'Brien, *A millennium project: constructing small groups*, Internat. J. Algebra Comput. **12** (2002), no. 5, 623–644, DOI 10.1142/S0218196702001115. MR1935567

[Bre00] Thomas Breuer, *Characters and automorphism groups of compact Riemann surfaces*, London Mathematical Society Lecture Note Series, vol. 280, Cambridge University Press, Cambridge, 2000. MR1796706

[CCN+85] J. H. Conway, R. T. Curtis, S. P. Norton, R. A. Parker, and R. A. Wilson, ATLAS *of finite groups*, Oxford University Press, Eynsham, 1985. Maximal subgroups and ordinary characters for simple groups; With computational assistance from J. G. Thackray. MR827219

[CH08] John J. Cannon and Derek F. Holt, *The transitive permutation groups of degree 32*, Experiment. Math. **17** (2008), no. 3, 307–314. MR2455702

[CHM98] John H. Conway, Alexander Hulpke, and John McKay, *On transitive permutation groups*, LMS J. Comput. Math. **1** (1998), 1–8, DOI 10.1112/S1461157000000115. MR1635715

[Con10] Marston Conder. Group actions on surfaces. `https://www.math.auckland.ac.nz/~conder/BigSurfaceActions-Genus2to101-ByGenus.txt`, 2010.

[DE05] Heiko Dietrich and Bettina Eick, *On the groups of cube-free order*, J. Algebra **292** (2005), no. 1, 122–137, DOI 10.1016/j.jalgebra.2004.12.001. MR2166799

[DE12] Heiko Dietrich and Bettina Eick, *Addendum to "On the groups of cube-free order" [J. Algebra 292 (1) (2005) 122–137] [MR2166799]*, J. Algebra **367** (2012), 247–248, DOI 10.1016/j.jalgebra.2012.06.014. MR2948221

[DEP22] Heiko Dietrich, Bettina Eick, and Xueyu Pan, *Groups whose orders factorise into at most four primes*, J. Symbolic Comput. **108** (2022), 23–40, DOI 10.1016/j.jsc.2021.04.005. MR4265711

[Dok] Tim Dokchitser. GroupNames. `http://groupnames.org`.

[Elk99] Noam D. Elkies, *The Klein quartic in number theory*, The eightfold way, Math. Sci. Res. Inst. Publ., vol. 35, Cambridge Univ. Press, Cambridge, 1999, pp. 51–101. MR1722413

[FKRS12] Francesc Fité, Kiran S. Kedlaya, Víctor Rotger, and Andrew V. Sutherland, *Sato-Tate distributions and Galois endomorphism modules in genus 2*, Compos. Math. **148** (2012), no. 5, 1390–1442, DOI 10.1112/S0010437X12000279. MR2982436

[GAP21] The GAP Group. *GAP – Groups, Algorithms, and Programming, Version 4.11.1*, 2021.

[HEO05] Derek F. Holt, Bettina Eick, and Eamonn A. O'Brien, *Handbook of computational group theory*, Discrete Mathematics and its Applications (Boca Raton), Chapman & Hall/CRC, Boca Raton, FL, 2005, DOI 10.1201/9781420035216. MR2129747

[HR20] Derek Holt and Gordon Royle, *A census of small transitive groups and vertex-transitive graphs*, J. Symbolic Comput. **101** (2020), 51–60, DOI 10.1016/j.jsc.2019.06.006. MR4109707

[HRT01] R. B. Howlett, L. J. Rylands, and D. E. Taylor, *Matrix generators for exceptional groups of Lie type*, J. Symbolic Comput. **31** (2001), no. 4, 429–445, DOI 10.1006/jsco.2000.0431. MR1823074

[Hul] Alexander Hulpke. Transitive groups library – a GAP package, version 3.0. `https://www.gap-system.org/Packages/transgrp.html`.

[Hul05] Alexander Hulpke, *Constructing transitive permutation groups*, J. Symbolic Comput. **39** (2005), no. 1, 1–30, DOI 10.1016/j.jsc.2004.08.002. MR2168238

[JR17] John W. Jones and David P. Roberts, *Artin L-functions of small conductor*, Res. Number Theory **3** (2017), Paper No. 16, 33, DOI 10.1007/s40993-017-0079-5. MR3669395

[NOVL04] M. F. Newman, E. A. O'Brien, and M. R. Vaughan-Lee, *Groups and nilpotent Lie rings whose order is the sixth power of a prime*, J. Algebra **278** (2004), no. 1, 383–401, DOI 10.1016/j.jalgebra.2003.11.012. MR2068084

[O'B90] E. A. O'Brien, *The p-group generation algorithm*, J. Symbolic Comput. **9** (1990), no. 5-6, 677–698, DOI 10.1016/S0747-7171(08)80082-X. Computational group theory, Part 1. MR1075431

[O'B91] E. A. O'Brien, *The groups of order* 256, J. Algebra **143** (1991), no. 1, 219–235, DOI 10.1016/0021-8693(91)90261-6. MR1128656

[OPS08] J. Opgenorth, W. Plesken, and T. Schulz. Carat, crystallographic algorithms and tables, Version 2.1b1. `https://github.com/lbfm-rwth/carat/`, Jul 2008.

[OVL05] E. A. O'Brien and M. R. Vaughan-Lee, *The groups with order p^7 for odd prime p*, J. Algebra **292** (2005), no. 1, 243–258, DOI 10.1016/j.jalgebra.2005.01.019. MR2166803

[Roe] David Roe. A fast nonisomorphism test for finite groups. In preparation.

[RSZB22] Jeremy Rouse, Andrew V. Sutherland, and David Zureick-Brown, *ℓ-adic images of Galois for elliptic curves over $\mathbb{Q}$ (and an appendix with John Voight)*, Forum Math. Sigma **10** (2022), Paper No. e62, 63, DOI 10.1017/fms.2022.38. With an appendix with John Voight. MR4468989

[RT98] L. J. Rylands and D. E. Taylor, *Matrix generators for the orthogonal groups*, J. Symbolic Comput. **25** (1998), no. 3, 351–360, DOI 10.1006/jsco.1997.0180. MR1615330

[Ser72] Jean-Pierre Serre, *Propriétés galoisiennes des points d'ordre fini des courbes elliptiques* (French), Invent. Math. **15** (1972), no. 4, 259–331, DOI 10.1007/BF01405086. MR387283

[Sta18] The Stacks Project Authors. *Stacks Project*. `https://stacks.math.columbia.edu`, 2018.

[Sut21] Andrew V. Sutherland, *Stronger arithmetic equivalence*, Discrete Anal., posted on 2021, Paper No. 23, 23, DOI 10.19086/da. MR4341956

[Tan11] O. Tange. GNU parallel - the command-line power tool. *;login: The USENIX Magazine*, 36(1):42–47, Feb 2011.

[Tay87] Don Taylor. Pairs of generators for matrix groups. I. Cayley Bulletin 3, 1987.

[VRD09] Guido Van Rossum and Fred L. Drake. *Python 3 Reference Manual*. CreateSpace, Scotts Valley, CA, 2009.

[WWT+] Robert Wilson, Peter Walsh, Jonathan Tripp, Ibrahim Suleiman, Richard Parker, Simon Norton, Simon Nickerson, Steve Linton, John Bray, and Rachel Abbott. Atlas of finite group representations - version 3. `http://brauer.maths.qmul.ac.uk/Atlas/v3/`.

[Zyw22] David Zywina. Explicit open images for elliptic curves over $\mathbb{Q}$. `arXiv:2206.14959`, 2022.

Lewis Combes, School of Mathematics and Statistics, University of Sheffield, UK, S3 7RH
Email address: `lmc577@proton.me`

John Jones, Department of Mathematics and Statistical Sciences, P.O. Box 871804, Arizona State University, Tempe, AZ, 85287
Email address: `jj@asu.edu`

Jennifer Paulhus, Department of Mathematics and Statistics, Mount Holyoke College, South Hadley, MA, 01075
Email address: `jpaulhus@mtholyoke.edu`

David Roe, Massachusetts Institute of Technology, Department of Mathematics, 77 Massachusetts Ave., Bldg. 2-336 Cambridge, MA 02139, United States of America
Email address: `roed@mit.edu`

Manami Roy, Department of Mathematics, Lafayette College, 730 High St, Easton, PA 18042
Email address: `royma@lafayette.edu`

Sam Schiavone, Massachusetts Institute of Technology, Department of Mathematics, 77 Massachusetts Ave., Bldg. 2-336 Cambridge, MA 02139, United States of America
Email address: `sam.schiavone@gmail.com`

Contemporary Mathematics
Volume **840**, 2026
https://doi.org/10.1090/conm/840/16812

Database-driven mathematical inquiry and the π-Base model for small semantic databases

Steven Clontz

ABSTRACT. Recent advances in computing have changed not only the nature of mathematical computation, but mathematical proof and inquiry itself. While artificial intelligence and formalized mathematics have been major topics within this conversation, this paper provides an informal survey of another class of tools for advancing mathematics research: databases of mathematical objects that enable semantic search. In addition to defining and exploring a few examples of these tools, we illustrate a particular line of research that was enabled by one such implementation: the π-Base community database of topological counterexamples.

1. Introduction

Much of the contemporary discussion on the use of computing to advance mathematical inquiry has been focused on the role of artificial intelligence (e.g. [**9**]) and formalization of mathematical proofs (e.g. [**8**]). In the former case, the aim is to emulate or substitute human mathematical reasoning, while the latter case replaces the need for human verification of mathematical claims.

In this paper, we will investigate instead how computing has the potential advance mathematics by augmenting the human act of mathematical research, specifically by the curation and adoption of **small semantic mathematical databases** which improve the findability and accessibility of mathematical knowledge that has been informally proven in the literature. We propose that this "formal modeling" of "informal mathematics" has the potential to advance mathematics today. This article thus aims to provide an informal survey of this landscape, explore a particular model for organizing small semantic databases, and finally examine a case study of an application implementing this model for general topology, the π**-Base** community database of topological counterexamples.

2. Formal modeling of mathematics

In his paper *What is the point of computers? A question for pure mathematicians* [**3**], Kevin Buzzard provides an excellent survey of the history of computationally-aided mathematical inquiry over the past few decades. While exhaustive search, as was utilized to prove the classic four-color theorem from graph theory once it was reduced down to 1,834 cases [**11**], is one example of using computers to tackle

2020 *Mathematics Subject Classification.* Primary 00A35, 54-02.

difficult problems constrained to a finite search space, recent years have increased the accessibility of so-called formal verification of mathematics, the main topic of Buzzard's survey. Such tools go under various names: "proof assistants", "theorem provers", "computer-verified mathematics", and so on; perhaps the most active modern example is Lean [**10**].

Such tools provide a mechanism for mathematicians to formalize their arguments as computer code that can be carefully checked by machine down to the finest detail. Of course, needing to refine down to first principles the usual hand waving of muddlesome technical details is a non-starter for most mathematicians; as such, these tools provide "tactics" that encode standard arguments into reusable instructions on how to build up the rigorous low-level proof. As an example, consider how one might prove that $(ab + ac) + ad = ba + (ca + da)$ for all $a, b, c, d \in \mathbb{R}$. While one might accomplish this through applications of associativity and commutativity (in Lean's `mathlib`, one might `rewrite` this goal using the theorems `add_assoc` and `mul_comm`), the single Lean tactic `ring` alternatively handles this proof in just four characters: *of course* the usual properties of `ring`s show that these expressions are equal to the other.

While such tactics have enabled the usual hand-waving of technicalities in many cases, it has not (to date) entirely brought formalized mathematics up to speed with the pace of classical mathematics. In [**23**], Terence Tao suggests that the difficulty ratio of writing a correct formal proof compared with writing a correct informal proof "is still well above one[, but] I believe there is no fundamental obstacle to dropping this ratio below one, especially with increased integration with AI [...] and other tools; this would be transformative to our field." Indeed, one may imagine a not-too-distant future where AI tools can predict formal mathematical proofs, which are then immediately checked by a computer verification systems, sending feedback to the AI, rinse and repeat, e.g. [**26**].

3. Semantic Mathematical Databases, Large and Small

However, there is an intermediate goal suggested by Buzzard in his survey: databases of mathematical concepts that enable semantic search, which can be used to query the corpus of peer-reviewed and/or formalized mathematics. Buzzard argues that "this sort of tool [...] has the potential to beat the techniques currently used by PhD students ('google hopefully,' 'page through a textbook/paper hopefully,' 'ask on a maths website and then wait,' 'ask another human') hands down".

But the benefits of curated databases of mathematical knowledge are not limited to students. A memory that has stuck with this author since graduate school is that of a particular topology seminar. The invited speaker noted at one point during his talk (paraphrasing), "and I will buy a bottle of wine a dinner tonight for whoever can help me remember how to construct an example of a normal T_0 space which is both arc connected and ultraconnected, but fails to be regular". This was the early 2010s, and many members of this audience had smart phones in their pocket, so one might expect that this question should have been answerable by a Google search. However, even today, the "trivial" example (take the reals with the topology of rays (x, ∞), example S42 of the π-Base [**24**]) is not revealed by a standard web search using those terms, and most of the trailheads that exist today and might lead one to this answer are websites and forum posts that did not exist at the time. Indeed, while Google has improved over the past decade or so in how it guesses semantics of words from their context in a search (or the user's search history), one might not be surprised that searching for the words "regular" and "normal" at that time could fail to specifically target mathematical results.

What stuck with this author primarily was the seeming consensus of the room that *of course* such an example existed, even though no one could quite articulate its details on the fly. And for a student in this audience, this could be a significant source of frustration, as this is the kind of knowledge that only comes with time immersed in a field. And even for experts, understanding that a result is "known" is not as useful as knowing "where" and "how", should the result or its details be needed for another application.[1]

To this end, let us define a **semantic mathematical database** as a database of mathematical objects or concepts that includes some technological measure to automatically connect its entries based upon their mathematical meaning. Dozens of examples of mathematical databases may be found by searching MathBases [**6**], but we wish to emphasize the need for *semantic* databases here. This distinction might be best illustrated by considering a common non-example: mathematics wikis. In addition to the wealth of mathematical knowledge curated by Wikipedia's dedicated contributors, there are several mathematics-specific wikis as well: ProofWiki.org, Math.Fandom.com, EnyclopediaOfMath.org, nCatLab.org, and more. While these certainly provide a great service by curating mathematical knowledge in an open and collaborative fashion, and the concepts they cover are indeed interconnected by relevant hyperlinks, all these connections must be made by human contributors. For example, ProofWiki doesn't "know" the difference between the rational numbers and the irrational numbers; each page is merely a human-curated document with human-authored hyperlinks pointing from one page to the next.

In contrast, while entries in a semantic mathematical database frequently contain narrative descriptions of the objects they represent, this exposition must be paired with **metadata** describing its mathematical features. An extreme example of this is the *L-functions and modular forms database*, or LMFDB for short [**13**]. To quote their homepage, "The LMFDB is a database of mathematical objects arising in number theory and arithmetic geometry that illustrates some of the mathematical connections predicted by the Langlands program." Like on a wiki, each object is given its own homepage (e.g. `https://www.lmfdb.org/L/2/31/31.30/c0/0/0` for the L-function given the ID 2-31-31.30-c0-0-0). And there are several curated subsets of the LMFDB that include exposition, for example, this list of "interesting" elliptic curves.

However, unlike a wiki, there is no exposition at all for the majority of entries. Instead, visitors of 2-31-31.30-c0-0-0 are provided a listing of its metadata: its Dirichlet series ($L(s) = 1 - 2^{-s} - 5^{-s} + -7^{-s} + 8^{-s} + 9^{-s} + \dots$), its functional equation ($\Lambda(s) = 31^{s/2}\Gamma_{\mathbb{C}}(s)L(s) = \Lambda(1-s)$), its degree (2), its conductor (31), its sign (1), and so on.

[1]During the preparation of this note, my colleague Katja Berčič shared this relevant (but perhaps apocryphal) anecdote concerning the graph theorists W.T. Tutte and Harold Coxeter: the latter mathematician asked the former in a phone call whether he knew of anyone studying a certain beautiful example of a 3-regular graph with 28 vertices and 42 edges. As the story goes, Tutte replied that of course he had, and that it was quite famously known by researchers in his circle as a Coxeter graph.

This lack of exposition is quite natural, as the LMFDB is an example of a **large** semantic mathematical database, wherein the vast majority of its entries are populated programmatically. For example, a sentence description is provided for `https://www.lmfdb.org/L/1/1/1.1/r0/0/0`, the well-known Riemann zeta function, but to populate the LMFDB with the other $1,448,483$ Dirichlet L-functions associated to primitive Dirichlet characters with conductor $N \leq 2800$, an exhaustive computation by algorithms described in David Platt's PhD thesis was required [**17**].

Given the computational intensity of generating so much data, the value of having such databases pre-computed is immediately seen. And with well-designed tools for navigation and **semantic** search (wherein we not only look for the presence/absence of certain keywords in a name or description, but are able to filter objects based upon the mathematical properties stored within their metadata), researchers and students have the potential to find the needles they need within such haystacks.

In contrast, we may also consider **small** semantic mathematical databases, where each entry is individually contributed and moderated by hand. While not small in a certain sense (with over $370,000$ entries to date), the On-Line Encyclopedia of Integer Sequences (OEIS) [**15**] is perhaps the best known small mathematics database. For example, the Fibonacci sequence $0,1,1,2,3,5,8,\ldots$ is given the ID A000045 and is available at `https://oeis.org/A000045`.

While search is supported (A000045 is the first result for the search `3,5,8,13,21` on its homepage, despite being modeled starting at `0`), one might not consider this database to be very semantic. For example, while it's noted that this sequence may be obtained from the formula $F(n) = F(n-1) + F(n-2)$ (with $F(0) = 0, F(1) = 1$), there's no direct way to obtain a list of all OEIS sequences obtained from the same formula but different base values.

In contrast, a small example that is quite semantic is the House of Graphs (HoG) [**7**]. A particularly unique feature of HoG is its search interface, which allows users to actually draw a graph of interest with their mouse, and it will query its database for a match. Of course, users are also able to search by specifying certain graph invariants; for example, obtaining over $4,000$ bipartite graphs of 1 average degree 3.

However, chief among its features for this author, as a casual graph theorist, is the following sentence paraphrased from its homepage, succinctly describing the utility of small semantic databases for researchers: **"Most [mathematicians] will agree that among the vast number of [objects in a given field] that exist, there are only a few thousand that can be considered really *interesting*."** So if the job of a large database is to help its users find a needle in a haystack, then the job of a small database is to serve as a well-curated museum, featuring the most important or interesting examples from its domain.

4. The π-Base model for small databases

The π-Base is another example of a small semantic mathematical database, founded by James Dabbs in 2014 [**24**]. Its implementation at `https://topology.pi-Base.org` is dedicated to results from general topology; in the next section we will dig deeper into some related research.

However, the goal at hand is to explore the "π-Base model" for organizing a small semantic mathematical database, appropriate for mathematicians interested in certain **objects**, the **properties** each object may or may not satisfy, and **theorems** that relate these properties.

4.1. Properties. Properties may be considered the atoms of the π-Base model. Each property is encapsulated in a single text file, illustrated in Listing 1 for the T_0 separation axiom. Each property is given an ID of the form `P######`, a short string to serve as its name (along with any alternative aliases the property may go by), and a longer description with appropriate references. In this sense, properties are quite similar to the pages of a wiki (albeit with a different mechanism of peer review, to be described in 4.4); it is how they are integrated into the objects and theorems that makes this database a semantic one.

LISTING 1. Source for property P1 on π-Base

```
---
uid: P000001
name: "$T_0$"
aliases:
  - Kolmogorov
  - T0
refs:
  - doi: 10.1007/978-1-4612-6290-9
    name: Counterexamples in Topology
---

Given any two distinct points, there is an open set containing
one but not the other.

Defined on page 11 of {{doi:10.1007/978-1-4612-6290-9}}.
```

4.2. Objects. Objects have the same features as properties: IDs (of the form `S######` at topology.pi-base.org, since it models topological Spaces), a name with aliases, and a longer description with references; see Listing 2.

LISTING 2. Source for space S1 on π-Base

```
---
uid: S000001
name: Discrete topology on a two-point set
aliases:
  - Finite Discrete Topology
refs:
  - doi: 10.1007/978-1-4612-6290-9
    name: Counterexamples in Topology
  - wikipedia: Discrete_space
    name: Discrete Space on Wikipedia
---

Let $X=2=\{0,1\}$ be the space containing two points and let
all subsets of $X$ be open.
```

```
Defined as counterexample #1 ("Finite Discrete Topology") in
{{doi:10.1007/978-1-4612-6290-9}}.
```

So while the π-Base doesn't "know" the mathematical content of these files, and isn't meant to automatically verify this content (a task left to its volunteer reviewers, see 4.4), it is aware of the boolean values asserted for specific space-property pairs. While space S1 is indeed P1, we will see soon why this is not manually asserted in the π-Base database; we instead illustrate in Listing 3 how S1 is asserted to be P52 (discrete).

LISTING 3. Source for space-property pair S1|P52 on π-Base

```
---
space: S000001 # The two-point discrete space
property: P000052 # The discrete property
value: true
refs:
- doi: 10.1007/978-1-4612-6290-9
  name: Counterexamples in Topology
---

All subsets of this space are open by definition.
```

4.3. Theorems. So why not assert that the two-point discrete space is T_0? The glue that holds the π-Base model together is its theorems. Each theorem asserts that if an object satisfies boolean values for one or more properties, it must also satisfy the given boolean value for another property.

Therefore, there's no need to assert that S1 is P1, given the assertion it is P52, along with the theorems in listings 4 and 5.

LISTING 4. Source for theorem T42 on π-Base

```
---
uid: T000042
if:
  P000052: true # The discrete property
then:
  P000002: true # The $T_1$ separation axiom
refs:
- doi: 10.1007/978-1-4612-6290-9
  name: Counterexamples in Topology
---

Asserted on Figure 9 of {{doi:10.1007/978-1-4612-6290-9}}.
```

LISTING 5. Source for theorem T119 on π-Base

```
---
uid: T000119
if:
  P000002: true # The $T_1$ separation axiom
then:
  P000001: true # The $T_0$ separation axiom
refs:
```

```
- doi: 10.1007/978-1-4612-6290-9
  name: Counterexamples in Topology
---

By definition, see page 11 of {{doi:10.1007/978-1-4612-6290-9}}.
```

In fact, only three properties are currently asserted for the two-point discrete space S1, and they characterize it uniquely: S52 (the space is discrete), P125 (the space has multiple points), and *not* P175 (the space has three or more points).

In addition to reducing the effort required to enter data for every space-property pair in the database, these theorems drastically reduce the room for error. For one, automated processes use these theorems to reject any submission to the π-Base that would violate a theorem. Furthermore, theorems are used to aid in search: see this result seeking examples of discrete spaces that fail to be T_0: not only are no spaces found, the π-Base application helpfully generates a proof for why they cannot exist, by chaining its theorems together.

Indeed, it's this **automated deduction** that distinguishes the π-Base model from most other mathematical database applications used by mathematicians today, semantic or otherwise. Buzzard cites this critical feature, enabled by the formal representation of mathematical *results* (without necessarily requiring formal verification of their *proofs*), as a powerful mechanism to enable "computer assisted learning", as well as to serve as an important stopgap in the absence of tooling that, as a practical matter, can support full formal verification of all mathematical results by the general audience of mathematics researchers [**3**].

4.4. Peer review. All the data in the π-Base is stored in these four types of text files. So how does a community of contributors collaborate on these?

Fortunately, such tools already exist: distributed version control systems such as Git have served software engineers since the early 2000s. While Git is not nearly as ubiquitous a tool for mathematicians, in recent years services such as GitHub have provided convenient ways for users to engage with these systems with little more effort than what's required to log into their web browser and start writing [**4**].

Thus, the π-Base uses these well-established and secure workflows for the business of peer review of database contributions, with the added bonus that the entire contents of its database is available as a free and open-source repository of text files. In particular, any proposed contribution to the database comes in the form of a "pull request" on GitHub. A discussion thread provided by GitHub for each pull request allows for contributors and reviewers to discuss the proposal, and ultimately the request is either "accepted and merged", or "closed".

GitHub provides two additional forums: a Discussions board where community members may freely discuss the project, and an Issues board where community members may suggest specific changes, without creating a formal pull request that would actually implement the proposed improvement. This allows for contributors to help curate the site, simply by engaging in conversations with other community members.

Finally, similar to Wikipedia, π-Base does not aim to be a home for original research. However, it was quickly discovered that limiting data to what's explicitly reflected in textbooks and peer-reviewed literature is quite constraining: so much of "known mathematics" is either never explicitly expressed in such manuscripts, or is presented in a context not immediately expressed within our model of objects/properties/theorems. As such, forums such as Math.StackExchange and MathOverflow are frequently used to ask and answer questions that fill in π-Base's gaps, relocating "original research" outside of the π-Base GitHub organization to venues with more eyes, as well as helping disseminate the π-Base resource to their broader audiences.

4.5. Applications to other fields, and limitations. In their 2013 *Notices* article [**2**], Billey and Tenner pose this scenario: "Suppose that M is a mathematician and that M has just proved theorem T. How is M to know if her result is truly new, or if T (or perhaps some equivalent reformulation of T) already exists in the literature?"[2] In fact, the π-Base software does exactly this, at least for for theorems of the form "all spaces with properties $\mathcal{P}$ also have properties $\mathcal{Q}$", and bounded above by the exhaustiveness of our community's contributions.

Given the abstractness of the π-Base model, one might suspect that its underlying free and open-source software could be used for a variety of mathematical disciplines. And indeed, the first direct fork of the π-Base aimed to serve topos theory is actively under development at `github.com/pi-base-topos`.

Additionally, other databases, many of which are tracked at MathBases.org, follow our model to some degree. A particular example is the Database of Ring Theory (DaRT) [**20**]. DaRT tracks two kinds of objects: rings and modules. Modules then have properties; however, they also have connections to rings (e.g. module M_{20} is given by $(x+(x,y)^2)$ over the ring $R_{23} = F_2[x,y]/(x,y)^2)$. Likewise, rings have properties, but for non-commutative rings these may be asymmetric, that is, hold differently for left or right multiplication. While π-Base's software could in theory be enhanced to allow for this flexibility, or forked to create a branch version of the software with this flexibility, this has not yet happened.

Another major blocker for the use of π-Base software for many databases is the need to not only model boolean properties, but valued properties. As an example in graph theory, consider regularity. A regular graph is a graph for which every vertex has the same degree (number of incident edges). However, many graph theoretic results need consider not only the regularity of the graph, but the type of regularity: an n-regular graph is a graph for which every vertex has degree n specifically. One could manually create separate properties for 1-regular, 2-regular, and so on; at least, the amount needed would be bounded above by the number of different graphs in the database. But of course, this is not a very elegant solution; the more sensible approach would perhaps be to store not only boolean values for object-property pairs, but also numerical values.

But this then reveals another limitation of the principle domain served by the π-Base today: properties from general topology typically aren't variable in terms of an integer or even real number value, but instead infinite cardinals. As a concrete example, consider the topological property of cardinality itself. As of Summer 2024, there were over twelve different properties in the π-Base that describe the cardinality of a space:

- $|X| = 0$ (P137),
- $|X| \geq 2$ (P125),
- $|X| \geq 3$ (P175),

[2]One might be reminded of Tai's 1994 rediscovery of the trapezoidal rule from calculus, published in the peer-reviewed medical journal *Diabetes Care*[**22**].

- $|X| \geq 4$ (P176),
- $|X| < \aleph_0$ (P78, i.e. finite),
- $|X| \leq \aleph_0$ (P57, i.e. countable),
- $|X| = \aleph_0$ (P181, i.e. countably infinite),
- $|X| = \aleph_1$ (P114, i.e. the smallest uncountable cardinality)
- $|X| < \mathfrak{c}$ (P58, i.e., smaller than the cardinality of the reals)
- $|X| \leq \mathfrak{c}$ (P163, i.e. no larger than the cardinality of the reals)
- $|X| = \mathfrak{c}$ (P65, i.e. the cardinality of the reals)
- $|X| \leq 2^{\mathfrak{c}}$ (P59, i.e. no larger than the cardinality of the power set of the reals)

The reader might reasonably judge this as quite a mess. Naively, one might consider a metaproperty of cardinality that could be assigned a cardinal value. But the π-Base is implemented in Javascript, whose built-in `Number` type merely supports a single non-finite value `Infinity`. Furthermore, the relationships of some of these properties depend on your model of ZFC: the π-Base does not have any theorem relating P114 and P58 above, as any such theorem would either imply the Continuum Hypothesis or its negation. Certainly, progress could be made here, but it will take the community time to find the most elegant and sustainable solution, particularly if the goal is for π-Base software to support areas besides general topology. (One possible solution currently being explored: using Lean as a programming language to assist in building the π-Base application from the plain text files described in Section 4.)

5. The π-Base as a vehicle for original research

We conclude by exploring how this author's attempt to model in the π-Base certain "well-known" results led to new mathematics research. A more technical article covering the results referenced here is available as [**5**]; here we instead focus on the story that led to these results.

The first iteration of the π-Base was a digital representation of Steen and Seebach's *Counterexamples in Topology* [**21**]. In this text, the usual topological separation axioms T_0 through T_6 were covered.[3] Here we explore two in particular.

DEFINITION 1 (P3 of π-Base). A space is T_2 (or Hausdorff) provided for each pair of distinct points x, y, there exist disjoint open sets U, V with $x \in U$ and $y \in V$.

DEFINITION 2 (P2 of π-Base). A space is T_1 provided for each pair of distinct points x, y, there exist (not necessarily disjoint) open sets U, V with $x \in U \setminus V$ and $y \in V \setminus U$.

In those early days, a theorem was created in π-Base to assert the immediate-from-the-definitions conclusion that $T_2 \Rightarrow T_1$. But after some time, the following two properties were contributed as well:

DEFINITION 3 (P99 of π-Base). A space is *US* ("Unique Sequential limits") provided that every convergent sequence has a unique limit.

DEFINITION 4 (P100 of π-Base). A space is *KC* ("Kompacts are Closed") provided that its compact subsets are closed.

[3]It's worth noting that the authors of *Counterexamples* did not follow the modern convention where $x \geq y$ if and only if $T_x \Rightarrow T_y$; another advantage of living semantic databases is to correct and disambiguate questionable notational decisions of the past.

At that time, not much justification was required to contribute a theorem, so it was asserted that $T_2 \Rightarrow KC \Rightarrow US \Rightarrow T_1$, based upon a links to a wiki called TopoSpaces. In 2017, the author received a small institutional grant to tidy up the π-Base and ensure it aligned with the peer-reviewed literature, so a more proper citation to the following theorem of Wilansky was provided instead.

THEOREM 5 (Thm 1 of [**25**]).

$$T_2 \Rightarrow KC \Rightarrow US \Rightarrow T_1$$

with no arrows reversing.

In order for π-Base to know none of these arrows can reverse, three counterexamples needed to be provided: one which was KC-not-T_2, one which was US-not-KC, and one which was T_1-not-US. But rather than describing such counterexamples here, the reader should recall that there's an app for that: Topology.pi-Base.org!

While this settled the situation at the time, in 2023 Patrick Rabau contributed to π-Base the following property defined in M. C. McCord's 1969 *Transactions* paper *Classifying spaces and infinite symmetric products* [**14**].

DEFINITION 6 (P143 of π-Base). A space is *weakly Hausdorff* or wH provided the continuous image of any compact Hausdorff space into the space is closed.

In that paper it was observed that this property lies strictly between T_2 and T_1 as well. While McCord was perhaps unfamiliar with the KC and US properties, one may show the following.

THEOREM 7.

$$T_2 \Rightarrow KC \Rightarrow wH \Rightarrow US \Rightarrow T_1$$

with no arrows reversing.

Since inclusion is a continuous map, $KC \Rightarrow wH$ follows immediately. To the best of this author's knowledge, the first result properly wiring up $wH \Rightarrow US$ was due to Rabau in Math.StackExchange 4267169: essentially, note that a copy of a converging sequence with its limit in $\mathbb{R}$ is compact Hausdorff, so given a converging sequence with its limit in the space, by wH this set is closed and therefore cannot have a second limit.

It was at this point this author decided to poke around the literature myself and look for any other significant investigations to properties within this spectrum of T_1-not-T_2. Indeed, there is another well-studied weakening of Hausdorff, along with the following two theorems.

DEFINITION? (Section 2 of [**12**] and Section 4 of [**19**]). A space X is said to be *k-Hausdorff* or kH provided that the diagonal $\Delta_X = \{\langle x, x\rangle : x \in X\}$ is k-closed in the product topology on X^2.

THEOREM? (Theorem 2.1 of [**12**]). $kH \Rightarrow KC$.

THEOREM? (Proposition 11.2 of [**19**]). $wH \Rightarrow kH$.

This leads to an immediate contradiction: we've already established that $KC \Rightarrow wH$, and that this arrow does not reverse, so what exactly is going on here?

As it turned there was no mistake made by the authors, even though they both operated using the same definition given above for k-Hausdorff. However, they were using different definitions for *k-closed.* So let us now proceed with a bit more care:

DEFINITION 8. A subset C of a space is k_1-*closed* provided for every compact subset K of the space, the intersection $C \cap K$ is closed in the subspace topology for K.

DEFINITION 9. A subset C of a space X is k_2-*closed* provided for every compact Hausdorff space K and continuous map $f : K \to X$, the preimage $f^{\leftarrow}[C]$ of C is closed in K.

DEFINITION 10. A space X is said to be k_i-*Hausdorff* or k_iH provided that the diagonal $\Delta_X = \{\langle x, x\rangle : x \in X\}$ is k_i-closed in the product topology on X^2.

With this mindful distinction made, we now may establish the following theorem.

THEOREM 11.

$$T_2 \Rightarrow k_1H \Rightarrow KC \Rightarrow wH \Rightarrow k_2H \Rightarrow US \Rightarrow T_1$$

with no arrows reversing.

PROOF. The first two arrows followed immediately from an alternative characterization of k_1H given in [**12**]: each compact subspace is closed and Hausdorff. The next arrow was noted in Theorem 7, and the fourth $wH \Rightarrow k_2H$ is from [**19**].

With the final arrow $US \Rightarrow T_1$ already established, it remained to be shown that $k_2H \Rightarrow US$. This was this author's personal contribution to the puzzle, first posted online at [**18**] and now detailed in [**5**]. The idea of this result comes from noting that the topological space which is simply a sequence converging to a unique limit, i.e. $K = \{1/n : n \in \mathbb{Z}^+\} \cup \{0\} \subseteq \mathbb{R}$, is a compact Hausdorff space. Then given a sequence a_n in a k_2H space X converging to the limits l_0, l_1, the map $f : K \to X^2$ defined by $f(1/n) = \langle a_n, a_n\rangle$ and $f(0) = \langle l_0, l_1\rangle$ is continuous. Then since Δ_X is k_2-closed, we have $f^{\leftarrow}[\Delta_X] \supseteq \{1/n : n \in \mathbb{Z}^+\}$ closed. It follows that $0 \in f^{\leftarrow}[\Delta_X]$, $f(0) = \langle l_0, l_1\rangle \in \Delta_X$, and thus the limit $l_0 = l_1$ is unique.

These arrows were each shown to not reverse by considering existing (or lightly modified) counterexamples, other than one: a space which is k_2H but not wH. This turned out to be what is now S165 of π-Base: the one-point compactification $X = Y \cup \{\infty\}$ of the Hausdorff "Arens-Fort space" $Y = \mathbb{Z} \cup \{p\}$ (itself S23 of π-Base). This followed from a new lemma: the one-point compactification of any KC space (and thus any Hausdorff space) is k_2H. But X is not wH: by removing the particular point p of the Arens-Fort space, the remainder $\mathbb{Z} \cup \{\infty\}$ is the one-point (Hausdorff) compactification of a countable discrete space (and thus the continuous image of a compact Hausdorff space), but is not closed in X. □

It's perhaps surprising that the properties k_1H, k_2H and wH investigated more or less independently all fell in line with Wilansky's original spectrum of T_1 weakenings of Hausdorff. One may consider various natural modifications of these to obtain more intermediate examples; for example, requiring unique limits of transfinite sequences, which would live between k_2H and US. And there are indeed several other T_1-not-T_2 properties found in the literature, including semi-Hausdorff (sH, P169 of π-Base), locally Hausdorff (lH, P84), and "has closed retracts" (RC, P101), which do not fall in line, even among themselves.

THEOREM 12.

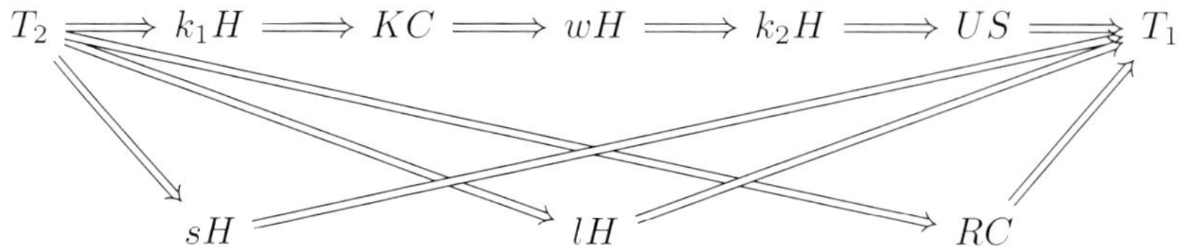

with no arrows reversing or missing.

The reader is again encouraged to use the π-Base application to obtain any desired details for the proof of Theorem 12.

We conclude with the following.

CLAIM. Semantic databases not only make "well-known" results from the literature easily accessible to any interested student or researcher of mathematics, but also serve as an excellent vehicle for driving future mathematical inquiry.

As of writing, there are six T_1 spaces on the π-Base for which the application cannot deduce the semi-Hausdorff property. Unless someone gets to it first, this would be a perfect graduate student project, and this is not the only piece of low-hanging fruit on the π-Base. But these things can be subtle, and simple-sounding problems in general topology frequently end up yielding numerous questions and answers at the research level [**16**].

So this author looks forward to welcoming more mathematicians to our community and others that maintain semantic mathematical databases. However, a major limitation is that while significant mathematical labor is involved in contributing to a semantic mathematical database, it generally doesn't "count" when it comes time for one to enter the job market, or submit one's annual faculty activity report. To that point, this author admits that before tenure, he mindfully avoided spending too much time on π-Base in order to focus on his traditional academic output. But after tenure, when this author decided to take his work on π-Base more seriously, several investigations such as the one just described have naturally kept him at least as active in authoring traditional publications as he was before. And with recognition from established mathematics organizations such as the American Institute of Mathematics [**1**], the visibility that comes with leading researchers such as Kevin Buzzard and Terence Tao promoting machine-assisted mathematics, and advocacy from newer organizations such as `code4math.org`, this author is optimistic that not only will the experience of mathematics research change dramatically in the coming decade, but as a community we will find ways to appropriately value the academic labor required to develop and maintain the shared digital infrastructure our research increasingly relies upon.

References

[1] AIMath, *Workshop on open-source cyberinfrastructure supporting mathematics research*, `https://aimath.org/workshops/upcoming/cyberinfrastructure/`.

[2] Sara C. Billey and Bridget E. Tenner, *Fingerprint databases for theorems*, Notices Amer. Math. Soc. **60** (2013), no. 8, 1034–1039, DOI 10.1090/noti1029. MR3113227

[3] Kevin Buzzard, *What is the point of computers? A question for pure mathematicians*, ICM—International Congress of Mathematicians. Vol. 2. Plenary lectures, EMS Press, Berlin, 2023, pp. 578–608. MR4680264

[4] Steven Clontz, *Github for mathematicians*.

[5] ———, *Non-hausdorff t_1 properties*, 2024.

[6] code4math Community, *MathBases — mathbases.org*, https://mathbases.org/, 2025, [Accessed 12-05-2025].
[7] Kris Coolsaet, Sven D'hondt, and Jan Goedgebeur, *House of Graphs 2.0: a database of interesting graphs and more*, Discrete Appl. Math. **325** (2023), 97–107, DOI 10.1016/j.dam.2022.10.013. MR4506063
[8] Susan D'Agostino, *Machine-human collaborations accelerate math research*, Notices Am. Math. Soc. **70** (2023), no. 9, 1490–1493 (English).
[9] Alex Davies, Petar Veličković, Lars Buesing, Sam Blackwell, Daniel Zheng, Nenad Tomašev, Richard Tanburn, Peter Battaglia, Charles Blundell, András Juhász, Marc Lackenby, Geordie Williamson, Demis Hassabis, and Pushmeet Kohli, *Advancing mathematics by guiding human intuition with ai*, Nature **600** (2021), no. 7887, 70–74.
[10] Leonardo de Moura, Soonho Kong, Jeremy Avigad, Floris van Doorn, and Jakob von Raumer, *The lean theorem prover (system description)*, Automated deduction—CADE 25, Lecture Notes in Comput. Sci., vol. 9195, Springer, Cham, 2015, pp. 378–388, DOI 10.1007/978-3-319-21401-6_26. MR3446905
[11] Georges Gonthier, *Formal proof—the four-color theorem*, Notices Amer. Math. Soc. **55** (2008), no. 11, 1382–1393. MR2463991
[12] J. Lawson and B. Madison, *Quotients of k-semigroups*, Semigroup Forum **9** (1974/75), no. 1, 1–18, DOI 10.1007/BF02194829. MR374322
[13] The LMFDB Collaboration, *The L-functions and modular forms database*, https://www.lmfdb.org, 2024.
[14] M. C. McCord, *Classifying spaces and infinite symmetric products*, Trans. Amer. Math. Soc. **146** (1969), 273–298, DOI 10.2307/1995173. MR251719
[15] OEIS Foundation Inc, *The on-line encyclopedia of integer sequences*, https://www.oeis.org, 2024.
[16] Elliott M Pearl, *Open problems in topology ii*, Elsevier, 2011.
[17] David J Platt, *Computing degree 1 l-functions rigorously*, 2011.
[18] Patrick Rabau, *Relationship between weak hausdorff and us properties*, Mathematics Stack Exchange, https://math.stackexchange.com/q/4267169 (version: 2023-03-06).
[19] Charles Rezk, *Compactly generated spaces*, nLab, https://ncatlab.org/nlab/files/Rezk_CompactlyGeneratedSpaces.pdf.
[20] Ryan C. Schwiebert, *The database of ring theory*.
[21] Lynn Arthur Steen and J. Arthur Seebach Jr., *Counterexamples in topology*, 2nd ed., Springer-Verlag, New York-Heidelberg, 1978. MR507446
[22] Mary M Tai, *A mathematical model for the determination of total area under glucose tolerance and other metabolic curves*, Diabetes care **17** (1994), no. 2, 152–154.
[23] Terence Tao, *Machine-assisted proof*, Notices Amer. Math. Soc. **72** (2025), no. 1, 6–13. MR4845676
[24] The π-Base Community, *π-base community database of topological counterexamples*, https://topology.pi-base.org.
[25] Albert Wilansky, *Between T_1 and T_2*, Amer. Math. Monthly **74** (1967), 261–266, DOI 10.2307/2316017. MR208557
[26] Kaiyu Yang, Aidan Swope, Alex Gu, Rahul Chalamala, Peiyang Song, Shixing Yu, Saad Godil, Ryan Prenger, and Anima Anandkumar, *LeanDojo: Theorem proving with retrieval-augmented language models*, Neural Information Processing Systems (NeurIPS), 2023.

Department of Mathematics and Statistics, 411 University Boulevard North, MSPB #314,Mobile, AL 36688-0002
Email address: sclontz@southalabama.edu

Contemporary Mathematics
Volume **840**, 2026
https://doi.org/10.1090/conm/840/16813

Local-global principle for 11-isogenies of elliptic curves is true over quadratic fields

Stevan Gajović, Jeroen Hanselman, and Angelos Koutsianas

ABSTRACT. In this paper, we prove that the local-global principle for 11-isogenies for elliptic curves over quadratic fields holds. This gives a positive answer to a conjecture by Banwait and Cremona [**5**, Conjecture 1.14]. The proof is based on the determination of the set of quadratic points on the modular curve $X_{D_{10}}(11)$.

1. Introduction

Let K be a number field and let E be an elliptic curve over K. If ℓ is a prime and E admits a K-rational ℓ-isogeny, then it is easy to show that the reduction $\tilde{E}_{\mathfrak{p}}/\mathbb{F}_{\mathfrak{p}}$ of E at a prime $\mathfrak{p}$ of good reduction of E also has a $\mathbb{F}_{\mathfrak{p}}$-rational ℓ-isogeny, where $\mathbb{F}_{\mathfrak{p}}$ is the residue field of $\mathfrak{p}$. It is natural to ask the converse question:

QUESTION 1.1. If $\tilde{E}_{\mathfrak{p}}/\mathbb{F}_{\mathfrak{p}}$ admits a $\mathbb{F}_{\mathfrak{p}}$-rational ℓ-isogeny for a density 1 set of primes $\mathfrak{p}$, then does E/K admit a K-rational ℓ-isogeny?

In [**20**] Sutherland studied the above question and explained that in most cases, we can expect the answer to be "yes". However, the most interesting case is when the answer is "no" and this local-global property is violated.

The existence of a K-rational ℓ-isogeny for E depends only on the j-invariant $j(E)$ of E when $j(E) \neq 0, 1728$; in other words, if E'/K is an elliptic curve with $j(E) = j(E')$ then the answer is "no" for E/K if and only if it is "no" for E'/K. Following Sutherland, a pair (ℓ, j_0) with $j_0 \neq 0, 1728$ and $j_0 \in K$ is called *exceptional for* K if there exists an elliptic curve E/K with $j(E) = j_0$ such that the answer to Question 1.1 is "no". The prime ℓ of an exceptional pair is called an *exceptional prime for* K.

We denote by $\rho_{E,\ell}$ the residual Galois representation that arises from the action of G_K on $E(\bar{K})[\ell]$. We denote by $G_{E,\ell} := \mathrm{Im}(\rho_{E,\ell}) \subseteq \mathrm{GL}_2(\mathbb{F}_\ell)$ and $H_{E,\ell}$ is the image of $G_{E,\ell}$ in $\mathrm{PGL}_2(\mathbb{F}_\ell)$. For $j(E) \neq 0, 1728$ the conjugacy class of $H_{E,\ell}$ in $\mathrm{PGL}_2(\mathbb{F}_\ell)$ depends only on $j(E)$.

For any number field K, it is known that the prime $\ell = 2$ is never an exceptional prime by [**1**, Remark 2.5]. Therefore, we assume that ℓ is odd and set $\ell^* = (-1)^{(\ell-1)/2}\ell$. We denote by D_{2n} the dihedral group of order $2n$.

2020 *Mathematics Subject Classification.* Primary 14G05, 11G18, Secondary 11G05.
Key words and phrases. Local-global principle, isogenies of elliptic curves, exceptional primes, modular curves, symmetric Chabauty.

THEOREM 1.2 ([**20**, Theorem 1, Lemma 1]). *Suppose $\sqrt{\ell^*} \notin K$ and (ℓ, j_0) is an exceptional pair for K. Let E/K be an elliptic curve with $j(E) = j_0$. Then, the following statements hold:*

(1) *The group $H_{E,\ell}$ is isomorphic to D_{2n}, where $n > 1$ is an odd divisor of $(\ell - 1)/2$.*
(2) $\ell \equiv 3 \pmod 4$.
(3) *The group $G_{E,\ell}$ is contained in the normaliser of a split Cartan subgroup of $\mathrm{GL}_2(\mathbb{F}_\ell)$.*
(4) *E obtains a rational ℓ-isogeny over $K(\sqrt{\ell^*})$.*

REMARK 1.3. As a consequence of Theorem 1.2, when $\sqrt{\ell^*} \notin K$, then any exceptional pair gives rise to a non-cuspidal K-rational point on the modular curve $X_{D_{2n}}(\ell)$ (see the definition in Section 2) for some odd divisor $n > 1$ of $\frac{\ell-1}{2}$.

Using Theorem 1.2 Sutherland proves that the only exceptional pair for $\mathbb{Q}$ is $(7, \frac{2268945}{128})$ [**20**, Theorem 2]. In [**5**, Proposition 1.3] the authors prove an analogous result to Theorem 1.2 for the case $\sqrt{\ell^*} \in K$.

THEOREM 1.4 ([**5**, Proposition 1.3]). *Suppose $\sqrt{\ell^*} \in K$. Then (ℓ, j_0) is exceptional for K if and only if one of the following holds for elliptic curves E/K with $j(E) = j_0$:*

- *$H_{E,\ell} \simeq A_4$ and $\ell \equiv 1 \pmod{12}$.*
- *$H_{E,\ell} \simeq S_4$ and $\ell \equiv 1 \pmod{24}$.*
- *$H_{E,\ell} \simeq A_5$ and $\ell \equiv 1 \pmod{60}$.*
- *$H_{E,\ell} \simeq D_{2n}$ and $\ell \equiv 1 \pmod 4$, $n > 1$ is a divisor of $(\ell - 1)/2$, and $G_{E,\ell}$ lies in a normaliser of a split Cartan subgroup.*

At the same time, Anni [**1**] focuses only on exceptional primes and proves that for a given K there are only finitely many.

THEOREM 1.5 ([**1**, Main Theorem]). *Let K be a number field of degree d over $\mathbb{Q}$ and discriminant Δ, and let $\ell_K := \max\{|\Delta|, 6d + 1\}$. The following holds:*

- *If (ℓ, j_0) is an exceptional pair for K then $\ell \leq \ell_K$.*
- *There are only finitely many exceptional pairs for K with $7 < \ell \leq \ell_K$.*

REMARK 1.6. The cases $\ell = 2, 3, 5, 7$ are also covered in [**1**]. In particular, the primes $\ell = 2, 3$ are not exceptional for any K, the prime $\ell = 5$ is exceptional if and only if $\sqrt{5} \in K$ and the prime $\ell = 7$ appears in infinitely many exceptional pairs for K if and only if the rank over K of the elliptic curve

$$y^2 = x^3 - 1715x + 33614,$$

is positive.

From Theorem 1.5 we understand that there are two important directions in which we can look for exceptional primes.

- Either we fix K and determine all the exceptional primes with $\ell \leq \ell_K$,
- or, we fix ℓ and a "suitable" family S of number fields, and determine all the number fields in S for which ℓ is an exceptional prime.

REMARK 1.7. According to Theorem 1.5 a natural choice of the family S is the set of all number fields of a fixed degree d. This choice becomes even more natural because, for odd degree extensions K, the bound $\ell_K = 6d + 1$ is a uniform bound with respect to d [**1**, Theorem 4.3].

Name	Coordinates
P_1	$(-3/4, 1/4, 0, \frac{\sqrt{77}}{2}, 0, 1)$
P_2	$(3/4, -5/4, 0, \frac{\sqrt{77}}{2}, 0, 1)$
P_3	$(1, 1, 1, \sqrt{-11}, \sqrt{-11}, 1)$
P_4	$(-1/3, 0, -1/3, \frac{\sqrt{22}}{3}, \frac{\sqrt{22}}{3}, 1)$
P_5	$(-2/5, 2/5, 1/5, \frac{\sqrt{209}}{5}, -\frac{\sqrt{209}}{5}, 1)$
P_6	$(-1, 7, 5, \sqrt{473}, -\sqrt{473}, 1)$

TABLE 1. The quadratic points of $X_{D_{10}}(11)$ up to conjugation.

Having all the above results in mind, in particular the fact that 11 is always in the range of potential exceptional primes when K is a quadratic field, Banwait and Cremona conjectured that the prime 11 is not an exceptional prime for any quadratic field K [**5**, Conjecture 1.14]. In this paper, we give a positive answer to the conjecture.

THEOREM 1.8. *The prime* 11 *is not an exceptional prime for any quadratic field.*

Theorem 1.8 is a consequence of Theorem 1.9 and [**5**, Proposition 10.1].

THEOREM 1.9. *The quadratic points of the modular curve $X_{D_{10}}(11)$ with respect to the model* (2.2) *are listed in Table* 1 *(up to conjugation).*

The key ingredients in the proof of Theorem 1.9 are the development of the (relative) symmetric Chabauty method and the Mordell-Weil sieve [**9**, **10**, **17**], the existence of Assaf's implementations of the space of modular forms for an arbitrary congruence subgroup Γ [**2**], the LMFDB modular curves database [**16**] and the fact that its quotient curve $X^+_{D_{10}}(11)$ is a genus 2 curve such that $\operatorname{rank}(J_{X_{D_{10}}}(K)) = \operatorname{rank}(J_{X^+_{D_{10}}}(K))$ for $K = \mathbb{Q}, \mathbb{Q}(\sqrt{-11})$.

A number of steps in the proofs were verified computationally using the Magma computer algebra system [**7**]. In our computations we used Magma V2.28-23. We rely on the Modular forms package by Eran Assaf [**2**] `https://github.com/assaferan/ModFrmGL2` and code by Samir Siksek [**18**] which is available on `https://github.com/samirsiksek/siksek.github.io/tree/main/progs/chabnf`. All of our computations were done on a machine running Ubuntu 22.04.1 with an 2 Intel Xeon E5-2660, 8 core CPUs at 2.2/3.0 GHz, 64 GB RAM. Most computations finish relatively quickly. Only computing the isomorphism between the model computed by Assaf's code and our own simplified equation might take more than an hour. The code used in this paper is available on `https://github.com/akoutsianas/local_global_isogenies`. In the paper we will clearly indicate whenever we rely on Magma. Instructions on how to reproduce the steps can be found in the repository.

2. The modular curve $X_{D_{10}}(11)$

Let N be a positive integer. Suppose G is a subgroup of $\mathrm{GL}_2(\mathbb{Z}/N\mathbb{Z})$ and $\Gamma_G = \{A \in \mathrm{SL}_2(\mathbb{Z}) : A \pmod N \in G_0\}$. Because $\Gamma_G \supset \Gamma(N)$ we have that Γ_G

is a congruence subgroup of $\mathrm{SL}_2(\mathbb{Z})$. We denote by X_G the modular curve that parametrizes elliptic curves over $\mathbb{C}$ whose residual representation modulo N lies in G up to conjugation. The curve X_G has a model defined over $\mathbb{Q}(\zeta_N)^{\det(G)}$ where $\det : G \to (\mathbb{Z}/N\mathbb{Z})^*$ is the determinant map. Over $\mathbb{C}$ the curve X_G is a compact Riemann surface and it holds $X_G(\mathbb{C}) \simeq \Gamma_G \setminus \mathbb{H}^*$.

Let $D_{10} \subset \mathrm{PGL}_2(\mathbb{F}_{11})$, $G_{D_{10}}$ the pullback of D_{10} to $\mathrm{GL}_2(\mathbb{F}_{11})$ and $\Gamma_{D_{10}} := \Gamma_{G_{D_{10}}}$. We define the modular curve[1] $X_{D_{10}}(11) := X_{\Gamma_{D_{10}}}$ which is, from the above, defined over $\mathbb{Q}$ because $\det(G) = \mathbb{F}_{11}^*$ [**20**, Proposition 3].

In [**14**] Galbraith described a method to compute a model of a modular curve $X(\Gamma)$ for a congruence subgroup Γ as long as we are able to compute a basis of $S_2(\Gamma)$. Galbraith's ideas have been used and extended by Banwait and Cremona[2] [**4**, **5**], Zywina [**21**] and Box [**8**]. Moreover, in [**2**] Assaf describes a general method of computing $S_2(\Gamma)$ and the algorithm has been implemented in Magma [**7**].

In our case we use Assaf's algorithm and his implementation to compute $S_2(\Gamma_{D_{10}})$. We pick an explicit subgroup H of $\mathrm{PGL}_2(\mathbb{F}_{11})$ isomorphic to D_{10}; in particular, the one that is generated by the following matrices

$$A = \begin{pmatrix} 4 & 0 \\ 0 & 3 \end{pmatrix}, \qquad B = \begin{pmatrix} 0 & 1 \\ 1 & 0 \end{pmatrix}.$$

With the terminology of [**2**, Definition 1.2.1] the group $\Gamma_{D_{10}}$ is of *real type*. Assaf's implementation computes the space $S_2(\Gamma_{D_{10}})$ which has dimension 6. This implies that the genus of $X_{D_{10}}(11)$ is 6. We note that one can compute the genus of $X_{D_{10}}(11)$ using known formulas, for example, in [**13**, Theorem 3.1.1].

REMARK 2.1. A different choice of the generators of D_{10} gives an isomorphic space of newforms which will not affect our computations of a model of $X_{D_{10}}(11)$.

The model we get for $X_{D_{10}}(11)$ by Assaf's implementation is given by equations with big coefficients. However, we know that $X_{D_{10}}(11)$ is isomorphic to $X_0(121)$ over $L := \mathbb{Q}(\sqrt{-11})$ [**4**, Lemma 3.3.7]. A model of $X_0(121)$ is given in Galbraith's thesis [**14**, p. 36] which we recall below.

(2.1)
$$X_0(121): \begin{cases} uw - 2vw + 2ux - 6vx + 2uy + 2vy + uz = 0, \\ uw + vw + 2ux - 2vx + 2uy - 10vy - 5uz + 11vz = 0, \\ -6u^2 + 6uv - 3v^2 - w^2 + 6wx - x^2 - 8wy + 10xy \\ -9y^2 - 4wz + 10xz = 0, \\ 6u^2 + 12uv + 12v^2 - 17wx - 2x^2 - 5wy + 4xy + 14y^2 \\ -6wz - 7xz - 11yz = 0, \\ -9v^2 - 8w^2 + wx + 9x^2 + 7wy - 10xy + y^2 - 7wz + 27xz - 11yz = 0, \\ -6u^2 - 12uv - 12v^2 - 3w^2 + 7wx - 6x^2 + 11wy + 12xy + 10y^2 + 4wz \\ +17xz - 11yz - 11z^2 = 0. \end{cases}$$

We search for a model of $X_{D_{10}}(11)$ so that it is isomorphic to $X_0(121)$ over L but not over $\mathbb{Q}$. Simultaneously, we test if it is isomorphic to the model obtained

---

[1] The curve $X_{D_{10}}(11)$ has label 11.132.6.b.1 in LMFDB beta version [**16**].

[2] The authors determine a model for the modular curve $X_{S_4}(13)$.

by Assaf's implementation using *IsIsomorphic* in Magma [**7**]. So, at the end we obtain the following model:
(2.2)

$$X_{D_{10}}(11):\begin{cases} uw-2vw+2ux-6vx+2uy+2vy+uz=0,\\ uw+vw+2ux-2vx+2uy-10vy-5uz+11vz=0,\\ -6u^2+6uv-3v^2+11w^2-66wx+11x^2+88wy-110xy+99y^2\\ +44wz-110xz=0,\\ 6u^2+12uv+12v^2+187wx+22x^2+55wy-44xy-154y^2+66wz\\ +77xz+121yz=0,\\ -9v^2+88w^2-11wx-99x^2-77wy+110xy-11y^2+77wz-297xz\\ +121yz=0,\\ -6u^2-12uv-12v^2+33w^2-77wx+66x^2-121wy-132xy-110y^2\\ -44wz-187xz+121yz+121z^2=0 \end{cases}$$

The isomorphism ϕ over L between the two models of $X_{D_{10}}(11)$ and $X_0(121)$ is given explicitly,

$$\begin{aligned} \phi:\ X_0(121)&\to X_{D_{10}}(11),\\ [x:y:z:u:v:w]&\mapsto[x:y:z:\sqrt{-11}u:\sqrt{-11}v:w]. \end{aligned} \tag{2.3}$$

Let w_{121} be the Atkin-Lehner involution of $X_0(121)$. We define

$$w_{11}:=\phi\circ w_{121}\circ\phi^{-1}. \tag{2.4}$$

We define $X_0^+(121):=X_0(121)/w_{121}$ and $X_{D_{10}}^+(11):=X_{D_{10}}(11)/w_{11}$.

LEMMA 2.2. *The Atkin-Lehner involution w_{121} on $X_0(121)$ and the involution w_{11} on $X_{D_{10}}(11)$ are both defined over $\mathbb{Q}$ and are given by $[x:y:z:u:v:w]\mapsto[x:y:z:-u:-v:w]$ with respect to the models defined by the equations given above.*

PROOF. We know that w_{121} is defined over $\mathbb{Q}$. From [**15**, Theorem 0.1], [**3**, Theorem 8] and [**6**, Proposition] we get that $\mathrm{Aut}_{\mathbb{Q}}(X_0(121))=\langle w_{121}\rangle\simeq C_2$. Finally, the statement for w_{11} is clear from its definition. □

Then[3] $C:=X_0^+(121)$ has genus $g(C)=2$, hence it is a hyperelliptic curve given by the equation

$$C:\ y^2=x^6-6x^5+11x^4-8x^3+11x^2-6x+1. \tag{2.5}$$

We also have explicitly computed the quotient maps $\phi_{X_0}:X_0(121)\to X_0^+(121)$ and $\phi_{X_{D_{10}}}:X_{D_{10}}(11)\to X_{D_{10}}^+(11)$ which can be found in our GitHub repository.

REMARK 2.3. In the computations in Section 3 we use the fact that both $\phi_{X_{D_{10}}}$ and $X_{D_{10}}^+(11)$ are defined over $\mathbb{Q}$.

LEMMA 2.4. *The curves $X_0^+(121)$ and $X_{D_{10}}^+(11)$ are isomorphic over $\mathbb{Q}$.*

[3] We recall that $X_0^+(121)\simeq X_{sp}^+(11)$.

PROOF. We know that $X_0(121)$ and $X_{D_{10}}(11)$ are isomorphic over L under the isomorphism ϕ above. From (2.3) and (2.4) we observe that

$$\phi^\sigma = \begin{cases} \phi, & \sigma \mid_L = id, \\ \phi \circ w_{121} = w_{11} \circ \phi, & \sigma \mid_L \neq id, \end{cases}$$

where $\phi^\sigma := \sigma^{-1} \circ \phi \circ \sigma$ and $\sigma \in \mathrm{Gal}(\bar{\mathbb{Q}}/\mathbb{Q})$. Together with the fact that ϕ_{X_0} and $\phi_{X_{D_{10}}}$ are defined over $\mathbb{Q}$ this implies that, when we take quotients by w_{121}, and w_{11}, the curves $X_0^+(121)$ and $X_{D_{10}}^+(11)$ become isomorphic over $\mathbb{Q}$. □

By abusing notation, let $\phi_{X_{D_{10}}}$ denote the composition of $\phi_{X_{D_{10}}}$ and the isomorphism between $X_{D_{10}}^+(11) \simeq C$ then we have the commutative diagram

$$\begin{array}{ccc} X_0(121) & \xrightarrow{\quad\phi\quad} & X_{D_{10}}(11) \\ & \searrow^{\phi_{X_0}} \quad \swarrow_{\phi_{X_{D_{10}}}} & \\ & C & \end{array} \tag{2.6}$$

where every map has been computed explicitly.

3. The Mordell-Weil group of $J_{D_{10}}$

We denote by J_0, $J_{D_{10}}$ and J_C the Jacobians of $X_0(121)$, $X_{D_{10}}(11)$ and C, respectively. It holds that

$$J_0 \sim E_{f_1} \oplus E_{f_2} \oplus E_{f_3} \oplus E_{f_4} \oplus E_{f_5}^2, \tag{3.1}$$

where f_i for $i = 1, 2, 3, 4$ are the four rational newforms of level 121, the ordering is according to the LMFDB, and f_5 is the unique rational newform of level 11. By modularity, f_i corresponds to the elliptic curve E_{f_i} over $\mathbb{Q}$ for each i. It holds that $\mathrm{rank}_{\mathbb{Q}}(E_{f_2}) = 1$ and $\mathrm{rank}_{\mathbb{Q}}(E_{f_i}) = 0$ for $i \neq 2$. Therefore, $\mathrm{rank}_{\mathbb{Q}}(J_0) = 1$.

Moreover, over L we have $\mathrm{rank}_L(E_{f_i}) = 0$ for $i \neq 2$ and $\mathrm{rank}_L(E_{f_2}) = 2$. Because the isogeny in (3.1) is defined over $\mathbb{Q}$, we get $\mathrm{rank}_L(J_0) = \mathrm{rank}_L(J_{D_{10}}) = 2$.

The curve C has two points at infinity with $\infty = (1, 1, 0)$ and $(-\infty) = (1, -1, 0)$ in the weighted projective plane $\mathbb{P}^2(1, 3, 1)$. Using the implementation of the method [**19**] in Magma, we prove that $J_C(\mathbb{Q}) \simeq \mathbb{Z}/5\mathbb{Z} \times \mathbb{Z}$ with generators $G_1 = [(0, -1) - (-\infty)]$ and $G_2 = [\infty - (-\infty)]$ of order 5 and infinite, respectively. Since Chabauty's rank condition is satisfied for C ($\mathrm{rank}(J_C(\mathbb{Q})) = 1 < 2 = g(C)$), we use the implementation of the method of Chabauty and Coleman in Magma to compute $C(\mathbb{Q})$.

PROPOSITION 3.1. *We have that* $C(\mathbb{Q}) = \{(1, \pm 2), (0, \pm 1), \pm\infty\}$.

For the computations and the proofs that follow it is important to explicitly compute $\phi_{X_{D_{10}}}^{-1}(C(\mathbb{Q}))$. Using Magma we get

$$\begin{aligned} \phi_{X_{D_{10}}}^{-1}((1, -2)) &= \{P_1, \bar{P}_1\}, & \phi_{X_{D_{10}}}^{-1}((1, 2)) &= \{P_2, \bar{P}_2\}, \\ \phi_{X_{D_{10}}}^{-1}(-\infty) &= \{P_3, \bar{P}_3\}, & \phi_{X_{D_{10}}}^{-1}(\infty) &= \{P_4, \bar{P}_4\}, \\ \phi_{X_{D_{10}}}^{-1}((0, -1)) &= \{P_5, \bar{P}_5\}, & \phi_{X_{D_{10}}}^{-1}((0, 1)) &= \{P_6, \bar{P}_6\}, \end{aligned}$$

where $\bar{P}_i$ is the conjugate of P_i.

We will use the commutative diagram (2.6) and the fact that $X_{D_{10}}(11)$ is isomorphic to $X_0(121)$ over L in order to determine a finite index subgroup of $J_{D_{10}}(\mathbb{Q})$.

The first step is to determine $J_C(L)$. Using the information of $J_C(L)$ and the fact that $\phi_{X_{D_{10}}}$ is defined over $\mathbb{Q}$ we are able to determine a finite index supgroup of $J_{D_{10}}(\mathbb{Q})$.

PROPOSITION 3.2. It holds that $J_C(L)_{\mathrm{Tor}} = \langle G_1 \rangle$ and $\mathrm{rank}(J_C(L)) = 2$. In particular, the group generated by $\langle G_1, G_2, G_3 \rangle$ is a finite index subgroup of $J_C(L)$ with

$$G_3 = \left[\left(\frac{-1+\sqrt{-3}}{2}, -\sqrt{-11} \right) + \left(\frac{-1-\sqrt{-3}}{2}, -\sqrt{-11} \right) - \infty - (-\infty) \right].$$

PROOF. By computing the order of the reduction of J_C modulo good primes and using the fact that the torsion subgroup injects into these groups (in Magma) we get that $\#J_C(L)_{\mathrm{Tor}} \leq 5$. Because $\#J_C(\mathbb{Q})_{\mathrm{Tor}} = 5$ and $J_C(\mathbb{Q})_{\mathrm{Tor}} \subset J_C(L)_{\mathrm{Tor}}$ we understand that $J_C(L)_{\mathrm{Tor}} = J_C(\mathbb{Q})_{\mathrm{Tor}} = \langle G_1 \rangle$.

Using the implementation of 2-descent [**19**] in Magma, we also get that $\mathrm{rank}(J_C(L)) \leq 2$. The point G_3 is a point of infinite order with $G_3 \notin J_C(\mathbb{Q})$. We prove that G_2 and G_3 are linearly independent elements in $J_C(L)$ by using Siksek's code. The idea is the following; if G_2, G_3 are linear dependent then the pairs $(\lambda, \mu) \in \mathbb{Z}^2$ such that $\lambda G_1 + \mu G_2 = 0$ is a subgroup Λ of $\mathbb{Z}^2$. We can also assume that there exists a pair (λ, μ) such that λ, μ are coprime away $\#J_C(L)_{\mathrm{Tor}}$. Using the reduction of $J_C(L)$ modulo small primes we prove that exist a prime $p \nmid \#J_C(L)_{\mathrm{Tor}}$ such that $\Lambda \subseteq p\mathbb{Z}^2$. □

PROPOSITION 3.3. It holds that $\mathrm{rank}(J_{D_{10}}(\mathbb{Q})) = 1$ and $J_{D_{10}}(\mathbb{Q})_{\mathrm{Tor}}$ is isomorphic to C_5 or $C_5 \times C_5$. In particular, the group $G = \langle D_1, D_2 \rangle$ where

$$D_1 = \left[P_6 + \bar{P}_6 - P_4 - \bar{P}_4 \right],$$
$$D_2 = \left[P_3 + \bar{P}_3 - P_4 - \bar{P}_4 \right],$$

with $5D_1 = 0$ and D_2 has infinite order, is a finite index subgroup of $J_{D_{10}}(\mathbb{Q})$ such that $10J_{D_{10}}(\mathbb{Q}) \subseteq G$.

PROOF. As in Proposition 3.2 we prove that $\#J_{D_{10}}(\mathbb{Q})_{\mathrm{Tor}} \mid 25$. At the same time we also prove that $J_{D_{10}}(\mathbb{Q})_{\mathrm{Tor}} \not\simeq C_{25}$, therefore the torsion part of $J_{D_{10}}(\mathbb{Q})$ is isomorphic to a subgroup of $C_5 \times C_5$. We have that $\phi^*_{X_{D_{10}}}$ injects $J_C(\mathbb{Q})_{\mathrm{Tor}}$ into $J_{D_{10}}(\mathbb{Q})_{\mathrm{Tor}}$ because $(\deg(\phi_{X_{D_{10}}}), \#J_{D_{10}}(\mathbb{Q})_{\mathrm{Tor}}) = 1$. Since, $J_C(\mathbb{Q})_{\mathrm{Tor}} \simeq C_5$ we understand that $J_{D_{10}}(\mathbb{Q})_{\mathrm{Tor}} \simeq C_5$ or $C_5 \times C_5$. Let $D_1 = \phi^*_{X_{D_{10}}}(G_1)$ then it holds $5D_1 = 0$.

The quotient map $\phi_{X_{D_{10}}} : X_{D_{10}}(11) \longrightarrow C$ induces an isogeny $J_{D_{10}} \sim J_C \times A$ where $A/\mathbb{Q}$ is some abelian variety and the isogeny is defined over $\mathbb{Q}$. Since $\mathrm{rank}(J_{D_{10}}(L)) = \mathrm{rank}(J_C(L)) = 2$, we have $\mathrm{rank}(A(L)) = 0$, so $\mathrm{rank}(A(\mathbb{Q})) = 0$. Hence, $\mathrm{rank}(J_{D_{10}}(\mathbb{Q})) = \mathrm{rank}(J_C(\mathbb{Q})) = 1$.

Let $D_2 = \phi^*_{X_{D_{10}}}(G_2) \in J_{D_{10}}(\mathbb{Q})$, then D_2 has an infinite order in $J_{D_{10}}(\mathbb{Q})$ because G_2 has an infinite order in $J_C(\mathbb{Q})$. Finally, from the above and [**9**, Proposition 3.1] we have that $10J_{D_{10}}(\mathbb{Q}) \subseteq G$. □

4. Symmetric Chabauty

In this section we apply the relative symmetric Chabauty as it is described in [**9**, **10**, **17**]. We also give a brief exposition of the idea of symmetric Chabauty following [**9**, Section 2.2].

4.1. Introduction. Let $X/\mathbb{Q}$ be a smooth projective curve of genus g that has good reduction at a prime $p > 2$. Let J be the Jacobian of X and r the rank of $J(\mathbb{Q})$. We also assume that $r \leq g-2$, i.e., the Chabauty condition holds for quadratic points. We denote by $\mathcal{X}$ the proper minimal regular model of X. In the space of global differential forms $\Omega_{X/\mathbb{Q}_p}(X)$, we have the $\mathbb{Z}_p$-submodule $\Omega_{\mathcal{X}/\mathbb{Z}_p}(\mathcal{X})$. In [**12**] Coleman defines the pairing

$$\Omega_{X/\mathbb{Q}_p}(X) \times J(\mathbb{Q}_p) \to \mathbb{Q}_p, \qquad \left(\omega, \left[\sum_i P_i - Q_i\right]\right) \mapsto \sum_i \int_{Q_i}^{P_i} \omega.$$

We denote by V the annihilator of $J(\mathbb{Q})$ under the above pairing, write $\mathcal{V} = V \cap \Omega_{\mathcal{X}/\mathbb{Z}_p}(\mathcal{X})$ and let $\tilde{V}$ be the image of $\mathcal{V}$ under the reduction map.

Let $\mathcal{Q} = \{Q_1, Q_2\} \in X^{(2)}(\mathbb{Q})$, where $X^{(2)}$ is the symmetric square of X. Suppose $\omega_1, \cdots, \omega_k$ is a basis of $\tilde{V}$. We fix a place v of $\mathbb{Q}(Q_1, Q_2)$ above p and denote by $\tilde{Q}_i$ the reduction of Q_i with respect to v. Let $t_{\tilde{Q}_i}$ be a uniformiser of $\tilde{Q}_i$. Then, we can expand ω_j around $\tilde{Q}_i$ with respect to $t_{\tilde{Q}_i}$ as a formal power series. In particular, it holds that

$$\omega_j = (a_0(\omega_j, t_{\tilde{Q}_i}) + a_1(\omega_j, t_{\tilde{Q}_i})t_{\tilde{Q}_i} + a_2(\omega_j, t_{\tilde{Q}_i})t_{\tilde{Q}_i}^2 + \cdots)dt_{\tilde{Q}_i} \tag{4.1}$$

If $Q_1 \neq Q_2$, we define

$$\mathcal{A} = \begin{pmatrix} a_0(\omega_1, t_{\tilde{Q}_1}) & a_0(\omega_1, t_{\tilde{Q}_2}) \\ \vdots & \vdots \\ a_0(\omega_k, t_{\tilde{Q}_1}) & a_0(\omega_k, t_{\tilde{Q}_2}) \end{pmatrix},$$

and when $Q_1 = Q_2$, we define

$$\mathcal{A} = \begin{pmatrix} a_0(\omega_1, t_{\tilde{Q}_1}) & a_1(\omega_1, t_{\tilde{Q}_1})/2 \\ \vdots & \vdots \\ a_0(\omega_k, t_{\tilde{Q}_1}) & a_1(\omega_k, t_{\tilde{Q}_1})/2 \end{pmatrix}.$$

THEOREM 4.1 ([**17**, Theorem 3.2]). *Suppose $p \geq 5$. If* $\operatorname{rank}(\mathcal{A}) = 2$, *then $\mathcal{Q}$ is the only point on $X^{(2)}(\mathbb{Q})$ in its residue class modulo p, i.e. for any $\mathcal{R} \in X^{(2)}(\mathbb{Q})$ such that $\mathcal{Q} \equiv \mathcal{R} \pmod{v}$, we have $\mathcal{R} = \mathcal{Q}$.*

Suppose $\rho : X \to C$ is a degree 2 map to a smooth non-singular curve C and $\mathcal{C}$ its proper minimal regular model. We also assume that C has good reduction at p. Suppose that ρ extends to a morphism $\mathcal{X} \to \mathcal{C}$ which corresponds to a degree 2 map $\tilde{X} = \tilde{\mathcal{X}} \to \mathcal{C}_{\mathbb{F}_p}$. The map $\rho : X \to C$ induces the trace map on the holomorphic differentials $\mathrm{Tr}\colon \Omega_{X/\mathbb{Q}_p}(X) \to \Omega_{C/\mathbb{Q}_p}(C)$, for more details, see [**10**, §2.6].

Let $\mathcal{V}_0 = \mathcal{V} \cap \ker\left(\Omega_{X/\mathbb{Q}_p}(X) \xrightarrow{\mathrm{Tr}} \Omega_{C/\mathbb{Q}_p}(C)\right)$ and $\tilde{V}_0$ the image of $\mathcal{V}_0$ under the reduction map. Let $\mathcal{Q} = \{Q_1, Q_2\} \in \rho^* C(\mathbb{Q})$, $\omega_1, \cdots, \omega_{k'}$ be a basis of $\tilde{V}_0$ and p, v, $\tilde{Q}_1$, $t_{\tilde{Q}_1}$ be as above.

THEOREM 4.2 ([**17**, Theorem 4.3]). *Suppose $p \geq 5$. If there exist ω_i for some $i \in \{1, \cdots, k'\}$ such that*

$$\frac{\omega_i}{dt_{\tilde{Q}_1}} \Big|_{t_{\tilde{Q}_1} = 0} \neq 0,$$

then every point in $X^{(2)}(\mathbb{Q})$ in the residue class of $\mathcal{Q}$ belongs to $\rho^*(C(\mathbb{Q}))$.

4.2. Relative Symmetric Chabauty for $X_{D_{10}}(11)$. Let W be the image of $1 - w_{11}^* : \Omega_{X_{D_{10}}(11)/\mathbb{Q}} \to \Omega_{X_{D_{10}}(11)/\mathbb{Q}}$.

PROPOSITION 4.3. The space W annihilates $J_{D_{10}}(\mathbb{Q})$ under the integration pairing, hence

$$\int_0^D \omega = 0,$$

for all $\omega \in W$ and $D \in J_{D_{10}}(\mathbb{Q})$. In addition, W lies in the kernel of the $\mathrm{Tr}_{\phi_{X_{D_{10}}}} : \Omega_{X_{D_{10}}(11)/\mathbb{Q}} \to \Omega_{C/\mathbb{Q}}$.

PROOF. The proof is similar to [**9**, Lemma 3.4] because $\mathrm{rank}(J_{D_{10}}(\mathbb{Q})) = \mathrm{rank}(J_C(\mathbb{Q}))$ and $1 + w_{11}^*$ is the trace map with respect to $\phi_{X_{D_{10}}}$. We also use [**10**, Lemma 2.2]. □

Let $\tilde{W}$ be the image of $W \cap \Omega_{\mathcal{X}_{D_{10}}/\mathbb{Z}_p}$ under the reduction map.

PROPOSITION 4.4. The space $\tilde{W}$ is the image of $1 - \tilde{w}_{11}^* : \Omega_{\tilde{\mathcal{X}}_{D_{10}}/\mathbb{F}_p} \to \Omega_{\tilde{\mathcal{X}}_{D_{10}}/\mathbb{F}_p}$.

PROOF. We use [**9**, Lemma 3.6] and the proof follows in the same way as the proof of [**9**, Proposition 3.5]. □

5. Mordell-Weil sieve

In this section, we briefly recall the Mordell-Weil sieve as discussed in [**9**, **10**, **17**] and which has its origin in [**11**]. Again we follow [**9**, Section 2.4].

Suppose $X/\mathbb{Q}$ be smooth projective curve with Jacobian J and $\rho : X \to C$ a degree 2 map (defined over $\mathbb{Q}$) where $C/\mathbb{Q}$ is another curve. We assume that we have the following data:

(1) $D_1, \cdots, D_r$ are divisors of $J(\mathbb{Q})$ that generate a finite index subgroup G of $J(\mathbb{Q})$,
(2) A number N such that $NJ(\mathbb{Q}) \subset G$,
(3) A rational degree 2 divisor which we denote by ∞,
(4) $\mathcal{L}'$ is a known finite subset of $X^{(2)}(\mathbb{Q})$. The set $\mathcal{L}'$ may also include points from $\rho^*(C(\mathbb{Q}))$,
(5) $p_1, \cdots, p_n$ are primes of good reduction for X.

We define $\mathcal{L} = \mathcal{L}' \cup \rho^* C(\mathbb{Q})$.

Let p be one of the primes $p_1, \cdots, p_r$. We define the maps $\phi : \mathbb{Z}^n \to G$ where $\phi(a_1, \cdots, a_n) = \sum_{i=1}^n a_i D_i$, $\iota : X^{(2)}(\mathbb{Q}) \to G$ where $\iota(\mathcal{Q}) = N[\mathcal{Q} - \infty]$ and $\iota_p : \tilde{X}^{(2)}(\mathbb{F}_p) \to J(\mathbb{F}_p)$ where $\iota_p(\mathcal{R}) = N[\mathcal{R} - \tilde{\infty}]$. In particular, we have the following diagram

$$\begin{array}{ccccccc}
\mathcal{L} & \xrightarrow{\ i\ } & X^{(2)}(\mathbb{Q}) & \xrightarrow{\ \iota\ } & G & \xleftarrow{\ \phi\ } & \mathbb{Z}^r \\
& \searrow & \downarrow{\scriptstyle \mathrm{red}} & & \downarrow{\scriptstyle \mathrm{red}} & \swarrow{\scriptstyle \phi_p} & \\
& & \tilde{X}^{(2)}(\mathbb{F}_p) & \xrightarrow{\ \iota_p\ } & J(\mathbb{F}_p) & &
\end{array}$$

where red is the reduction map and $\phi_p := \mathrm{red} \circ \phi$.

Let $\mathcal{M}_p \subset \tilde{X}^{(2)}(\mathbb{F}_p)$ be the subset of points $\mathcal{R} \in \iota_p^{-1}(\mathrm{Im}(\phi_p))$ which satisfy one of the following:

- $\mathcal{R} \notin \mathrm{red}(\mathcal{L}')$,

- $\mathcal{R} = \tilde{\mathcal{Q}}$ for some point $\mathcal{Q} \in \mathcal{L}' \setminus \rho^* C(\mathbb{Q})$ not satisfying the conditions of Theorem 4.1,
- $\mathcal{R} = \tilde{\mathcal{Q}}$ for some $\rho^* C(\mathbb{Q})$ not satisfying the conditions of Theorem 4.2.

THEOREM 5.1 (Mordell-Weil sieve [**17**, Theorem 5.1], [**9**, Theorem 2.6]). *Suppose*

$$\bigcap_{j=1}^{r} \phi_{p_j}^{-1}\left(\iota_{p_j}\left(\mathcal{M}_{p_j}\right)\right) = \emptyset,$$

then $X^{(2)}(\mathbb{Q}) = \mathcal{L}$.

6. Proof of Theorem 1.9

In this section we give the proof of Theorem 1.9. With the notation of Section 5 we set $\mathcal{L}' = \emptyset$ and $\mathcal{L} = \phi^*_{X_{D_{10}}} C(\mathbb{Q}) \subset X_{D_{10}}(11)^{(2)}(\mathbb{Q})$.

PROOF OF THEOREM 1.9. We apply relative symmetric Chabauty for $p = 5, 7, 13, 17, 19, 23$ and we prove that the elements in $\mathcal{L}$ are the only elements in their residue disks modulo p. Because $\mathcal{L}' = \emptyset$ we only had to apply Theorem 4.2. Finally, it is enough to apply the Mordell-Weil sieve, Theorem 5.1, where G is the finite-index subgroup of $J_{D_{10}}(\mathbb{Q})$ from Proposition 3.3, and $N = 10$, for the above choice of primes and get $\mathcal{L} = X_{D_{10}}(11)^{(2)}(\mathbb{Q})$. □

PROOF OF THEOREM 1.8. We may assume that $K \neq \mathbb{Q}(\sqrt{-11})$ by Theorem 1.4. For the other quadratic fields, we may apply Theorem 1.2 and Remark 1.3, which explains that it is enough to compute the quadratic points on $X_{D_{10}}(11)$.

From Theorem 1.9 we have determined the quadratic points of $X_{D_{10}}(11)$. All the quadratic points of $X_{D_{10}}(11)$ are pullbacks of the rational points of C. The image of the set of rational points of C under the j-map has been computed in the LMFDB modular curves database[4] and is equal to

$$J = [\infty, -3375, 8000, -884736, 16581375, -884736000].$$

Let $\Phi_{11}(x, y)$ be the class modular polynomial. With a short script we show that $\Phi_{11}(x, j_0)$ has a linear factor over $\mathbb{Q}[x]$, hence it also has a linear factor over any quadratic extension $K/\mathbb{Q}$, for all possible $j_0 \in J$ with $j_0 \neq \infty$, which by [**5**, Proposition 10.1] is enough to conclude the proof. □

7. Acknowledgments

The second and third authors would like to thank the first author and Charles University for their hospitality when this collaboration took place in Prague. The first and the third author want to thank the University of Groningen for their hospitality when they collaborated in Groningen. The first author was supported by the Czech Science Foundation GAČR, grant 21-00420M, Junior Fund grant for postdoctoral positions at Charles University and by the MPIM guest postdoctoral fellowship program during various stages of this project. The second author is supported by MaRDI, funded by the Deutsche Forschungsgemeinschaft (DFG), project number 460135501, NFDI 29/1. The third author is supported by the Special Account for Research Funds, AUTH research grant *"Solving Diophantine Equations-10349"*. Moreover, the third author wants to thank Professor Imin Chen

[4]Since $C \simeq X^+_{sp}(11)$ the j-map of C can be found in `https://beta.lmfdb.org/ModularCurve/Q/11.66.2.a.1/`.

for providing access to the servers of the Mathematics Department of Simon Fraser University, where the code was written and tested. We are especially grateful to the anonymous referees for their careful reports and many suggestions and improvements to this paper. We want to thank Imin Chen, Diana Mocanu, Steffen Müller and Lazar Radičević for useful conversations about the project.

References

[1] Samuele Anni, *A local-global principle for isogenies of prime degree over number fields*, J. Lond. Math. Soc. (2) **89** (2014), no. 3, 745–761, DOI 10.1112/jlms/jdu002. MR3217647

[2] Jennifer S. Balakrishnan, Noam Elkies, Brendan Hassett, Bjorn Poonen, Andrew V. Sutherland, and John Voight (eds.), *Arithmetic geometry, number theory, and computation*, Simons Symposia, Springer, Cham, 2021, DOI 10.1007/978-3-030-80914-0. MR4427958

[3] A. O. L. Atkin and J. Lehner, *Hecke operators on* $\Gamma_0(m)$, Math. Ann. **185** (1970), 134–160, DOI 10.1007/BF01359701. MR268123

[4] Barinder S. Banwait, *On some local to global phenomena for abelian varieties*, ProQuest LLC, Ann Arbor, MI, 2013. Thesis (Ph.D.)–University of Warwick (United Kingdom). MR3389327

[5] Barinder S. Banwait and John E. Cremona, *Tetrahedral elliptic curves and the local-global principle for isogenies*, Algebra Number Theory **8** (2014), no. 5, 1201–1229, DOI 10.2140/ant.2014.8.1201. MR3263141

[6] Francesc Bars, *The group structure of the normalizer of* $\Gamma_0(N)$ *after Atkin-Lehner*, Comm. Algebra **36** (2008), no. 6, 2160–2170, DOI 10.1080/00927870801949682. MR2418382

[7] Wieb Bosma, John Cannon, and Catherine Playoust, *The Magma algebra system. I. The user language*, J. Symbolic Comput. **24** (1997), no. 3-4, 235–265, DOI 10.1006/jsco.1996.0125. Computational algebra and number theory (London, 1993). MR1484478

[8] Josha Box, *Computing models for quotients of modular curves*, Res. Number Theory **7** (2021), no. 3, Paper No. 51, 34, DOI 10.1007/s40993-021-00276-8. MR4294165

[9] Josha Box, *Quadratic points on modular curves with infinite Mordell-Weil group*, Math. Comp. **90** (2021), no. 327, 321–343, DOI 10.1090/mcom/3547. MR4166463

[10] Josha Box, Stevan Gajović, and Pip Goodman, *Cubic and quartic points on modular curves using generalised symmetric Chabauty*, Int. Math. Res. Not. IMRN **7** (2023), 5604–5659, DOI 10.1093/imrn/rnab358. MR4565697

[11] Nils Bruin and Michael Stoll, *Deciding existence of rational points on curves: an experiment*, Experiment. Math. **17** (2008), no. 2, 181–189. MR2433884

[12] Robert F. Coleman, *Torsion points on curves and p-adic abelian integrals*, Ann. of Math. (2) **121** (1985), no. 1, 111–168, DOI 10.2307/1971194. MR782557

[13] Fred Diamond and Jerry Shurman, *A first course in modular forms*, Graduate Texts in Mathematics, vol. 228, Springer-Verlag, New York, 2005. MR2112196

[14] Steven Galbraith. *Equations for modular curves.* PhD thesis, University of Oxford, 1996. `https://www.math.auckland.ac.nz/~sgal018/thesis.pdf`.

[15] M. A. Kenku and Fumiyuki Momose, *Automorphism groups of the modular curves* $X_0(N)$, Compositio Math. **65** (1988), no. 1, 51–80. MR930147

[16] The LMFDB Collaboration. The L-functions and modular forms database. `https://www.lmfdb.org`, 2024. [Online; accessed 18 October 2024].

[17] Samir Siksek, *Chabauty for symmetric powers of curves*, Algebra Number Theory **3** (2009), no. 2, 209–236, DOI 10.2140/ant.2009.3.209. MR2491943

[18] Samir Siksek, *Explicit Chabauty over number fields*, Algebra Number Theory **7** (2013), no. 4, 765–793, DOI 10.2140/ant.2013.7.765. MR3095226

[19] Michael Stoll, *Implementing 2-descent for Jacobians of hyperelliptic curves*, Acta Arith. **98** (2001), no. 3, 245–277, DOI 10.4064/aa98-3-4. MR1829626

[20] Andrew V. Sutherland, *A local-global principle for rational isogenies of prime degree* (English, with English and French summaries), J. Théor. Nombres Bordeaux **24** (2012), no. 2, 475–485, DOI 10.5802/jtnb.807. MR2950703

[21] David Zywina. Computing actions on cusp forms. 2021. `arXiv:2001.07270`.

Max Planck Institute for Mathematics in Bonn, Charles University Prague

Faculty of Mathematics and Physics, Charles University, Ke Karlovu 3, 121 16 Praha 2, Czech Republic
Email address: stevangajovic@gmail.com

AG Algebra, Geometrie und Computeralgebra, RPTU Kaiserslautern-Landau
Email address: hanselm@rptu.de

Department of Mathematics, Aristotle University of Thessaloniki, 54124, Thessaloniki, Greece.
Email address: akoutsianas@math.auth.gr

Contemporary Mathematics
Volume **840**, 2026
https://doi.org/10.1090/conm/840/16814

Quadratic Chabauty experiments on genus 2 bielliptic modular curves in the LMFDB

Kate Finnerty

ABSTRACT. We present results of quadratic Chabauty experiments on genus 2 bielliptic modular curves of Jacobian rank 2 that have recently been added to the LMFDB. We apply quadratic Chabauty methods over both the rationals and quadratic imaginary fields. In a number of cases, the experiments yielded algebraic irrational points among the set of mock rational points. We highlight specific notable examples, including the non-split Cartan modular curve $X_{ns}^{+}(15)$. Lastly, we offer a conjecture relating the level of the modular curve to the potential number fields over which points can arise.

1. Introduction

1.1. Background and Motivation. For a projective, smooth, absolutely integral curve X over $\mathbb{Q}$, let $X(\mathbb{Q})$ denote the set of its rational points. If X is of genus 0, then $X(\mathbb{Q})$ is empty or infinite (and X is isomorphic to $\mathbb{P}^1$). If the genus g of X is 1 and $X(\mathbb{Q})$ is nonempty, then X is an elliptic curve, and Mordell's theorem states that $X(\mathbb{Q})$ is a finitely generated group [**Mor23**]. When the genus is 2 or greater, Faltings' theorem assures us that $X(\mathbb{Q})$ is finite [**Fal83**]. However, the proof does not provide an effective construction to *find* these points. It is an ongoing question to provably compute $X(\mathbb{Q})$.

A classical method to compute $X(\mathbb{Q})$ comes from the work of Chabauty [**Cha41**] and Coleman [**Col85a**]. Given a prime p of good reduction for X and a known $P \in X(\mathbb{Q})$, if the rank of the Jacobian variety J of X is less than the genus of the curve, we can construct a finite set $X(\mathbb{Q}_p)_1$ that contains $X(\mathbb{Q})$. In practice, one can often succeed in provably computing all of $X(\mathbb{Q})$ by checking to see which p-adic points easily lift to rational points and then using a tool such as the Mordell-Weil sieve to rule out the remaining points from being rational.

In the case where the rank of the curve is not less than the genus, this fails to provide finite bounds. Kim developed a nonabelian Chabauty program using Selmer varieties and their iterated p-adic integrals [**Kim09**]. Just as the classical method creates the set $X(\mathbb{Q}_p)_1$, nonabelian Chabauty produces analogous sets:

$$X(\mathbb{Q}_p)_1 \supset X(\mathbb{Q}_p)_2 \supset X(\mathbb{Q}_p)_3 \supset \cdots \supset X(\mathbb{Q}_p)_n \supset X(\mathbb{Q}).$$

Kim conjectured that for n sufficiently large, $X(\mathbb{Q}_p)_n$ is finite. He further conjectured that for sufficiently large n, we achieve the equality $X(\mathbb{Q}_p)_n = X(\mathbb{Q})$.

2020 *Mathematics Subject Classification.* 11D45, 11Y50.

This work has led to a wave of development already in the first nonabelian case, $n = 2$, known as quadratic Chabauty. The quadratic Chabauty set $X(\mathbb{Q}_p)_2$ is known to be finite when the rank of the Jacobian of the curve is less than $g - 1 + rk(NS(J))$, where $NS(J)$ denotes the Néron-Severi group of the Jacobian over $\mathbb{Q}$. Balakrishnan and Dogra establish this bound for finiteness in [**BD18**] and together with Müller compute some explicit examples. Developments have led to computational tools for new and more complicated examples in recent years. In [**Bia20**], Bianchi further developed quadratic Chabauty by using the p-adic sigma function instead of the double Coleman integral to compute the quadratic Chabauty locus for genus 2 bielliptic curves. Balakrishnan, Besser, Bianchi, and Müller later generalized this to rational points on genus 2 bielliptic curves over number fields [**BBBM21**]. Bianchi and Padurariu then implemented and scaled these methods to compute the set of rational points on bielliptic curves of Jacobian rank 2 from the LMFDB's database of genus 2 curves [**BP24, Cola**].

1.2. Aim of this paper. After Bianchi and Padurariu's computations, the beta version of the LMFDB added a collection of modular curves satisfying the following [**Colb**]:

LMFDB CONDITIONS. The LMFDB beta database contains data on modular curves X_H with $\det(H)$ equal to $\hat{\mathbb{Z}}^\times$ such that at least one of the following holds:

(1) H has level at most 70.
(2) H has prime-power level at most 335.
(3) H contains $-I$ and has genus at most 24 and level at most 335.
(4) H does not contain $-I$ and has genus at most 8 and level at most 335.

The database includes information on rational cusps and rational CM points for every X_H. Rational and low degree points data are available for points that correspond to elliptic curves already in the LMFDB, but these points are not yet provably computed for all curves in the database.

In this manuscript, we consider genus 2 bielliptic curves from this database with Jacobian rank 2 as well as curves of Jacobian rank 1 that achieve Jacobian rank 2 over a quadratic imaginary number field with an eye towards understanding the set of *mock rational points*, which we describe below.

Recall that Balakrishnan and Dogra [**BD18**] define a finite set $X(\mathbb{Q}_p)_Z \subset \mathbb{Q}_p$, dependent on a choice of correspondence $Z \subset X \times X$, that contains $X(\mathbb{Q}_p)_2$. Our work will use the simplification developed in [**BP24**]: the set of rational points is contained in the finite set

$$A = \{z \in X(\mathbb{Q}_p) : \widetilde{\rho}(z) \in \Omega\}. \tag{1.1}$$

Here, $\widetilde{\rho}$ is an explicitly computable locally analytic function on $X(\mathbb{Q}_p)$. We will define it precisely in the next section. Additionally, we can explicitly describe and compute Ω as a finite subset of $\mathbb{Q}_p$ (see Theorem 2.3). In this paper, we will refer to the points in $A \setminus X(\mathbb{Q})$ as mock rational points. We aim to study this set with a focus on the question:

QUESTION 1.1. Are there algebraic points among the mock rational points for a given curve?

The question of algebraic points on modular curves has been studied in recent years by Özman, Box, and others (see [**OS19**],[**AKMJ$^+$24**],[**Box21**],[**BGG23**]).

We have also seen algebraic points appear as mock rational points before. For example, when computing the set of points $X_0(37)(\mathbb{Q}(i))$, Balakrishnan, Dogra, and Müller also discovered points on $X_0(37)$ defined over $\mathbb{Q}(\sqrt{-3})$ [**BD18**]. This example invites several questions. Under what conditions do these points emerge? Over what fields do these points live? What characteristics do these points share?

To explore these questions, we carry out computations on the genus 2 bielliptic modular curves described above. We analyze genus 2 curves with non-simple Jacobian of rank 1 and 2: 209 curves with Jacobian rank 2 and 1237 curves with Jacobian rank 1.

For the genus 2 curves of Jacobian rank 2 over $\mathbb{Q}$, we can use quadratic Chabauty. On the other hand, for curves of Jacobian rank 1 over $\mathbb{Q}$, we consider them over quadratic imaginary fields over which the rank of the Jacobian of the curve increases to 2 and carry out quadratic Chabauty with respect to that field. In particular, since the Jacobian of a bielliptic curve is isogenous to a product of elliptic curves $E_1 \times E_2$, the rank is the sum of the ranks of the elliptic curves. We focus on the case where both E_1 and E_2 have rank 1 over $\mathbb{Q}(\sqrt{-D})$.

In the rank 2 case, we vary the prime used for quadratic Chabauty to observe differences in what points appear. We take p a prime of good, ordinary reduction for X with $3 < p < 100$. In the rank 1 case, we vary not only the prime but also the quadratic imaginary field used. We determine over which fields $\mathbb{Q}(\sqrt{-D})$ with $1 < D < 20$ and $D \in \{43, 67, 163\}$ the rank of the Jacobian of the curve is 2 (while we sought a tractable bound for D, adding the three larger values allows us to consider *all* imaginary quadratic fields of class number one at minimal additional cost). For each such field $\mathbb{Q}(\sqrt{-D})$, we restrict to primes p of good, ordinary reduction that split in $\mathbb{Q}(\sqrt{-D})$.

We provide code and data on GitHub to demonstrate computations in the paper [**Fin25**]. The code is modified from [**Bal**], which in turn is modified from [**BPb**] and uses SageMath [**The24**]. We will describe the new modifications in Sections 3 and 4. From these computations, we obtain the following:

PROPOSITION 1.2. Among the genus 2 Jacobian rank 2 modular curves satisfying LMFDB Conditions (1.2), Tables 4 and 6[1] contain curves that have points over the listed number fields through quadratic Chabauty for some prime $p < 100$ of good, ordinary reduction, along with the smallest prime for which they appear, up to precision $O(p^{25})$.

PROPOSITION 1.3. Among the genus 2 Jacobian rank 1 modular curves satisfying LMFDB Conditions (1.2), Tables 5 and 7 contain curves that have points over the listed number fields through quadratic Chabauty for some prime $p < 100$ of good, ordinary reduction, along with the smallest prime for which they appear, up to precision $O(p^{25})$.

We note that this list of algebraic points is not provably exhaustive. We can, however, make a conjecture based on the results of the explicit computations. It holds for the genus 2 modular curves satisfying LMFDB Conditions (Section 1.2).

CONJECTURE 1.4. Let X be a bielliptic modular curve of genus 2 and level N. If $Jac(X)$ is of rank 2 over $\mathbb{Q}$ and X has a mock rational point over $\mathbb{Q}(\sqrt{D})$ then D must be a divisor of N.

[1]Tables 6 and 7 are on the Supplementary Tables file on GitHub [**Fin25**].

We might hope for an analogous conjecture in the rank one case. Suppose X is a bielliptic modular curve of genus 2 and level N with $Jac(X)$ of rank 1. If X has a mock rational point over $\mathbb{Q}(\sqrt{D})$ arise through quadratic Chabauty when computing over $\mathbb{Q}(\sqrt{D'})$, we frequently observe one of the following:

(1) D divides N.
(2) D divides D'.

Note that in order to use $\mathbb{Q}(\sqrt{D'})$ for quadratic Chabauty contributions for the curve X, we need the rank of the Jacobian of X to be 2 over this field.

However, we have observed examples for which this is not the case. The curve `24.36.2.bo.1` has $\mathbb{Q}(\sqrt{-3})$-points appear when using $p = 13$ and working over $\mathbb{Q}(\sqrt{-10})$. Also, the curve $X_0(37)$ has the aforementioned points over $\mathbb{Q}(\sqrt{-3})$ appear when using $p = 13$ and working over $\mathbb{Q}(i)$. We suspect there is a systematic explanation for these not captured in the above divisibility conditions on D.

1.3. Acknowledgements. Many thanks are due to Francesca Bianchi and Oana Padurariu for their virtual correspondence related to their previous work. We also thank Jennifer Balakrishnan for proposing the project and for helpful and insightful conversations throughout. The third modular curves workshop for the Simons Collaboration in Arithmetic Geometry, Number Theory, and Computation in March 2024, with support from the Simons Foundation, was also instrumental. Lastly, we are grateful to the reviewers of this article for their insight and attention to detail, both of which greatly improved the paper. The author was partially supported by NSF Grant DMS-1945452 and US-Israel BSF Grant 2022393.

2. Quadratic Chabauty for Bielliptic Curves

Our main source for this expository section is [**BP24**]. Let X be a non-singular genus 2 bielliptic curve

$$X : y^2 = a_6x^6 + a_4x^4 + a_2x^2 + a_0, a_i \in \mathbb{Z}, \tag{2.1}$$

and consider the two elliptic curves

$$E_1 : y^2 = x^3 + a_4x^2 + a_2a_6x + a_0a_6^2, \tag{2.2}$$

$$E_2 : y^2 = x^3 + a_2x^2 + a_4a_0x + a_6a_0^2. \tag{2.3}$$

We have degree 2 maps $\phi_i : X \to E_i$ with

$$\phi_1(x, y) = (a_6x^2, a_6y), \quad \phi_2(x, y) = (a_0x^{-2}, a_0yx^{-3}). \tag{2.4}$$

Remark 2.1. The genus 2 curves of our dataset do not initially come in this form. However, any genus 2 curve with a degree 2 map to an elliptic curve has a model of this form [**Wet97**].

Recall our goal is to define a function $\widetilde{\rho}$ so that we can compute the set A in Equation (1.1). Let log denote the p-adic logarithm extended to $\mathbb{Q}_p^\times$ by setting $\log(p) = 0$. While some summands of $\widetilde{\rho}$ will depend on the choice of branch of log, the final function is independent of the value of $\log(p)$. On E_i (although we could replace this discussion with any $E/\mathbb{Q}$) with good reduction at p, for $P \in E_i(\mathbb{Q}_p)$, we let $Log(P) = \int_\infty^P \omega_i$, where $\omega_i = \frac{dx}{2y}$ and the integral is Coleman integration [**Col85b**, Theorem 2.8].

If P_1 and P_2 are in the same residue disk, then $Log(P_1) - Log(P_2) = \int_{P_1}^{P_2} \omega_i$ and the integral can be calculated via formal antidifferentiation. Since ω_i is holomorphic, the function *Log* is locally analytic. Furthermore, *Log* vanishes at a point P if and only if that point is torsion [**Col85b**, Proposition 3.1].

We consider the decomposition of the global p-adic height into local heights λ_q. Given a prime q, we let ord_q denote the q-adic valuation on $\mathbb{Q}_q$, normalized to be surjective onto the integers and let $|\cdot|_q$ denote the standard absolute value on $\mathbb{Q}_q$. With this, the following results give us the desired objects:

PROPOSITION 2.2 ([**BP24**, Proposition 2.2]). Let $q \neq p$. If $P \in E(\mathbb{Q}_q)$ reduces to a nonsingular point modulo q, then $\lambda_q(P) = \log(max\{1, |x(P)|_q\})$.

Let W_q^E be the set of values attained by λ_q on points in $E(\mathbb{Q}_q)$ of the form (x, y) with $x, y \in \mathbb{Z}_q$. Then W_q^E is finite, explicitly computable, and equal to $\{0\}$ for all but finitely many q.

In particular, $W_q^E \subseteq \{0\}$ at all primes of good reduction for the given model of the elliptic curve E.

Note that in the above proposition, we require the model for E to be minimal, see [**Bia20**]. Next, we use the following additional notation in the following theorem and remark. For a prime q, we define

$$Z_q = X(\mathbb{Q}_q) \setminus \{P : x(P) \in \{0, \infty\}\}.$$

THEOREM 2.3 ([**BP24**, Theorem 2.3]). *Suppose that each of E_1 and E_2 have rank 1 over $\mathbb{Q}$, and let p be a prime of good reduction for the equation of X. For each $i \in \{1, 2\}$, fix a choice of subspace of $H^1_{dR}(E_i/\mathbb{Q}_p)$ complementary to the space of holomorphic forms, and consider the corresponding global height h_p and local Néron functions λ_q at every q. Let $P_i \in E_i(\mathbb{Q})$ be a point of infinite order. Define $\alpha_i = \frac{h_p(P_i)}{Log^2(P_i)}$. Then:*

(1) *The constant α_i is independent of choice of P_i.*
(2) *The function $\rho : Z_p \to \mathbb{Q}_p$ given by*

$$\rho(z) = \lambda_p(\phi_1(z)) - \lambda_p(\phi_2(z)) - 2\log(x(z)) - \alpha_1 Log^2(\phi_1(z)) + \alpha_2 Log^2(\phi_2(z))$$

can be continued to a locally analytic function $\widetilde{\rho} : X(\mathbb{Q}_p) \to \mathbb{Q}_p$.

(3) *For a prime $q \neq p$, let*

$$\Omega_q = (-W_q^{E_1} + W_q^{E_2} + \{-n \log q : -ord_q(a_6) \leq n \leq ord_q(a_0), n \equiv 0 \bmod 2\}) \cup (\log |a_0|_q - W_q^{E_1}) \cup (-\log |a_6|_q + W_q^{E_2})$$

and set $\Omega = \{\sum_{q\ bad} w_q : w_q \in \Omega_q\}$. Then Ω is finite and contains the image $\widetilde{\rho}(X(\mathbb{Q}))$.

REMARK 2.4. In the development of the theory and the first examples of quadratic Chabauty for bielliptic curves, Balakrishnan and Dogra also define a set Ω [**BD18**, Theorem 1.2]. In this paper, the set of points whose image under a certain quadratic Chabauty function lands in Ω is a superset rather than strictly equal to $X(\mathbb{Q}_p)_2$ (note that Balakrishnan and Dogra worked over general number fields). In turn, the set of Theorem 2.3 is a superset of the set of Balakrishnan and Dogra, so the set A of Equation (1.1) is a superset of $X(\mathbb{Q}_p)_2$. This method of giving a superset that is explicitly computable is standard: in addition to [**BD18**], see [**BDM+19**], [**BD21**]. The authors of [**BDM+19**] remark that this can be thought of as an analogue to working with a suitable quotient of the Jacobian as in [**Maz77**] and [**Mer96**].

3. Computations over the Rationals

We begin with 209 curves of Jacobian rank 2. Many have the same model, and we use the coarse curves only to obtain a unique list of models. This strategy is permissible because any fine modular curve in our list shares a model with a coarse modular curve. This fact can be checked by hand: for example, the fine curve `14.96.2-14.g.1.1` has Weierstrass model

$$y^2 = x^6 - 3x^5 - x^4 + 7x^3 - x^2 - 3x + 1,$$

which is the same as that of the coarse curve `14.48.2.g.1`. By excluding fine modular curves, we obtain a list of 76 equations defining bielliptic curves. We also exclude curves with no known rational points. We then note that many modular curves have exceptional isomorphisms. We consider only one representative from each isomorphism class, giving us 30 equations $y^2 = f(x)$ to consider [2]. We convert each to a bielliptic model by using Magma to obtain Richelot isogenous surfaces that are of the form $E_1 \times E_2$. Performing changes of variables on these elliptic curves allows us to find a_i that determine a model for X in the form of Equation (2.1). We run the analysis for each prime p less than or equal to 97 of good, ordinary reduction.

3.1. On Choice of Bielliptic Model. A given curve does not have a unique representation of the form in Equation (2.1). Most immediately, a change of variables $[x : y : z] \mapsto [-z : y : x]$ gives an isomorphism between $y^2 = a_6x^6 + a_4x^4 + a_2x^2 + a_0$ and $y^2 = a_0x^6 + a_2x^4 + a_4x^2 + a_6$. However, there are often other possibilities. For example, the following are isomorphic:

$$y^2 = -x^6 + 83x^4 - 19x^2 + 1,$$
$$y^2 = 64x^6 - 304x^4 + 332x^2 - 1.$$

A model with a larger discriminant will produce a larger set of p-adic points when completing quadratic Chabauty. We are therefore interested in finding a model of minimal discriminant to streamline the analysis.

The following result describes how discriminants of isomorphic curves are related to each other. For more, one can refer to [**Liu94**].

Lemma 3.1 ([**BSS+16**, Section 2.1]). *Let F be a perfect field and X a curve of genus 2 over F. Suppose X has a Weierstrass equation $y^2 + h(x)y = f(x)$. If $(y')^2 + h'(x')y' = f'(x')$ is another Weierstrass model for the curve X, then it is related to the original via a change of variables of the form*

$$x' = \frac{ax+b}{cx+d}, \qquad y' = \frac{ey + j(x)}{(cx+d)^3}, \qquad \text{with } ad - bc, e \in F^\times, j \in F[x], \deg j \le 3.$$

Furthermore, we have the discriminants $\Delta(f', h')$ and $\Delta(f, h)$ are related as follows:

$$\Delta(f', h') = e^{20}(ad - bc)^{-30}\Delta(f, h).$$

[2] While we checked for these isomorphisms by noting when Weierstrass models on the LMFDB were identical or off by a sign difference, checks for exceptional isomorphisms were also performed by David Roe in October 2024 [**Roe24**], and our results agree for the curves under consideration here.

The last statement of the lemma immediately gives us the following:

COROLLARY 3.2 ([**BSS+16**, Section 2.1]). The discriminant is invariant as an element of $F^\times/F^{\times 10}$.

We now describe how to attain a model of minimal discriminant in our case. We can represent transformations between models with matrices of the form

$$M(a,b,c,d,e) := \begin{pmatrix} a & 0 & b \\ 0 & e & 0 \\ c & 0 & d \end{pmatrix}.$$

By the lemma, this transformation changes the discriminant by a factor of $e^{20}(ad-bc)^{-30}$. The following is a result of direct computation with the previous lemma and the definition of $M(a,b,c,d,e)$.

PROPOSITION 3.3. Let $X : y^2 = a_6x^6 + a_4x^4 + a_2x^2 + a_0$ with a_i all integers. For $M(a,b,c,d,e)$ to yield an isomorphic curve of the form in Equation (2.1) with integer coefficients, it is necessary and sufficient to have the following conditions:

- The following are all equal to zero:
  - $a_0(3ac^5) + a_2(2ac^3d^2 + bc^4d) + a_4(acd^4 + 2bc^2d) + a_6(3bd^5)$.
  - $a_0(5a^3c^3) + a_2(a^3cd + 3a^2bc^2d + ab^2c^3) + a_4(a^2bd^3 + 3ab^2cd^2 + b^3c^2) + a_6(5b^3d^3)$.
  - $a_0(3a^5c) + a_2(a^4bd + 2a^3b^2c) + a_4(2a^2b^3d + ab^4c) + a_6(3b^5d)$.
- The following are all congruent to zero modulo $(ad-bc)^6$:
  - $e^2(c^6a_0 + c^4d^2a_2 + c^2d^4a_4 + d^6a_6)$
  - $e^2(15a^2c^4a_0 + 6a^2c^2d^2a_2 + a^2d^4a_4 + 8abc^3da_2 + 8abcd^3a_4 + b^2c^4a_2 + 6b^2c^2d^2a_4 + 15b^2d^4a_6)$
  - $e^2(15a^4c^2a_0 + a^4d^2a_2 + 8a^3bcda_2 + 6a^2b^2c^2a_2 + 6a^2b^2d^2a4 + 8ab^3cda_4 + b^4c^2a_4 + 15b^4d^2a_6)$
  - $e^2(a^6a_0 + a^4b^2a_2 + a^2b^4a_4 + b^6a_6)$

We can think of the first set as vanishing conditions: these ensure that the odd-power terms have zero coefficient. The latter are integrality conditions that ensure the even-power terms are elements of $\mathbb{Z}$.

Let $\prod_i p_i^{e_i}$ be the prime factorization of the discriminant of a given model for the curve X. We compute the potential discriminant-lowering effects $p^k < 1$ and, for each such p^k, the set of $M(a,b,c,d,e)$ such that $e^{20}(ad-bc)^{-30} = p^k$ subject to the condition that the absolute values of a,b,c,d,e are less than or equal to the maximum p_i. In addition to considering integral a,b,c,d,e, we also consider numbers of the form $1/p$ for p prime to ensure that we can obtain transformations with effects that reduce the discriminant by a prime power that is a negative even multiple of 10.

Note that our bounds on the matrix coefficients are sufficient because we can obtain a reduction in the discriminant by a factor of p^{10} for any p dividing the discriminant, which is sufficient by Corollary 3.2. While matrices that reduce the discriminant by a given factor are not unique, we offer the following as an example of a transformation to reduce the discriminant by p^{-10}:

$$M(p,0,0,1,p) = \begin{pmatrix} p & 0 & 0 \\ 0 & p & 0 \\ 0 & 0 & 1 \end{pmatrix}.$$

For each potential factor by which we might lower the discriminant, we apply Proposition 3.3 to obtain a matrix $M(a,b,c,d,e)$ that will yield an isomorphic model of the desired form with integer coefficients, if such a matrix exists. If it exists, we apply it and obtain a new model. If not, we continue to the next factor. We repeat until there is no more allowable transformation matrix that would reduce the discriminant while keeping it in the desired form. In this way, we obtain optimal models to use in the analysis.

For example, initial transformation of the curve `30.60.2.i.1` into a bielliptic model yielded

$$y^2 = -2700x^6 + 8000,$$

which has discriminant $2^{54}3^{21}5^{25}$. Applying the algorithm produces the model

$$y^2 = -108x^6 + 5,$$

of discriminant $2^{24}3^{21}5^5$, removing a factor of $2^{30}5^{20}$.

3.2. Execution of the Analysis. A description of the algorithm is as follows. Our input is: f a polynomial over $\mathbb{Q}$ of the form in Equation (2.1) such that the two elliptic curves of Equations (2.2) and (2.3) both have rank 1 over $\mathbb{Q}$, p an odd prime such that E_1 and E_2 have good reduction and a fixed p-adic precision n. In our computations, we set $n = 25$, sufficient for all curves under the precision analysis in [**BP24**]. Our output is a list of rational points on $H : y^2 = f(x)$ and a list of p-adic points on H that are not recognized as rational points. We compute this by the following:

(1) For each elliptic curve E_i, compute the following: $E_{i,m}$ (the minimal models), Δ_i (the discriminant), h_i (the canonical p-adic height), P_i (a generator of the free part of $E_i(\mathbb{Q})$), $E_{i,K}$ (the elliptic curve over $K = \mathbb{Q}_p$ up to precision n), $E_{i,p}$ (the elliptic curve over $\mathbb{F}_p$).
(2) Set H_p as $H : y^2 = f(x)$ over $\mathbb{F}_p$. Let D be the set of points of H_p up to automorphism.
(3) Compute the set Ω per Theorem 2.3.
(4) For every point in D, compute $\widetilde{\rho}$ per Theorem 2.3. Then, for each element $\omega \in \Omega$, compute the zeroes of $\widetilde{\rho}-\omega$. Note that these will be p-adic numbers. Find the points on H over $\mathbb{Q}_p$ with these zeroes as x-coordinates and save them in a list L.
(5) For each element of L, attempt to lift to a rational point. Return a list of rational points and a list of candidate algebraic points for further analysis.

We focus on p-adic points that do not obviously lift to rational points. We use SageMath's algdep functionality to see if either of the points' coordinates satisfy relations of degree up to four. We distinguish between Weierstrass points, which are unsurprising to encounter, and other non-Weierstrass points.

REMARK 3.4. Since we work with precision $n = 25$, one might wonder whether the points truly appear via this analysis or if they are only transcendental points that are p-adically very close. In all cases, we can immediately confirm that the algebraic point satisfies the defining equation for the curve. Then, to be more robust, we can apply the Weierstrass preparation theorem. We consider the power series given by $\widetilde{\rho} - \omega$ and factor out the polynomial part that has a root corresponding to the point in question. If we can evenly factor out that polynomial part with coefficients in $\mathbb{Q}$ up to our working precision and the root is simple, then we have the desired result.

For example, consider the curve `16.48.2.a.1`, denoted X, with model given by $y^2 = x^6 - 5x^4 - 5x^2 + 1$ of Jacobian rank 2. Working with both $p = 11$ and $p = 17$, we find p-adic points that appear to correspond to $(\pm 1, \pm 2\sqrt{-2})$. Without loss of generality, we let $Q = (1, 2\sqrt{-2})$, since the other three points are redundant up to automorphism. One can immediately see that Q is in $X(\mathbb{Q}(\sqrt{-2}))$ (as are its images under automorphism).

Let t_0 represent the root of $\widetilde{\rho} - \omega$ that was recognized as Q for $p = 17$. It turns out that $(t - t_0)$ evenly divides $\widetilde{\rho} - \omega$. This is a simple root with a linear coefficient of just 1, so we can conclude that the point is algebraic. Although the terms in the case of $p = 11$ are 11-adic instead of 17-adic, the result of being able to evenly factor out a simple linear term is the same.

Another change to the code relates to the computation of the set Ω. When computing the local height at a given bad prime q, one can check if the curve has potential good reduction at q. If so, then the contribution at that prime can be understood entirely by information about the local height at a single $\mathbb{Q}_q$-point on the curve [**BP24**]. The original implementation assumes that such a point will exist. We remove this assumption by first checking in MAGMA [**BCP97**] if the curve is locally soluble at q . In the event that the code fails to obtain a $\mathbb{Q}_q$-point, we proceed to the general case to compute the height contribution at q.

The last change is concerning the point at infinity. At each residue disc, we need to compute the function $\widetilde{\rho}$. For the given point P, which is a $\mathbb{F}_p$-point, we let Q denote a lift to the curve over $\mathbb{Q}_p$. We let f_1 and f_2 denote the images of Q to E_1 and E_2 over $\mathbb{Q}_p$. If, for one f_i, we have f_i multiplied by the order of E_i over $\mathbb{F}_p$ is the point at infinity, then we must compute the Coleman integral from the point at infinity to f_i. This step can be prohibitively computationally expensive for high precision. Therefore, we take advantage of the fact that $\int_{\infty}^{P_1} \omega_i = \frac{1}{2}\int_{-P_1}^{P_1} \omega_i$ to circumvent difficulties with the point at infinity. This is a consequence of extending the following lemma to tangential basepoints, as was done on page 289 of [**BKK11**]:

LEMMA 3.5 ([**BBK10**, Lemma 16]). *Let ω_i be an odd, everywhere meromorphic differential on X. Choose P, Q as points in $X(\mathbb{C}_p)$ which are not poles of ω_i, with P Weierstrass. Then for ι the hyperelliptic involution, we have $\int_P^Q \omega_i = \frac{1}{2}\int_{\iota(Q)}^{Q} \omega_i$.*

After the analysis described in this section, we were also able to replicate the strategy of Bianchi and Padurariu to use the Mordell-Weil sieve to confirm the complete sets of rational points for these curves. For information about the methodology, one can refer to Sections 4.3 and 4.4 of [**BP24**]. This allowed us to compute the rational points on more curves, but since the technology behind the computation is not new, it is not a focus of this paper. It does, however, give the following:

PROPOSITION 3.6. The analysis of [**Fin25**] produces the set of rational points for genus 2 bielliptic modular curves of Jacobian rank 1 or 2 that satisfy LMFDB Conditions (Section 1.2).

3.3. Algebraic Weierstrass Mock Rational Points. Weierstrass points form a convenient set of examples when considering in which cases an algebraic point of a curve will appear in the analysis. We obtain precise criteria by comparing the splitting behavior of the defining polynomials with the results in Table 6, which can be found in the Supplementary Tables file of the GitHub [**Fin25**].

PROPOSITION 3.7. Let $P = (x_0, 0)$ be a Weierstrass point of a genus 2 bielliptic modular curve X with rank 2 Jacobian satisfying LMFDB Conditions (1.2) with $x_0 \notin \mathbb{Q}$. Then for all $p < 100$, all nonrational Weierstrass points P in $A \setminus X(\mathbb{Q})$ (see Equation (1.1)), up to precision $O(p^{25})$, satisfy the following properties:

(1) The prime p is of good and ordinary reduction for both elliptic curves E_i, where $Jac(X) \sim E_1 \times E_2$, so it satisfies the conditions for quadratic Chabauty for bielliptic curves.
(2) The defining polynomial $f(x)$ for X factors as $(ax^2 + bx + c)g(x)$, and x_0 is a root of $ax^2 + bx + c$.
(3) The point P lives in $\mathbb{Q}(\sqrt{D})$ and D is a quadratic residue modulo p.

Consider the curve X with label `24.72.2.iy.1`. It has a model given by $y^2 = 6x^6 + 12x^4 + 12x^2 + 6$. The defining polynomial for X factors over the rationals as $6(x^2 + 1)(x^2 - x + 1)(x^2 + x + 1)$, so we have three eligible terms in the factorization, and we expect to observe all 6 Weierstrass points. Note E does not have ordinary reduction at 7 or 47, so we cannot use these primes for quadratic Chabauty. All other primes p with $3 < p < 100$ are of good, ordinary reduction. Since the Weierstrass points live over $\mathbb{Q}(i)$, and $\mathbb{Q}(\sqrt{-3})$, we can check the Legendre symbol of these primes with respect to -1 and -3. The behavior we see is as expected: the points $(\pm i, 0)$ appear exactly when $p \equiv 1 \bmod 4$ and the points $(\pm\frac{1}{2} \pm \frac{\sqrt{-3}}{2}, 0)$ appear exactly when $p \equiv 1 \bmod 3$.

Next, consider the modular curve `16.48.2.a.1`. We use $y^2 = x^6 - 5x^5 - 5x^2 + 1$, which factors over the rationals as $(x^2 + 1)(x^2 - 2x - 1)(x^2 + 2x - 1)$. The first term has roots $\pm i$, and the latter terms have roots living over $\mathbb{Q}(\sqrt{2})$. The primes p that are 1 modulo 4 are those such that -1 is a quadratic residue modulo p, so we consider the list $\{5, 13, 17, 29, 37, 41, 53, 61, 73, 89, 97\}$. We can check which primes are of good, ordinary reduction for the corresponding elliptic curves and obtain the list $\{17, 41, 73, 89, 97\}$. These are exactly the primes for which we observe the points $(\pm i, 0)$. However, we do not observe the points over $\mathbb{Q}(\sqrt{2})$ for any prime < 100. The conditions for a Weierstrass point to appear are necessary but not sufficient.

4. Computations over Quadratic Number Fields

We perform a similar process as the start of the analysis of curves of Jacobian rank 2. Although we start with 1237 curves with Jacobian rank 1, we obtain a list of 116 unique defining equations of curves with a known rational point. We once again convert them to the desired bielliptic form $y^2 = a_6x^6 + a_4x^4 + a_2x^2 + a_0$. In this section, we use code from Bianchi and Padurariu [**BPa**] to assist in some of the transformations.

As stated in the introduction, an eligible choice of number field $K = \mathbb{Q}(\sqrt{-D})$ and prime p is one where the Jacobian of the curve is rank 2 over K and p is a prime of good, ordinary reduction for the curve that splits in K. For each p between 3 and 100 of good, ordinary reduction, we check potential values of D until we have obtained a list of at least five eligible pairs or run out of primes. We obtain a list of eligible pairs (D, p) for each curve and run quadratic Chabauty on each.

Code modifications from the previous section were also used for this analysis. In addition, we need to consider the height contributions of primes that divide the discriminant of the number field K. We incorporate this into the computation of Ω. More precisely, we can modify Theorem 2.3 as follows:

THEOREM 4.1. *Let X be a bielliptic genus 2 curve of the form in Equation (2.1) whose Jacobian splits into a product of two elliptic curves as $J \sim E_1 \times E_2$ such that $rk(E_1(\mathbb{Q})) + rk(E_2(\mathbb{Q})) = 1$.*

Suppose that each of E_1 and E_2 has rank 1 over an imaginary quadratic number field $K/\mathbb{Q}$, and let p be a prime of good reduction for the equation of X. For each $i \in \{1, 2\}$, fix a choice of subspace of $H^1_{dR}(E_i/\mathbb{Q}_p)$ complementary to the space of holomorphic forms, and consider the corresponding global height h_p and local Néron functions λ_q at every q.

Let $P_i \in E_i(K)$ be a point of infinite order. Define $\alpha_i = \frac{h_p(P_i)}{[K:\mathbb{Q}]Log^2(P_i)}$. Then:

(1) *The constant α_i is independent of choice of P_i.*
(2) *The function $\rho : Z_p \to \mathbb{Q}_p$ given by*

$$\rho(z) = \lambda_p(\phi_1(z)) - \lambda_p(\phi_2(z)) - 2\log(x(z)) - \alpha_1 Log^2(\phi_1(z)) + \alpha_2 Log^2(\phi_2(z))$$

can be continued to a locally analytic function $\widetilde{\rho} : X(\mathbb{Q}_p) \to \mathbb{Q}_p$.

(3) *For a prime $q \neq p$, let*

$$\Omega_q = (-W_q^{E_1} + W_q^{E_2} + \{-n \log q : -ord_q(a_6) \leq n \leq ord_q(a_0), n \equiv 0 \bmod 2\})$$
$$\cup(\log|a_0|_q - W_q^{E_1}) \cup (-\log|a_6|_q + W_q^{E_2})$$

*and set $\Omega = \{\sum_{q\in S} w_q : w_q \in \Omega_q\}$, where S consists of the bad primes for each E_i, the prime factors of a_0 and a_6 for X, **and** the prime divisors of the discriminant of K. Then Ω is finite and contains $\widetilde{\rho}(X(\mathbb{Q}))$.*

Many of the arguments of Bianchi and Padurariu for Theorem 2.3 in the case of $K = \mathbb{Q}$ apply, see [**BP24**]. We now discuss the necessary steps to make this extended result hold. The addition of the factor $\frac{1}{[K:\mathbb{Q}]}$ in the definition of α_i is a necessary normalization factor for p-adic heights over a quadratic field, see Corollary 8.1 ii) of [**BD18**]. With this normalization complete, the only difference with the previous Theorem is the elements of the set S that contribute to the definition of Ω. In particular, the set S encapsulates the possible nontrivial local height contributions. When working over a number field K, it is possible that the ramification of primes can lead to nontrivial height contributions, meaning we need to consider them. However, since there are only finitely many such primes, Ω still remains finite.

4.1. Appearance of Weierstrass Points in the Rank 1 Case. We checked the rank 1 results to compare the appearance of Weierstrass points to that in the rank 2 case, keeping in mind the results described in Proposition 3.7. Since we did not run the analysis on Jacobian rank 1 curves for all primes between 3 and 100, we cannot directly claim an analogue to Proposition 3.7. However, for primes that were tested, Weierstrass points that appeared satisfied the conditions, regardless of the number field used.

Consider the curve `60.48.2.c.1` with model $y^2 = x^6 + 20x^4 + 25x^2 - 750$ as an example. This factors as $-2(x^2 - 5)(x^2 + 10)(x^2 + 15)$. The Weierstrass points live over $\mathbb{Q}(\sqrt{5}), \mathbb{Q}(\sqrt{-10})$, and $\mathbb{Q}\sqrt{-15})$. The primes used were 11, 13, 19, 23, and 37. As expected, we saw the points over $\mathbb{Q}(\sqrt{5})$ when using $p = 11$ and 19, the points over $\mathbb{Q}(\sqrt{-10})$ using all five primes, and the points over $\mathbb{Q}(\sqrt{-15})$ using $p = 19$ and 23.

5. Results

Table 4 lists those curves of Jacobian rank 2 over $\mathbb{Q}$ for which the analysis yielded non-Weierstrass algebraic **non-rational** points. For each curve, the table states:

(1) The identifying label of the curve, linked to its page on the Beta LMFDB.
(2) The model used in the analysis.
(3) The smallest prime that realized the points.
(4) The number field over which the points live.
(5) The set of discovered algebraic points with respect to the model used.

Table 6 enumerates the curves of Jacobian rank 2 with detected Weierstrass points. Table 5 and Table 7 do the same for the Jacobian rank 1 curves. However, instead of simply stating the minimal prime, we also state the value D for which the computation was performed over $\mathbb{Q}(\sqrt{-D})$. Table 1 summarizes the results. Note that the sum of each row is greater than the number of unique models because some curves had points in multiple categories. Tables 4 and 5 are in the appendix and Tables 6 and 7 are on the Supplementary Tables file on GitHub [**Fin25**].

As discussed in the computations sections, some modular curves with different labels share a model. We remark on labeling conventions here [**Cola**]. For coarse modular curves X_H, labels have the form N.i.g.c.n, which denote respectively the level, index, genus, Gassmann class identifier, and an integer to distinguish non-conjugate subgroups where the first four identifiers are the same. Table 8 in the Supplementary Tables on GitHub lists each label whose model realized quadratic points with any other modular curves in the database that share a model [**Fin25**]. This includes both exceptional isomorphisms and fine modular curves.

Any such curve has level divisible by that of the curve used in the main results. For example, our first curve is `14.48.2.g.1`. In our analysis, we used the model $y^2 = -x^6 + 83x^4 - 19x^2 + 1$. This model can also represent several other curves, including `28.96.2-14.g.1.1`, `42.96.2-14.g.1.1`, `56.96.2-14.g.1.1`, and `70.96.2-14.g.1.1`. We can add these as evidence for Conjecture 1.4.

TABLE 1. Count of genus 2 curves for which a point with the given characteristic was found, by rank of the Jacobian

Rank	x-coordinate irrational, $y \neq 0$	x-coordinate rational, y-coordinate irrational	Weierstrass	None
2	6	3	14	10
1	42	8	39	65

6. Examples

6.1. $X_{ns}^{+}(15)$**.** We highlight points found on the modular curve $X_{ns}^{+}(15)$, which has LMFDB label `15.60.2.d.1`. The model we use is $y^2 = -45x^6 + 75x^4 - 15x^2 + 1$. The primes less than 100 of good, ordinary reduction for the curve are 7, 13, 19, 31, 37, 43, 61, 67, 73, and 97. For most of these primes, we recover no notable mock rational points. However, for $p \in \{19, 31, 61\}$ we observe four points: $\{(\pm\frac{\sqrt{5}}{5}, \pm\frac{4}{5})\}$. The sizes of the relevant outputs are in Table 2.

6.2. A Rank 1 Example. We highlight points over different number fields found on the modular curve `16.24.2.f.1`, using model $y^2 = 4x^6 + 6x^4 + 4x^2 + 1$.

TABLE 2. Analysis Data for $X_{ns}^{+}(15)$

prime	# recovered rational points	# mock rational p-adic points
19	14	284
31	14	404
61	14	780

TABLE 3. Analysis Data for Curve `16.24.2.f.1`

prime	number field	# mock rational p-adic points	Algebraic points
5	$\mathbb{Q}(\sqrt{-6})$	66	$(\pm i, \pm i)$
37	$\mathbb{Q}(\sqrt{-7})$	566	$(\pm i, \pm i), (\pm\frac{2\sqrt{-7}}{7}, \pm\frac{5\sqrt{-7}}{49})$

The Jacobian of this curve has rank 1. The results for two iterations of the analysis are in Table 3. Both cases recovered the two known rational points. Note that the points $(\pm i, \pm i)$ appeared both times, but the points over $\mathbb{Q}(\sqrt{-7})$ only appeared in the latter. A given viable field and prime will not necessarily capture all algebraic points.

6.3. A Return to $X_0(37)$. We conclude this section by revisiting an example from the beginning of this paper. A motivating example for performing the analysis of curves with Jacobian rank 1 was $X_0(37)$, which has label `37.38.2.a.1`. Recall it was found to have points over both $\mathbb{Q}(i)$ and $\mathbb{Q}(\sqrt{-3})$. In this work, we found further points over $\mathbb{Q}(\sqrt{-7})$.

Using the model $y^2 = -x^6 - 9x^4 - 11x^2 + 37$, we can run the analysis with $p = 11$ over $\mathbb{Q}(\sqrt{-7})$. In this case, we observe the following set of points:

$$\{(\pm\sqrt{-7}, \pm 4), (\pm\frac{\sqrt{-7}}{3}, \pm\frac{172}{27})\}.$$

The interested reader can find more about the quadratic points on $X_0(37)$ in Balakrishnan and Mazur's 2024 paper regarding Ogg's Torsion Conjecture [**BM25**] and an earlier paper of Box [**Box21**].

7. Future Work

We have gathered extensive data regarding quadratic points that appear during quadratic Chabauty computations for modular curves. We conclude with some remarks regarding additional analyses and questions to explore. These questions emerged through conversations with Kiran Kedlaya and the anonymous referees.

First, a potential counterexample to the conjecture could be found in the following setup. Given a rational Heegner point on one of the elliptic curves E_1 or E_2 in the decomposition of the Jacobian of the genus 2 curve, this could pull back to a quadratic point on X. If this point satisfies the local height conditions required to be a mock rational point, it could be possible to appear in the quadratic Chabauty analysis while not having the properties of the points in the conjecture. These examples, if they exist, would be rare. It is worth exploring if they are possible, and to find them if so.

We also have two questions immediately from looking at the tables:

QUESTION 7.1. All points found were over quadratic or biquadratic fields. Does this generalize to modular curves of different genus or ranks of the Jacobian, or to other hyperelliptic curves in general?

QUESTION 7.2. It is also the case that upon observing one algebraic point, we always find the other points in its Galois orbit. Once again, how does this generalize?

Another interesting set of questions question pertains to the Mordell-Weil group of the Jacobian.

QUESTION 7.3. Could some of the algebraic points found be linearly equivalent to those in the Mordell-Weil group of the Jacobian over the base field? If they have the same heights away from p, could this explain their appearance in the analysis?

A related but not entirely equivalent question is the following:

QUESTION 7.4. If we find algebraic points over number fields for a given curve, how does the Mordell-Weil group change when computed over these fields or their compositum versus the rationals?

8. Appendix: Tables

TABLE 4. Rank 2 Non-Weierstrass Points

Label	Model $y^2 = f(x)$	p	Number Field	Points
14.48.2.g.1	$-x^6 + 83x^4 - 19x^2 + 1$	37	$\mathbb{Q}(\sqrt{-3}, \sqrt{21})$	$(\pm\frac{\sqrt{-3}}{3}, \pm\frac{8\sqrt{21}}{9})$
15.60.2.d.1	$-45x^6 + 75x^4 - 15x^2 + 1$	19	$\mathbb{Q}(\sqrt{5})$	$(\frac{\sqrt{5}}{5}, \pm\frac{4}{5})$
16.48.2.a.1	$x^6 - 5x^4 - 5x^2 + 1$	11	$\mathbb{Q}(\sqrt{-2})$	$(\pm 1, \pm 2\sqrt{2})$
18.72.2.f.1	$9x^6 - 99x^4 + 27x^2 - 1$	7	$\mathbb{Q}(\sqrt{-3})$	$(\pm\frac{\sqrt{-3}}{3}, \pm\frac{8\sqrt{-3}}{3})$
28.48.2.j.1	$x^6 - 83x^4 + 19x^2 - 1$	19	$\mathbb{Q}(\sqrt{-3}, \sqrt{7})$	$(\pm\frac{\sqrt{-3}}{3}, \pm\frac{8\sqrt{-21}}{9})$
36.72.2.d.1	$-9x^6 + 99x^4 - 27x^2 + 1$	13	$\mathbb{Q}(\sqrt{-3}, \sqrt{3})$	$(\pm\frac{\sqrt{-3}}{3}, \pm\frac{8\sqrt{3}}{3})$
48.48.2.dk.1	$12x^6 + 54x^4 + 48x^2 + 12$	73	$\mathbb{Q}(\sqrt{3})$	$(0, \pm 2\sqrt{3})$
48.48.2.dm.1	$24x^6 + 54x^4 - 24x^2 + 3$	73	$\mathbb{Q}(\sqrt{3})$	$(0, \pm\sqrt{3})$
48.48.2.dw.1	$27x^6 - 72x^4 + 54x^2 - 8$	73	$\mathbb{Q}(\sqrt{6})$	$(\pm\frac{\sqrt{6}}{9}, \pm 2)$

TABLE 5. Rank 1 Non-Weierstrass Points

Label	Model $y^2 = f(x)$	(D, p)	Number Field	Points
11.66.2.a.1	$11x^6 + 11x^4 - 7x^2 + 1$	[7,23]	$\mathbb{Q}(\sqrt{-7})$	$(\pm\frac{\sqrt{-7}}{3}, \frac{76}{27})$
16.24.2.f.1	$4x^6 + 6x^4 + 4x^2 + 1$	[6,5]	$\mathbb{Q}(i)$	$(\pm i, \pm i)$
16.24.2.f.1	$4x^6 + 6x^4 + 4x^2 + 1$	[7,37]	$\mathbb{Q}(\sqrt{-7})$	$(\pm\frac{2\sqrt{-7}}{7}, \pm\frac{5\sqrt{-7}}{49})$
16.48.2.bm.1	$8x^6 + 16x^4 + 9x^2 + 1$	[5,41]	$\mathbb{Q}(\sqrt{-5})$	$(\pm\frac{3\sqrt{-5}}{7}, \pm\frac{62}{343})$
16.48.2.bx.1	$-x^6 + 7x^4 - 7x^2 + 1$	[1,5]	$\mathbb{Q}(i)$	$(\pm i, \pm 4)$
16.48.2.bx.1	$-x^6 + 7x^4 - 7x^2 + 1$	[17,7]	$\mathbb{Q}(\sqrt{-17})$	$(\pm\sqrt{-17}, \pm 84)$
16.48.2.c.1	$-x^6 - 5x^4 + 5x^2 + 1$	[7,11]	$\mathbb{Q}(\sqrt{-7})$	$(\pm\sqrt{-7}, \pm 8)$
16.48.2.c.1	$-x^6 - 5x^4 + 5x^2 + 1$	[1,17]	$\mathbb{Q}(i, \sqrt{2})$	$(\pm i, \pm 2\sqrt{-2})$
16.48.2.c.2	$-2x^6 - 10x^4 + 10x^2 + 2$	[19,11]	$\mathbb{Q}(\sqrt{-19})$	$(\pm\frac{9\sqrt{-19}}{19}, \pm\frac{680\sqrt{-19}}{361})$
16.48.2.c.2	$-2x^6 - 10x^4 + 10x^2 + 2$	[1,17]	$\mathbb{Q}(i)$	$(\pm i, \pm 4i)$
16.48.2.y.1	$-8x^6 + 18x^4 - 8x^2 + 1$	[15,17]	$\mathbb{Q}(\sqrt{2})$	$(\pm\frac{\sqrt{2}}{2}, \pm\frac{\sqrt{2}}{2})$
18.54.2.e.1	$-3x^6 + 18x^4 - 3x^2 + 4$	[3,7]	$\mathbb{Q}(\sqrt{-3})$	$(\pm\frac{\sqrt{-3}}{3}, \pm\frac{8}{3})$, $(\pm\sqrt{-3}, \pm 16)$
20.30.2.1.1	$256x^6 + 113x^4 + 18x^2 + 1$	[7,11]	$\mathbb{Q}(\sqrt{-7})$	$(\pm\frac{\sqrt{-7}}{7}, \pm\frac{2\sqrt{-7}}{49})$
24.36.2.a.1	$8x^6 + 1$	[7,37]	$\mathbb{Q}(\sqrt{-7})$	$(\pm\frac{2\sqrt{-7}}{7}, \pm\frac{13\sqrt{-7}}{49})$
				Continued on next page

Table 5 – continued from previous page

Label	Model	(D,p)	Number Field	Points
24.36.2.bo.1	$72x^6+48x^4+12x^2+1$	[10,13]	$\mathbb{Q}(\sqrt{-3})$	$(\pm\frac{\sqrt{-3}}{3},\pm\frac{\sqrt{-3}}{3})$
24.36.2.bo.1	$72x^6+48x^4+12x^2+1$	[2,19]	$\mathbb{Q}(\sqrt{-2})$	$(\pm\frac{\sqrt{-2}}{2},\pm\sqrt{-2})$
24.36.2.bv.1	$-2x^6+2$	[1,13]	$\mathbb{Q}(i)$	$(\pm i,\pm 2)$
24.36.2.ev.1	$x^6-13x^4+39x^2-27$	[3,13]	$\mathbb{Q}(\sqrt{-3})$	$(0,\pm 3\sqrt{-3})$
24.36.2.ev.1	$x^6-13x^4+39x^2-27$	[3,19]	$\mathbb{Q}(\sqrt{-3},\sqrt{-2})$	$(\pm\sqrt{-3},\pm 12\sqrt{-2})$
24.36.2.gc.1	$-6x^6+6$	[1,13]	$\mathbb{Q}(\sqrt{3},i)$	$(\pm i,\pm 2\sqrt{3})$
24.36.2.gh.1	$-6x^6-78x^4+78x^2+6$	[1,5]	$\mathbb{Q}(i)$	$(\pm i,\pm 12i)$
24.36.2.gj.1	$-9x^6-21x^4+21x^2+9$	[1,5]	$\mathbb{Q}(i,\sqrt{-6})$	$(\pm i,\pm 2\sqrt{-6})$
24.36.2.gk.1	$x^6+13x^4-13x^2-1$	[1,5]	$\mathbb{Q}(i,\sqrt{6})$	$(\pm i,\pm 2\sqrt{6})$
24.36.2.gk.1	$x^6+13x^4-13x^2-1$	[1,13]	$\mathbb{Q}(i)$	$(0,\pm i)$
24.36.2.gm.1	$6x^6+14x^4-14x^2-6$	[1,5]	$\mathbb{Q}(i)$	$(\pm i,\pm 4)$
24.36.2.gm.1	$6x^6+14x^4-14x^2-6$	[11,5]	$\mathbb{Q}(\sqrt{-11})$	$(\pm\frac{\sqrt{-11}}{5},\pm\frac{192}{125})$
24.36.2.x.1	$2x^6-2$	[1,13]	$\mathbb{Q}(i)$	$(\pm i,\pm 2i)$
24.36.2.x.1	$2x^6-2$	[17,7]	$\mathbb{Q}(\sqrt{-17})$	$(\pm\frac{15\sqrt{-17}}{17},\pm\frac{4774\sqrt{-17}}{289})$
24.72.2.ba.1	$24x^6+12x^4+18x^2+9$	[8,43]	$\mathbb{Q}(\sqrt{-2})$	$(\pm\frac{\sqrt{-2}}{4},\pm\frac{21}{8})$
24.72.2.cm.1	$-9x^6+3x^4-3x^2+1$	[1,13]	$\mathbb{Q}(i)$	$(\pm i,\pm 4)$
24.72.2.da.1	$-72x^6+12x^4-6x^2+1$	[8,43]	$\mathbb{Q}(\sqrt{-2})$	$(\pm\frac{\sqrt{-2}}{2},\pm 4)$
24.72.2.db.1	$9x^6-6x^4+12x^2-8$	[10,13]	$\mathbb{Q}(\sqrt{-10})$	$(\pm\frac{\sqrt{-10}}{15},\pm\frac{208\sqrt{-10}}{225})$
24.72.2.db.1	$9x^6-6x^4+12x^2-8$	[2,19]	$\mathbb{Q}(\sqrt{-2})$	$(0,\pm 2\sqrt{-2})$, $(\pm\sqrt{-2},\pm 8\sqrt{-2})$
24.72.2.fy.1	$216x^6+72x^4+12x^2+1$	[17,13]	$\mathbb{Q}(\sqrt{-3})$	$(\pm\frac{\sqrt{-3}}{3},\pm\sqrt{-3})$
24.72.2.hy.1	$-6x^6+12x^4-12x^2+6$	[1,5]	$\mathbb{Q}(i)$	$(\pm i,\pm 6)$
24.72.2.ja.1	$24x^6-24x^4+12x^2-3$	[2,11]	$\mathbb{Q}(\sqrt{-2})$	$(\pm\frac{\sqrt{-2}}{2},\pm 3\sqrt{-2})$ $(\pm\frac{1}{2},\pm\frac{3\sqrt{-2}}{4})$
24.72.2.t.1	$-24x^6+12x^4-18x^2+9$	[6,31]	$\mathbb{Q}(\sqrt{-6})$	$(\pm\frac{\sqrt{-6}}{2},\pm 12)$
24.72.2.u.1	$9x^6+9x^4+3x^2+3$	[1,13]	$\mathbb{Q}(i)$	$(\pm 2i,\pm 21i)$
28.32.2.a.1	$2x^6+46x^4+30x^2-14$	[19,5]	$\mathbb{Q}(\sqrt{-19})$	$(\pm\sqrt{-19},\pm 48)$
28.32.2.a.1	$2x^6+46x^4+30x^2-14$	[2,17]	$\mathbb{Q}(\sqrt{-2})$	$(\pm\frac{3\sqrt{-2}}{2},\pm\frac{49}{2})$
28.48.2.g.1	$49x^6+70x^4+25x^2+4$	[17,13]	$\mathbb{Q}(\sqrt{-17})$	$(\pm\frac{\sqrt{-17}}{7},\pm\frac{64}{49})$
28.48.2.g.1	$49x^6+70x^4+25x^2+4$	[67,19]	$\mathbb{Q}(\sqrt{7})$	$(\pm\frac{\sqrt{7}}{7},\pm\frac{8\sqrt{7}}{7})$
30.30.2.b.1	$150x^6-690x^4+546x^2-6$	[19,17]	$\mathbb{Q}(\sqrt{-15})$	$(\pm\frac{\sqrt{-15}}{5},\pm\frac{32\sqrt{-15}}{5})$
36.24.2.a.1	x^6+108	[15,19]	$\mathbb{Q}(\sqrt{-3})$	$(\pm\sqrt{-3},\pm 9)$
36.54.2.i.1	$x^6+6x^4+9x^2+16$	[3,13]	$\mathbb{Q}(\sqrt{-3})$	$(\pm\sqrt{-3},\pm 4)$
37.38.2.a.1	$-x^6-9x^4-11x^2+37$	[3,7]	$\mathbb{Q}(\sqrt{-3})$	$(\pm\sqrt{-3},\pm 4)$, $(\pm 3\sqrt{-3},\pm 116)$
37.38.2.a.1	$-x^6-9x^4-11x^2+37$	[7,11]	$\mathbb{Q}(\sqrt{-7})$	$(\pm\sqrt{-7},\pm 4)$ $(\pm\frac{\sqrt{-7}}{3},\pm\frac{172}{27})$
38.40.2.b.1	$4x^6+64x^4+304x^2+361$	[3,13]	$\mathbb{Q}(\sqrt{-3})$	$(\pm\frac{19\sqrt{-3}}{12},\pm\frac{437\sqrt{-3}}{288})$
40.30.2.e.1	$25x^6-115x^4+91x^2-1$	[1,13]	$\mathbb{Q}(i)$	$(0,\pm i)$
40.30.2.f.1	$x^6-91x^4+115x^2-25$	[1,13]	$\mathbb{Q}(i)$	$(0,\pm 5i)$
45.54.2.c.1	$5x^6+21x^4-9x^2-1$	[1,13]	$\mathbb{Q}(i)$	$(0,\pm i)$
48.48.2.ez.1	$3x^6-21x^4+21x^2-3$	[1,13]	$\mathbb{Q}(\sqrt{-3},i)$	$(\pm i,\pm 4\sqrt{-3})$
56.48.2.l.1	$2x^6+25x^4+140x^2+196$	[17,13]	$\mathbb{Q}(\sqrt{14})$	$(\pm\sqrt{14},\pm 112)$
60.30.2.p.1	$-3x^6-5x^4-225x^2-3375$	[17,7]	$\mathbb{Q}(\sqrt{-5})$	$(\pm\sqrt{-5},\pm 20\sqrt{-5})$
60.30.2.p.1	$-3x^6-5x^4-225x^2-3375$	[15,19]	$\mathbb{Q}(\sqrt{-15})$	$(\pm\frac{\sqrt{-15}}{6},\pm\frac{355\sqrt{-15}}{24})$
60.48.2.f.1	$9x^6+60x^4+25x^2-250$	[13,17]	$\mathbb{Q}(\sqrt{-15})$	$(\pm\frac{\sqrt{-15}}{3},\pm\frac{10\sqrt{-15}}{3})$
70.30.2.b.1	$x^6-98x^4+833x^2-10976$	[7,11]	$\mathbb{Q}(\sqrt{-7})$	$(\pm\sqrt{-7},\pm 56\sqrt{-7})$

References

[AKMJ+24] Nikola Adžaga, Timo Keller, Philippe Michaud-Jacobs, Filip Najman, Ekin Ozman, and Borna Vukorepa, *Computing quadratic points on modular curves $X_0(N)$*, Math. Comp. **93** (2024), no. 347, 1371–1397, DOI 10.1090/mcom/3902. MR4708039

[Bal] Jennifer Balakrishnan. Sagemath code. URL: `https://github.com/jbalakrishnan/QC_bielliptic`.

[BBBM21] Jennifer S. Balakrishnan, Amnon Besser, Francesca Bianchi, and J. Steffen Müller. Explicit quadratic Chabauty over number fields. *Israel J. Math.*, 243(1):185–232, 2021. DOI 10.1007/s11856-021-2158-5.

[BBK10] Jennifer S. Balakrishnan, Robert W. Bradshaw, and Kiran S. Kedlaya. Explicit Coleman integration for hyperelliptic curves. In *Algorithmic number theory*, volume 6197 of *Lecture Notes in Comput. Sci.*, pages 16–31. Springer, Berlin, 2010. DOI 10.1007/978-3-642-14518-6_6

[BCP97] Wieb Bosma, John Cannon, and Catherine Playoust, *The Magma algebra system. I. The user language*, J. Symbolic Comput. **24** (1997), no. 3-4, 235–265, DOI 10.1006/jsco.1996.0125. Computational algebra and number theory (London, 1993). MR1484478

[BD18] Jennifer S. Balakrishnan and Netan Dogra, *Quadratic Chabauty and rational points, I: p-adic heights*, Duke Math. J. **167** (2018), no. 11, 1981–2038, DOI 10.1215/00127094-2018-0013. With an appendix by J. Steffen Müller. MR3843370

[BD21] Jennifer S. Balakrishnan and Netan Dogra, *Quadratic Chabauty and rational points II: Generalised height functions on Selmer varieties*, Int. Math. Res. Not. IMRN **15** (2021), 11923–12008, DOI 10.1093/imrn/rnz362. MR4294137

[BDM+19] Jennifer Balakrishnan, Netan Dogra, J. Steffen Müller, Jan Tuitman, and Jan Vonk, *Explicit Chabauty-Kim for the split Cartan modular curve of level 13*, Ann. of Math. (2) **189** (2019), no. 3, 885–944, DOI 10.4007/annals.2019.189.3.6. MR3961086

[BGG23] Josha Box, Stevan Gajović, and Pip Goodman, *Cubic and quartic points on modular curves using generalised symmetric Chabauty*, Int. Math. Res. Not. IMRN **7** (2023), 5604–5659, DOI 10.1093/imrn/rnab358. MR4565697

[Bia20] Francesca Bianchi, *Quadratic Chabauty for (bi)elliptic curves and Kim's conjecture*, Algebra & Number Theory **14** (2020), no. 9, 2369–2416, DOI 10.2140/ant.2020.14.2369. MR4172711

[BKK11] Jennifer S. Balakrishnan, Kiran S. Kedlaya, and Minhyong Kim, *Appendix and erratum to "Massey products for elliptic curves of rank 1" [MR2629986]*, J. Amer. Math. Soc. **24** (2011), no. 1, 281–291, DOI 10.1090/S0894-0347-2010-00675-3. MR2726605

[BM25] Jennifer S. Balakrishnan and Barry Mazur, *Ogg's torsion conjecture: fifty years later*, Bull. Amer. Math. Soc. (N.S.) **62** (2025), no. 2, 235–268, DOI 10.1090/bull/1851. With an appendix by Netan Dogra. MR4885858

[Box21] Josha Box, *Quadratic points on modular curves with infinite Mordell-Weil group*, Math. Comp. **90** (2021), no. 327, 321–343, DOI 10.1090/mcom/3547. MR4166463

[BPa] Francesca Bianchi and Oana Padurariu. MAGMA code. URL: `https://github.com/oana-adascalitei/MWSieveForDatabase`.

[BPb] Francesca Bianchi and Oana Padurariu. SageMath code. URL: `https://github.com/bianchifrancesca/QC_bielliptic`.

[BP24] Francesca Bianchi and Oana Padurariu, *Rational points on rank 2 genus 2 bielliptic curves in the LMFDB*, LuCaNT: LMFDB, computation, and number theory, Contemp. Math., vol. 796, Amer. Math. Soc., Providence, RI, 2024, pp. 215–242, DOI 10.1090/conm/796/16003. MR4732689

[BSS+16] Andrew R. Booker, Jeroen Sijsling, Andrew V. Sutherland, John Voight, and Dan Yasaki, *A database of genus-2 curves over the rational numbers*, LMS J. Comput. Math. **19** (2016), no. suppl. A, 235–254, DOI 10.1112/S146115701600019X. MR3540958

[Cha41] Claude Chabauty, *Sur les points rationnels des courbes algébriques de genre supérieur à l'unité* (French), C. R. Acad. Sci. Paris **212** (1941), 882–885. MR4484

[Cola] The LMFDB Collaboration. The L-functions and modular forms database. URL: `http://www.lmfdb.org`.

[Colb] The LMFDB Collaboration. The L-functions and modular forms database, Beta version. URL: `http://beta.lmfdb.org`.

[Col85a] Robert F. Coleman, *Effective Chabauty*, Duke Math. J. **52** (1985), no. 3, 765–770, DOI 10.1215/S0012-7094-85-05240-8. MR808103

[Col85b] Robert F. Coleman, *Torsion points on curves and p-adic abelian integrals*, Ann. of Math. (2) **121** (1985), no. 1, 111–168, DOI 10.2307/1971194. MR782557

[Fal83] G. Faltings, *Endlichkeitssätze für abelsche Varietäten über Zahlkörpern* (German), Invent. Math. **73** (1983), no. 3, 349–366, DOI 10.1007/BF01388432. MR718935

[Fin25] Kate Finnerty. SageMath code, 2025. URL: `https://github.com/kfinnerty19/QC_bielliptic`.

[Kim09] Minhyong Kim, *The unipotent Albanese map and Selmer varieties for curves*, Publ. Res. Inst. Math. Sci. **45** (2009), no. 1, 89–133, DOI 10.2977/prims/1234361156. MR2512779

[Liu94] Qing Liu, *Conducteur et discriminant minimal de courbes de genre* 2 (French), Compositio Math. **94** (1994), no. 1, 51–79. MR1302311

[Maz77] B. Mazur, *Modular curves and the Eisenstein ideal*, Inst. Hautes Études Sci. Publ. Math. **47** (1977), 33–186 (1978). With an appendix by Mazur and M. Rapoport. MR488287

[Mer96] Loïc Merel, *Bornes pour la torsion des courbes elliptiques sur les corps de nombres* (French), Invent. Math. **124** (1996), no. 1-3, 437–449, DOI 10.1007/s002220050059. MR1369424

[Mor23] L. J. Mordell, *On the rational solutions of the indeterminate equations of the third and fourth degrees*, Proc. Cambridge Philos. Soc. **21** (1922/23), 179–192. MR4656011

[OS19] Ekin Ozman and Samir Siksek, *Quadratic points on modular curves*, Math. Comp. **88** (2019), no. 319, 2461–2484, DOI 10.1090/mcom/3407. MR3957901

[Roe24] David Roe. Private communication. Harvard University CMSA Math and Machine Learning Program Zulip Server, 2024.

[The24] The Sage Developers. *SageMath, the Sage Mathematics Software System (Version 9.3)*, 2024. `https://www.sagemath.org`.

[Wet97] Joseph Loebach Wetherell, *Bounding the number of rational points on certain curves of high rank*, ProQuest LLC, Ann Arbor, MI, 1997. Thesis (Ph.D.)–University of California, Berkeley. MR2696280

KATE FINNERTY
Current address: Boston University, 665 Commonwealth Avenue, Boston, MA 02215
Email address: `ksfinn@bu.edu`

Contemporary Mathematics
Volume **840**, 2026
https://doi.org/10.1090/conm/840/16815

Gluing genus 1 and genus 2 curves along ℓ-torsion

Pitchayut Saengrungkongka and Noah Walsh

ABSTRACT. Let Y be a genus 2 curve over $\mathbb{Q}$. We provide a method to systematically search for possible candidates of a prime $\ell \geq 3$ and a genus 1 curve X for which there exists a genus 3 curve Z over $\mathbb{Q}$ whose Jacobian is, up to quadratic twist, (ℓ, ℓ, ℓ)-isogenous to the product of Jacobians of X and Y, building on the work by Hanselman, Schiavone, and Sijsling for $\ell = 2$. We find several such pairs (X, Y) for prime ℓ up to 13. We also improve their numerical gluing algorithm, allowing us to successfully glue genus 1 and genus 2 curves along their 13-torsion.

1. Introduction

The study of curves is a central topic in arithmetic geometry. Exhaustive lists of curves have long been a useful tool in number theory, starting with the Antwerp tables that contain elliptic curves up to conductor 200 [**BK75**], continuing with Cremona's tables of elliptic curves [**Cre06**], and persisting today with the collections of elliptic curves and genus 2 curves in the LMFDB [**LMF24**].

A large number of higher genus curves can be constructed by gluing curves of smaller genera. For any two curves X and Y of genera g_X and g_Y, the process of ***gluing*** produces a curve Z of genus $g_X + g_Y$ such that the Jacobian of Z, denoted $\mathrm{Jac}(Z)$, is isogenous to the product $\mathrm{Jac}(X) \times \mathrm{Jac}(Y)$. In other words, $\mathrm{Jac}(Z)$ is a quotient of $\mathrm{Jac}(X) \times \mathrm{Jac}(Y)$ by a finite subgroup G. If G is a subgroup of the product of the n-torsion subgroups $\mathrm{Jac}(X)[n]$ and $\mathrm{Jac}(Y)[n]$ for a positive integer n, we say that this gluing is ***along n-torsion***.

Given generic curves X and Y, there are many possible choices of subgroups G that give rise to gluings over $\mathbb{C}$, and one can construct such gluings analytically by considering $\mathrm{Jac}(X)$ and $\mathrm{Jac}(Y)$ as complex tori. However, for arithmetic applications, we are interested in curves defined over non-algebraically closed fields such as $\mathbb{Q}$. If we take curves X and Y randomly, then it is very likely there will be no gluing Z that is defined over $\mathbb{Q}$. We are interested in systematically producing pairs of curves that admit a gluing over $\mathbb{Q}$. These curves often have interesting properties that deviate from generic behavior, such as a special torsion structure, a nontrivial endomorphism ring, or an interesting Sato-Tate group.

1.1. Previous Work. The simpler case of gluing two curves of genus 1 (i.e., $1 + 1 = 2$) along n-torsion has been studied in many aspects.

2020 *Mathematics Subject Classification.* Primary 11G30, 14H25, 14H40.

(i) **Criteria for gluings to exist.** Frey and Kani [**FK91**] derived the necessary conditions for two curves to admit a gluing along n-torsion, namely, if E_1 and E_2 are gluable, then there exists an antisymplectic, $\mathrm{Gal}(\overline{\mathbb{Q}}/\mathbb{Q})$-equivariant isomorphism $E_1[n] \to E_2[n]$.

(ii) **Find all gluable pairs.** Given a list of elliptic curves (e.g., all elliptic curves in the L-functions and modular forms database, LMFDB [**LMF24**]), we are interested in systematically searching for all pairs that are gluable. When $n = \ell$ is prime, Cremona and Freitas [**CF22**] gave a complete algorithm for detecting all such antisymplectic isomorphisms and applied their methods to all pairs of elliptic curves in the LMFDB.

(iii) **Parametrize all gluable pairs.** Building on (ii), we ask if one can parameterize all pairs of elliptic curves that are gluable. For a fixed elliptic curve E_1, the space of all elliptic curves E_2 that are gluable is one-dimensional and is a twist of the modular curve $X(n)$. For $n \in \{2, 3, 4, 5\}$, the space is isomorphic to $\mathbb{P}^1$. The details have been worked out by Rubin and Silverberg in [**RS01**,**RS95**,**Sil97**]. Fisher [**Fis14**] derived explicit formulas for $n \in \{7, 11\}$, but in that case there are finitely many E_2's for a fixed E_1. There are also efforts to parametrize *pairs* of gluable elliptic curve along n-torsion for higher n: Fisher [**Fis20**, Cor. 1.3] showed that there are infinitely many pairs of gluable curves for all $n \leq 10$.

(iv) **Computing gluing.** One method of computing gluings over $\mathbb{Q}$ is via analytic techniques over $\mathbb{C}$. In simpler cases, we also have explicit formulas. Howe, Leprévost, and Poonen [**HLP00**, Prop. 4] gave an explicit formula for $n = 2$, and Bröker et. al. [**BHLS15**, §A.1] gave a formula for $n = 3$. As n grows larger, the explicit formulas quickly become unwieldy.

Genus 1 plus genus 2 gluing (i.e., $1 + 2 = 3$) has not been studied as widely. Ritzenthaler and Romagny [**RR18**] gave a formula for recovering the equation of the genus 2 factor of a genus 1 plus genus 2 gluing along 2-torsion, provided that the genus 1 factor is known. Then Hanselman, Schiavone, and Sijsling [**HSS21**] gave a comprehensive algorithm and explicit formula for gluing along 2-torsion. They answer some of the questions above for gluing genus 1 and 2 curves as follows.

- To answer (i), they [**HSS21**, §1] provided a concrete criterion for a gluing along ℓ-torsion to exist for each prime ℓ, which we recall in Section 2.
- To answer (iv), they [**HSS21**, §2.1] outlined the analytic algorithm to compute the gluing along ℓ-torsion, which was implemented in [**HSS20**]. In the case of $\ell = 2$, they also provided an explicit formula to glue genus 1 and 2 curves along 2-torsion.
- To answer (iii), if $\ell = 2$, for a genus two curve Y, Hanselman [**Han20**, §2.2] constructed an infinite family of genus 3 curves resulting from gluing Y with an elliptic curve along their 2-torsion.

  More generally, for a fixed genus 2 curve Y such that there exists an elliptic curve X that admits a gluing with Y along ℓ-torsion, one can reduce the problem of finding all gluable elliptic curves X' to finding elliptic curves X' whose ℓ-torsion is symplectically isomorphic (as a $\mathrm{Gal}(\overline{\mathbb{Q}}/\mathbb{Q})$-module) to the ℓ-torsion of X. See Section 6.2 for more details.

Our work attempts to answer (ii) and improves existing methods in (iv).

1.2. Our Results. This paper considers the problem of gluing two curves of genus 1 and genus 2 along their ℓ-torsion to obtain a curve of genus 3 for a prime ℓ. We focus on the case of $\ell \geq 3$ that has not been as widely studied. There are two key results.

The first key result is that, given a curve Y of genus 2, we demonstrate how to systematically and efficiently search for genus 1 curves X in the LMFDB and corresponding primes ℓ such that X and Y are gluable along ℓ-torsion. More specifically, given a genus 2 curve Y, we first eliminate all but finitely many ℓ's. Then for each such ℓ, we search for elliptic curves X such that there exists a $\mathrm{Gal}(\overline{\mathbb{Q}}/\mathbb{Q})$-stable subgroup $G \subset \mathrm{Jac}(X)[\ell] \times \mathrm{Jac}(Y)[\ell]$ such that the quotient

$$Q = \frac{\mathrm{Jac}(X) \times \mathrm{Jac}(Y)}{G} \tag{1.1}$$

is a principally-polarized abelian variety. The search process culminates in the workflow described in Section 6. We also provide an algorithm to rigorously verify the existence of such G in the generic case, as detailed in Algorithms 4.16 and 5.9. If Q is isomorphic to a Jacobian of some genus 3 curve Z (which happens generically, when Q is not a product of two or more Jacobians), then curves X and Y produce a gluing Z. We use this result to glue a large number of curves in the LMFDB and discover curves with interesting geometric endomorphism rings in Section 7.2.

The second key result is an improvement of the numerical gluing algorithm in [**HSS21**, §2.1] to obtain a more efficient Algorithm 6.2. With this algorithm, we compute a gluing along 13-torsion in 24 minutes in Example 7.2.

1.3. Organization of the Article. In Section 2, we describe the condition under which two curves are gluable, following [**HSS21**]. Then in Section 3, we explain how to use Frobenius elements to filter pairs of gluable curves efficiently. Section 4 explains how to utilize Serre's modularity conjecture to rigorously verify (in the generic case) a part of the gluability condition. Section 5 adapts the symplectic test in [**FK22**] to verify the other part of the gluability condition. We put all the pieces together and describe our current workflow in Section 6, which includes the speedup of the gluing algorithm. Finally, Section 7 lists some examples produced from running our workflow on curves in the LMFDB.

1.4. Notations. For any prime p and rational number $r \neq 0$, let $\nu_p(r)$ be the p-adic valuation of r, i.e., the exponent of p in the prime factorization of r.

For any abelian variety A over a field k and a positive integer n, $A[n]$ denotes the set of n-torsion points over the algebraic closure $\overline{k}$. For any curve X, let N_X denote the conductor of X, and for any prime p not dividing N_X, let $a_{p,X}$ denote the trace of Frobenius at p of X. Additionally, we define $a_{p,X}$ to be 1, -1, or 0 if X is an elliptic curve with split multiplicative, non-split multiplicative, or additive reduction modulo p, respectively.

2. Gluability Conditions

Let $n \geq 2$ be an integer. Let X and Y be smooth curves of any genus over a base field k with characteristic not dividing n. Informally, a gluing of X and Y is a curve Z with an isomorphism $\mathrm{Jac}(Z) \simeq (\mathrm{Jac}(X) \times \mathrm{Jac}(Y))/G$, where G is a subgroup of $(\mathrm{Jac}(X) \times \mathrm{Jac}(Y))[n]$ for some n. Such an isomorphism does not exist for all subgroups G; in what follows, we consider some necessary conditions for $(\mathrm{Jac}(X) \times \mathrm{Jac}(Y))/G$ to be isomorphic to a Jacobian and consider how they translate to conditions on G.

First, the Jacobian of any curve C comes with a ***principal polarization***. This is an isomorphism $\lambda : \mathrm{Jac}(C) \to \mathrm{Jac}(C)^\vee$, satisfying technical conditions laid out in [**Mil08**, §1.11]. We will consider more generally necessary conditions for $(A_1 \times A_2)/G$ to be a Jacobian, where A_1 and A_2 are arbitrary principally polarized abelian varieties. Let $\lambda_1 : A_1 \to A_1^\vee$ and $\lambda_2 : A_2 \to A_2^\vee$ be these polarizations.

For any abelian variety A, a polarization $\lambda : A \to A^\vee$ induces for each n the ***Weil Pairing***

$$e_n^\lambda \colon A[n] \times A[n] \to \mu_n, \tag{2.1}$$

an alternating bilinear form taking values in the n-th roots of unity. (See [**Mil08**, §1.13] for the exact definition.)

The pairings $e_n^{\lambda_1}$ and $e_n^{\lambda_2}$ on A_1 and A_2 induce a pairing

$$\begin{aligned} e_n \colon (A_1 \times A_2)[n] \times (A_1 \times A_2)[n] &\to \mu_n \\ ((P_1, P_2), (Q_1, Q_2)) &\mapsto e_n^{\lambda_1}(P_1, Q_1)\, e_n^{\lambda_2}(P_2, Q_2) \end{aligned} \tag{2.2}$$

coming from the product polarization $\lambda_1 \times \lambda_2 : A_1 \times A_2 \to A_1^\vee \times A_2^\vee \simeq (A_1 \times A_2)^\vee$.

We only consider cases in which the polarization on the quotient comes from the n-th power of this product polarization. (See [**Han20**, Rmk. 1.1.4, Thm. 1.1.10] for why generically one does not gain anything by considering polarizations other than the n-th power of the product.) In order for this to yield a polarization on the quotient, G must be ***isotropic***, i.e., the pairing must vanish on $G \times G$. Furthermore, by [**BCCK24**, Lem. 2.1], G must be maximal with respect to this property in order for the polarization to be principal.

The polarization of a Jacobian is indecomposable, so it is also necessary for the polarization on the quotient to be indecomposable. Therefore, it is also necessary that G be ***indecomposable***, i.e., not of the form $G_1 \times G_2$ for $G_1 \subset A_1$, $G_2 \subset A_2$, as otherwise the polarization will be a product of polarizations on A_1/G_1 and A_2/G_2.

To summarize, G must be an ***indecomposable maximal isotropic subgroup***. Let us now specialize to the case where $A_1 = \mathrm{Jac}(X)$ and $A_2 = \mathrm{Jac}(Y)$ for curves X and Y of genera 1 and 2, respectively. Suppose that G is an indecomposable maximal isotropic subgroup. Then $(\mathrm{Jac}(X) \times \mathrm{Jac}(Y))/G$ is an abelian variety of dimension 3 defined over some extension k' of k. If the polarization of $(\mathrm{Jac}(X) \times \mathrm{Jac}(Y))/G$ is indecomposable, then by [**OU73**], $(\mathrm{Jac}(X) \times \mathrm{Jac}(Y))/G$ is isomorphic to the Jacobian of some curve Z, where Z and the isomorphism are defined over some further extension k'' of k'. Serre proved in the appendix to [**Lau01**] that one may in fact take k'' to be a quadratic extension of k'.

We can now formally define a gluing.

Definition 2.3. An ***(n, n)-gluing*** of the curves X and Y over k is a triple (Z, ψ, G), where Z is a smooth curve over k and a subgroup $G \subseteq \mathrm{Jac}(X)[n] \times \mathrm{Jac}(Y)[n]$ along with an isomorphism of principally polarized abelian varieties

$$\psi : \frac{\mathrm{Jac}(X) \times \mathrm{Jac}(Y)}{G} \xrightarrow{\sim} \mathrm{Jac}(Z). \tag{2.4}$$

In the situation when n equals ℓ, a prime number, [**HSS21**, §1] gives the following description of such subgroups.

PROPOSITION 2.5. [**HSS21**, Prop. 1.18] *A subgroup* $G \subseteq \mathrm{Jac}(X)[\ell] \times \mathrm{Jac}(Y)[\ell]$ *is an indecomposable maximal isotropic subgroup if and only if*

$$G = \{(x, y) \mid \phi(x) = y + H\} \tag{2.6}$$

for some one-dimensional subgroup $H \subset G$ *and some antisymplectic isomorphism* $\phi : \mathrm{Jac}(X)[\ell] \to H^{\perp}/H$.

For most choices of G, the resulting abelian variety $(\mathrm{Jac}(X) \times \mathrm{Jac}(Y))/G$ will not be defined over k. In order for $(\mathrm{Jac}(X) \times \mathrm{Jac}(Y))/G$ to be defined over k, it is necessary that G be Galois-stable. [**HSS21**, Prop. 1.39] worked out what this means in terms of (G, ϕ) in Proposition 2.5.

THEOREM 2.7. [**HSS21**, Prop. 1.39] *Following the notations of Proposition* 2.5, G *is Galois stable if and only if both of the following hold*

(i) H *is Galois-stable; and*
(ii) ϕ *is Galois-equivariant.*

Conversely, from the paragraph preceding Definition 2.3, we have the following converse.

THEOREM 2.8. *Suppose that*

(i) G *is Galois-stable (i.e., satisfies both conditions of Theorem* 2.7*); and*
(ii) *the quotient* $(\mathrm{Jac}(X) \times \mathrm{Jac}(Y))/G$ *is not a product of two or more Jacobians.*

Then there exists a curve Z *such that* $\mathrm{Jac}(Z) \simeq (\mathrm{Jac}(X) \times \mathrm{Jac}(Y))/G$, *and the isomorphism is defined over a quadratic extension of* k.

Condition (ii) of Theorem 2.8 can be verified numerically in terms of a period matrix of $(\mathrm{Jac}(X)\times\mathrm{Jac}(Y))/G$: at most one of its even theta values can vanish. For most of the paper, we focus on searching for pairs of curves (X, Y) over $\mathbb{Q}$ for which there exists a Galois-stable maximal isotropic subgroup $G \subset \mathrm{Jac}(X)[\ell] \times \mathrm{Jac}(Y)[\ell]$ (equivalently, (H, ϕ) satisfying the conditions of Theorem 2.7). For this to hold, we need both of the following to be true:

(i) there exists a $\mathrm{Gal}(\overline{\mathbb{Q}}/\mathbb{Q})$-stable subgroup of $\mathrm{Jac}(Y)[\ell]$.
(ii) $H^{\perp}/H$ and $\mathrm{Jac}(X)[\ell]$ are isomorphic as $\mathrm{Gal}(\overline{\mathbb{Q}}/\mathbb{Q})$-modules, and the isomorphism is antisymplectic.

Fix a genus 2 curve Y. In Section 3, we describe how to rule out elliptic curves that do not satisfy (i) or the isomorphic part of (ii). Then in Section 5, we discuss how to test the remaining condition of (ii): that the isomorphism is antisymplectic.

REMARK 2.9. Even if there exists a Galois-stable maximal isotropic subgroup of $\mathrm{Jac}(X)[\ell] \times \mathrm{Jac}(Y)[\ell]$, there might still not be a gluing. Consider the genus 2 curve Y given by `2646.b.71442.1` on the LMFDB, the elliptic curve X given by `21.a6` on the LMFDB, and $\ell = 3$. In this case, we can show that there exists such subgroup G. However, we have strong numerical evidence that there does not exist a gluing between X and Y. More specifically, the Jacobian of the genus 2 curve is 2-isogenous to a product of two elliptic curves, and quotienting out by G splits the $\mathrm{Jac}(Y)$ factor into a product of two elliptic curves. We have yet to investigate a condition to tell a priori whether $(\mathrm{Jac}(X) \times \mathrm{Jac}(Y))/G$ will be a Jacobian or not.

3. Rational ℓ-torsion Subgroups

Throughout this section, we fix a genus 2 curve Y and study for what primes ℓ there might exist an (ℓ, ℓ)-gluing, and if so, whether there exists such a genus 1 curve X for which X and Y admit an (ℓ, ℓ)-gluing.

It turns out that for a fixed Y, the condition (i) of Theorem 2.7 already rules out the possible primes ℓ to a finite set. We explain the reason in Section 3.1. Once we know ℓ, we present an algorithm that computes the trace of a Frobenius element acting on $H^\perp/H$ in Section 3.2. We explain how this constrains X in Section 3.3.

We now introduce the concept of a Frobenius polynomial. For any prime $p \neq \ell$ and genus two curve Y for which Y has good reduction $\overline{Y}$ modulo p, let K be a number field such that $\mathrm{Jac}(Y)$ has a full ℓ-torsion over K. Pick any prime ideal $\mathfrak{p}$ above p, and let $\mathrm{Frob}_p \in \mathrm{Gal}(K/\mathbb{Q})$ be the Frobenius element corresponding to $\mathfrak{p}$[1]. Then $\mathbb{F}_\mathfrak{p} := \mathcal{O}_K/\mathfrak{p}$ is a finite field, and Frob_p acts on points of $\mathrm{Jac}(Y)$ in a way that

$$\mathrm{Frob}_p((x:y:z)) \equiv (x^p : y^p : z^p) \pmod{\mathfrak{p}} \tag{3.1}$$

Thus, we get an action of Frob_p on $\mathrm{Jac}(\overline{Y})[\ell] \simeq \mathbb{F}_\ell^4$ by a linear map in $\mathrm{GL}_4(\mathbb{F}_\ell)$ with characteristic polynomial congruent modulo ℓ to the ***Frobenius polynomial***, which is of the form

$$F_{Y,p}(T) := T^4 - a_{p,Y}T^3 + a'_{p,Y}T^2 - pa_{p,Y}T + p^2 \in \mathbb{Z}[x], \tag{3.2}$$

independent of ℓ. The coefficients $a_{p,Y}$ and $a'_{p,Y}$ are determined by the number of points of Y over $\mathbb{F}_p$ and $\mathbb{F}_{p^2}$. See [**Mil08**, Chapter II] for more details.

3.1. Finding Possible Primes. The condition that $\mathrm{Jac}(Y)[\ell]$ must have a 1-dimensional Galois-stable subspace H already restricts the possible values of ℓ to a finite set, even without knowing the curve X. Let $\mathcal{L}$ be a set of primes for which H exists. An algorithm to determine a finite superset of $\mathcal{L}$ is studied in [**BCCK24**, §3.1], using Dieulefait's criterion in [**Die02**, §3.1]. Their algorithm only filters out ℓ's not dividing the conductor N_Y and does not do anything for those ℓ's that divide N_Y. Thus, we modify their algorithm so that it can rule out some primes ℓ dividing N_Y as well.

The action of an element $\sigma \in \mathrm{Gal}(\overline{\mathbb{Q}}/\mathbb{Q})$ on $\mathrm{Jac}(Y)[\ell] \in \mathbb{F}_\ell^4$ can be expressed as a matrix in $\mathrm{GL}_4(\mathbb{F}_\ell)$. This induces a representation $\overline{\rho}_Y : \mathrm{Gal}(\overline{\mathbb{Q}}/\mathbb{Q}) \to \mathrm{GL}(\mathrm{Jac}(Y)[\ell])$ of dimension 4. The 1-dimensional $\mathrm{Gal}(\overline{\mathbb{Q}}/\mathbb{Q})$-stable subspace H induces two 1-dimensional representations $\mathrm{Gal}(\overline{\mathbb{Q}}/\mathbb{Q}) \to \mathbb{F}_\ell^\times$:

- on H itself; we let this representation correspond to the character
$$\varepsilon_1 : \mathrm{Gal}(\overline{\mathbb{Q}}/\mathbb{Q}) \to \mathrm{GL}(H) \simeq \mathbb{F}_\ell^\times.$$
- on $\mathrm{Jac}(Y)[\ell]/H^\perp$ (by Galois-equivariance of the Weil pairing); we let this representation correspond to the character
$$\varepsilon_2 : \mathrm{Gal}(\overline{\mathbb{Q}}/\mathbb{Q}) \to \mathrm{GL}(\mathrm{Jac}(Y)[\ell]/H^\perp) \simeq \mathbb{F}_\ell^\times.$$

By Galois-equivariance of the Weil pairing, we have $\varepsilon_1\varepsilon_2 = \chi_\ell$, where χ_ℓ is the mod-ℓ cyclotomic character (the unique character such that $\sigma(\zeta_\ell) = \zeta_\ell^{\chi_\ell(\sigma)}$). Because $\mathbb{F}_\ell^\times$ is abelian, by the Kronecker-Weber theorem, ε_i $(i = 1, 2)$ must factor through $\mathrm{Gal}(\mathbb{Q}(\zeta_e)/\mathbb{Q}) \simeq (\mathbb{Z}/e\mathbb{Z})^\times$ for some positive integer e (where ζ_e is a primitive e-th root of unity). The smallest such e is the ***conductor*** of ε_i.

[1] Although Frob_p is defined only up to conjugacy class, the choice of which Frob_p we pick will not matter.

PROPOSITION 3.3. *Let N_Y be the conductor of Y, and let d be the largest integer such that d^2 divides $N_Y/\ell^{\nu_\ell(N_Y)}$. Define*

$$D = \begin{cases} \ell d & \text{if } \ell \text{ divides } N_Y, \\ d & \text{if } \ell \text{ does not divide } N_Y. \end{cases} \tag{3.4}$$

Then the conductor of at least one of ε_1 and ε_2 divides D.

PROOF. First, we show in general that the conductor of ε_1 and ε_2 both divide ℓd. Let $I_\ell = \mathrm{Gal}(\overline{\mathbb{Q}}_\ell/\mathbb{Q}_\ell^{\mathrm{unr}})$ be the inertia group. By [**Ser87**, §2.3], there exist α and β such that if $\varepsilon_1' = \varepsilon_1\chi_\ell^{-\alpha}$ and $\varepsilon_2' = \varepsilon_2\chi_\ell^{-\beta}$, then both ε_1' and ε_2' are characters from $\mathrm{Gal}(\overline{\mathbb{Q}}/\mathbb{Q}) \to \mathbb{F}_\ell^\times$ unramified at ℓ. Since $\varepsilon_1\varepsilon_2 = \chi_\ell$, we must have $\alpha + \beta \equiv 1 \pmod{\ell - 1}$ and $\varepsilon_1'\varepsilon_2'$ is the trivial character. In particular, ε_1' and ε_2' have the same conductor; let it be k.

Therefore, k^2 divides the conductor of the mod-ℓ representation $\overline{\rho}_Y$, so k^2 divides $N_Y/\ell^{\nu_\ell(N_Y)}$, so k divides d. Thus, the conductors of ε_1' and ε_2' both divide d, which implies that the conductors of both ε_1 and ε_2 divide ℓd.

Now, assume ℓ does not divide N_Y. By Raynaud's result [**Ray74**, Cor. 3.4.4], we have $\{\alpha, \beta\} = \{0, 1\}$. Therefore, at least one of ε_1 and ε_2 is unramified away from ℓ, implying that the conductor of ε_1 or ε_2 is the same as that of ε_1' or ε_2'. □

Proposition 3.3 shows that at least one of ε_1 and ε_2, say ε, factors through $\mathrm{Gal}(\mathbb{Q}(\zeta_D)/\mathbb{Q}) \simeq (\mathbb{Z}/D\mathbb{Z})^\times$, inducing a Dirichlet character $\chi : (\mathbb{Z}/D\mathbb{Z})^\times \to \mathbb{F}_\ell^\times$. This narrows down the possible candidates for χ to a finite set.

Furthermore, for any prime p, the image of Frob_p in $\mathrm{Gal}(\mathbb{Q}(\zeta_D)/\mathbb{Q}) \simeq (\mathbb{Z}/D\mathbb{Z})^\times$ is p. Thus, if f is the order of p in $(\mathbb{Z}/D\mathbb{Z})^\times$, then

$$\varepsilon(\mathrm{Frob}_p)^f = \varepsilon\big(\mathrm{Frob}_p^f\big) = \varepsilon\left(p^f\right) = \varepsilon(1) = 1. \tag{3.5}$$

Moreover, $\varepsilon(\mathrm{Frob}_p)$ is an eigenvalue of Frob_p (with eigenvector in H). This means that the polynomial $T - \varepsilon(\mathrm{Frob}_p)$ divides $F_{Y,p}(T)$. In other words, $F_{Y,p}(T)$ and $T^f - 1$ have a common root in $\mathbb{F}_\ell$, and so the resultant $\mathrm{Res}(F_{Y,p}(T), T^f - 1)$ is divisible by ℓ. This gives rise to the following algorithm.

ALGORITHM 3.6. *Input:* a curve Y of genus 2, its conductor N_Y, and a finite nonempty set of primes $\mathcal{P}$, each of which gives good reduction of Y.

Output: a finite superset of $\mathcal{L}$, the set of primes ℓ for which there exists a one-dimensional $\mathrm{Gal}(\overline{\mathbb{Q}}/\mathbb{Q})$-stable subgroup $H \subset \mathrm{Jac}(X)[\ell]$.

(1) Let d be the largest positive integer such that $d^2 \mid N_Y$.
(2) For each prime $p \in \mathcal{P}$, do the following:
 - compute the Frobenius polynomial $F_{Y,p}(T)$; and
 - compute the order f_p of p in $(\mathbb{Z}/d\mathbb{Z})^\times$.
(3) Let $\mathcal{L}_{\mathrm{good}}$ be the set of primes dividing $\gcd_{p\in\mathcal{P}} \mathrm{Res}(F_{Y,p}(T), T^{f_p} - 1)$.
(4) For each prime ℓ dividing N_Y, we run the following test for each prime $p \in \mathcal{P}$:
 - compute the Frobenius polynomial $F_{Y,p}(T)$ and the order g_p of p in $(\mathbb{Z}/\ell d\mathbb{Z})^\times$; and
 - check whether $F_{Y,p}(T)$ and $T^{g_p} - 1$ have a common root in $\mathbb{F}_\ell$.

 Let $\mathcal{L}_{\mathrm{bad}}$ be the set of primes ℓ dividing N_Y that pass the above test for all primes p.
(5) Return $\mathcal{L} = \mathcal{L}_{\mathrm{good}} \cup \mathcal{L}_{\mathrm{bad}}$.

The discussion preceding this algorithm shows that Algorithm 3.6 indeed returns a valid superset of $\mathcal{L}$. Such a superset is always finite: by the Riemann hypothesis for curves, all roots of $F_{Y,p}$ have absolute value $\sqrt{p}$, so $\operatorname{Res}(F_{Y,p}(T), T^{f_p}-1) \neq 0$ for all primes p, and hence $\mathcal{L}_{\text{good}}$ is always finite.

3.2. The Frobenius Action on $H^\perp/H$. Assuming that H exists, we now proceed to study the condition (ii) of Theorem 2.7. We defer the study of the symplectic condition to Section 5 and focus on the condition that $H^\perp/H$ and $\operatorname{Jac}(X)[\ell]$ are isomorphic as $\operatorname{Gal}(\overline{\mathbb{Q}}/\mathbb{Q})$-modules.

For any prime p at which Y has good reduction, the Frobenius element Frob_p acts on $H^\perp/H \simeq \mathbb{F}_\ell^2$ by a matrix in $\operatorname{GL}_2(\mathbb{F}_\ell)$. We define

$$b_p = \text{trace of } \operatorname{Frob}_p \text{ on } H^\perp/H. \tag{3.7}$$

The traces b_p are important data describing the Galois action on $H^\perp/H$, which will allow us to search for a suitable elliptic curve X. See Proposition 3.10 for more details. In this subsection, we will explain how to narrow down the possible values of b_p.

Recall from Section 3.1 that the action of $\operatorname{Gal}(\overline{\mathbb{Q}}/\mathbb{Q})$ on H is determined by a character $\varepsilon : \operatorname{Gal}(\overline{\mathbb{Q}}/\mathbb{Q}) \to \mathbb{F}_\ell^\times$, and ε factors through $\operatorname{Gal}(\mathbb{Q}(\zeta_D)/\mathbb{Q}) \simeq (\mathbb{Z}/D\mathbb{Z})^\times$, where D is defined in (3.4). Thus, ε corresponds to a Dirichlet character $\chi : (\mathbb{Z}/D\mathbb{Z})^\times \to \mathbb{F}_\ell^\times$, and we have $\varepsilon(\operatorname{Frob}_p) = \chi(p)$.

For any prime p, the action of Frob_p onto the one-dimensional subspace H and $\operatorname{Jac}(Y)[\ell]/H^\perp$ must be multiplication by $\chi(p)$ and $p/\chi(p)$ in some order, which are two eigenvalues of Frob_p acting on $\operatorname{Jac}(Y)$. The trace on $H^\perp/H$ must be the sum of the remaining two eigenvalues. Thus,

$$b_p = a_{p,Y} - \chi(p) - \tfrac{p}{\chi(p)} \pmod{\ell}. \tag{3.8}$$

Hence, if one knows χ, one can determine b_p for all p. For a fixed Y, there are finitely many possibilities for $\chi : (\mathbb{Z}/D\mathbb{Z})^\times \to \mathbb{F}_\ell^\times$. We rule out χ by checking that, for any prime q not dividing N_Y, the number $\chi(q)$ is a root of $F_{Y,q}(T)$. This leads to the following algorithm.

ALGORITHM 3.9. *Input.* A genus 2 curve Y, its conductor N_Y, a prime ℓ, and a finite set of primes $\mathcal{Q}$.

Output. A finite list of candidate functions that take a prime number $p \neq \ell$ and output b_p (defined in (3.7)) modulo ℓ.

For each one-dimensional Galois-stable subgroup $H \subset \operatorname{Jac}(Y)[\ell]$, the function that outputs the trace of Frob_p on $H^\perp/H$ must be in the output list (but there might be extraneous functions).

(1) Compute D as defined in (3.4).
(2) Enumerate the set $\mathcal{X}$ of all Dirichlet characters $(\mathbb{Z}/D\mathbb{Z})^\times \to \mathbb{F}_\ell^\times$.
(3) For each prime $q \in \mathcal{Q}$, compute $F_{Y,q}(T)$. Remove from $\mathcal{X}$ any character χ such that $\chi(q)$ is not a root of $F_{Y,q}(T)$.
(4) For each function $\chi \in \mathcal{X}$, we return the function

$$p \mapsto a_{p,Y} - \chi(p) - \tfrac{p}{\chi(p)} \pmod{\ell}.$$

From (3.8), we can see that the output of this algorithm includes at least one function for each possible H. If there are multiple possible H's, the algorithm returns multiple functions, each corresponding to a candidate H.

3.3. Frobenius Traces of Gluable Elliptic Curve. For this entire section, fix a genus 2 curve Y and a prime $\ell \geq 3$ such that there exists a one-dimensional $\mathrm{Gal}(\overline{\mathbb{Q}}/\mathbb{Q})$-stable subgroup $H \subset \mathrm{Jac}(Y)[\ell]$. For any prime p at which Y has good reduction, let b_p denote the trace of Frob_p in $H^\perp/H$ (which we have narrowed down the possibilities in Algorithm 3.9).

Suppose that X is an elliptic curve such that $\mathrm{Jac}(X)[\ell]$ and $H^\perp/H$ are isomorphic as $\mathrm{Gal}(\overline{\mathbb{Q}}/\mathbb{Q})$-modules. By comparing b_p with the trace of Frobenius at p of X (which we denoted $a_{p,X}$), we can rule out some of such X, as detailed in the following proposition.

PROPOSITION 3.10. *Let X be an elliptic curve such that* $\mathrm{Jac}(X)[\ell]$ *and* $H^\perp/H$ *are isomorphic as* $\mathrm{Gal}(\overline{\mathbb{Q}}/\mathbb{Q})$*-modules. Let $p \neq \ell$ be a prime such that Y has good reduction modulo p.*

(i) *If X has good reduction modulo p, then $b_p \equiv a_{p,X} \pmod{\ell}$.*
(ii) *If X has split multiplicative reduction modulo p, then $b_p \equiv 1+p \pmod{\ell}$.*
(iii) *If X has non-split multiplicative reduction modulo p, then $b_p \equiv -(1+p) \pmod{\ell}$.*
(iv) *If $\ell \geq 5$, then X cannot have additive reduction modulo p.*

PROOF.

(i) The Frobenius element Frob_p acts on $H^\perp/H$ and on $X_{\mathbb{F}_p}[\ell]$ by the same matrix in $\mathrm{GL}_2(\mathbb{F}_\ell)$ up to a change of basis. Thus, they have the same trace modulo ℓ.
(ii) By the theory on Tate's curve [**Sil94**, Thm. V.5.3.], there exists $q \in \mathbb{Q}_p$ such that $X_{\overline{\mathbb{Q}}_p} \simeq \overline{\mathbb{Q}_p}^\times/q^{\mathbb{Z}}$ as groups. This isomorphism is Galois equivariant. By considering the ℓ-torsion of both sides, we find that $X_{\overline{\mathbb{Q}}_p}[\ell] \simeq \langle q^{1/\ell}, \zeta_\ell\rangle$ (where ζ_ℓ is the ℓ-root of unity). Therefore, the Frobenius element Frob_p acts on $X_{\mathbb{F}_p}[\ell]$ by matrix $\left(\begin{smallmatrix}1 & * \\ 0 & p\end{smallmatrix}\right)$, so it must act on $H^\perp/H$ by the same matrix in modulo ℓ. Hence, we conclude that $b_p \equiv 1+p \pmod{\ell}$.
(iii) There exists a quadratic twist X' of X such that X' has a split multiplicative reduction modulo ℓ. By (ii), the trace of Frob_p acting on $X'_{\mathbb{F}_p}[\ell]$ is $1+p$. Thus, the trace of Frob_p acting on $X_{\mathbb{F}_p}[\ell]$ is $-(1+p)$, which must be equal to b_p in modulo ℓ.
(iv) If $p \neq \ell \geq 5$ and X has additive reduction modulo p, then we claim that the torsion field $\mathbb{Q}(X[\ell])$ is ramified at p. To see this, following the notations of [**Sil09**, Chapter VII], let X_0 and X_1 denote the points in $X_{\mathbb{Q}_p}$ that reduce to non-singular point and the additive identity $\widetilde{O}$ in $X_{\mathbb{F}_p}$, respectively. By [**Sil09**, Thm. VII.6.1], the size of the group $X_{\mathbb{Q}_p}/X_0$ is at most 4. Thus, for $\ell \geq 5$, we have $X_{\mathbb{Q}_p}[\ell] \subseteq X_0$. However, if $\mathbb{Q}(\mathrm{Jac}(X)[\ell])$ were unramified at p, X_1 would not have any ℓ-torsion, so by [**Sil09**, Prop. 2.1], one would have $(\mathbb{Z}/\ell\mathbb{Z})^2$ as a subgroup of $(X_{\overline{\mathbb{F}}_p})_{\mathrm{ns}} \simeq (\overline{\mathbb{F}}_p, +)$, a contradiction.

   Thus, the Galois representation $\overline{\rho}_X : \mathrm{Gal}(\overline{\mathbb{Q}}/\mathbb{Q}) \to \mathrm{GL}_2(\mathbb{F}_\ell)$ is ramified at p. However, the Galois representation $\overline{\rho}_{H^\perp/H} : \mathrm{Gal}(\overline{\mathbb{Q}}/\mathbb{Q}) \to \mathrm{GL}_2(\mathbb{F}_\ell)$ is isomorphic to $\overline{\rho}_X$ but is unramified at p, a contradiction. □

COROLLARY 3.11. *Suppose $p \neq \ell$ and $\ell \geq 5$. If*

$$b_p \notin \{-(p+1), p+1\} \cup [-2\sqrt{p}, 2\sqrt{p}], \tag{3.12}$$

then there is no elliptic curve X such that $H^\perp/H$ and $\mathrm{Jac}(X)[\ell]$ are isomorphic as $\mathrm{Gal}(\overline{\mathbb{Q}}/\mathbb{Q})$-modules.

PROOF. Follows from Proposition 3.10 and Hasse-Weil bound. □

Corollary 3.11 can be used to prove that some genus 2 curves have no gluable elliptic curves. We provide an example for $\ell \in \{11, 13, 19\}$.

EXAMPLE 3.13. When $\ell = 11$, The curve Y with LMFDB label `353.a.353.1` and equation $y^2 + (x^3 + x + 1)y = x^2$ has rational 11-torsion point (so H exists), and its Frobenius polynomial at 2 is

$$F_{Y,2}(T) = T^4 + T^3 + 3T^2 + 2T^3 + 4 = (T-1)(T-2)(T^2 + 4T + 2).$$

Following Algorithm 3.9, we have $\chi \equiv 1$, so we can compute $b_2 = -4 \pmod{11}$. Thus, Y has no gluable elliptic curves.

Similarly, the curves with LMFDB labels `349.a.349.1` and `169.a.169.1` have no gluable elliptic curves with $\ell = 13$ and $\ell = 19$, respectively.

REMARK 3.14. When $\ell \in \{3, 5\}$, by [**Rub97**, Thm. 3], there always exists an elliptic curve X such that $H^\perp/H$ and $\mathrm{Jac}(X)[\ell]$ are isomorphic as $\mathrm{Gal}(\overline{\mathbb{Q}}/\mathbb{Q})$-modules. Furthermore, one can make it isomorphic with a correct symplectic type [**Fis06**, §13]. Thus, there always exists a gluable elliptic curve provided that $\ell \in \{3, 5\}$ and H exists.

We do not yet know whether there is an example of a genus 2 curve for which H exists when $\ell = 7$, but there is no gluable elliptic curve.

4. Proving Isomorphism of Galois Representations

In this section, we utilize Serre's modularity conjecture to study the action of $\mathrm{Gal}(\overline{\mathbb{Q}}/\mathbb{Q})$ on $H^\perp/H$ more closely. The result of this is that, if the Galois representation associated to $H^\perp/H$ is irreducible, then there is an algorithm to rigorously prove that $H^\perp/H$ and $\mathrm{Jac}(X)[\ell]$ are isomorphic as $\mathrm{Gal}(\overline{\mathbb{Q}}/\mathbb{Q})$-modules.

We begin by reviewing the statement of Serre's modularity conjecture in Section 4.1 and use it to show that the representation corresponding to $H^\perp/H$ is modular in Section 4.2. After this, the algorithm is divided into two steps.

(1) Proving that there indeed exists a one-dimensional $\mathrm{Gal}(\overline{\mathbb{Q}}/\mathbb{Q})$-stable subgroup $H \subset \mathrm{Jac}(Y)[\ell]$. We do this by utilizing a numerical algorithm from analytic description of Y. This is detailed in Section 4.3.
(2) Proving the isomorphism between $H^\perp/H$ and $\mathrm{Jac}(X)[\ell]$. We do this by combining modularity of the representation corresponding to $H^\perp/H$ and Sturm's bound. This gives an explicit bound on how many Frobenius traces to check, which is explained in Section 4.4.

Finally, in Section 4.5, we provide some comments about the reducible case.

4.1. Serre's Modularity Conjecture. In this section, we briefly review the statement of Serre's modularity conjecture, which was proven by Khare and Wintenberger in [**KW09**].

For any positive integer N, we have the congruence subgroup

$$\Gamma_0(N) = \left\{ \left(\begin{smallmatrix} a & b \\ c & d \end{smallmatrix}\right) \in \mathrm{SL}_2(\mathbb{Z}) : c \equiv 0 \pmod{N} \right\} \subset \mathrm{SL}_2(\mathbb{Z}). \tag{4.1}$$

For any Dirichlet character $\varepsilon : (\mathbb{Z}/N\mathbb{Z})^\times \to \mathbb{C}^\times$ and any positive integer k, we have the (finite-dimensional) $\mathbb{C}$-vector space $S_k(N, \varepsilon)$ of cusp forms in $\Gamma_0(N)$ with weight

k and nebentypus ε. We also denote $S_2(N) := S_2(N, \chi_{\mathrm{triv}})$, where $\chi_{\mathrm{triv}}(a) = 1$ for all a.

Each function $f \in S_2(N, \chi)$ has a ***q*-expansion** $f(z) = \sum_{n\geq 1} a_n q^n$, where $q = e^{2\pi i z}$. For any prime ℓ, one can reduce modular forms modulo ℓ by taking any mapping from $\overline{\mathbb{Q}}$ to $\overline{\mathbb{F}}_\ell$ and reducing the q-expansion coefficients along the mapping. This gives a ***modulo ℓ-modular form*** $\overline{f} = \sum_{n\geq 1} \overline{a}_n q^n$ where $\overline{a}_n \in \overline{\mathbb{F}}_\ell$.

We now state Serre's modularity conjecture.

THEOREM 4.2 (Serre's modularity conjecture, [**KW09**]). *Suppose* $\overline{\rho} : \mathrm{Gal}(\overline{\mathbb{Q}}/\mathbb{Q}) \to \mathrm{GL}_2(\mathbb{F}_\ell)$ *satisfies*

- $\overline{\rho}$ *is irreducible; and*
- $\overline{\rho}$ *is odd, i.e.,* $\det \overline{\rho}(c) = -1$, *where c is the complex conjugation map.*

Then there exist positive integers $N = N(\overline{\rho})$, $k = k(\overline{\rho})$, *a character* $\varepsilon = \varepsilon(\overline{\rho}) : (\mathbb{Z}/N\mathbb{Z})^\times \to \mathbb{C}^\times$, *and a modulo ℓ modular form* $f \in \mathbb{F}_\ell[[q]]$ *arising from* $S_k(N, \varepsilon)$ *with coefficients in* $\mathbb{F}_\ell$ *such that* $a_p \equiv \mathrm{tr}\,\rho(\mathrm{Frob}_p) \pmod{\ell}$ *for all p not dividing N.*

[**Ser87**] gives an explicit recipe for determining $N(\overline{\rho})$, $\varepsilon(\overline{\rho})$, and $k(\overline{\rho})$. We will not reproduce the full definitions here, but it is important to note that they satisfy $\det \overline{\rho} = \varepsilon(\chi_\ell)^{k-1}$ and $k \in [2, \ell^2 + 1]$.

4.2. Modularity of the Representation. Let Y be a genus 2 curve and ℓ be a prime such that there exists a Galois-stable subgroup $H \subseteq \mathrm{Jac}(Y)[\ell]$. Let $V_Y = \mathrm{Jac}(Y)[\ell]$. We now use Serre's modularity conjecture to study the $\mathrm{Gal}(\overline{\mathbb{Q}}/\mathbb{Q})$-module structure of $H^\perp/H$.

Since H is $\mathrm{Gal}(\overline{\mathbb{Q}}/\mathbb{Q})$-stable, by Galois-equivariance of the Weil pairing, we deduce that $H^\perp$ is $\mathrm{Gal}(\overline{\mathbb{Q}}/\mathbb{Q})$-stable. Thus, the action of $\mathrm{Gal}(\overline{\mathbb{Q}}/\mathbb{Q})$ onto $H^\perp/H$ is well-defined and thus induces the 2-dimensional mod ℓ-representation $\overline{\rho}_{H^\perp/H} : \mathrm{Gal}(\overline{\mathbb{Q}}/\mathbb{Q}) \to \mathrm{GL}(H^\perp/H)$. First, we have the following result about the determinant.

PROPOSITION 4.3. *For any* $\sigma \in \mathrm{Gal}(\overline{\mathbb{Q}}/\mathbb{Q})$, *we have*

$$\det(\overline{\rho}_{H^\perp/H}(\sigma)) = \chi_\ell(\sigma), \tag{4.4}$$

where $\chi_\ell : \mathrm{Gal}(\overline{\mathbb{Q}}/\mathbb{Q}) \to \mathbb{F}_\ell$ *is the cyclotomic character, determined by the action of the primitive ℓ-th root of unity.*

PROOF. Consider the $\mathrm{Gal}(\overline{\mathbb{Q}}/\mathbb{Q})$-invariant filtration $0 \subset H \subset H^\perp \subset V_Y$, which has successive quotients are H, $H^\perp/H$, and $V_Y/H^\perp$. Therefore, we deduce that

$$\det(\sigma|_H)\det(\sigma|_{H^\perp/H})\det(\sigma|_{V_Y/H^\perp}) = \det(\sigma|_{V_Y}) = \chi_\ell(\sigma)^2. \tag{4.5}$$

However, by Galois equivariance of the Weil pairing, $\det(\sigma|_H)\det(\sigma|_{V_Y/H^\perp}) = \chi_\ell(\sigma)$, so it follows that $\det(\overline{\rho}_{H^\perp/H}(\sigma)) = \det(\sigma|_{H^\perp/H}) = \chi_\ell(\sigma)$. □

Plugging the complex conjugation map c into Proposition 4.3, we see that $\det \overline{\rho}_{H^\perp/H}(c) = \chi_\ell(c) = -1$, so $\overline{\rho}_{H^\perp/H}$ is odd. Thus, if we assume that $\overline{\rho}_{H^\perp/H}$ is irreducible, then by Serre's modularity conjecture and the previous lemma, $\overline{\rho}_{H^\perp/H}$ is modular with trivial nebentypus. The next proposition then restricts the level.

PROPOSITION 4.6. *If* $\overline{\rho}_{H^\perp/H}$ *is irreducible, then the level* $N(\overline{\rho}_{H^\perp/H})$ *divides* N_Y.

PROOF. We claim that the conductor divides the conductor of the representation $\overline{\rho} : \mathrm{Gal}(\overline{\mathbb{Q}}/\mathbb{Q}) \to \mathrm{GL}(V_Y)$, which in turn divides the conductor of Y.

To prove this, it suffices to show that $\nu_p(N(\overline{\rho}_{H^\perp/H})) \leq \nu_p(N(\overline{\rho}_Y))$ for all primes p. Recall from [**Ser87**] that the exponents of p in the conductors are defined by

$$\nu_p(N(\overline{\rho}_{H^\perp/H})) = \sum_{i\geq 0} \frac{|G_i|}{|G_0|} \dim(H^\perp/H)^{G_i} \tag{4.7}$$

$$\nu_p(N(\overline{\rho}_Y)) = \sum_{i\geq 0} \frac{|G_i|}{|G_0|} \dim(V_Y)^{G_i}, \tag{4.8}$$

where G_i are p-adic ramification groups (with lower numbering) and V^G is the subspace of V fixed by G.

Thus, it suffices to show that for any subgroup $G \subset \mathrm{Gal}(\overline{\mathbb{Q}}/\mathbb{Q})$, we have $\dim(H^\perp/H)^G \leq \dim(V_Y)^G$.

Consider the $\mathrm{Gal}(\overline{\mathbb{Q}}/\mathbb{Q})$-invariant filtration $0 \subset H \subset H^\perp \subset V_Y$, which has successive quotients H, $H^\perp/H$, and $V_Y/H^\perp$, respectively. For any subgroup $G \subset \mathrm{Gal}(\overline{\mathbb{Q}}/\mathbb{Q})$, we have

$$\dim H^G + \dim(H^\perp/H)^G + \dim(V_Y/H^\perp)^G = \dim(V_Y)^G, \tag{4.9}$$

so $\dim(H^\perp/H)^G \leq \dim(V_Y)^G$, as desired. □

We now determine the weight of the representation.

THEOREM 4.10. *Assume that $\overline{\rho}_{H^\perp/H}$ is irreducible (hence modular) and ℓ does not divide N_Y. Then $\overline{\rho}_{H^\perp/H}$ is modular with weight* 2.

PROOF. By [**Ser87**, Prop. 4], it suffices to show that $H^\perp/H$ is a finite $\mathbb{Z}_\ell$-group scheme. Since $\mathrm{Jac}(Y)$ has good reduction modulo ℓ, it follows that $\mathrm{Jac}(Y)[\ell]$ (and hence both H and $H^\perp$) is a finite $\mathbb{Z}_\ell$-group scheme. Finally, by [**Ray74**, Cor. 3.3.6 (1)] (with $e = 1$), it follows that $H^\perp/H$ is a finite $\mathbb{Z}_\ell$-group scheme. □

COROLLARY 4.11. *Assume that $\overline{\rho}_{H^\perp/H}$ is irreducible. Let*

$$k = \begin{cases} 2 & \ell \text{ does not divide } N_Y \\ \ell^2 + 1 & \ell \text{ divides } N_Y. \end{cases} \tag{4.12}$$

Then there exists a modular form $f \in S_k(N_Y)$ modulo ℓ such that $\overline{\rho}_{f,\ell} \simeq \overline{\rho}_{H^\perp/H}$.

In particular, if $f = \sum_{n=1}^\infty c_n q^n$ is the q-expansion of f, then $c_p \equiv b_p \pmod{\ell}$ for all primes p not dividing N_Y.

PROOF. If ℓ does not divide N_Y, this is immediate by Theorem 4.10.

Otherwise, if ℓ divides N_Y, then the (untwisted) weight of $\overline{\rho}_{H^\perp/H}$ is in the interval $[2, \ell^2 + 1]$ and is $\equiv 2 \pmod{\ell - 1}$. By von Staudt–Clausen theorem, $\nu_\ell(B_{\ell-1}) = -1$ (where $B_{\ell-1}$ denotes Bernoulli numbers). Thus, the q-expansion of Eisenstein series $E_{\ell-1} = 1 + \frac{2(\ell-1)}{B_{\ell-1}} \sum_{n=1}^\infty \sigma_{\ell-1}(n) q^n$ is congruent to 1 modulo ℓ. Thus, for any modular form $f \in S_k(N_Y)$ modulo ℓ, the modular form $E_{\ell-1} f \in S_{k+\ell-1}(N_Y)$, whose q-expansion in modulo ℓ is the exactly the same as f in modulo ℓ. Thus, regardless of what the weight of $\overline{\rho}_{H^\perp/H}$ is, one can increment the weight by $\ell - 1$ at a time without changing the coefficients modulo ℓ. Hence, there is a corresponding modular form of weight $\ell^2 + 1$. This concludes the proof. □

4.3. Finding rational H. In this subsection, we describe a numerical algorithm that can compute the approximate coordinates of H, if it exists.

Recall that the Jacobian of a curve of genus 2 over $\mathbb{C}$ is isomorphic to a complex torus $\mathbb{C}^2/\Lambda$ for some four-dimensional lattice Λ. This isomorphism can be constructed explicitly. To do so, recall that points on the $\operatorname{Jac}(Y)$ can be parametrized by $\{P_1, P_2\} \in \operatorname{Sym}^2 Y$. For two chosen base points $Q_1, Q_2 \in \operatorname{Jac}(Y)$, we define the ***Abel-Jacobi map***

$$(4.13)\qquad \begin{aligned} \mathrm{AJ}_{Q_1,Q_2} : \operatorname{Sym}^2 Y &\to H^0(Y,\Omega)/H_1(Y,\mathbb{Z}) = \mathbb{C}^2/\Lambda \\ \{P_1, P_2\} &\mapsto \left(\int_{Q_1}^{P_1} \omega_1 + \int_{Q_2}^{P_2} \omega_1, \quad \int_{Q_1}^{P_1} \omega_2 + \int_{Q_2}^{P_2} \omega_2 \right), \end{aligned}$$

where one picks a basis for ω_1, ω_2 for differentials $H^0(Y, \Omega)$; a working basis is $\omega_1 = \frac{dx}{y}$ and $\omega_2 = \frac{x\,dx}{y}$ if $y^2 = f(x)$ is a hyperelliptic model of Y. This map does not depend on the choice of a path, as we are considering their values modulo $\Lambda = H_1(Y, \mathbb{Z})$. For more details about this map, we refer the reader to [**Mil08**, §1.18]. For our purposes, we take Q_1 and Q_2 to be ∞ and ∞ or ∞_1 and ∞_2 according to whether we are using an odd model or an even model. This ensures that the set $\{P_1, P_2\}$ is Galois-invariant.

On this torus, the ℓ-torsion is the ℓ^4 points in $\frac{1}{\ell}\Lambda/\Lambda$. For a candidate subgroup H we may now calculate coordinates and check whether it is rational as follows.

ALGORITHM 4.14. *Input.* A one-dimensional subgroup of the ℓ-torsion of $\operatorname{Jac}(Y)$, as represented by its coordinates in $\mathbb{C}^2/\Lambda$.

Output. If the algorithm recognizes H as Galois-stable, `true`. If the algorithm recognizes H as Galois-nonstable, `false`. Otherwise, `unknown`.

(1) Let the nonzero points in H be $P_1, P_2, \ldots, P_{\ell-1}$.
(2) For each i from 1 to $\ell - 1$, pull back P_i to a point on $\operatorname{Sym}^2 Y$. Define

$$(4.15)\quad f_i(T) := \begin{cases} (T - x_{i,1})(T - x_{i,2}) & \text{if } P_i \mapsto \{(x_{i,1}, y_{i,1}), (x_{i,2}, y_{i,2})\} \in \operatorname{Sym}^2 Y \\ T - x_i & \text{if } P_i \mapsto \{(x_i, y_i)\} \in \operatorname{Sym}^2 Y. \end{cases}$$

(3) Compute $f(T) := \prod_{i=1}^{\ell-1} f_i(T)$.
(4) If any coordinate has a recognizably nonzero complex part, output `false`.
(5) Run a rational recognition algorithm over the coefficients of f. If the coefficients are successfully recognized as rational numbers, then we verify that this is correct by computing coordinate of P_1 (which will lie in an field extension of $\mathbb{Q}$ of degree at most $4(\ell - 1)$) and check that $\ell P_1 = 0$. If this test passes, output `true`. Otherwise, output `unknown`.

This algorithm works because if H is Galois-stable, the list of x-coordinates collected in step 2 must be Galois-stable as well, meaning $f(T) \in \mathbb{Q}[T]$.

This algorithm does not always return `true` or `false`, even if the precision is sufficiently high. If the precision is sufficiently high and the algorithm still returns `unknown`, then H is probably Galois-nonstable.

4.4. Proving Isomorphism. One notable property of modular forms is that if the first few coefficients (up to the Sturm bound [**Stu87**]) of two modular forms are congruent modulo ℓ, then all coefficients of those modular forms are congruent. Combining this fact with the modularity of $\overline{\rho}_{H^\perp/H}$ gives an algorithm to deterministically prove the isomorphism between two mod-ℓ Galois representations $\overline{\rho}_{H^\perp/H}$

and $\overline{\rho}_X$. The algorithm mirrors [**KO92**, Prop. 4], which is used to prove mod-ℓ congruences between elliptic curves. We now describe a version of this algorithm in our setting of genus 2 curves.

ALGORITHM 4.16. *Input.*

- An elliptic curve X (with conductor N_X),
- a prime $\ell \geq 3$,
- a genus 2 curve Y (with conductor N_Y), and
- the coordinates of a generator of one-dimensional $\mathrm{Gal}(\overline{\mathbb{Q}}/\mathbb{Q})$-stable subgroup $H \subset \mathrm{Jac}(Y)[\ell]$ (given by their minimal polynomials) such that $\overline{\rho}_{H^\perp/H}$ is irreducible.

Output. `true` if $\overline{\rho}_{H^\perp/H} \simeq \overline{\rho}_X$ and `false` otherwise.

(1) Compute the level M and the Sturm's bound B by

$$M = \mathrm{lcm}\left(N_X,\ N_Y \prod_{\substack{\text{prime } p \\ p \mid N_Y}} p\right) \quad \text{and} \quad B = \frac{k}{12} M \prod_{\substack{\text{prime } p \\ p \mid M}} \left(1 - \frac{1}{p}\right), \tag{4.17}$$

where k is defined in (4.12).

(2) For each prime $p \leq B$ not dividing N_Y, do the following.
   - Compute the action of Frob_p acting on H, which will be a mutiplication by some $t \in \mathbb{F}_\ell^\times$. This can be done in a finite extension of $\mathbb{F}_p$.
   - Compute b_p (defined in (3.7)) by (cf. (3.8)) $b_p = a_{p,Y} - t - \frac{p}{t}$.
   - If p does not divide N_X, check whether $b_p \equiv a_{p,X} \pmod{\ell}$.
   - If X has a multiplicative reduction modulo p, check whether $b_p a_{p,X} \equiv p + 1 \pmod{\ell}$.

Return `true` if the condition holds for all $p \leq B$ and `false` otherwise.

This algorithm is only feasible when the conductor of Y is very small: if the conductor N_Y is squarefree, then the algorithm requires checking at least N_Y^2 traces. Most genus two curves have conductor at least 1000, so this algorithm is only feasible for curves with very small conductor.

REMARK 4.18. To certify irreducibility of $\overline{\rho}_{H^\perp/H}$, we do the following: first enumerate the set $\mathcal{X}$ of characters $\varepsilon : \mathrm{Gal}(\overline{\mathbb{Q}}/\mathbb{Q}) \to \mathbb{F}_\ell^\times$ that has conductor dividing D (where D is in Proposition 3.3). Then for each prime $p \leq B$ for which Y has good reduction, eliminate any character $\varepsilon \in \mathcal{X}$ such that $b_p \not\equiv \varepsilon(p) + \frac{p}{\varepsilon(p)} \pmod{\ell}$. If it happens that $\mathcal{X} = \emptyset$, we can conclude that H is irreducible.

PROOF OF CORRECTNESS OF ALGORITHM 4.16. If $\overline{\rho}_{H^\perp/H} \simeq \overline{\rho}_X$, then by Proposition 3.10, the conditions in step (4) are all satisfied, so the algorithm returns `true`. Thus, it suffices to show that if the conditions in step (4) are all satisfied, then $\overline{\rho}_{H^\perp/H} \simeq \overline{\rho}_X$.

By the Modularity Theorem [**BCDT01**, Thm. A], let $\overline{f} \in S_2(\Gamma_0(N_X))$ be the modular form corresponding to ρ_X. By the weight-incrementing argument in the proof of Corollary 4.11, there is a modular form $f \in S_k(\Gamma_0(N_X))$ that has the same reduction modulo ℓ as $\overline{f}$. By Corollary 4.11, let $g \in S_k(\Gamma_0(N_Y))$ be the modular form corresponding to $\rho_{H^\perp/H}$.

For any modular form h, we let $N(h)$ be the level of h (i.e., the smallest positive integer such that $h \in S_k(\Gamma_0(N(h)))$). Recall that the corresponding L-function of

h is $L_h(s) = \sum_{n\geq 1} c_n n^{-s}$. Define $F_{p,h}(T) = \sum_{n=0}^{\infty} c_{p^n} T^n$. If h is an eigenform, then by properties of Hecke operators (see [**DS05**, Thm. 5.9.2]), then we have

$$L_h(s) = \prod_{p \text{ prime}} F_{p,h}(T), \quad F_{p,h}(T) = \begin{cases} (1 - c_p T)^{-1} & \text{if } p \mid N \\ (1 - c_p T + pT^2)^{-1} & \text{if } p \nmid N. \end{cases} \tag{4.19}$$

Let $R_d : S_k(\Gamma_0(N)) \to S_k(\Gamma_0(Nd))$ denote the operator that takes a function h to another function $\tau \mapsto h(d\tau)$. Note that the L-function corresponding to $R_d h$ is $d^{-s} L_h(s)$. Thus, for any polynomial P, the L-function corresponding to $P(R_d)h$ is $P(d^{-s})L_h(s)$. By comparing individual L-factors, we have

$$L_{P(R_p)h} = \prod_{q \text{ prime}} F_{q,P(R_p)h}(T), \quad \text{and} \tag{4.20}$$

$$F_{q,P(R_p)h}(T) = \begin{cases} F_{q,h}(T) & \text{if } q \neq p \\ P(T) \cdot F_{p,h}(T) & \text{if } q = p \end{cases}. \tag{4.21}$$

For each prime $p \leq B$, we define operators U_p that will be applied to f and V_p that will be applied to g so that

(i) the L-factors of $U_p f$ and $V_p g$ at p are equal,
(ii) the L-factors of $U_p f$ and $V_p g$ at all primes $q \neq p$ remain unchanged, and
(iii) for any prime p, $\nu_p(U_p f)$ and $\nu_p(V_p g)$ are at most $\nu_p(M)$.

(U_p and V_p can depend on f and g.) We define those by splitting into four cases.

(1) **If p divides N_Y,** then we define

$$U_p = F_{p,f}(R_p) \quad \text{and} \quad V_p = F_{p,g}(R_p),$$

so by applying (4.21) twice, we have

$$F_{p,U_p f}(T) = F_{p,f}(T) \cdot F_{p,f}(T)^{-1} = 1,$$
$$F_{p,V_p g}(T) = F_{p,g}(T) \cdot F_{p,g}(T)^{-1} = 1,$$

verifying (i). Moreover, $\deg F_{p,f} = \max(2 - \nu_p(N_X), 0)$ by considering reduction types of X, so

$$\nu_p(N(U_p f)) \leq \nu_p(N_X) + \deg F_{p,f} = \max(2, \nu_p(N_X)).$$

Similarly, $\deg F_{p,g} = 1$, so we have

$$\nu_p(N(V_p g)) \leq \nu_p(N_Y) + 1.$$

Therefore, both $\nu_p(N(U_p f))$ and $\nu_p(N(V_p g))$ are at most $\nu_p(M)$, verifying (iii).

(2) **If p does not divide either N_X or N_Y.** Then we have $F_{p,f}(T) = (1 - a_{p,X}T + pT^2)^{-1}$ and $F_{p,g}(T) = (1 - b_p T + pT^2)^{-1}$, which are automatically congruent modulo ℓ. Thus, we can define $U_p = V_p = 1$, so the levels of $U_p f$ and $V_p g$ are not divisible by p. Both (i) and (iii) are then satisfied.

(3) **If p does not divide N_Y and $\nu_p(N_X) = 1$,** then we have $F_{p,f}(T) = (1 - a_{p,X}T)^{-1}$ and $F_{p,g}(T) = (1 - b_p T + T^2)^{-1}$. The fact that $a_{p,X} = \pm 1$, and the $a_{p,X} b_p \equiv p+1 \pmod{\ell}$ implies that $1 - a_{p,X}T$ divides $1 - b_p T + pT^2$ in $\mathbb{F}_\ell[T]$. Let the quotient be $1 - cT$ for $c \in \mathbb{F}_\ell$. Then we define

$$U_p = 1 \quad \text{and} \quad V_p = 1 - cR_p,$$

so that by (4.21), we have $F_{p,U_pf}(T) = G_{p,U_pf}(T) = (1 - a_{p,X}T)^{-1}$, verifying (i). Moreover, $\nu_p(N(U_pf)) = \nu_p(N_X) = 1$ and $\nu_p(N(V_pg)) \le 1$, verifying (iii).

(4) **If p does not divide N_Y and $\nu_p(N_X) \ge 2$,** then the L-factors at p of f and g are 1 and $(1 - b_pT + T^2)^{-1}$. Thus, we define

$$U_p = 1 \quad \text{and} \quad V_p = 1 - b_pR_p + R_p^2,$$

so from (4.21), we have $F_{p,U_pf}(T) = F_{p,U_pg} = 1$, verifying (i). Moreover, $\nu_p(N(U_pf)) = \nu_p(N_X)$ and $\nu_p(N(U_pg)) \le 2$, both are at most $\nu_p(M)$.

Now, let $U = \prod_{p \le B} U_p$ and $V = \prod_{p \le B} V_p$. Hence, we have that

- For all primes $p \le B$, the L-functions corresponding to Uf and Vg can be expressed as an Euler product, and the L-factors at p of Uf and Vg are equal.
- Both Uf and Vg are in $S_k(\Gamma_0(M))$ for M defined in (4.17).

Thus, the coefficients of $1, q, q^2, \dots, q^B$ in the q-expansions of the modular forms Uf and Vg are congruent modulo ℓ. Hence, by Sturm's theorem [**Stu87**, Thm. 1], we get that Uf and Vg are congruent modulo ℓ. This means that $a_p \equiv b_p \pmod{\ell}$ for *every* prime $p \nmid M$.

Thus, for any prime p, the matrices $\overline{\rho}_{H^\perp/H}(\mathrm{Frob}_p)$ and $\overline{\rho}_X(\mathrm{Frob}_p)$ are conjugate. By the Chebotarev's density theorem, for any $\sigma \in \mathrm{Gal}(\overline{\mathbb{Q}}/\mathbb{Q})$, one can select p such that σ and Frob_p are conjugate, and so $\overline{\rho}_{H^\perp/H}(\sigma)$ and $\overline{\rho}_X(\sigma)$ are conjugate for any $\sigma \in \mathrm{Gal}(\overline{\mathbb{Q}}/\mathbb{Q})$. Thus, by the Brauer–Nesbitt theorem, $\overline{\rho}_{H^\perp/H}$ and $\overline{\rho}_X$ are isomorphic up to semisimplification. However, since we assumed $\overline{\rho}_{H^\perp/H}$ is irreducible, it follows that $\overline{\rho}_{H^\perp/H} \simeq \rho_X$. □

4.5. The Reducible Case. If that $\overline{\rho}_{H^\perp/H}$ is reducible, we cannot use the algorithm above because $\overline{\rho}_{H^\perp/H}$ is not necessarily modular. Even if $\overline{\rho}_{H^\perp/H}$ were modular, we can only use Algorithm 4.16 to show that $\overline{\rho}_{H^\perp/H}$ and ρ_X are isomorphic up to semisimplification. An algorithm using modular polynomials similar to [**CF22**, §3.6.] does not extend well to genus 2 curves due to the unwieldy nature of modular polynomials in genus 2. One might need to resort to explicitly computing $\overline{\rho}_{H^\perp/H}$.

5. Checking the Antisymplectic Condition

Throughout this section, let X and Y be curves of genus 1 and genus 2, respectively, and let ℓ be a prime such that

- there is a Galois-stable 1-dimensional subspace $H \subseteq \mathrm{Jac}(Y)[\ell]$; and
- there is a $\mathrm{Gal}(\overline{\mathbb{Q}}/\mathbb{Q})$-module isomorphism $\phi : H^\perp/H \to \mathrm{Jac}(X)[\ell]$.

Then the only condition left to verify before we can glue X and Y is that ϕ is antisymplectic, i.e., for any $P, Q \in H^\perp/H$, we have $e_\ell(\phi(P), \phi(Q)) = e_\ell(P, Q)^{-1}$. If $\ell = 2$, then this condition is tautological. Hence, we assume ℓ is odd for the remainder of this section.

If the image of the mod-ℓ representation $\overline{\rho}_{H^\perp/H} : \mathrm{Gal}(\overline{\mathbb{Q}}/\mathbb{Q}) \to \mathrm{GL}(H^\perp/H)$ is sufficiently nice, we may still be able to determine whether this condition holds even before trying to glue those curves. We adapt the method in [**FK22**] to do so.

5.1. Symplectic Type. If $\phi : H^\perp/H \to \mathrm{Jac}(X)[\ell]$ is a Galois module isomorphism, then there exists a constant $\alpha \in \mathbb{F}_\ell^\times$ such that for any $P, Q \in H^\perp/H$, we have $e_\ell(\phi(P), \phi(Q)) = e_\ell(P, Q)^\alpha$ because α is simply the determinant of ϕ under a suitable basis. Note that if we replace ϕ by $[t] \circ \phi$ where $t \in \mathbb{Z}$ (which is still a Galois module isomorphism), then α becomes αt^2.

We thus define the ***symplectic type*** of ϕ to be the image of α in $\mathbb{F}_\ell^\times/(\mathbb{F}_\ell^\times)^2 \simeq \{\pm 1\}$. If the image is $+1$, then ϕ has ***positive symplectic type***, and if the image is -1, then ϕ has ***negative symplectic type***.[2] This terminology is not to be confused with symplectic and antisymplectic.

PROPOSITION 5.1. *Suppose* $\ell \equiv 1 \pmod 4$ *(resp.* $\ell \equiv 3 \pmod 4$*). Then there exists an antisymplectic* $\mathrm{Gal}(\overline{\mathbb{Q}}/\mathbb{Q})$*-module isomorphism* $\phi : H^\perp/H \to \mathrm{Jac}(X)[\ell]$ *if and only if there exists a* $\mathrm{Gal}(\overline{\mathbb{Q}}/\mathbb{Q})$*-module isomorphism* $\psi : H^\perp/H \to \mathrm{Jac}(X)[\ell]$ *of positive (resp. negative) symplectic type.*

PROOF. Both facts follow from: $-1 \in (\mathbb{F}_\ell^\times)^2$ if and only if $\ell \equiv 1 \pmod 4$. □

From Proposition 5.1, it suffices to determine whether a $\mathrm{Gal}(\overline{\mathbb{Q}}/\mathbb{Q})$-module isomorphism we have is of positive or negative symplectic type. It is possible that there is a $\mathrm{Gal}(\overline{\mathbb{Q}}/\mathbb{Q})$-module isomorphism of positive symplectic type and a $\mathrm{Gal}(\overline{\mathbb{Q}}/\mathbb{Q})$-module isomorphism of negative symplectic type at the same time. By [**FK22**, Thm. 15], this happens if and only if the image of mod ℓ-representation $\overline{\rho}_{X,\ell}(\mathrm{Gal}(\overline{\mathbb{Q}}/\mathbb{Q}))$ is abelian and not the subgroup generated by a conjugate of $\left(\begin{smallmatrix} a & 1 \\ 0 & a \end{smallmatrix}\right) \in \mathrm{GL}_2(\mathbb{F}_\ell)$ for some $a \in \mathbb{F}_\ell^\times$.

5.2. A Local Test for Symplectic Type. We do not have a general algorithm for determining the symplectic type given $H^\perp/H$ and X. However, if there exists $\sigma \in \mathrm{Gal}(\overline{\mathbb{Q}}/\mathbb{Q})$ which acts on $\mathrm{Jac}(X)[\ell]$ by a non-diagonalizable matrix (i.e., conjugate of $\left(\begin{smallmatrix} a & 1 \\ 0 & a \end{smallmatrix}\right)$ for some $a \in \mathbb{F}_\ell^\times$), then this element σ could be used to determine the symplectic type, as detailed in the following proposition, which is a variant of [**FK22**, Thm. 16].

PROPOSITION 5.2. *Let* $\langle \bullet, \bullet \rangle : \mathbb{F}_\ell^2 \times \mathbb{F}_\ell^2 \to \mathbb{F}_\ell$ *be a non-degenerate alternating bilinear pairing. Let* $M \in \mathrm{GL}_2(\mathbb{F}_\ell)$ *be a non-diagonalizable matrix. Then as* v *varies through* $\mathbb{F}_\ell^2$*, then* $\langle v, Mv \rangle$ *does not depend on* v *up to multiplication by a square.*

PROOF. Without loss of generality, change the basis so that $M = \left(\begin{smallmatrix} a & 1 \\ 0 & a \end{smallmatrix}\right)$ and, scale the inner product by a constant so that $\left\langle \left(\begin{smallmatrix} x \\ y \end{smallmatrix}\right), \left(\begin{smallmatrix} x' \\ y' \end{smallmatrix}\right) \right\rangle = xy' - yx'$. Thus, $\langle v, Mv \rangle = -y^2$, where $v = \left(\begin{smallmatrix} x \\ y \end{smallmatrix}\right)$. □

In the case of the ℓ-torsion of an elliptic curve X, if there exists $\sigma \in \mathrm{Gal}(\overline{\mathbb{Q}}/\mathbb{Q})$ acting on $\mathrm{Jac}(X)[\ell]$ by a non-diagonalizable matrix, then by Proposition 5.2, the Weil pairing $e_\ell(P, \sigma(P))$ is either 1 or does not depend on $P \in \mathrm{Jac}(X)[\ell]$ (up to multiplication by a square in $\mathbb{F}_\ell^\times$).

Similarly, σ acts on $H^\perp/H$ by the same matrix, so by Proposition 5.2 again, the Weil pairing $e_\ell(Q, \sigma(Q))$ is either 1 or does not depend on $Q \in H^\perp/H$ (up to multiplication by a square in $\mathbb{F}_\ell^\times$). If $P = \phi(Q)$, then

$$(5.3) \qquad e_\ell(P, \sigma(P)) = e_\ell(\phi(Q), \sigma(\phi(Q))) = e_\ell(\phi(Q), \phi(\sigma(Q))) = e_\ell(Q, \sigma(Q))^\alpha,$$

[2] [**FK22**] and [**CF22**] call positive and negative symplectic type symplectic and antisymplectic isomorphism, respectively. We choose a different terminology to avoid confusion.

where α is the symplectic type. Thus, comparing the nontrivial values of $e_\ell(P, \sigma(P))$ and $e_\ell(Q, \sigma(Q))$ for any $P \in \mathrm{Jac}(X)[\ell]$ and $Q \in H^\perp/H$ determines the symplectic type.

To compute this, we note that, by the Chebotarev's density theorem, there exists a prime p for which Frob_p and σ are in the same conjugacy class. Thus, we may take the reduction of both curves modulo p and consider their equations in $\overline{\mathbb{F}}_p$. The following Lemma 5.4 shows that all torsion points are contained in a field extension of degree only $O(\ell^2)$.

LEMMA 5.4. *Suppose that p is a prime such that the action of Frob_p on $\mathrm{Jac}(X)[\ell]$ (and hence on $H^\perp/H$) is a non-diagonalizable matrix. Then we have*

$$\mathrm{Jac}(X)_{\overline{\mathbb{F}}_p}[\ell] = \mathrm{Jac}(X)_{\mathbb{F}_{p^{\ell(\ell-1)}}}[\ell] \quad \textit{and} \tag{5.5}$$

$$\mathrm{Jac}(Y)_{\overline{\mathbb{F}}_p}[\ell] = \begin{cases} \mathrm{Jac}(Y)_{\mathbb{F}_{p^{\ell(\ell-1)}}}[\ell] & \textit{if } \ell \neq 3 \\ \mathrm{Jac}(Y)_{\mathbb{F}_{p^{18}}}[\ell] & \textit{if } \ell = 3 \end{cases} \tag{5.6}$$

PROOF. Suppose that the action of Frob_p on $\mathrm{Jac}(X)_{\overline{\mathbb{F}}_p}[\ell]$ is conjugate to $\left(\begin{smallmatrix} \gamma & 1 \\ 0 & \gamma \end{smallmatrix}\right)$ for some $\gamma \in \mathbb{F}_\ell$. We compute the order of Frob_p in both torsion fields.

- For X, by the condition, Frob_p is conjugate to $\left(\begin{smallmatrix} \gamma & 1 \\ 0 & \gamma \end{smallmatrix}\right)$. The n-th power of this is $\left(\begin{smallmatrix} \gamma^n & n\gamma^{n-1} \\ 0 & \gamma^n \end{smallmatrix}\right)$, which is congruent modulo ℓ to the identity matrix when $n = \ell(\ell-1)$.
- For Y, we have that Frob_p must act on $\mathrm{Jac}(Y)_{\overline{\mathbb{F}}_p}[\ell]$ by matrix with eigenvalues α, β, γ, γ, all in $\mathbb{F}_\ell$, where α and β are the eigenvalues corresponding to H and $\mathrm{Jac}(Y)_{\overline{\mathbb{F}}_p}[\ell]/H^\perp$. In particular, we deduce that $(\mathrm{Frob}_p)^{\ell-1} - 1$ has all eigenvalues 0, and hence is a nilpotent matrix.

  In particular, if no Jordan block of Frob_p has size greater than ℓ, then

  $$0 = \left((\mathrm{Frob}_p)^{\ell-1} - 1\right)^\ell = (\mathrm{Frob}_p)^{\ell(\ell-1)} - 1, \tag{5.7}$$

  where the second equality holds because we are working in modulo ℓ. The only case that the previous sentence does not cover is when $k = 4$ and $\ell = 3$ (i.e., Frob_p is a single Jordan block), in which case the order is 18.

Thus, every element in $\mathrm{Jac}(X)_{\overline{\mathbb{F}}_p}[\ell]$ and $\mathrm{Jac}(Y)_{\overline{\mathbb{F}}_p}[\ell]$ is fixed by $(\mathrm{Frob}_p)^{\ell(\ell-1)}$ (or $(\mathrm{Frob}_p)^{18}$ for $\mathrm{Jac}(Y)_{\overline{\mathbb{F}}_p}[\ell]$ and $\ell = 3$), completing the proof. □

REMARK 5.8. In the version of Algorithm 5.9 given below, we will consider only the case in which α and β are both distinct from γ, in which case one needs to consider only $\mathbb{F}_{p^{\ell(\ell-1)}}$ even for $\ell = 3$.

5.3. Algorithm for Determining Symplectic Type. With all the tools developed in Section 5.2, we now describe an algorithm to determine the symplectic type on some of the curves.

ALGORITHM 5.9. *Input.* Two curves X and Y of genus 1 and genus 2 for which a one-dimensional Galois-stable $H \subseteq \mathrm{Jac}(Y)[\ell]$ exists and there exists a $\mathrm{Gal}(\overline{\mathbb{Q}}/\mathbb{Q})$-module isomorphism $\phi : H^\perp/H \to \mathrm{Jac}(X)[\ell]$.

Output. Either `positive` or `negative` symplectic type of ϕ or `fail`, which occurs if the image $\overline{\rho}_{H^\perp/H}$ does not contain a non-diagonalizable matrix.

(1) Find a prime p such that $a_{X,p}^2 \equiv 4p \pmod{\ell}$. Then check that the matrix $\mathrm{Frob}_p|_{\mathrm{Jac}(X)[\ell]}$ is not diagonalizable by checking that the order of $\mathrm{Frob}_p|_{\mathrm{Jac}(X)[\ell]}$ does not divide $\ell - 1$.

If one cannot find p after a sufficiently many trials, then the image of Galois representation likely does not have a non-diagonalizable element, so return `fail`.

Once we find p, consider curves X and Y over $\mathbb{F}_{p^{\ell(\ell-1)}}$.

(2) Pick a random point $P \in \mathrm{Jac}(X)[\ell]$ and then compute

$$w_1 := e_\ell(P, \mathrm{Frob}_p(P)).$$

Repeat until this result is not 1.

(3) Determine the characteristic polynomial of $\mathrm{Frob}_p|_{\mathrm{Jac}(Y)[\ell]}$. Write it in the form $(T-\alpha)(T-\beta)(T-\gamma)^2 \in \mathbb{F}_\ell[T]$ such that $(T-\gamma)^2$ is the characteristic polynomial of $\mathrm{Frob}_p|_{\mathrm{Jac}(X)[\ell]}$.

If $\alpha = \beta = \gamma$, repeat (1) again with larger primes.

(4) Pick a random point $R \in \mathrm{Jac}(Y)[\ell]$. Compute

$$Q = (\mathrm{Frob}_p - \alpha)(\mathrm{Frob}_p - \beta)(R).$$

Then compute $w_2 := e_\ell(Q, \mathrm{Frob}_p(Q))$. Repeat until this result is not 1.

(5) Return `positive` if $w_1 = w_2^{t^2}$ for some $t \in \mathbb{F}_\ell^\times$, `negative` otherwise.

REMARK 5.10. In step (1), by the Chebotaraev density theorem, the density of such p is at least $(\ell-1)/|\,\mathrm{GL}_2(\mathbb{F}_\ell)| = \Omega(1/\ell^3)$. Since ℓ is generally small (practically, $\ell \leq 19$), it is not difficult to obtain p by trying the first few primes.

REMARK 5.11. One can show that the probability that each attempt of both steps (2) and (4) fails is $\frac{1}{\ell}$. (For step (4), note that $\alpha\beta = \gamma^2 = p$, so both α and β are distinct from γ.)

PROPOSITION 5.12. *Algorithm* 5.9 *correctly determines the symplectic type (if it does not return* `fail`*).*

PROOF. From (5.3), it suffices to show that $Q \in H^\perp$. Assume without loss of generality that Frob_p acts on H by multiplication by α. Let

$$V = \mathrm{Ker}(\mathrm{Frob}_p - \beta)(\mathrm{Frob}_p - \gamma)^2.$$

Since V is a span of the three columns corresponding to β, γ, γ in the Jordan block decomposition, we deduce that $\dim V = 3$. Moreover, the action of Frob_p on $H^\perp$ has characteristic polynomial $(T-\beta)(T-\gamma)^2$, so we have $H^\perp \subseteq V$. Comparing dimensions gives $H^\perp = V$. Finally,

$$(\mathrm{Frob}_p - \beta)(\mathrm{Frob}_p - \gamma)^2 Q = (\mathrm{Frob}_p - \alpha)(\mathrm{Frob}_p - \beta)^2(\mathrm{Frob}_p - \gamma)^2 R = 0, \tag{5.13}$$

so $Q \in H^\perp$ as desired. □

This algorithm is generally able to handle $\ell \in \{3, 5, 7\}$ in a few seconds.

6. Gluing Curves

Given a genus 2 curve Y, we wish to find a prime ℓ and a genus 1 curve X to which it can be glued, and then compute the resulting gluing. In this section, we will give a concrete description of our workflow for finding X and computing the gluing. For simplicity, we only look for curves Y such that the Galois representation corresponding to $H^\perp/H$ is irreducible. We then give a demonstration of how the steps pan out on a particular curve.

Our workflow starts with a genus 2 curve Y. Because attempting to compute gluings is time-consuming, we first narrow down the possibilities for ℓ and X as described in sections Sections 6.1 and 6.2. We then attempt to analytically construct gluings for each of the remaining possibilities in Section 6.3. The implementation of the workflow described in this section is available at [**SW25b**].

6.1. Determining ℓ. As specified by Item (i) we must find ℓ for which some $H \subset \mathrm{Jac}(Y)[\ell]$ is Galois-stable. We use Algorithm 3.6 to find all possible ℓ for which this holds, possibly along with some spurious ℓ which do not work.

6.2. Determining X. Once ℓ is restricted to some finite set, we fix some ℓ. We assume that H as in Item (i) exists. We wish to find X such that some ψ as in Item (ii) exists.

Each of the $\mathrm{Gal}(\overline{\mathbb{Q}}/\mathbb{Q})$-stable subgroups H corresponds to a trace function that takes a prime number p and output b_p, the trace of $H^\perp/H$. We use Algorithm 3.9 to recover all possible trace functions. (If it reports no trace function, this means that no such H exists.) For each such test function $p \mapsto b_p$, we query the LMFDB all elliptic curves X over $\mathbb{Q}$ that satisfies the conditions of Proposition 3.10 for all primes $p \leq 100$. This leaves us with a list of potential elliptic curves.

Next, given a potential X and Y, one can rule out most cases where ϕ has the wrong symplectic type by applying Algorithm 5.9.

Once we have at least one elliptic curve X for which there exists an antisymplectic $\mathrm{Gal}(\overline{\mathbb{Q}}/\mathbb{Q})$-module isomorphism $\phi : H^\perp/H \to \mathrm{Jac}(X)[\ell]$, it is possible to characterize all of them. Indeed, we note that for any elliptic curve X', a symplectic $\mathrm{Gal}(\overline{\mathbb{Q}}/\mathbb{Q})$-module isomorphism $\psi : \mathrm{Jac}(X)[\ell] \to \mathrm{Jac}(X')[\ell]$ gives an antisymplectic $\mathrm{Gal}(\overline{\mathbb{Q}}/\mathbb{Q})$-module isomorphism $\phi' = \psi \circ \phi$. Further, all antisymplectic $\mathrm{Gal}(\overline{\mathbb{Q}}/\mathbb{Q})$-module isomorphisms $\phi' : H^\perp/H \to \mathrm{Jac}(X')[\ell]$ arise in this manner, specifically from $\psi = \phi' \circ \phi^{-1}$. Thus, given such an X, the problem of finding all such X' is reduced to finding an elliptic curve ℓ-congruent to X with positive symplectic type. When $\ell \in \{2, 3, 5\}$, the moduli space of all such X' is a curve of genus 0, which has been worked out in [**RS01**] and [**RS95**]. When $\ell \geq 7$, the moduli space of X' is a curve of genus at least 3. By Falting's theorem, any curve of genus greater than 1 has finitely many rational points. Thus, for $\ell \geq 7$, there will be only finitely many such X' defined over $\mathbb{Q}$. Still, the equation for the curve of all possible X' has been worked out for $\ell \in \{7, 11\}$ in [**Fis14**].

Remark 6.1 (The reducible case). If the Galois representation corresponding to $H^\perp/H$ is reducible, following the above steps above is not enough to filter the list of elliptic curves and reduce it to a list of manageable size. For example, starting with the curve `961.a.961.1` in the LMFDB and $\ell = 5$ leaves 3083 potential elliptic curves. This is because Frobenius traces can only prove that $\overline{\rho}_X$ and $\overline{\rho}_{H^\perp/H}$ are isomorphic up to semisimplification.

There are two possible strategies that we can use to narrow this list down further. Suppose that X is an elliptic curve such that $\overline{\rho}_X \simeq \overline{\rho}_{H^\perp/H}$.

(1) **Discriminant.** Suppose that p is a prime not dividing N_Y such that X has a multiplicative reduction modulo p. Thus, the Galois representation $\overline{\rho}_X$ should be unramified at p, and so by using the theory of Tate's curve,

one can deduce that ℓ divides $\nu_p(\Delta_{\min}(X))$ (where $\Delta_{\min}$ denotes the minimal discriminant). Thus, we can rule out a large number of potential elliptic curves for which this condition does not hold.

(2) **Diagonal Matrix.** Let p be a prime such that $\overline{\rho}_Y(\mathrm{Frob}_p)$ is a diagonal matrix in $\mathrm{GSp}_4(\mathbb{F}_\ell)$, then $\overline{\rho}_X(\mathrm{Frob}_p)$ is also a diagonal matrix in $\mathrm{GL}_2(\mathbb{F}_\ell)$. Thus, one can find such primes p and use them to rule of the elliptic curves further.

Running these two strategies on the curve `961.a.961.1` (with 12 primes 1301, 2351, 4211, 5171, 16001, 17881, 24371, 31181, 35531, 36451, 37361, and 45751 in Step (2)) reduces the number of candidates to only 9. However, this does not work very well on some other curves, especially with $\ell = 3$, due to the sheer number of candidates and the rarity of primes p in Step (2).

6.3. Computing a Gluing. The computation of gluing is based on the numerical algorithm in [**HSS21**, §2.1] and the code from [**HSS20**]. However, we made several optimizations. First, in the step of computing the curve invariants from lattice (Step (4) in Algorithm 6.2 below), we replace the original theta function algorithm with a faster implementation from FLINT [**EK25**]. Second, the original code repeatedly tests a random maximal isotropic subgroup of $\mathrm{Jac}(X)[\ell] \times \mathrm{Jac}(Y)[\ell]$. However, this proves to be inefficient since there are $\Theta(\ell^6)$ such subgroups [**HSS21**, Cor. 1.20].

To optimize this, we propose the following two-step approach to search for the desired subgroup.

(1) Determine which one-dimensional subgroups H are rational.
(2) Determine which antisymplectic isomorphisms $\phi : \mathrm{Jac}(X)[\ell] \to H^\perp/H$ give rise to a gluing.

This gives the following algorithm.

ALGORITHM 6.2. *Input.* An elliptic curve X, a genus 2 curve Y, and a prime ℓ.

Output. A (possibly empty) list of all genus 3 curves Z defined over $\mathbb{Q}$ such that $\mathrm{Jac}(Z) \sim (\mathrm{Jac}\, X \times \mathrm{Jac}\, Y)/G$ for some maximal isotropic subgroup $G \subset \mathrm{Jac}(X)[\ell] \times \mathrm{Jac}(Y)[\ell]$.

(1) Compute the period matrix of Y, giving a basis $\{P_1, P_2, P_3, P_4\}$ of the lattice Λ such that $\mathrm{Jac}(Y) \simeq \mathbb{C}^2/\Lambda$.
(2) For each ratio $(a_1 : a_2 : a_3 : a_4) \in \mathbb{P}^3(\mathbb{F}_\ell)$, test whether the subgroup H generated by the torsion point $\frac{a_1}{\ell}P_1 + \frac{a_2}{\ell}P_2 + \frac{a_3}{\ell}P_3 + \frac{a_4}{\ell}P_4$ is $\mathrm{Gal}(\overline{\mathbb{Q}}/\mathbb{Q})$-stable by using Algorithm 4.14.
(3) For each H determined to be $\mathrm{Gal}(\overline{\mathbb{Q}}/\mathbb{Q})$-stable in (2), enumerate all antisymplectic isomorphisms $\phi : \mathrm{Jac}(X)[\ell] \to H^\perp/H$.
(4) For each such ϕ, use (ϕ, H) to generate the subgroup G according to [**HSS21**, Prop. 1.18]. Compute a period matrix of the lattice $\mathrm{Jac}(X) \times \mathrm{Jac}(Y)/G$.
(5) Compute its Diximier–Ohno invariants [**KLL+18**] or Shioda invariants [**BILV16**] and test whether they are defined over $\mathbb{Q}$ or not.
(6) For any set of rational invariants, reconstruct the curve Z using [**HSS21**, Alg. 2.21] for plane quartics or [**BILV16**] for hyperelliptic curves.

In Step (2), there are $\#\mathbb{P}^3(\mathbb{F}_\ell) = \ell^3 + \ell^2 + \ell + 1 = \Theta(\ell^3)$ possible H's to check. In Step (3), there are $\#\,\mathrm{SL}_2(\mathbb{F}_\ell) = \ell(\ell^2 - 1) = \Theta(\ell^3)$ isomorphisms to check. Thus, this algorithm reduces checking $\Theta(\ell^6)$ subgroups to checking at most $\Theta(\ell^3)$ torsion points and isomorphisms, making it much more efficient

We implemented this algorithm in Magma V.2.28-16. The timing of this algorithm against simply enumerating all maximal isotropic subgroups is shown in Table 1. The timing was taken on CPU 12th Gen Intel i9-12900K (24) @5.100GHz on five different test cases in which one can find a gluing when working with 500 decimal digits. Note that the timing only measures time to compute the (Dixmier-Ohno or Shioda) invariants and does not include time to reconstruct the curve.

TABLE 1. Time to compute invariants of all possible gluings 500 digits.

	$\ell = 3$	$\ell = 5$	$\ell = 7$
Enumerate all subgroups	144.5 s	5760 s	(not attempted)
Using Algorithm 6.2	11.8 s	71.9 s	241 s

6.4. Example. We provide a rundown of our algorithms on a particular curve.

EXAMPLE 6.3. Let Y be the genus 2 curve $y^2 + (x^3 + x^2 + x + 1)y = -x^2 - x$ (`277.a.277.1` in the LMFDB). Running Algorithm 3.6 on this curve yields the result $\{3, 5\}$, meaning that the only possible choices of ℓ are 3 and 5. From now, suppose we are looking for a $(5, 5)$-gluing, i.e., $\ell = 5$.

We next run Algorithm 3.9 on the input $(Y, 5)$. Since the conductor of Y is 277, which is prime, the algorithm concludes immediately that χ must be trivial, i.e., $\chi(p) = 1$ for all p. (This is also reflected by the fact that Y has a rational 5-torsion subgroup.) Thus, we may compute the Frobenius traces b_p of $H^\perp/H$ from the Frobenius traces $a_{p,Y}$ of Y. For example,

$$\begin{aligned} F_{Y,13}(T) &= T^4 - 3T^3 + 7T^2 - 39T + 169, \\ a_{Y,13} &= 3, \quad \text{and} \quad b_{13} = 3 - 1 - \tfrac{13}{1} = 4 \pmod 5). \end{aligned} \tag{6.4}$$

We can repeat the above process to compute b_p for all primes $p \leq 100$. Then we search for all elliptic curves in the LMFDB satisfying the trace constraints detailed in Proposition 3.10. The result is the four curves shown in Table 2.

TABLE 2. Potential elliptic curves gluable to curve `277.a.277.1`

	LMFDB	Equation	Symplectic Type
X_1	`1939.b1`	$y^2 + y = x^3 - 1916x - 32281$	positive
X_2	`18559.a1`	$y^2 + y = x^3 + 11734x - 21208$	negative
X_3	`21883.b1`	$y^2 + y = x^3 - 86x - 11120$	positive
X_4	`32963.c1`	$y^2 + y = x^3 - 77866x + 8364065$	positive

We run the symplectic test in Algorithm 5.9 to test the antisymplectic condition. All curves can be tested using $p = 19$, and the results are shown in Table 2. Since $\ell \equiv 1 \pmod 4$, by Proposition 5.1, the symplectic test rules out X_2

Thus, our algorithm found the curves X_1, X_3, and X_4. This is *not* a proof that these curves are gluable to Y. We can either use Algorithm 4.16 or computing the gluing explicitly to prove that they actually form a Galois stable maximal isotropic subgroup G.

For $i \in \{1, 3, 4\}$, running the code referenced in Section 6.3 on X_i and Y shows that there are indeed gluings Z_i. The invariants of Z_1 and Z_3 were obtained by computing at 500-digit precision, while the invariants of Z_4 were obtained at 1000-digit precision. It took about 30 seconds to compute the coordinates a nonzero point in H and less than 60 seconds to compute the invariants for each of Z_1, Z_3, and Z_4. However, by far dominating every computation we have done is minimizing the equations of Z_4, which took just over an hour. The minimized equations of Z_1, Z_3, and Z_4 are given below.

$$\begin{aligned} Z_1 : 88189x^4 &- 398531x^3y + 7700x^3z - 678120x^2y^2 + 1444780x^2yz + 231034x^2z^2 \\ &+ 238603xy^3 - 1620885xy^2z - 218291xyz^2 - 420855xz^3 + 82587y^4 \\ &- 2912900y^3z + 333537y^2z^2 - 959874yz^3 - 281678z^4 = 0 \end{aligned}$$

$$\begin{aligned} Z_3 : y^2 = 448x^8 &+ 3584x^7 + 2016x^6 - 476x^5 \\ &- 13020x^4 - 16408x^3 - 18340x^2 - 8988x - 4025 \end{aligned}$$

$$\begin{aligned} Z_4 : 19351616x^4 &+ 136748535x^3y + 106394158x^3z - 235515177x^2y^2 \\ &- 46043175x^2yz + 67674485x^2z^2 - 549641282xy^3 + 36999650xy^2z \\ &- 160500711xyz^2 - 36439076xz^3 + 272167382y^4 + 488584945y^3z \\ &- 488728851y^2z^2 + 152950443yz^3 - 115190535z^4 = 0 \end{aligned}$$

7. Examples

We now report on some examples resulting from the search process in Section 6. As in Section 6.3, the amount of time to compute a gluing is defined to be the time taken to compute Diximier-Ohno invariants or Shioda invariants. It does not include time to reconstruct or minimize the curve. All timings were done on CPU 12th Gen Intel i9-12900K (24) @5.100GHz.

7.1. Gluing Along Large Torsion. Our work allows us to look for gluings along ℓ-torsion for larger ℓ.

We first note that for $\ell \in \{7, 11, 13, 17, 19, 37, 43, 67, 163\}$ (i.e., all $\ell \geq 7$ for which there exists an ℓ-isogeny of elliptic curves, by Mazur's isogeny theorem [**Maz78**, Thm. 1]), one can construct infinitely many gluable candidates of genus 1 and genus 2 curves. Here, a ***gluable candidate*** is a pair of curves (X, Y) such that X has genus 1, Y has genus 2, and there exists a Galois stable maximal isotropic subgroup $G \subset \mathrm{Jac}(X)[\ell] \times \mathrm{Jac}(Y)[\ell]$. As noted in Remark 2.9, a gluable candidate does not necessary produce a gluing.

There is an easy (and uninteresting) way to construct those gluable candidates. We will construct pairs of elliptic curve X and genus 2 curve Y such that

- Y is isogenous to a product of two elliptic curves $Y_1 \times Y_2$;
- Y_1 has an ℓ-isogeny; and
- $\mathrm{Jac}(X)[\ell]$ and $\mathrm{Jac}(Y_2)[\ell]$ are isomorphic as $\mathrm{Gal}(\overline{\mathbb{Q}}/\mathbb{Q})$-modules.

Call a gluable candidate (X, Y) ***uninteresting*** if it is of the above form and ***interesting*** otherwise.

Proposition 7.1. *For $\ell \in \{7, 11, 13, 17, 19, 37, 43, 67, 163\}$, there exists infinitely many uninteresting gluable candidates (X, Y) along ℓ-torsion.*

Proof. Let $p = 2$ if $\ell \equiv 1$ or $3 \pmod 8$ and $p = 3$ if $\ell \in \{7, 13, 37\}$. Let Y_1 be an elliptic curve with ℓ-isogeny, and let Y_2 be an elliptic curve such that $Y_2[p]$ and

$Y_1[p]$ are antisymplectically isomorphic as $\mathrm{Gal}(\overline{\mathbb{Q}}/\mathbb{Q})$-modules. There are infinitely many such Y_2 because when $p=2$, one can take elliptic curves whose 2-torsion field is isomorphic to that of Y_1, and when $p=3$, this follows from [**Fis06**, §13].

The curves Y_1 and Y_2 are gluable along p-torsion, producing a curve Y for which there exists a p-isogeny $\phi : Y_1 \times Y_2 \to \mathrm{Jac}(Y)$. Since $\ell \neq p$, the map ϕ induces an isomorphism $\phi : \mathrm{Jac}(Y_1)[\ell] \times \mathrm{Jac}(Y_2)[\ell] \xrightarrow{\sim} \mathrm{Jac}(Y)[\ell]$. This isomorphism is antisymplectic because it has degree p (so by properties of Weil pairing, $e_\ell(\phi(P), \phi(Q)) = e_\ell(P,Q)^p$), and by our choice of ℓ, the number $-p$ is a quadratic residue modulo ℓ.

Let X be any elliptic curve such that $\mathrm{Jac}(X)[\ell]$ and $\mathrm{Jac}(Y_2)[\ell]$ are isomorphic as $\mathrm{Gal}(\overline{\mathbb{Q}}/\mathbb{Q})$-modules. In particular, we may take $X = Y_2$. We now show that X and Y are gluable candidates.

- Since Y_1 has an ℓ-isogeny, there exists a one-dimensional subgroup $G \subset \mathrm{Jac}(Y_1)[\ell]$. Then the image $H := \phi(G \times \{0\})$ is a one-dimensional subgroup of $\mathrm{Jac}(Y)$.
- By Galois equivariance of the Weil pairing, we have $H^\perp/H = \phi(\{0\} \times Y_2)$. Since ϕ is antisymplectic, $H^\perp/H$ and $\mathrm{Jac}(Y_2)[\ell]$ are antisymplectic as $\mathrm{Gal}(\overline{\mathbb{Q}}/\mathbb{Q})$-modules. □

We thus look for interesting gluable candidates. For $\ell \geq 11$, we considered all genus 1 and 2 curves in the LMFDB and concluded that there are no interesting gluable candidates in the LMFDB. Therefore, we use a larger dataset of genus 2 curves provided by Sutherland [**Sut22**] to obtain the following examples. We check that the gluing is interesting by computing the geometric endomorphism algebra of Y using [**CMSV19**].

EXAMPLE 7.2 (Interesting gluing along 13-torsion). Let X be an elliptic curve $y^2 + y = x^3 + x^2 - 208x - 1256$ (`75.a1` in the LMFDB). Let Y be a genus 2 curve $y^2 + x^3 y = -5x^4 + 45x^2 + 9x$, which has conductor 151 875 and minimal discriminant 2 883 251 953 125. Our code computes a gluing along 13-torsion of X and Y in 24 minutes under with 1000-digit precision, and minimizing the model gives $42y^2 = 24x^8 - 96x^7 + 8x^6 + 192x^5 + 230x^4 - 117x^3 - 177x^2 - 59x + 49$.

EXAMPLE 7.3 (Interesting gluing along 11-torsion). Let X be an elliptic curve $y^2 + xy = x^3 + 9096x + 224832$ (`966.k1` in the LMFDB). Let Y be a genus 2 curve $y^2 + (x^2 + x)y = x^6 - 3x^5 + 9x^4 - 5x^3 + 12x^2 - 6x$. The resulting gluing when $\ell = 11$ is

$$\begin{aligned} Z : 39753x^4 + 89236x^3y - 76006x^3z - 3537x^2y^2 - 469x^2yz + 46697x^2z^2 - 2200xy^3 + 42003xy^2z \\ + 29597xyz^2 - 58478xz^3 - 4883y^4 - 12000y^3z - 9287y^2z^2 - 398yz^3 + 6544z^4 = 0. \end{aligned}$$

7.2. Curves with Interesting Geometric Endomorphism Rings. For any abelian variety A, the ***geometric endomorphism ring*** $\mathrm{End}(A_{\overline{\mathbb{Q}}})$ is the ring of all endomorphisms $A \to A$ defined over $\overline{\mathbb{Q}}$. For any curve C, its geometric endomorphism ring is defined as the geometric endomorphism ring of its Jacobian $\mathrm{End}(C_{\overline{\mathbb{Q}}}) := \mathrm{End}(\mathrm{Jac}\, C_{\overline{\mathbb{Q}}})$. The ***geometric endomorphism algebra*** of a curve C is defined as $\mathrm{End}(C_{\overline{\mathbb{Q}}}) \otimes_{\mathbb{Z}} \mathbb{Q}$. If Z is a gluing of X and Y, then the geometric endomorphism algebra of Z can be easily determined by the geometric endomorphism algebra of X and Y. Furthermore, one can verify the endomorphism algebra by the code in [**CMSV19**].

EXAMPLE 7.4. Let X be the elliptic curve $y^2 = x^3 + x^2 - 3x + 1$ (256.a2 in the LMFDB). Let Y be the genus 2 curve $y^2 + y = 6x^5 + 9x^4 - x^3 - 3x^2$ (20736.l.373248.1 in the LMFDB).

There are two gluings between X and Y with $\ell = 3$, whose minimized equations are given below

$$\begin{aligned} Z &: y^2 = -210x^7 - 630x^6 + 245x^5 + 1155x^4 - 70x^3 - 700x^2 + 140 \\ Z' &: 8x^4 - 26x^3y + 26x^2y^2 + 32xy^3 - 3y^4 + 29x^3z - 26x^2yz + 234xy^2z \\ &\quad - 26y^3z + 153x^2z^2 + 22xyz^2 + 17y^2z^2 + 65xz^3 - 250yz^3 + 25z^4 = 0 \end{aligned}$$

One can compute $\mathrm{End}(X_{\overline{\mathbb{Q}}}) \otimes_{\mathbb{Z}} \mathbb{Q} \simeq \mathbb{Q}[\sqrt{-2}]$ and $\mathrm{End}(Y_{\overline{\mathbb{Q}}}) \otimes_{\mathbb{Z}} \mathbb{Q} \simeq B_{2,3}$, so

$$\mathrm{End}(Z_{\overline{\mathbb{Q}}}) \otimes_{\mathbb{Z}} \mathbb{Q} \simeq \mathrm{End}(Z'_{\overline{\mathbb{Q}}}) \otimes_{\mathbb{Z}} \mathbb{Q} \simeq \mathbb{Q}[\sqrt{-2}] \times B_{2,3},$$

where $B_{2,3}$ is the unique quaternion algebra over $\mathbb{Q}$ ramified at 2 and 3.

7.3. Gluing Curves in the LMFDB. We have applied our algorithms to every genus 2 curve Y in the LMFDB to identify triples (X, Y, ℓ) such that X and Y are potentially gluable along ℓ-torsion where one of the following criteria are met:

- $\ell = 3$ and $H^\perp/H$ is an irreducible Galois module. (If $H^\perp/H$ is reducible, our filtering methods described in Remark 6.1 result in a large number of false positives due to sheer number of candidate pairs of curves (X, Y).) Our methods identified 3009 genus 2 curves Y for which at least one candidate X exists.
- $\ell \geq 5$. We have identified 704 pairs of (Y, ℓ) for which at least one candidate X exists. The distribution by ℓ is shown in Table 3. All gluings for $\ell \geq 11$ are uninteresting in the sense of Proposition 7.1. Of these pairs, 44 curve Y's have $H^\perp/H$ reducible, all of which comes from $\ell = 5$.

TABLE 3. Number of Y for which there exists at least one candidate of gluable elliptic curves along ℓ-torsion.

ℓ	5	7	11	13	19	37	67
Number of Y's	649	35	11	2	4	2	1

For every such (Y, ℓ) such that $H^\perp/H$ is irreducible and we can run the symplectic test, we attempt to glue Y to the candidate X which has minimal conductor. For $\ell = 3$ we attempt this even when we cannot run the symplectic test. The result is shown in Table 4. In all but two cases that we did not find a gluing, the Jacobian of curve Y splits, in which believe that the cause of failure comes from that the condition (ii) of Theorem 2.8 does not hold (cf. Remark 2.9). One further case comes from a pair with the wrong symplectic type, but the symplectic test did not produce a result. The only other case that we did not find a gluing is when $\ell = 3$, Y is the genus 2 curve labeled 471900.a.943800.1 in the LMFDB, and X is the elliptic curve 298800.ff.1 in the LMFDB. In this case, the Galois representations $H^\perp/H$ and $X[3]$ are not isomorphic because the former is unramified at 83 but the latter is not.

The output of our gluings can be found in [**SW25a**].

TABLE 4. Gluing Results.

	$\ell = 3$	$\ell = 5$	$\ell = 7$
$H^{\perp}/H$ irreducible and (for $\ell \geq 5$) passes symplectic test	3009	595	32
Successful gluing	2575	536	19
Gives an error	5	9	1
Did not find gluing	429	50	12

Acknowledgements. This research was conducted through the MIT Department of Mathematics's Summer Program for Undergraduate Research (SPUR). The authors would like to thank our mentors Edgar Costa and Sam Schiavone for their guidance throughout the program. We also thank Eran Assaf and Shiva Chidambaram for helpful discussions. We also thank Prof. David Jerison and Jonathan Bloom for organizing SPUR and for their thoughtful comments about our research. Finally, we thank anonymous referees who provide helpful comments to this paper.

References

[BCCK24] Raymond van Bommel, Shiva Chidambaram, Edgar Costa, and Jean Kieffer, *Computing isogeny classes of typical principally polarized abelian surfaces over the rationals*, LuCaNT: LMFDB, computation, and number theory, Contemp. Math., vol. 796, Amer. Math. Soc., [Providence], RI, [2024] ©2024, pp. 187–214, DOI 10.1090/conm/796/16002. MR4732688

[BCDT01] Christophe Breuil, Brian Conrad, Fred Diamond, and Richard Taylor, *On the modularity of elliptic curves over* **Q***: wild 3-adic exercises*, J. Amer. Math. Soc. **14** (2001), no. 4, 843–939, DOI 10.1090/S0894-0347-01-00370-8. MR1839918

[BHLS15] Reinier Bröker, Everett W. Howe, Kristin E. Lauter, and Peter Stevenhagen, *Genus-2 curves and Jacobians with a given number of points*, LMS J. Comput. Math. **18** (2015), no. 1, 170–197, DOI 10.1112/S1461157014000461. MR3349314

[BILV16] Jennifer S. Balakrishnan, Sorina Ionica, Kristin Lauter, and Christelle Vincent, *Constructing genus-3 hyperelliptic Jacobians with CM*, LMS J. Comput. Math. **19** (2016), no. suppl. A, 283–300, DOI 10.1112/S1461157016000322. MR3540961

[BK75] B. J. Birch and W. Kuyk (eds.), *Modular functions of one variable. IV*, Lecture Notes in Mathematics, vol. Vol. 476, Springer-Verlag, Berlin-New York, 1975. Edited by B. J. Birch and W. Kuyk. MR0376533

[CF22] John E. Cremona and Nuno Freitas, *Global methods for the symplectic type of congruences between elliptic curves*, Rev. Mat. Iberoam. **38** (2022), no. 1, 1–32, DOI 10.4171/rmi/1269. MR4382462

[CMSV19] Edgar Costa, Nicolas Mascot, Jeroen Sijsling, and John Voight, *Rigorous computation of the endomorphism ring of a Jacobian*, Math. Comp. **88** (2019), no. 317, 1303–1339, DOI 10.1090/mcom/3373. MR3904148

[Cre06] John Cremona, *The elliptic curve database for conductors to 130000*, Algorithmic number theory, Lecture Notes in Comput. Sci., vol. 4076, Springer, Berlin, 2006, pp. 11–29, DOI 10.1007/11792086_2. MR2282912

[Die02] Luis V. Dieulefait, *Explicit determination of the images of the Galois representations attached to abelian surfaces with* $\mathrm{End}(A) = \mathbb{Z}$, Experiment. Math. **11** (2002), no. 4, 503–512 (2003). MR1969642

[DS05] Fred Diamond and Jerry Shurman, *A first course in modular forms*, Graduate Texts in Mathematics, vol. 228, Springer-Verlag, New York, 2005. MR2112196

[EK25] Noam D. Elkies and Jean Kieffer, *Fast evaluation of riemann theta functions in any dimension*, 2025, Preprint.

[Fis06] Tom Fisher, *The Hessian of a genus one curve*, Proc. Lond. Math. Soc. (3) **104** (2012), no. 3, 613–648, DOI 10.1112/plms/pdr039. MR2900238

[Fis14] Tom Fisher, *On families of 7- and 11-congruent elliptic curves*, LMS J. Comput. Math. **17** (2014), no. 1, 536–564, DOI 10.1112/S1461157014000059. MR3356045

[Fis20] Tom Fisher, *Explicit moduli spaces for congruences of elliptic curves*, Math. Z. **295** (2020), no. 3-4, 1337–1354, DOI 10.1007/s00209-019-02392-9. MR4125692

[FK91] Gerhard Frey and Ernst Kani, *Curves of genus* 2 *covering elliptic curves and an arithmetical application*, Arithmetic algebraic geometry (Texel, 1989), Progr. Math., vol. 89, Birkhäuser Boston, Boston, MA, 1991, pp. 153–176, DOI 10.1007/978-1-4612-0457-2_7. MR1085258

[FK22] Nuno Freitas and Alain Kraus, *On the symplectic type of isomorphisms of the p-torsion of elliptic curves*, Mem. Amer. Math. Soc. **277** (2022), no. 1361, v+105, DOI 10.1090/memo/1361. MR4403927

[Han20] Jeroen Hanselman, *Gluing curves of genus 2 and genus 1 along their 2-torsion*, Ph.D. thesis, Univeristät Ulm, 2020.

[HLP00] Everett W. Howe, Franck Leprévost, and Bjorn Poonen, *Large torsion subgroups of split Jacobians of curves of genus two or three*, Forum Math. **12** (2000), no. 3, 315–364, DOI 10.1515/form.2000.008. MR1748483

[HSS20] Jeroen Hanselman, Sam Schiavone, and Jeroen Sijsling, *`gluing`, a magma package for gluing curves of low genus along their torsion*, 2020.

[HSS21] Jeroen Hanselman, Sam Schiavone, and Jeroen Sijsling, *Gluing curves of genus 1 and 2 along their 2-torsion*, Math. Comp. **90** (2021), no. 331, 2333–2379, DOI 10.1090/mcom/3627. MR4280304

[KLL+18] Pınar Kılıçer, Hugo Labrande, Reynald Lercier, Christophe Ritzenthaler, Jeroen Sijsling, and Marco Streng, *Plane quartics over $\mathbb{Q}$ with complex multiplication*, Acta Arith. **185** (2018), no. 2, 127–156, DOI 10.4064/aa170227-16-3. MR3856187

[KO92] A. Kraus and J. Oesterlé, *Sur une question de B. Mazur* (French), Math. Ann. **293** (1992), no. 2, 259–275, DOI 10.1007/BF01444715. MR1166121

[KW09] Chandrashekhar Khare and Jean-Pierre Wintenberger, *Serre's modularity conjecture. I*, Invent. Math. **178** (2009), no. 3, 485–504, DOI 10.1007/s00222-009-0205-7. MR2551763

[Lau01] Kristin Lauter, *Geometric methods for improving the upper bounds on the number of rational points on algebraic curves over finite fields*, J. Algebraic Geom. **10** (2001), no. 1, 19–36. With an appendix in French by J.-P. Serre. MR1795548

[LMF24] The LMFDB Collaboration, *The L-functions and modular forms database*, `https://www.lmfdb.org`, 2024, [Online; accessed 30 July 2024].

[Maz78] B. Mazur, *Rational isogenies of prime degree (with an appendix by D. Goldfeld)*, Invent. Math. **44** (1978), no. 2, 129–162, DOI 10.1007/BF01390348. MR482230

[Mil08] James S. Milne, *Abelian varieties (v2.00)*, `www.jmilne.org/math/CourseNotes/av.html`, 2008, pp. 166+vi.

[OU73] Frans Oort and Kenji Ueno, *Principally polarized abelian varieties of dimension two or three are Jacobian varieties*, J. Fac. Sci. Univ. Tokyo Sect. IA Math. **20** (1973), 377–381. MR364265

[Ray74] Michel Raynaud, *Schémas en groupes de type* $(p, \dots, p)$ (French), Bull. Soc. Math. France **102** (1974), 241–280. MR419467

[RR18] Christophe Ritzenthaler and Matthieu Romagny, *On the Prym variety of genus 3 covers of genus 1 curves* (English, with English and French summaries), Épijournal Géom. Algébrique **2** (2018), Art. 2, 8, DOI 10.46298/epiga.2018.volume2.3663. MR3781951

[RS95] K. Rubin and A. Silverberg, *Families of elliptic curves with constant mod p representations*, Elliptic curves, modular forms, & Fermat's last theorem (Hong Kong, 1993), Ser. Number Theory, I, Int. Press, Cambridge, MA, 1995, pp. 148–161. MR1363500

[RS01] K. Rubin and A. Silverberg, *Mod 2 representations of elliptic curves*, Proc. Amer. Math. Soc. **129** (2001), no. 1, 53–57, DOI 10.1090/S0002-9939-00-05539-8. MR1694877

[Rub97] Karl Rubin, *Modularity of mod* 5 *representations*, Modular forms and Fermat's last theorem (Boston, MA, 1995), Springer, New York, 1997, pp. 463–474. MR1638489

[Ser87] Jean-Pierre Serre, *Sur les représentations modulaires de degré* 2 *de* $\mathrm{Gal}(\overline{\mathbf{Q}}/\mathbf{Q})$ (French), Duke Math. J. **54** (1987), no. 1, 179–230, DOI 10.1215/S0012-7094-87-05413-5. MR885783

[Sil94] Joseph H. Silverman, *Advanced topics in the arithmetic of elliptic curves*, Graduate Texts in Mathematics, vol. 151, Springer-Verlag, New York, 1994, DOI 10.1007/978-1-4612-0851-8. MR1312368

[Sil97] Alice Silverberg, *Explicit families of elliptic curves with prescribed mod N representations*, Modular forms and Fermat's last theorem (Boston, MA, 1995), Springer, New York, 1997, pp. 447–461. MR1638488

[Sil09] Joseph H. Silverman, *The arithmetic of elliptic curves*, 2nd ed., Graduate Texts in Mathematics, vol. 106, Springer, Dordrecht, 2009, DOI 10.1007/978-0-387-09494-6. MR2514094

[Stu87] Jacob Sturm, *On the congruence of modular forms*, Number theory (New York, 1984), Lecture Notes in Math., vol. 1240, Springer, Berlin, 1987, pp. 275–280, DOI 10.1007/BFb0072985. MR894516

[Sut22] Andrew Sutherland, *A database of genus 2 curves over $\mathbb{Q}$ with conductor less than 2^{20}.*, 2022.

[SW25a] Pitchayut Saengrungkongka and Noah Walsh, *gluing_data*, 2025. https://github.com/ndwrobotics/gluing_data

[SW25b] ———, *gluing_utilities*, 2025. https://github.com/ndwrobotics/gluing_utilities

Department of Mathematics, Massachusetts Institute of Technology, Cambridge, MA 02139, USA

Email address: psaeng@mit.edu

Department of Mathematics, Massachusetts Institute of Technology, Cambridge, MA 02139, USA

Email address: nwalsh21@mit.edu

Contemporary Mathematics
Volume **840**, 2026
https://doi.org/10.1090/conm/840/16816

A census of genus 6 curves over $\mathbb{F}_2$

Yongyuan Huang, Kiran S. Kedlaya, and Jun Bo Lau

Abstract. We compile a complete list of isomorphism class representatives of curves of genus 6 over $\mathbb{F}_2$. We use explicit descriptions of canonical curves in each stratum of the Brill–Noether stratification of the moduli space $\mathcal{M}_6$, due to Mukai in the generic case. Our computed value of $\#\mathcal{M}_6(\mathbb{F}_2)$ agrees with the Lefschetz trace formula as recently computed by Bergstrom–Canning–Petersen–Schmitt. We also report progress on compiling a corresponding list in genus 7.

1. Introduction

For $g > 1$, let $\mathcal{M}_g$ denote the moduli stack of curves of genus g. (All "curves" herein are smooth, projective, and geometrically irreducible unless otherwise specified.) For each prime power q, the set $\mathcal{M}_g(\mathbb{F}_q)$ of $\mathbb{F}_q$-valued points of $\mathcal{M}_g$ is finite; it is naturally identified with the set of isomorphism classes of curves of genus g over $\mathbb{F}_q$. We equip $\mathcal{M}_g(\mathbb{F}_q)$ with the measure which gives the isomorphism class of a curve C the weight $\frac{1}{\#\mathrm{Aut}(C)}$, as in the Lefschetz trace formula for Deligne–Mumford stacks [**Beh93**]. (Here we count automorphisms of C over $\mathbb{F}_q$ itself, not its base extension to an algebraic closure.) The resulting stacky point count $\#M_g(\mathbb{F}_q)$ can also be interpreted as the number of geometric isomorphism classes of genus g curves which admit a model over $\mathbb{F}_q$ (see Lemma 4.2 for the relationship between these two counting problems).

Since $\mathcal{M}_g$ has relative dimension $3g-3$ over $\mathbb{Z}$, it is feasible to compute the set $\mathcal{M}_g(\mathbb{F}_q)$ for small values of g and q, especially for $q = 2$ where this has been done previously for $g \leq 5$ [**Xar20, Dra24**]. In this paper, we extend the computation to the case $g = 6$.

Theorem 1.1. *We obtain an explicit list of isomorphism class representatives for $\mathcal{M}_6(\mathbb{F}_2)$: it consists of* 72227 *elements, and for the weighting by automorphisms we have*

$$\#\mathcal{M}_6(\mathbb{F}_2) = 68615. \qquad (1)$$

2020 *Mathematics Subject Classification.* 11G20.

Kedlaya was supported by NSF (grant DMS-2053473), UC San Diego (Warschawski professorship), and the Simons Foundation (Simons Fellowship in Mathematics, 2023–24). Lau was supported by the Simons Foundation grant #550023 for the Collaboration on Arithmetic Geometry, Number Theory, and Computation. Computational resources (including the Diophantus cluster at MIT) were provided by the Simons Collaboration in Arithmetic Geometry, Number Theory, and Computation and by Boston University's Department of Mathematics and Statistics.

A list of isomorphism class representatives, as well as the SageMath [**Sag23**] and Magma [**BCP97**] code used to generate it, can be found at

`https://github.com/junbolau/genus-6.`

The list is also available via the table of isogeny classes of abelian varieties over finite fields in the *L-functions and modular forms database* (LMFDB) [**LMF24**]. We observe that 38327 of the 164937 isogeny classes of abelian sixfolds over $\mathbb{F}_2$ contain at least one Jacobian, representing all 20 of the possible Newton polygons, and that the maximum number of Jacobians in a single isogeny class is 20.

Our approach to Theorem 1.1 follows the partial census carried out in [**Ked24b**] (which was limited to curves with particular zeta functions, see below): for each stratum in the Brill–Noether stratification of $\mathcal{M}_6$, we use the descriptions of general canonical curves in each stratum (due to Mukai [**Muk93**] for the generic stratum) to construct a covering set for the isomorphism classes of curves over $\mathbb{F}_2$ in that stratum. We then make extensive use of Magma's implementation of function fields to identify isomorphic curves and compute automorphism groups; the only groups that occur are

$$C_1, C_2, C_3, C_4, C_2 \times C_2, C_5, C_6, S_3, C_{10}, D_5, D_{10}, A_5.$$

We have two main applications in mind for Theorem 1.1. One is to identify curves with a given zeta function; for example, the following statements can now be verified by database queries in LMFDB.

- The maximum number of $\mathbb{F}_2$-points on a curve of genus 6 is 10, achieved by exactly two curves [**Rig10**].
- There are 70 supersingular curves of genus 6 over $\mathbb{F}_2$, with 28 distinct zeta functions.
- There is no curve of genus 6 over $\mathbb{F}_2$ with any of the three possible zeta functions for an *excessive* genus 6 curve allowed by [**FGH**, Theorem 3.2]. We thus recover [**FGH**, Theorem 5.1]: the maximum gonality of a curve of genus 6 over $\mathbb{F}_2$ is 6.
- There is no curve C of genus 6 over $\mathbb{F}_2$ with $\#C(\mathbb{F}_{2^4}) = 0$. This recovers the previous assertion as well as a nonexistence statement made in [**Ked22**, §6].
- There is a unique curve C of genus 6 over $\mathbb{F}_2$ with $(\#C(\mathbb{F}_{2^i}))_{i=1}^6 = (0, 0, 0, 20, 15, 90)$ [**Ked23**, Lemma 10.2].
- As reported in [**Ked24b**, Table 1], there are 52 curves with zeta functions matching one of the options in [**Ked24b**, Table 2]. As shown in [**Ked23**], this list includes every curve of genus 6 admitting an étale double cover with trivial relative class group.

The other intended application is to the computation of the rational cohomology of $\mathcal{M}_g$. There has been much progress in this direction recently; for instance, it was shown in [**CL24b**, Corollary 1.6] that for $g \leq 6$, $\#\mathcal{M}_g(\mathbb{F}_q)$ is a polynomial in q (this has been subsequently improved, see below). More precisely, this follows from the Lefschetz trace formula and the fact that in these cases, the rational cohomology of $\mathcal{M}_g$ can be computed using the tautological Chow ring. The latter can be computed using the SageMath package described in [**DSvZ21**]; by so doing, one can recover the explicit polynomials for $g = 4, 5$, and 6 (see [**BT07**, §4] or [**BFP24**, Theorem 1.5] for $g = 4$ and [**BCPS**] for $g = 5, 6$). The resulting formula for $g = 6, q = 2$ agrees with (1); while Theorem 1.1 is in principle logically independent of this agreement, admitting it allows for an alternate justification of the correctness of our result (see §4).

On the other hand, a tabulation of curves of genus g also yields, for every positive integer n, a point count for the stack $\mathcal{M}_{g,n}$ of n-pointed genus g curves (where the points are distinct and distinguishable) or more generally any quotient of $\mathcal{M}_{g,n}$ by a subgroup of S_n. For example, Theorem 1.1 yields the following.

COROLLARY 1.2. *We have*

$$\#\mathcal{M}_{6,1}(\mathbb{F}_2) = 223317, \tag{2}$$

$$\#(\mathcal{M}_{6,2}/S_2)(\mathbb{F}_2) = 471210, \tag{3}$$

$$\#\mathcal{M}_{6,2}(\mathbb{F}_2) = 650838, \tag{4}$$

$$\#(\mathcal{M}_{6,3}/S_3)(\mathbb{F}_2) = 927153, \tag{5}$$

$$\#\mathcal{M}_{6,3}(\mathbb{F}_2) = 1679646, \tag{6}$$

$$\#(\mathcal{M}_{6,4}/S_4)(\mathbb{F}_2) = 1794569. \tag{7}$$

In all but the last case, [**CLPW24**, Theorem 1.5] implies that the point count over $\mathbb{F}_q$ is a polynomial in q, but as of now the computation of these polynomials remains infeasible using [**DSvZ21**]. Our computation provides one linear constraint on the coefficients of the polynomial, and thus reduces by one the number of rational cohomology groups that need to be computed in order to determine the polynomial. (One could adapt our methods to perform a census over $\mathbb{F}_3$ and thus obtain a second linear constraint; we do not plan to do this.)

We observe that [**Ked24b**] also includes a partial census of genus 7 curves over $\mathbb{F}_2$, which it should be possible to similarly upgrade to a full census; we report some progress at the end of this paper. For comparison, we note that the existence of polynomial formulas for $\#\mathcal{M}_7(\mathbb{F}_q)$ and $\#\mathcal{M}_{7,1}(\mathbb{F}_q)$ has recently been established in [**CLPW24**, Theorem 1.1, Theorem 1.5]; the explicit formula for $\#\mathcal{M}_7(\mathbb{F}_q)$ will appear in [**BCPS**]. Combining the latter with Corollary 1.2 and known polynomial formulas for $\#M_{g,n}(\mathbb{F}_q)$ for $g \leq 6$ will yield the value of $\#\mathcal{M}_7(\mathbb{F}_2)$. (The latter is computed by first computing the point count on the compactification $\overline{\mathcal{M}}_7$ parametrizing stable curves; removing the boundary contribution requires some point counts on $M_{g,n}$ for $g < 7$.)

It is unclear whether one can push this further, say to genus 8 or even 9. On one hand, the expected number of curves (approximately 2^{3g-3} in genus g) is manageable, and we again have explicit descriptions of canonical curves in these genera [**IM03, Muk10, Muk22**]. On the other hand, these descriptions are currently only available over an algebraically closed based field; moreover, it is known that $\#\mathcal{M}_8(\mathbb{F}_q)$ admits a polynomial formula (see again [**CLPW24**, Theorem 1.1]) but it is unclear whether it is feasible to compute this polynomial using current technology. Moreover, it is known that $\#\mathcal{M}_g(\mathbb{F}_q)$ does not admit a polynomial formula for any $g \geq 9$ (see yet again [**CLPW24**, Theorem 1.1]), although for each fixed g it is expected that one can express $\#\mathcal{M}_g(\mathbb{F}_q)$ in terms of Fourier coefficients of certain automorphic forms, such as the discriminant modular form Δ.

2. The Brill–Noether stratification of $\mathcal{M}_6$

We first recall some relevant terminology and facts about $\mathcal{M}_6$. Throughout this discussion, let C be a curve of genus g over a finite field k and let $\overline{k}$ be an algebraic closure of k. Let K be the canonical divisor on C, and $|K|$ be the canonical linear system.

A g^r_d on C is a linear system of dimension r and degree d, which if basepoint-free defines a degree d morphism $C \to \mathbf{P}^r_k$. We call C *hyperelliptic* if there is a finite morphism $C \to \mathbf{P}^1_k$ of degree 2, or equivalently if C admits a g^1_2 (which is automatically basepoint-free if $g \geq 1$). We call the morphism $\iota\colon C \to \mathbf{P}^{g-1}_k$ defined by $|K|$ the *canonical morphism*. For $g \geq 2$, $|K|$ is very ample if and only if C is not hyperelliptic. Thus if C is nonhyperelliptic, the canonical morphism ι is an embedding, and if C is hyperelliptic, ι factors as a degree 2 morphism $C \to \mathbf{P}^1_k$ followed by the Veronese embedding $\mathbf{P}^1_k \to \mathbf{P}^{g-1}_k$. In particular, every genus two curve is hyperelliptic.

For $g \geq 4$, we say C is *trigonal* if C admits a g^1_3 but not a g^1_2; let $\mathcal{T}_g$ be the stack of smooth trigonal curves. The moduli space $\mathcal{T}_g$ admits a stratification by locally closed substacks $\mathcal{T}_{g,n}$ where n runs over integers with $0 \leq n \leq \frac{g+2}{3}$ and $n \equiv g \pmod 2$. The integer n denotes the *Maroni invariant* of a trigonal curve C, defined as the unique nonnegative integer n such that the trigonal cover $C \to \mathbf{P}^1_k$ factors through a closed embedding

$$C \to \mathbf{F}_n := \mathbf{P}_{\mathbf{P}^1_k}(\mathcal{O}_{\mathbf{P}^1_k} \oplus \mathcal{O}(n)_{\mathbf{P}^1})$$

in such a way that the structure map $\mathbf{F}_n \to \mathbf{P}^1_k$ restricts to the trigonal cover $C \to \mathbf{P}^1_k$. Note that $\mathbf{F}_n$ is ruled by the fibers of the structure map; it is in fact the nth Hirzebruch surface (which for $n = 0$ degenerates to $\mathbf{P}^1_k \times_k \mathbf{P}^1_k$), and can also be represented as an $(n,1)$-hypersurface in $\mathbf{P}^1_k \times_k \mathbf{P}^2_k$.

Last but not least, we say C is *bielliptic* if it admits a degree 2 map to a genus 1 curve over k. Any such map gives rise to a g^1_4, but not conversely.

Due to work of Petri and Mukai [**Muk93**], we have the following classification of genus 6 curves over finite fields.

THEOREM 2.1. *Let C be a curve of genus 6 over a finite field k. Then one (and only one) of the following holds.*

1. *The curve C is hyperelliptic.*
2. *The curve C is bielliptic.*
3. *The curve C occurs as a smooth quintic in $\mathbf{P}^2_k$.*
4. *The curve C is trigonal of Maroni invariant 0. In this case, C occurs as a curve of bidegree $(3,4)$ in $\mathbf{P}^1_k \times \mathbf{P}^1_k$.*
5. *The curve C is trigonal of Maroni invariant 2. In this case, C occurs as a complete intersection of type $(2,1) \cap (1,3)$ in $\mathbf{P}^1_k \times_k \mathbf{P}^2_k$, where the $(2,1)$-hyperplane is isomorphic to the Hirzebruch surface $\mathbf{F}_2$.*
6. *The curve C occurs as a transverse intersection of four hyperplanes, a quadric hypersurface, and the 6-dimensional Grassmannian $\mathrm{Gr}(2,5)$ in $\mathbf{P}^9_k$.*

PROOF. Most of the above follows from Petri's theorem. For the details in the last case, see [**Ked24b**, Theorem 3.1] and the references therein. □

REMARK 2.2. Curves as in case 6 of Theorem 2.1 are known as *Brill–Noether-general* curves (c.f. [**PV15**, Theorem 4.1]). We will henceforth refer to them as *generic* curves of genus 6.

REMARK 2.3. As stated in [**PV15**, Theorem 4.1], the space $\mathcal{M}_6$ can be stratified into locally closed substacks consisting of the loci corresponding to each of the cases in Theorem 2.1. In particular:

(1) The locus $\mathcal{H}_6$ of hyperelliptic curves of genus 6 has dimension 11.
(2) The locus $\mathcal{B}_6$ of bielliptic curves of genus 6 has dimension 10.
(3) The locus $\mathcal{Q}_6$ of smooth plane quintic curves of genus 6 has dimension 12.
(4) The locus $\mathcal{T}_{6,0}$ of trigonal curves of genus 6 with Maroni invariant 0 has dimension 13.
(5) The locus $\mathcal{T}_{6,2}$ of trigonal curves of genus 6 with Maroni invariant 2 has dimension 12.
(6) The locus $\mathcal{M}_6^{\mathrm{BN}}$ of generic curves of genus 6 has dimension 15 (it is open in $\mathcal{M}_6$).

3. Tabulation of data

We begin by recording a few convenient facts that allow us to more efficiently search and filter putative genus 6 curves.

(1) Using an analogue of the explicit formula from analytic number theory, Serre (c.f. [**Ser20**, Theorem 5.3.2, 7.1 Table 1] shows that a curve of genus 6 has at most 10 $\mathbb{F}_2$-points, which is a notable refinement from the Hasse-Weil bound 15.
(2) LMFDB contains a complete list of isogeny classes of abelian varieties of dimension 6 over $\mathbb{F}_2$ and their corresponding L-polynomials. Using the fact that a curve and its Jacobian have the same Weil polynomial, we recover a finite set containing the tuple $(\#C(\mathbb{F}_{2^i}))_{i=1}^6$ for any curve C of genus 6 over $\mathbb{F}_2$. The relevant code written in SageMath can be found in `./Census/Shared/weil_poly_utils.sage` in our code base (taken from [**Ked24b**]). We make use of this list when it would presumptively speed up our tabulation process.

In several cases, we use the *orbit lookup trees* introduced by the second author (see [**Ked24b**, Appendix A]) to efficiently compute orbit representatives for the action of a group G on n-element subsets of a finite set S equipped with a left G-action for small values of n. The implementation of this algorithm in SAGEMATH can be found in the file `./Census/Shared/orbits.sage` in our code base (again taken from [**Ked24b**]). The orbit lookup tree algorithm and implementation have subsequently been improved [**Ked24a**], and we have been using the updated version in our census-making in the genus 7 case.

To simplify the code somewhat, initially we only construct a finite set of genus 6 curves over $\mathbb{F}_2$ which meets every isomorphism class with "limited redundancies". We use a separate postprocessing step to remove these redundancies (see §3.7).

3.1. Hyperelliptic curves. Here we follow the strategy used in [**Xar20**, **Dra24**] where the enumerations were done in cases $g = 4, 5$. This strategy is adapted to characteristic 2; for a good approach in odd characteristic, see [**How25**].

Any hyperelliptic curve of genus g over $\mathbb{F}_2$ can be represented as $y^2 + q(x)y = p(x)$ with $p(x), q(x) \in \mathbb{F}_2[x]$ and $2g + 1 \leq \max\{2\deg(q(x)), \deg(p(x))\} \leq 2g + 2$.

Xarles presented a method to determine the isomorphism class of a hyperelliptic curve using the action of $\mathrm{PGL}_2(\mathbb{F}_2)$ on $\mathbb{F}_2[x]_{\leq g+1}$.

LEMMA 3.1. *([**Xar20**], Lemma 1) Let H_1, H_2 be hyperelliptic curves represented by the equations $y^2 + q_i(x)y = p_i(x)$ for $i \in \{1, 2\}$ respectively as above. Suppose that $H_1 \cong H_2$. Then there exists $A \in \mathrm{PGL}_2(\mathbb{F}_2)$ such that $q_2(x) = \psi_{g+1}(A)(q_1(x))$, where the action of $\mathrm{PGL}_2(\mathbb{F}_2)$ on $\mathbb{F}_2[x]_{\leq n}$ is given by*

$$\psi_n(A)(q(x)) := (cx + d)^n q\left(\frac{ax + b}{cx + d}\right), \qquad A = \begin{pmatrix} a & b \\ c & d \end{pmatrix} \in \mathrm{PGL}_2(\mathbb{F}_2).$$

We compute orbit representatives for this action, test for pairwise isomorphism, and record the resulting curves. The implementation of this method can be found in the folder `./Census/hyperelliptic/` in our code base.

3.2. Bielliptic curves. Here we follow a construction from [**Ked24b**] for enumerating bielliptic curves of genus 7 over $\mathbb{F}_2$. As the core elements of the argument exhibit no dependence at all on the genus, it is straightforward to replicate in genus 6; this case was not needed in [**Ked24b**] because in that setting (where the zeta function is heavily restricted) one can use a more conceptual argument to rule out bielliptic curves of genus 6.

By Riemann–Hurwitz plus the fact that double covers in characteristic 2 have only wild ramifications, the map from a bielliptic curve C of genus 6 to its elliptic quotient E has ramification divisor of the form $2D$ where D is an effective divisor of degree $g - 1 = 5$ on E. We may thus generate all bielliptic curves by enumerating over a set of isomorphism class representatives of elliptic curves E over $\mathbb{F}_2$ (there are 5 of them). For each E, we use MAGMA to enumerate over all effective divisors D of degree 5. For each D, we enumerate over all order-2 quotients of the ray class group of $2D$, form the corresponding abelian extension, then check to see if it indeed has genus 6 (and if so record the resulting curve). The implementation of this method can be found in `./Census/bielliptic/` in our code base.

3.3. Smooth plane quintic curves. Since the space of quintic polynomials over $\mathbb{F}_2$ has dimension $\binom{7}{2} = 21$, we can skip the removal of redundancy using the action of $\mathrm{GL}(3, \mathbb{F}_2)$; we simplify identify all of the nonsingular polynomials and record the resulting smooth curves. The implementation of this method can be found in `./Census/plane_quintic/` in our code base.

3.4. Trigonal curves of Maroni invariant 0. In this case, we are looking for $(3, 4)$-curves in $\mathbf{P}^1 \times \mathbf{P}^1$, and we follow the strategy used in [**Ked24b**]. We first compute orbit representatives for the action of $\mathrm{PGL}_2(\mathbb{F}_2) \times \mathrm{PGL}_2(\mathbb{F}_2)$ on *all* subsets of $(\mathbf{P}^1 \times \mathbf{P}^1)(\mathbb{F}_2)$. For each orbit representative, we identify the $(3, 4)$-polynomials which vanish on the points in the chosen subset and do not vanish elsewhere; since we are working over $\mathbb{F}_2$, this is an affine subspace of the vector space of $(3, 4)$-polynomials. We then pick out the nonsingular polynomials and record the resulting smooth curves. The implementation of this method can be found in `./Census/trigonal_maroni_0/` in our code base.

3.5. Trigonal curves of Maroni invariant 2. In this case, we are looking for complete intersections of type $(2,1) \cap (1,3)$ in $\mathbf{P}^1 \times \mathbf{P}^2$. More specifically, if we write $\mathbf{P}^1 \times \mathbf{P}^2 = \operatorname{Proj} \mathbb{F}_2[x_0, x_1; y_0, y_1, y_2]$, then we may take the $(2,1)$-hypersurface X_1 to be

$$(x_0^2 + x_1^2)y_1 + x_0 x_1 y_2 = 0. \tag{8}$$

Indeed, over a field of characteristic 0, the equation of the Hirzebruch surface $\mathbf{F}_n$ is isomorphic to the hypersurface defined by $x_0^n y_1 - x_1^n y_2$ in $\mathbf{P}^1 \times \mathbf{P}^2$ (c.f. [**Huy04**] Exercise 2.4.5), and we obtain (8) by taking $n = 2$ and making a change of variables to get an equation with smooth mod-2 reduction.

The hypersurface (8) is fixed by the group G generated by the three involutions

$$x_0 \leftrightarrow x_1; \qquad y_0 \mapsto y_0 + y_1; \qquad y_0 \mapsto y_0 + y_2.$$

We now proceed as in the previous case: we compute orbit representatives for the action of G on all subsets of X_1; for each orbit representative, we identify the $(1,3)$-polynomials which vanish on the points in the chosen subset and do not vanish elsewhere; we then pick out the nonsingular polynomials and record the resulting smooth curves. The implementation of this method can be found in `./Census/trigonal_maroni_2/` in our code base.

3.6. Generic curves. Here we follow a modified version of the strategy used in [**Ked24b**]. This is the most computationally intensive case. We first identify orbit representatives for the action of $\mathrm{PGL}_5(\mathbb{F}_2)$ on 4-tuples of points in $\mathbf{P}^{9^\vee}(\mathbb{F}_2)$. Each 4-tuple defines 4 linear forms and hence 4 hyperplanes on $\mathbf{P}^9$; we next compute representatives for the linear action of $\mathrm{PGL}_4(\mathbb{F}_2)$ on such 4-tuples preserving the intersection of the 4 hyperplanes. We record all cases where the intersection of the 4 hyperplanes with the Grassmannian $\mathrm{Gr}(2,5)$ is irreducible with singular locus of codimension greater than 1; there are 17 such intersections, of which 7 are smooth, corresponding to the fact that quintic del Pezzo surfaces over a finite field are indexed by conjugacy classes in S_5 (e.g., see [**Tre20**, Table 1]).

For each of these 17 intersections, we first record all the quadrics defined on the span of the 4 linear forms, which reduces the enumeration of quadrics from a $\binom{10}{2} = 55$-dimensional space to a $\binom{7}{2} = 21$-dimensional space); we then record the cases where the intersection is smooth of genus 6. The implementation of this method can be found in `./Census/generic/` in our code base.

3.7. Postprocessing. For each stratum, the computation described above yields a finite set of curves of genus 6 over $\mathbb{F}_2$ lying in that stratum and including at least one representative of each isomorphism class. It then remains to remove redundant representatives.

For this, we first hash the curves by their zeta function, or equivalently by the function $C \mapsto (\#C(\mathbb{F}_{2^i}))_{i=1}^6$. Within each hash class, we use MAGMA to construct the function field of each curve, then use `Isomorphisms` to test whether any pair of curves is isomorphic. Once this is done, we compute the automorphism group of each curve that remains.

For the record, we mention some bugs in MAGMA that we had to work around. These were fixed in subsequent releases based on our reports.

- For two function fields, the function `Isomorphisms` returns a list of all isomorphisms between the two fields, but in some cases with repeated entries. This caused `AutomorphismGroup` to yield errors in certain cases, for which we computed the group structure directly from the output of `Isomorphisms`.

- For two function fields, the function `IsIsomorphic` sometimes returned `false` even when the two fields are isomorphic. We instead tested whether `Isomorphisms` returns a nonempty list.

4. Consistency checks

The proof of Theorem 1.1 implicitly depends on the correctness both of the relevant features of the underlying computational systems (in particular SageMath and Magma) and of our implementation of the search strategies described above. It is thus highly desirable to perform some logically independent consistency checks of the resulting data. We describe several such checks here.

4.1. Point counting on $\mathcal{M}_6$. We first verify the numerical assertion (1). By [**CL24b**, Corollary 1.6], there exists a monic polynomial $P(T) \in \mathbb{Z}[T]$ of degree 15 such that $\#\mathcal{M}_6(\mathbb{F}_q) = P(q)$ for every prime power q. On account of the Lefschetz trace formula for Deligne–Mumford stacks [**Beh93**, Theorem 3.1.2], it is a feasible but challenging computation to extract the exact polynomial by computing in the tautological ring of $\mathcal{M}_6$ as indicated (and implemented) in [**DSvZ21**].

Theorem 4.1. *For every prime power q, we have*

$$\#\mathcal{M}_6(\mathbb{F}_q) = q^{15} + q^{14} + 2q^{13} + q^{12} - q^{10} + q^3 - 1.$$

In particular, $\#\mathcal{M}_6(\mathbb{F}_2) = 68615$ as asserted in Theorem 1.1.

Proof. See [**BCPS**]. □

Given Theorem 4.1, one can give an alternate proof of Theorem 1.1 by independently checking the following two concrete assertions:

- For each tabulated curve C, the order of $\#\mathrm{Aut}(C)$ is no greater than the reported value.
- No two of the tabulated curves lying in the same stratum are isomorphic. (For an extra consistency check, we tested this in Magma also for pairs of curves not lying in the same stratum.)

Given these assertions, one may then directly verify from our data that $\#\mathcal{M}_6(\mathbb{F}_q) \geq 68615$ with equality if and only if our census is complete. Combining with Theorem 4.1 then yields Theorem 1.1.

4.2. Point counts with marked points. As noted earlier, given Theorem 1.1 one can count the $\mathbb{F}_2$-points of any moduli stack corresponding to genus 6 curves with some additional marked structure, as in Corollary 1.2. This count will always yield an integer thanks to the following fact.

Lemma 4.2. *Let $\mathcal{X}$ be a Deligne–Mumford stack over a finite field $\mathbb{F}_q$ admitting a coarse moduli space X. Then $\#\mathcal{X}(\mathbb{F}_q) = \#X(\mathbb{F}_q)$.*

Proof. See [**BFP24**, Proposition 1.3(3)]. □

4.3. Point counts in strata. Point counts of some strata of $\mathcal{M}_6$ are also known, and can be used to check the corresponding sections of our table. See Table 1 for a summary of this discussion.

- For hyperelliptic curves, it is straightforward to compute that
$$\#\mathcal{H}_6(\mathbb{F}_q) = q^{11};$$
see [**Ber09**] for much stronger results.
- For plane quintics, Gorinov [**Gor05**] showed that $\mathcal{Q}_6$ has trivial rational cohomology, yielding
$$\#\mathcal{Q}_6(\mathbb{F}_q) = q^{12}.$$
This has been rederived by elementary means by Wennink [**Wen**].

We are not aware of any prior computation of $\#\mathcal{T}_{6,n}(\mathbb{F}_q)$. Comparing the values for $q = 2$ with Zheng's results on the stable cohomology of $\mathcal{T}_{g,n}$ [**Zhe24**] suggests that
$$\#\mathcal{T}_{6,0}(\mathbb{F}_q) \approx q^{13} - q^{10}, \qquad \#\mathcal{T}_{6,2}(\mathbb{F}_q) \approx q^{12} + q^{11}.$$
While there is no reason to expect a dearth of lower-order terms, for $q = 2$ these expressions come astonishingly close to the computed counts (7168 and 6144 in place of the actual values 7166 and 6148).

For $\#\mathcal{B}_6(\mathbb{F}_q)$, we have the following result for odd primes that does not appear to have been reported previously, but which does not yield a correct prediction for $q = 2$ (see below).

LEMMA 4.3. *For $g \geq 6$, let $H_{2g-2} \to \mathcal{M}_1$ be the representable morphism of Artin stacks whose fiber over a point $[E]$ is the Hilbert scheme of reduced zero-dimensional closed subschemes of E of length $2g - 2$. Then for every odd prime power q, $\#\mathcal{B}_g(\mathbb{F}_q) = \#H_{2g-2}(\mathbb{F}_q)$.*

PROOF. For any bielliptic curve C of genus $g \geq 6$, by Castelnuovo–Severi the map from C to its genus-1 quotient is unique up to composition by an automorphism of the target. In particular, the bielliptic involution ι of C is central in $\mathrm{Aut}(C)$.

For a given elliptic curve E over $\mathbb{F}_q$ (which as usual has a marked point O) and a given g, every bielliptic covering $C \to E$ of genus g gives rise to a pair $(D, \mathcal{L})$ in which D is an effective squarefree divisor on E of degree $2g - 2$ (the branch locus) and $\mathcal{L}$ is a square root of the line bundle $\mathcal{O}(D)$. In particular, such a pair can only exist if the sum over D yields an element of $2E(\mathbb{F}_q)$; when this condition does hold, the square roots of $\mathcal{O}(D)$ form a torsor for the group $E(\mathbb{F}_q)[2]$. Moreover, the bielliptic covering is determined by the pair up to a relative quadratic twist. This yields the claim. □

REMARK 4.4. *Lemma 4.3 breaks down for q even because the bielliptic covering is no longer tame, so its branch locus is no longer reduced. See Remark 4.6 for further discussion.*

PROPOSITION 4.5. *For $6 \leq g \leq 11$, for every odd prime q,*
$$\#\mathcal{B}_g(\mathbb{F}_q) = \frac{q^{2g} - q^{2g-4} - q^{2g-5} + (-1)^{g+1} q}{q^2 + 1}. \tag{9}$$

PROOF. For E an elliptic curve over $\mathbb{F}_q$, let E° denote the set of closed points of E (of arbitrary degree) and let $a(E) := q + 1 - \#E(\mathbb{F}_q)$ be the trace of Frobenius of E. For $n \geq 0$, let $d_n(E)$ denote the number of effective squarefree divisors of degree n on E. View $\mathcal{M}_{1,1}(\mathbb{F}_q)$ as a measure space by weighting the isomorphism class of E by $\frac{1}{\#\mathrm{Aut}(E)}$. By Lemma 4.3,

$$\#\mathcal{B}_g(\mathbb{F}_q) = \int_{\mathcal{M}_{1,1}(\mathbb{F}_q)} \frac{d_{2g-2}(E)}{q - a(E) + 1}.$$

We compute the generating series for $d_n(E)$ by writing

$$\begin{aligned}\sum_{n=0}^{\infty} d_n(E)T^n &= \prod_{x \in E^\circ} (1 + T^{\deg(x)}) = \prod_{x \in E^\circ} \frac{1 - T^{2\deg(x)}}{1 - T^{\deg(x)}} \\ &= \frac{Z(X,T)}{Z(X,T^2)} = \frac{(1-T^2)(1-qT^2)(1-a(E)T+qT^2)}{(1-T)(1-qT)(1-a(E)T^2+qT^4)} \\ &= 1 + (q - a(E) + 1)T\frac{1-qT^3}{(1-qT)(1-a(E)T^2+qT^4)}.\end{aligned}$$

In other words, writing $[T^n]f$ for the coefficient of T^n in the power series f, we have

$$\frac{d_n(E)}{q - a(E) + 1} = [T^{n-1}]\frac{1-qT^3}{(1-qT)(1-a(E)T^2+qT^4)} \qquad (n > 0)$$

and hence

$$\#\mathcal{B}_g(\mathbb{F}_q) = \int_{\mathcal{M}_{1,1}(\mathbb{F}_q)} [T^{2g-3}]\frac{1-qT^3}{(1-qT)(1-a(E)T^2+qT^4)}.$$

Since we are only extracting odd coefficients, we may rewrite this as

$$\begin{aligned}\#\mathcal{B}_g(\mathbb{F}_q) &= \frac{1}{2}\int_{\mathcal{M}_{1,1}(\mathbb{F}_q)} [T^{2g-3}]\left(\frac{1-qT^3}{1-qT} - \frac{1+qT^3}{1+qT}\right)\frac{1}{1-a(E)T^2+qT^4} \\ &= q\int_{\mathcal{M}_{1,1}(\mathbb{F}_q)} [T^{2g-3}]\frac{T(1-T^2)}{(1-q^2T^2)(1-a(E)T^2+qT^4)} \\ &= q\int_{\mathcal{M}_{1,1}(\mathbb{F}_q)} [T^{g-2}]\frac{1-T}{(1-q^2T)(1-a(E)T+qT^2)} \\ &= q[T^{g-2}]\frac{1-T}{1-q^2T}\int_{\mathcal{M}_{1,1}(\mathbb{F}_q)} \frac{1}{1-a(E)T+qT^2}.\end{aligned}$$

At this point, we begin to write a instead of $a(E)$ to shorten the notation. To evaluate the integral, we first recall that $\mathcal{M}_{1,1}(\mathbb{F}_q)$ has total measure q. We next recall that elliptic curves over $\mathbb{F}_q$ come in quadratic twist pairs whose Frobenius traces differ by a sign, so $\int_{\mathcal{M}_{1,1}(\mathbb{F}_q)} a^{2n+1} = 0$ for all $n \geq 0$. In particular,

$$\begin{aligned}\int_{\mathcal{M}_{1,1}(\mathbb{F}_q)} \frac{1}{1-aT+qT^2} &= \int_{\mathcal{M}_{1,1}(\mathbb{F}_q)} \frac{1}{2(1-aT+qT^2)} + \frac{1}{2(1+aT+qT^2)} \\ &= \int_{\mathcal{M}_{1,1}(\mathbb{F}_q)} \frac{1+qT^2}{1-(a^2-2q)T^2+q^2T^4} \\ &\equiv \int_{\mathcal{M}_{1,1}(\mathbb{F}_q)} (1 + (a^2-q)T^2 + (a^4 - 3a^2q + q^2)T^4 \\ &\quad + (a^6 - 5a^4q + 6a^2q^2 - q^3)T^6 \\ &\quad + (a^8 - 7a^6q + 15a^4q^2 - 10a^2q^3 + q^4)T^8) \pmod{T^{10}}.\end{aligned}$$

We finally invoke a result of Birch [**Bir68**] (as reformulated in [**KP17**, Theorem 1]): for q an odd prime,

$$\int_{\mathcal{M}_{1,1}(\mathbb{F}_q)} a^2 = q^2 - 1$$
$$\int_{\mathcal{M}_{1,1}(\mathbb{F}_q)} a^4 = 2q^3 - 3q - 1$$
$$\int_{\mathcal{M}_{1,1}(\mathbb{F}_q)} a^6 = 5q^4 - 9q^2 - 5q - 1$$
$$\int_{\mathcal{M}_{1,1}(\mathbb{F}_q)} a^8 = 14q^5 - 28q^3 - 20q^2 - 7q - 1.$$

This yields

$$\int_{\mathcal{M}_{1,1}(\mathbb{F}_q)} \frac{1}{1 - aT + qT^2} \equiv q - T^2 - T^4 - T^6 - T^8 \pmod{T^{10}};$$

hence for $6 \leq g \leq 11$,

$$\begin{aligned}
\#\mathcal{B}_g(\mathbb{F}_q) &= q[T^{g-2}]\frac{1-T}{1-q^2T}\left(q - \frac{T^2}{1-T^2}\right) \\
&= q[T^{g-2}]\frac{q-(q+1)T^2}{(1+T)(1-q^2T)} \\
&= \frac{q}{q^2+1}[T^{g-1}]\left(\frac{q-(q+1)T^2}{1-q^2T} - \frac{q-(q+1)T^2}{1+T}\right) \\
&= \frac{q}{q^2+1}(qq^{2g-2} - (q+1)q^{2g-6} - q(-1)^{g-1} + (q+1)(-1)^{g-3})
\end{aligned}$$

which simplifies to the stated expression. □

REMARK 4.6. *There is a mild misattribution in the proof of Proposition* 4.5*: Birch proved his theorem assuming* $q \geq 5$*, whereas we stated our result for* $q \geq 3$*. However, Ihara's extension of Birch's formula works uniformly in all characteristics and provides an extension to prime powers (reflecting the fact that* $\overline{\mathcal{M}}_{1,2g-2}$ *is smooth over* $\mathbb{Z}$*); see* [**KP17**, Theorem 2] *for a compact statement.*

Even with this, the proof of Proposition 4.5 *does not cover* $q = 2$ *(Remark* 4.4*), and indeed the formula* (9) *breaks down: it predicts* $\#\mathcal{B}_6(\mathbb{F}_2) = 742$*, whereas the computed value is* 744.

REMARK 4.7. *It is also shown in* [**Bir68**] *that* $\int_{\mathcal{M}_{1,1}(\mathbb{F}_q)} a(E)^{10}$ *includes a nonzero contribution from the* Δ *modular form, as then do* $\#\mathcal{M}_{1,11}(\mathbb{F}_q)$ *and* $\#\mathcal{B}_{12}(\mathbb{F}_q)$*; in particular, neither of these is a polynomial in* q*. This state of affairs corresponds to the fact that the bielliptic locus of* $\mathcal{M}_g$ *has only tautological cycle classes for* $g \leq 11$ [**CL24a**] *but not for* $g = 12$ [**vZ18**].

TABLE 1. Point counts (unweighted and weighted) over $\mathbb{F}_2$ of the various strata of $\mathcal{M}_6$. When known, the weighted point count over $\mathbb{F}_q$ is also listed.

Stratum	Unweighted	Weighted	Weighted count over $\mathbb{F}_q$
$\mathcal{H}_6$	4134	2048	q^{11}
$\mathcal{B}_6$	1530	744	$q^{10} - q^8 - q^5 + q^3 - q$ (q odd)
$\mathcal{Q}_6$	4204	4096	q^{12}
$\mathcal{T}_{6,0}$	7282	7166	?
$\mathcal{T}_{6,2}$	6181	6148	?
$\mathcal{M}_6^{\mathrm{BN}}$	48896	48413	?
$\mathcal{M}_6$	72227	68615	$q^{15} + q^{14} + 2q^{13} + q^{12} - q^{10} + q^3 - 1$

5. Progress in genus 7

As in [**Ked23**], one can also describe the Brill–Noether stratification of $\mathcal{M}_7$ explicitly.

THEOREM 5.1. *Let C be a curve of genus 7 over a finite field k. Then one (and only one) of the following holds.*

(1) *The curve C is hyperelliptic.*
(2) *The curve C is trigonal of Maroni invariant 3. In this case, C occurs as a hypersurface of degree 9 in $\mathbf{P}(1:1:3)_k$.*
(3) *The curve C is trigonal of Maroni invariant 1. In this case, C occurs as a complete intersection of type $(1,1) \cap (3,3)$ in $\mathbf{P}^1_k \times_k \mathbf{P}^2_k$.*
(4) *The curve C is bielliptic.*
(5) *The curve C is not bielliptic but admits a g^2_6 which is self-adjoint (squares to the canonical class). In this case, C is a complete intersection of type $(3) \cap (4)$ in $\mathbf{P}(1:1:1:2)_k$, where the degree 3 hypersurface can be taken to be defined by $x_0x_3 + P_3(x_1, x_2) = 0$ for some separable cubic P_3.*
(6) *The curve C admits a pair of distinct g^2_6's. In this case, C occurs as a complete intersection of type $(1,1) \cap (1,1) \cap (2,2)$ in $\mathbf{P}^2_k \times_k \mathbf{P}^2_k$.*
(7) *The curve C does not admit a g^2_6 but $C_{\overline{k}}$ does. In this case, C occurs as a complete intersection of type $(1,1) \cap (1,1) \cap (2,2)$ in the quadratic twist of $\mathbf{P}^2_k \times_k \mathbf{P}^2_k$.*
(8) *The curve C admits a g^1_4 but $C_{\overline{k}}$ does not admit a g^2_6. In this case, C occurs as a complete intersection of type $(1,1) \cap (1,2) \cap (1,2)$ in $\mathbf{P}^1_k \times_k \mathbf{P}^3_k$ in which the $(1,1)$-hypersurface is a $\mathbf{P}^2$-bundle over $\mathbf{P}^1$.*
(9) *The curve C does not admit a g^1_4. In this case, C occurs as a transverse intersection of 9 hyperplanes and the orthogonal Grassmannian $\mathrm{OG}^+(5,10)$ in $\mathbf{P}^{15}_k$.*

PROOF. Again, most of the above follows from Petri's theorem. For the rest, see [**Ked24b**, Theorem 3.2]. □

We list the dimensions of the corresponding strata in moduli in Table 2. Note that cases (6) and (7) of Theorem 5.1 together correspond to a single stratum in $\mathcal{M}_7$, as the distinction between them is not stable under base change; the listed value for the dimension in these cases corresponds to this single stratum.

TABLE 2. Preliminary point counts (unweighted and weighted) over $\mathbb{F}_2$ of the various strata of $\mathcal{M}_7$. The cases "rational g^2_6" and "irrational g^2_6" together constitute a single stratum.

Stratum	Dimension	Unweighted	Weighted
$\mathcal{H}_7$	13	16544	8192
$\mathcal{B}_7$	12	6205	2970
$\mathcal{T}_{7,3}$	13	8340	8264
$\mathcal{T}_{7,1}$	15	42857	42725
self-adjoint g^2_6	15	24925	24580
rational g^2_6	16	43012	42240.5
irrational g^2_6		50791	49982.5
tetragonal	17	171102	169850
$\mathcal{M}_7^{\mathrm{BN}}$	18	?	?

Using the methods of [**Ked24b**], we have enumerated $\mathbb{F}_2$-points of $\mathcal{M}_7$ in all but the generic stratum; preliminary results are reported in Table 2, but these should be treated with some caution as we have not yet been able to run many consistency checks. That said, we can identify some features of the data that comport with theoretical predictions.

- We have $\#\mathcal{H}_7(\mathbb{F}_2) = 2^{13}$.
- We have $\#\mathcal{B}_7(\mathbb{F}_2) = 2^{12} - 2^{10} - 2^7 + 2^5 - 2^3 - 2$ which agrees with the formula from (9) for q odd, in contrast with Remark 4.6.
- Each stratum has integral weighted point count, as predicted by Lemma 4.2. In particular, the sum of the rational and irrational g^2_6 counts is integral.

We also observe that all 31 possible Newton polygons for abelian varieties of dimension 7 over $\mathbb{F}_2$ occur for Jacobians.

Acknowledgments

Thanks to Samir Canning for discussions about [**BCPS**] and to David Roe for importing our data into the abelian varieties over finite fields section of the LMFDB [**LMF24**].

References

[BCP97] Wieb Bosma, John Cannon, and Catherine Playoust, *The Magma algebra system. I. The user language*, J. Symbolic Comput. **24** (1997), no. 3-4, 235–265, DOI 10.1006/jsco.1996.0125. Computational algebra and number theory (London, 1993). MR1484478

[BCPS] Jonas Bergström, Samir Canning, Dan Petersen, and Johannes Schmitt, in preparation.

[Beh93] Kai A. Behrend, *The Lefschetz trace formula for algebraic stacks*, Invent. Math. **112** (1993), no. 1, 127–149, DOI 10.1007/BF01232427. MR1207479

[Ber09] Jonas Bergström, *Equivariant counts of points of the moduli spaces of pointed hyperelliptic curves*, Doc. Math. **14** (2009), 259–296. MR2538614

[BFP24] Jonas Bergström, Carel Faber, and Sam Payne, *Polynomial point counts and odd cohomology vanishing on moduli spaces of stable curves*, Ann. of Math. (2) **199** (2024), no. 3, 1323–1365, DOI 10.4007/annals.2024.199.3.7. MR4740541

[Bir68] B. J. Birch, *How the number of points of an elliptic curve over a fixed prime field varies*, J. London Math. Soc. **43** (1968), 57–60, DOI 10.1112/jlms/s1-43.1.57. MR230682

[BT07] Jonas Bergström and Orsola Tommasi, *The rational cohomology of $\overline{\mathcal{M}}_4$*, Math. Ann. **338** (2007), no. 1, 207–239, DOI 10.1007/s00208-006-0073-z. MR2295510

[CL24a] Samir Canning and Hannah Larson, *The bielliptic locus in genus 11*, Michigan Math. J. **75** (2025), no. 3, 623–638, DOI 10.1307/mmj/20226306. MR4929116

[CL24b] ———, *On the Chow and cohomology rings of moduli spaces of stable curves*, Journal of the European Mathematical Society (2024), arXiv:2208.02357.

[CLPW24] Samir Canning, Hannah Larson, Sam Payne, and Thomas Willwacher, *Moduli spaces of curves with polynomial point counts*, arXiv:2410.19913.

[Dra24] Dušan Dragutinović, *Computing binary curves of genus five*, J. Pure Appl. Algebra **228** (2024), no. 4, Paper No. 107522, 19, DOI 10.1016/j.jpaa.2023.107522. MR4642980

[DSvZ21] Vincent Delecroix, Johannes Schmitt, and Jason van Zelm, *admcycles—a Sage package for calculations in the tautological ring of the moduli space of stable curves*, J. Softw. Algebra Geom. **11** (2021), no. 1, 89–112, DOI 10.2140/jsag.2021.11.89. MR4387186

[FG22] Xander Faber and Jon Grantham, *Binary curves of small fixed genus and gonality with many rational points*, J. Algebra **597** (2022), 24–46, DOI 10.1016/j.jalgebra.2022.01.008. MR4372127

[FGH] Xander Faber, Jon Grantham, and Everett W. Howe, *On the maximum gonality of a curve over a finite field*, Algebra Number Theory **19** (2025), no. 8, 1637–1662, DOI 10.2140/ant.2025.19.1637. MR4930589

[Gor05] Alexei G. Gorinov, *Real cohomology groups of the space of nonsingular curves of degree 5 in* $\mathbf{CP}^2$ (English, with English and French summaries), Ann. Fac. Sci. Toulouse Math. (6) **14** (2005), no. 3, 395–434. MR2172585

[How25] Everett W. Howe, *Enumerating hyperelliptic curves over finite fields in quasi-linear time*, Res. Number Theory **11** (2025), no. 1, Paper No. 26, 24, DOI 10.1007/s40993-024-00594-7. MR4852290

[Huy04] Daniel Huybrechts, *Complex geometry*, Universitext, Springer-Verlag, Berlin, 2005. An introduction. MR2093043

[Iha67] Yasutaka Ihara, *Hecke Polynomials as congruence ζ functions in elliptic modular case*, Ann. of Math. (2) **85** (1967), 267–295, DOI 10.2307/1970442. MR207655

[IM03] Manabu Ide and Shigeru Mukai, *Canonical curves of genus eight*, Proc. Japan Acad. Ser. A Math. Sci. **79** (2003), no. 3, 59–64. MR1967047

[Ked22] Kiran S. Kedlaya, *The relative class number one problem for function fields, I*, Res. Number Theory **8** (2022), no. 4, Paper No. 79, 21, DOI 10.1007/s40993-022-00369-y. MR4493405

[Ked23] Kiran S. Kedlaya, *The relative class number one problem for function fields, I*, Res. Number Theory **8** (2022), no. 4, Paper No. 79, 21, DOI 10.1007/s40993-022-00369-y. MR4493405

[Ked24a] ———, *Orbit lookup trees*, https://github.com/kedlaya/orbits/blob/main/orbits.pdf.

[Ked24b] Kiran S. Kedlaya, *The relative class number one problem for function fields, III*, LuCaNT: LMFDB, computation, and number theory, Contemp. Math., vol. 796, Amer. Math. Soc., Providence, RI, 2024, pp. 55–73, DOI 10.1090/conm/796/15997. MR4732683

[KP17] Nathan Kaplan and Ian Petrow, *Elliptic curves over a finite field and the trace formula*, Proc. Lond. Math. Soc. (3) **115** (2017), no. 6, 1317–1372, DOI 10.1112/plms.12069. MR3741853

[LMF24] The LMFDB Collaboration, *The L-functions and modular forms database*, https://www.lmfdb.org, 2024, [Online; accessed 6 January 2024].

[Muk93] Shigeru Mukai, *Curves and Grassmannians*, Algebraic geometry and related topics (Inchon, 1992), Conf. Proc. Lecture Notes Algebraic Geom., I, Int. Press, Cambridge, MA, 1993, pp. 19–40. MR1285374

[Muk10] Shigeru Mukai, *Curves and symmetric spaces, II*, Ann. of Math. (2) **172** (2010), no. 3, 1539–1558, DOI 10.4007/annals.2010.172.1539. MR2726093

[Muk22] Shigeru Mukai, *Curves and symmetric spaces III: BN-special vs. 1-PS degeneration*, Proc. Indian Acad. Sci. Math. Sci. **132** (2022), no. 2, Paper No. 57, 9, DOI 10.1007/s12044-022-00702-6. MR4493391

[PV15] Nikola Penev and Ravi Vakil, *The Chow ring of the moduli space of curves of genus six*, Algebr. Geom. **2** (2015), no. 1, 123–136, DOI 10.14231/AG-2015-006. MR3322200

[Rig10] Alessandra Rigato, *Uniqueness of low genus optimal curves over* $\mathbb{F}_2$, Arithmetic, geometry, cryptography and coding theory 2009, Contemp. Math., vol. 521, Amer. Math. Soc., Providence, RI, 2010, pp. 87–105, DOI 10.1090/conm/521/10276. MR2744036

[Sag23] The Sage Developers, *SageMath, the Sage Mathematics Software System*, 2023, DOI 10.5281/zenodo.6259615.

[Ser20] Jean-Pierre Serre, *Rational points on curves over finite fields*, Documents Mathématiques (Paris) [Mathematical Documents (Paris)], vol. 18, Société Mathématique de France, Paris, 2020. With contributions by Everett Howe, Joseph Oesterlé and Christophe Ritzenthaler; Edited by Alp Bassa, Elisa Lorenzo García, Christophe Ritzenthaler and René Schoof. MR4242817

[Tre20] Andrey Trepalin, *Del Pezzo surfaces over finite fields*, Finite Fields Appl. **68** (2020), 101741, 32, DOI 10.1016/j.ffa.2020.101741. MR4149657

[vZ18] Jason van Zelm, *Nontautological bielliptic cycles*, Pacific J. Math. **294** (2018), no. 2, 495–504, DOI 10.2140/pjm.2018.294.495. MR3770123

[Wen] Tom Wennink, *Counting the number of trigonal curves of genus five over finite fields*, thesis, Utrecht University, 2016, `https://studenttheses.uu.nl/bitstream/handle/20.500.12932/23652/ThesisTomWennnink.pdf`.

[Xar20] Xavier Xarles, *A census of all genus 4 curves over the field with 2 elements*, `arXiv:2007.07822`.

[Zhe24] Angelina Zheng, *Stable cohomology of the moduli space of trigonal curves*, Int. Math. Res. Not. IMRN **2** (2024), 1123–1153, DOI 10.1093/imrn/rnad011. MR4692368

Department of Mathematics, University of California San Diego, 2985 Muir Ln, La Jolla, CA 92093, USA
Email address: `yoh011@ucsd.edu`

Department of Mathematics, University of California San Diego, 2985 Muir Ln, La Jolla, CA 92093, USA
Email address: `kedlaya@ucsd.edu`

Department of Electrical Engineering, KU Leuven, Kasteelpark Arenberg 10/2452, 3001 Leuven (Heverlee), Belgium
Email address: `laujunbo@gmail.com`

Contemporary Mathematics
Volume **840**, 2026
https://doi.org/10.1090/conm/840/16817

The Kodaira dimension of Hilbert modular threefolds

Adam Logan

ABSTRACT. Following a method introduced by Thomas-Vasquez and developed by Grundman, we prove that many Hilbert modular threefolds of geometric genus 0 and 1 are of general type, and that some are of nonnegative Kodaira dimension. The new ingredient is a detailed study of the geometry and combinatorics of totally positive integral elements x of a fractional ideal I in a totally real number field K with the property that $\operatorname{tr} xy < \min I \operatorname{tr} y$ for some $y \gg 0 \in K$.

1. Introduction

This paper is about the birational geometry, and in particular the Kodaira dimension, of Hilbert modular varieties. We begin by recalling the definition of these varieties as complex orbifolds, as in [**28**, Chapter 1], [**6**, §2.2]. Let K be a totally real number field of degree d. The standard formula $\left(\begin{smallmatrix} a & b \\ c & d \end{smallmatrix}\right) z = \frac{az+b}{cz+d}$ gives an action of $\mathrm{GL}_2^+(\mathbb{R})$ on the upper half-plane $\mathcal{H}$. The product of the d embeddings $K \hookrightarrow \mathbb{R}$ gives an injection $\mathrm{PSL}_2(\mathcal{O}_K) \hookrightarrow \mathrm{GL}_2^+(\mathbb{R})^d$ whose image is discrete and acts on $\mathcal{H}^d$ properly discontinuously and with finite stabilizers; thus the quotient inherits a structure of complex orbifold from $\mathcal{H}^d$.

It is sometimes important to consider a quotient by $\mathrm{PGL}_2^+(\mathcal{O}_K)$, the subgroup of $\mathrm{GL}_2(\mathcal{O}_K)$ consisting of matrices whose determinant is a totally positive unit, rather than PSL_2. These groups are the same when the class number and narrow class number of K are equal, but if ε is a totally positive unit that is not a square then $\left(\begin{smallmatrix} \varepsilon & 0 \\ 0 & 1 \end{smallmatrix}\right)$ represents an element of PGL_2^+ not in PSL_2. Again the quotient is a complex orbifold of dimension d.

Let A be a fractional ideal of $\mathcal{O}_K$ and consider $\mathrm{Aut}(\mathcal{O}_K \oplus A)$. This can be viewed as a group of matrices $\left(\begin{smallmatrix} a & b \\ c & d \end{smallmatrix}\right)$, where $a, d \in \mathcal{O}_K$, $b \in A^{-1}$, $c \in A$ and where the determinant belongs to $\mathcal{O}_K^\times$. We thus write $\mathrm{PSL}(\mathcal{O}_K \oplus A), \mathrm{PGL}^+(\mathcal{O}_K \oplus A)$ for the subgroups of $\mathrm{Aut}(\mathcal{O}_K \oplus A)$ for which the determinant is the square of a unit, respectively a totally positive unit, modulo the scalar matrices.

We state some results from [**28**, I.4] on these groups. If A is generated by a totally positive element $x_+ \in K$, then $\mathrm{PSL}(\mathcal{O}_K \oplus A)$ is conjugate to PSL_2 by $\left(\begin{smallmatrix} x_+ & 0 \\ 0 & 1 \end{smallmatrix}\right)$, which identifies the action of PSL_2 on $\mathcal{H}^d$ with that of $\mathrm{PSL}(\mathcal{O}_K \oplus A)$. Similarly, if $A = I^2$, then the two groups are conjugate by a matrix of determinant 1 whose two columns belong to I, I^{-1}. On the other hand, the groups $\mathrm{PSL}(\mathcal{O}_K \oplus A), \mathrm{PSL}(\mathcal{O}_K \oplus$

2020 *Mathematics Subject Classification.* Primary 14G35, 11F41, 14E15; Secondary 11Y40, 32S45.

B) are not conjugate in $\mathrm{GL}_2^+(K)$ if AB^{-1} is not of the form TI^2, where T is a narrowly principal ideal. Thus the conjugacy classes of groups of this form are in bijection with the genera of K. Similarly for PGL^+.

DEFINITION 1.1 ([**2**, (2.1.4)]). Let I be a nonzero ideal of $\mathcal{O}_K$ and let A be a fractional ideal. Let $\Gamma_0(I;A)$ and $\hat{\Gamma}_0(I;A)$ be the subgroups of $\mathrm{PSL}_2(\mathcal{O}_K \oplus A)$ and $\mathrm{PGL}_2^+(\mathcal{O}_K \oplus A)$ respectively consisting of matrices whose lower left entry belongs to AI. These are subgroups of finite index, so we again obtain a complex orbifold as the quotient. This parametrizes abelian d-folds with a particular kind of action of $\mathcal{O}_K$ and a distinguished $\mathcal{O}_K$-submodule of the torsion isomorphic to $\mathcal{O}_K/I$.

DEFINITION 1.2. Let K be a totally real number field and Γ a subgroup commensurable with $\mathrm{PGL}_2^+(\mathcal{O}_K)$ as above. The quotient $\Gamma\backslash\mathcal{H}^d$ is the *Hilbert modular variety* for the group Γ.

A priori this is only a complex orbifold, but in fact it has an algebraic structure by [**28**, Theorem II.7.1].

NOTATION 1.3. The Hilbert modular variety for the group $\Gamma_0(I;A)$, where I is an ideal in $\mathcal{O}_K$, will be denoted $H_{K,I;A}$. When A is (1) we will generally omit it from the notation, writing $H_{K,I}$. Similarly if $I = (1)$ we may write $H_{K;A}$, and if both are (1) we simply write H_K.

These quotients are not compact; they are compactified by adjoining finitely many points called cusps. The cusps of H_K are in bijection with $\mathrm{PSL}_2(\mathcal{O}_K)\backslash\mathbb{P}^1(K)$, or equivalently with ideal classes of $\mathcal{O}_K$. Giving a resolution of the cusps is essentially equivalent to decomposing a fundamental domain for the action of $\mathcal{O}_{K,+}^\times$, the totally positive units of $\mathcal{O}_K$, on the totally positive cone in $K \otimes \mathbb{R}$, into cones spanned by bases for ideals. In the case $[K:\mathbb{Q}] = 2$ this is treated in great detail in [**28**, Chapter 2]; in [**23**] the problem is discussed briefly for arbitrary degree in Section 1 and studied much more closely in the rest of the paper. In contrast to the case $[K:\mathbb{Q}] = 2$ where there is a canonical way to subdivide a cone in $\mathbb{R}^2$, there are many different ways to resolve the cusp singularities in dimension greater than 2, none of which is obviously the most natural. As noted in [**28**, p. 36], this is related to the lack of a single minimal model for varieties of dimension greater than 2.

In addition, there are singularities that arise from the isolated fixed points of elements of Γ on $\mathcal{H}^d$. For $[K:\mathbb{Q}]$ fixed there are only finitely many such types and they are much more tractable than the cusps.

1.1. Previous results. Recall the definitions of Kodaira dimension κ_V and general type [**18**, Definition 1-5-2, Proposition 1-5-3]:

NOTATION 1.4. Let V be a smooth variety. The canonical line bundle of V will be denoted K_V; since we are not concerned with specific divisors D with $\mathcal{O}(D) \cong K_V$, we write its powers as $K_V^{\otimes n}$ rather than nK_V.

DEFINITION 1.5. Let V be an irreducible variety and V' a smooth variety birationally equivalent to V (the choice does not matter). If $|K_{V'}^{\otimes n}| = 0$ for all $n > 0$, then we define $\kappa_V = -\infty$. If $|K_{V'}^{\otimes n}| \leq 1$ for all $n > 0$ with equality for at least one n, then define $\kappa_V = 0$. Otherwise, there exist positive integers k, n and positive real numbers c, C such that $cm^k < \dim |K_{V'}^{\otimes mn}| < Cm^k$ for sufficiently large m, and we define $\kappa_V = k$. If $\kappa_V = \dim V$ then V is *of general type*.

We now introduce two well-studied families of cubic fields.

DEFINITION 1.6 ([**23**, §3.I (*), (3.2)]). Let $1 \le r \le s-2$. Define $K_{r,s}$ to be the field $\mathbb{Q}(\alpha_{r,s})$, where $\alpha_{r,s}$ satisfies $x(x-r)(x-s)-1=0$. Let $n>0$. Define K_n to be $\mathbb{Q}(\alpha_n)$, where α_n is a root of $x^3+(n+1)x^2+(n-2)x-1$.

These fields have easily constructed sets of units of maximal rank: the K_n are Galois and so α_n and α_n^σ are units where $\sigma \in \mathrm{Aut}(K_n)$, while $\alpha_{r,s}, \alpha_{r,s}-r \in \mathcal{O}_{K_{r,s}}^\times$. This implies in some cases that the singularities of the Hilbert modular varieties associated to them are relatively tractable: see Remark 2.4.

Knöller proves the following [**13**, Satz 1, Satz 2]:

THEOREM 1.7. (1) *If $\mathcal{O}_{K_n}$ is generated by α_n (e.g., if n^2-n+7 is squarefree) and has class number* 1 *and $n>4$, then H_{K_n} is of general type.*
(2) *If $\mathcal{O}_{K_{r,s}}$ is generated by $\alpha_{r,s}$ and has class number* 1 *and $3\zeta_{K_{r,s}}(2)D_{K_{r,s}} > 4\pi^6$, and $r>1$, then $H_{K_{r,s}}$ is of general type.*

Every H_K that Knöller proves to be of general type satisfies $p_g(H_K) > 1$. Grundman proved the following statement:

THEOREM 1.8 ([**8**, Theorem 3], [**9**, Theorem 3]). *Let $K \in \{K_{1,7}, K_{2,5}, K_{3,5}\}$. Then H_K is of geometric genus* 0 *and positive Kodaira dimension.*

This was surprising because a Hilbert modular surface for level 1 is rational if and only if its geometric genus is 0 [**28**, Proposition VII.6.1].

1.2. Summary of results.

We now state our main result.

THEOREM 6.6, 6.7. (1) *Let K be a cubic field with discriminant ≥ 473 such the geometric genus of H_K is* 0 *or* 1. *Then H_K is of general type, unless possibly $K \cong \mathbb{Q}[x]/(x^3-7x-5)$ is the cubic field with LMFDB label 3.3.697.1 or $K \cong \mathbb{Q}[x]/(x^3-x^2-7x-3)$ is the cubic field with LMFDB label 3.3.788.1.*
(2) *Let K be a cubic field such that the geometric genus of H_K is* 0 *or* 1. *Let A be an ideal that is not narrowly principal (this implies that $h_K^+=2$). Then $H_{K;A}$ is of general type, except possibly for the field 3.3.229.1 defined by x^3-4x-1.*

We are able to prove something about four more fields:

PROPOSITION 6.8, 6.10. (1) *Let K be the cubic field* 3.3.404.1 $= \mathbb{Q}(\alpha)$, *where α is a root of x^3-x^2-5x-1. Then $\kappa_{H_K} \ge 0$.*
(2) *Let K be the cubic field* 3.3.469.1 $= \mathbb{Q}(\alpha)$, *where α is a root of x^3-x^2-5x+4. Then $\kappa_{H_K} > 0$.*
(3) *Let K be the cubic field* 3.3.229.1 $= \mathbb{Q}(\alpha)$, *where α is a root of x^3-4x-1, and let A represent the nonprincipal genus. Then $\kappa_{H_{K;A}} \ge 0$.*
(4) *Let K be the cubic field* 3.3.788.1 $= \mathbb{Q}(\alpha)$, *where α is a root of x^3-x^2-7x-3, and let A represent the principal genus. Then $K_{H_{K;A}} > 0$.*

It would seem more natural to study the $H_{K;A}$ with $p_g \le 1$ rather than the $H_{K;A}$ for which $p_g(H_K) \le 1$. In [**17**], which will be published as part of a subsequent paper, it is proved that when K is a cubic field the dimensions of the spaces of modular forms for all of the $\Gamma_0(I;A)$ are equal. Since the number of cusps is the same as well, the geometric genera are equal, as we will see in Remark 1.21. So these conditions are equivalent.

REMARK 1.9. Our method determines two constants $c_1(K) = -2\zeta_K(-1), c_2(K)$ associated to a field such that if $c_1(K) > c_2(K)$ then H_K is of general type. Both of these grow without extreme oscillations (see Tables 2, 3), and for the cubic field 3.3.1937.1 of maximal discriminant for which $p_g(H_K) \leq 1$ we already have $c_1(K) = 28, c_2(K) = 397/30$. Therefore we believe that if $p_g(H_K) > 1$ (which already implies that $\kappa_{H_K} > 0$) then all $H_{K;A}$ are of general type, and that this can be verified for all such K by the methods of this paper. In order to prove this it would be necessary to verify that $c_1 > c_2$ for all cubic fields violating the inequality $\frac{D_K\zeta_K(2)}{hR} \geq 2^{3-2}\pi^3$ of [**7**, Corollary 10] and all choices of A. The amount of computer time required is substantial but not prohibitive, especially since it appears that this inequality fails for only 421 cubic fields. See Section 6.1 for more details.

Previous work on the Kodaira dimension of Hilbert modular varieties of dimension greater than 2 has largely been restricted to level 1, perhaps because it is easier to compute invariants under this assumption and because $\kappa_{H_{K,I;A}} \geq \kappa_{H_{K;A}}$ (in characteristic 0 this is a general statement about dominant maps). In Propositions 7.10, 7.12 we classify the $H_{K,I}$ for which the geometric genus p_g is at most 1. We can show that many of the $H_{K,I}$ are of general type even when we cannot prove this for H_K. Here is a sample of our results. We use $\mathfrak{p}_q$ to refer to a prime of norm q; it does not matter which one we choose because of the Galois automorphism. The full result is presented in Propositions 7.10 to 7.13. See also Tables 6, 7.

THEOREM 7.6. Let $K = \mathbb{Q}(\zeta_7)^+$. The Hilbert modular variety $H_{K,I}$ is of geometric genus 0 if and only if $I \in \{(1), \mathfrak{p}_7, (2), \mathfrak{p}_{13}, \mathfrak{p}_{29}, \mathfrak{p}_{43}\}$, and 1 if and only if $I \in \{(3), \mathfrak{p}_{41}, \mathfrak{p}_7^2, 2\mathfrak{p}_7, (4), \mathfrak{p}_{71}, \mathfrak{p}_7\mathfrak{p}_{13}, \mathfrak{p}_{97}, \mathfrak{p}_{113}, \mathfrak{p}_{127}, \mathfrak{p}_{13}^2\}$.

PROPOSITION 7.11, 7.16. All of these varieties are of general type, with the possible exceptions of $I \in \{(1), \mathfrak{p}_7, (2)\}$. We have $\kappa_{H_{K,(2)}} > 0$.

We would like to study the varieties not shown to be of general type and understand their geometry explicitly as in [**16**]. However, this seems to be a very difficult problem, and we have no new results in this direction.

1.3. Method for studying the Kodaira dimension of Hilbert modular threefolds. In order to prove that Hilbert modular threefolds are of general type, we will show that powers of the canonical bundle have many sections. Here we describe these sections as Hilbert modular forms; later, in Section 2, we explain which ones belong to $H^0(nK)$, and in Sections 3–5 we will give methods for exact and asymptotic calculation of the dimensions. In Sections 6, 7 we present the results. The statements of this section and Section 2 are not new, going back to [**13**]. However, results like those stated in Section 1.2 and proved in subsequent sections have previously appeared only in special cases.

NOTATION 1.10. Fix an ordering of the d real embeddings $\pi_i : K \hookrightarrow \mathbb{R}$. For $x \in K$ and $1 \leq i \leq d$, let $x_i = \pi_i(x)$. For $z \in \mathcal{H}^d$, let z_i be the ith component.

DEFINITION 1.11 ([**28**, Definition I.6.1]). Let k be a nonnegative even integer. A *Hilbert modular form* of (parallel) weight $2k$ for a group Γ on an open subset $U \subseteq \mathcal{H}^d$ is a holomorphic function satisfying

$$f(\gamma z) = \left(\prod_{i=1}^{d}(c_i z_i + d_i)^{2k}(\det\gamma)_i^{-k}\right) f(z)$$

for all pairs (γ, z) with $\gamma \in \Gamma$ and $z, \gamma z \in U$. We denote the space of Hilbert modular forms of weight $2k$ by $M_{2k}(U;\Gamma)$, omitting U or Γ when these are clear.

REMARK 1.12. An easy calculation [**28**, Lemma III.3.1] shows that if f is a Hilbert modular form of weight $2k$ then $f(z_1, \ldots, z_n)(dz_1 \ldots dz_n)^{\otimes k}$ is invariant under Γ. In particular, if $\Gamma\backslash U$ is nonsingular then $\mathrm{H}^0(\Gamma\backslash U, K_U^{\otimes k})$ is naturally identified with $M_{2k}(U;\Gamma)$. However, this does not imply that the space $\mathrm{H}^0(H_K, K_U^{\otimes k})$ of modular forms of weight k can be identified with $M_{2k}(\Gamma)$, because there are additional conditions at the cusps and elliptic points for a modular form to give a section of $K_U^{\otimes k}$. Rather, as in the classical case, the space of modular forms of weight $2k$ is identified with $\mathrm{H}^0(H_K, (K(\log \text{cusps}))^{\otimes k})$, and Proj of the ring of modular forms is the *Baily-Borel compactification* of H_K [**28**, Theorem II.7.1]. In particular H_K is always of log general type.

We recall some facts about the rate of growth of the dimension of M_{2k}. Although an exact formula for the dimension requires a detailed examination of the elliptic points (Definition 1.15), the asymptotic formula does not require this. The starting point is a result of Siegel:

THEOREM 1.13 ([**28**, Theorem IV.1.1]). *Relative to the standard hyperbolic metric on $\mathcal{H}^d$, the volume of* $\mathrm{PSL}_2(\mathcal{O}_K)\backslash\mathcal{H}^d$ *is equal to* $2\zeta_k(-1)$.

Specializing [**20**, Theorem 11], Thomas and Vasquez give a formula for the dimension of the space of modular forms for $\mathrm{PSL}_2(\mathcal{O}_K)$ and its torsion-free finite-index subgroups. The following well-known result follows from this:

PROPOSITION 1.14. *Let Γ be a group commensurable with* $\mathrm{PSL}_2(\mathcal{O}_K)$, *where K is a totally real field of degree d. Then the dimension of the space of cusp forms of weight $2k$ is asymptotically equal to* $2\zeta_K(-1)(-k)^d[\mathrm{PSL}_2(\mathcal{O}_K) : \Gamma]$.

PROOF. In the case where $\Gamma = \mathrm{PSL}_2(\mathcal{O}_K)$, this follows from [**25**, (2.2)], since in their terminology $\mathrm{PSL}_2(\mathcal{O}_K)$ is always of modular type. To pass to a commensurable subgroup we use [**6**, Theorem II.3.5] to see that the leading coefficient is proportional to $[\mathrm{PSL}_2(\mathcal{O}_K) : \Gamma]$, it being clear that the terms in the formula other than $\mathrm{vol}(\Gamma\backslash\mathcal{H}^n)(2r-1)^n$ are of lower order. □

We now consider the special points of Hilbert modular varieties, namely fixed points of nonidentity elements of Γ and the cusps that compactify $\Gamma\backslash\mathcal{H}^d$. The first of these must be studied to obtain an exact formula for the dimension of the space of modular forms, while the second are vital for estimating the difference between this dimension and the dimension of H^0 of powers of the canonical line bundle.

DEFINITION 1.15 ([**28**, p. 15]). An *elliptic point* of $H_{K,I;A}$ is the image of the fixed point of an element of $\Gamma_0(I;A)$ of finite order greater than 1.

NOTATION 1.16. Let $C_{K,I;A}, E_{K,I;A}$ be the sets of cusps and elliptic points of $H_{K,I;A}$ respectively.

DEFINITION 1.17 ([**28**, Definition I.6.2]). Let U be an open subset of $\mathcal{H}^d \setminus E_{K,I;A}$. Let C_U be the subset of $C_{K,I;A}$ consisting of cusps such that U contains a punctured neighbourhood of some representative of C, and similarly for E_U and elliptic points. A *Hilbert cusp form* on U is a Hilbert modular form on U that extends to a holomorphic function on $U \cup \Gamma\backslash(C_U \cup E_U)$ that is 0 on all cusps of C_U.

THEOREM 1.18 ([**12**, Section 3.6, Lemma]). *For a resolution of singularities* $\tilde{H}_{K,I;A}$ *of* $H_{K,I;A}$, *the sections of* $K_{\tilde{H}_{K,I;A}}$ *are precisely the Hilbert cusp forms of weight* 2.

Hirzebruch only states this for quadratic fields, but the proof does not use this assumption. This theorem does not extend to powers of $K_{\tilde{H}_{K,I;A}}$. To describe the situation, we introduce some further notation.

NOTATION 1.19. Let P be a cusp or elliptic point for H. Let U_P be a small punctured neighbourhood of P (a complex manifold) and let $V_P = U_P \cup \{P\}$. Let $\tilde{V}_P$ be any resolution of the singular point P of V_P (nothing depends on the choice). The natural injection $\mathrm{H}^0(\tilde{V}_P, K_{\tilde{V}_P}^{\otimes q}) \hookrightarrow \mathrm{H}^0(U_P, K_{U_P}^{\otimes q})$ will be denoted $e_{P;q}$.

DEFINITION 1.20 ([**13**, p. 6]). The *qth defect* $\delta_P(q)$ of P is $\dim \operatorname{coker} e_{P;q}$.

REMARK 1.21. We have $\delta_P(0) = 0$ for all P; for elliptic points this is the case $m = 0$ of [**13**, Satz 2.0], while for cusps it is a direct consequence of [**13**, Satz 2.4] (see Definition 2.1 for an explanation of the notation). For $q = 1$, Theorem 1.18 shows that $\delta_P(1) = 1$ if P is a cusp, at least if $[K : \mathbb{Q}] > 1$. In addition, if P is an elliptic point then $\delta_P(1) = 0$. This was first shown by Freitag; it is clearly explained on [**28**, p. 56].

Now let $S = C_{K,I;A} \cup E_{K,I;A}$ and let $U = H_{K,I;A} \setminus S$, which is a smooth complex manifold. For each $P \in S$, let $\tilde{V}_P$ be a resolution of $U \cup P$, and let $\tilde{H}_{K,I;A}$ be a simultaneous resolution of all the singularities, obtained by identifying the $\tilde{V}_P$ along the open subsets identified with U. We then have $\mathrm{H}^0(\tilde{H}_{K,I;A}, K^{\otimes q}) = \bigcap_{P\in S} \operatorname{im} e_{P;q}$. It follows that $\mathrm{h}^0(\tilde{H}_{K,I;A}, K^{\otimes q}) \geq \dim M_{2q} - \sum_{P\in S} \delta_P(q)$, since $M_{2q} = \mathrm{h}^0(H_{K,I;A} \setminus S, K^{\otimes q})$ (in contrast to the case $K = \mathbb{Q}$, where we need an additional condition to ensure that modular forms are holomorphic at the cusps; see [**28**, Section I.6]). We thus obtain a lower bound for $\mathrm{h}^0(K^{\otimes q})$, namely

$$\mathrm{h}^0(K^{\otimes q}) \geq \dim M_{2q} - \sum_{P\in S} \delta_P(q). \tag{1.1}$$

If we can prove that this lower bound is positive (resp. greater than 1) for some q, then we have shown that the Kodaira dimension of the Hilbert modular variety is nonnegative (resp. positive). If we can estimate the lower bound as being at least cq^d for some $c > 0$, the variety is of general type. Grundman did this and essentially proved in [**8**, Theorem 3], [**9**, Theorem 3] that the component of the Hilbert modular threefolds corresponding to the principal genus for 3.3.697.1 is of positive Kodaira dimension, while for 3.3.761.1, 3.3.985.1 it is of general type. Grundman states only that the plurigenera are not equal to 0 for the principal genus, concluding that the varieties are not rational. However, the statement that H_K is of general type for the second and third fields follows from her results. We will explain this in Example 6.2 and Remark 6.3.

2. Defects

Thomas and Vasquez give a formula [**25**, Theorem 3.10] for the dimension of M_{2q} for certain subgroups of $\mathrm{PSL}_2(K)$. A general formula for $\Gamma_0(I)$ does not seem to appear in one single place in the literature, but can be extracted from [**29**, 39.10] and [**2**, 5.1]. The calculation is implemented in a Magma script [**15**] (I thank John

Voight for many clarifying comments). This leaves the problem of computing the defects. We recall the method of Thomas [**26**, Section 1] (citing [**13**]) for evaluating the defect in terms of a count of elements with multiples of bounded trace.

DEFINITION 2.1 ([**13**]). Let $M \subset K$ be a free abelian subgroup of rank $d = [K : \mathbb{Q}]$ and $V \subseteq \mathcal{O}_{K,+}^\times$ a group of totally positive units preserving M. Following Knöller, we consider the cusp singularity at the image of ∞ in the quotient of $\mathcal{H}^d \cup \{\infty\}$ by the group of matrices $\{(\begin{smallmatrix} v & g \\ 0 & 1 \end{smallmatrix}) : v \in V, g \in M\}$: we refer to this as a cusp of *type* (M, V). Let $\delta_{(M,V)}(q)$ be the qth defect of a cusp of type (M, V). Let $\hat{M} = \{x \in K : \operatorname{tr} xm \in \mathbb{Z} \text{ for all } m \in M\}$ be the dual of M with respect to the trace. For $q \in \mathbb{N}$, let $\Lambda_q(M) = \{x \in \hat{M}_+ : \operatorname{tr} xm < q \text{ for some } m \in M_+\} \cup \{0\}$.

For any such M, a suitable subgroup V exists:

LEMMA 2.2. *Let $M \subset K$ be a free abelian subgroup of rank d. The subgroup of $\mathcal{O}_{K,+}^\times$ preserving M is of finite index in $\mathcal{O}_{K,+}^\times$.*

PROOF. By multiplying by a positive integer we may assume that $M \subset \mathcal{O}_K$. Then $\mathcal{O}_{K,+}^\times$ acts on subgroups $N \subseteq \mathcal{O}_K$ with $\mathcal{O}_K/M \cong \mathcal{O}_K/N$. There are finitely many such subgroups, so the stabilizer of any one of them is of finite index. □

THEOREM 2.3 ([**13**, Satz 2.4]). *The qth defect of a cusp of type (M, V) is equal to $\#(\Lambda_q(M)/V)$.*

In particular, we always have $\Lambda_1 = \{0\}$ and so $\delta_{M,V}(1) = 1$. This is closely related to Theorem 1.18.

REMARK 2.4. Thomas states that it is difficult to calculate $\delta_{M,V}(q)$ in general and imposes an additional restriction that implies that every element of $\Lambda_q(\mathcal{O}_K)/V$ is represented by $x \in \hat{\mathcal{O}}_{K,+}$ with $\operatorname{tr} x < q$. Later works of Thomas-Vasquez [**23**] and Grundman [**8**, **9**] follow Thomas in imposing this restriction. This holds for the $K_{r,s}$ and K_n (Definition 1.6) under the additional hypothesis that a root of the given defining polynomial generates the ring of integers [**23**, Section 3]. In these cases they construct explicit cusp resolutions: see [**23**, Section 3], [**8**, Section 6].

In this work we will describe practical methods for computing $\delta_{M,V}$ without an explicit cusp resolution. This allows us to prove results analogous to those of [**9**, **23**] for general totally real cubic fields.

REMARK 2.5. The definition of the fields $K_{r,s}$ can be extended to arbitrary degree, setting $K_{r_1,\dots,r_n}$ to be the field of degree $n+1$ defined by the polynomial $x\prod_{i=1}^n (x - r_i) - 1$. For $n = 1$, we obtain the quadratic fields $\mathbb{Q}(\sqrt{n^2+4})$. However, these do not necessarily have the property that every element of $\Lambda_q(\mathcal{O}_K)/V$ is represented by an element of trace less than q. For example, with $n = 16$ this field is $\mathbb{Q}(\sqrt{65})$, in which $\frac{9+\sqrt{65}}{2} \cdot \frac{9-\sqrt{65}}{2} = 4$. One easily concludes that $\frac{9+\sqrt{65}}{2}$ is an element of $\Lambda_9(\mathcal{O}_K)$ having no multiple by a totally positive unit of trace less than 9. In Section 4 we will present systematic methods to find such elements or prove that they do not exist.

We now return to describe the defects of cyclic quotient singularities, following [**28**, II.6, III.3]. Van der Geer only treats the case of dimension 2, but his methods apply without change to the general situation. Thus we only state the results.

NOTATION 2.6. Let $a_1, \dots, a_n$ be positive integers relatively prime to m and less than m. Let L be the sublattice of $\mathbb{Z}^n$ consisting of vectors $(x_1, \dots, x_n)$ with $m|\sum_{i=1}^n a_i x_i$. The dual lattice, which is generated over $\mathbb{Z}^n$ by $(a_1, \dots, a_n)/m$, is denoted M. For $k = 1, \dots, m-1$ let $P_k = \frac{1}{m}(ka_j \bmod m)_{j=1}^n \in M$, and define a simplex T_k by the conditions $x_1, \dots, x_n \geq 1$ and $\sum_i P_{ki} x_i < 1$. (Note that $T_k = \emptyset$ if $\sum_i P_{ki} \geq 1$.) Let $T = \cup_i T_i$ and let $T(q)$ be T scaled up by q.

DEFINITION 2.7. Let $m \in \mathbb{Z}^+$ and let $a_1, \dots, a_n$ be positive integers less than m and relatively prime to m. Let the cyclic group C_m of order m act on $\mathbb{A}^n$ such that a generator acts by $(x_1, \dots, x_n) \to (\mu_m^{a_1} x_1, \dots, \mu_m^{a_n} x_n)$. A singularity locally isomorphic to that of $\mathbb{A}^n/C_m$ at the origin is called a *cyclic quotient singularity of type* $(a_1, \dots, a_n; m)$.

Such a singularity is isolated by [**19**, Corollary 2.2].

THEOREM 2.8 ([**28**, p. 56]). *The qth defect of a cyclic quotient singularity of type $(a_1, \dots, a_n; m)$ is the number of L-points of $T(q)$.*

COROLLARY 2.9. *The qth defect is asymptotic to* $\frac{\operatorname{vol} T}{m} q^d + O(q^{d-1})$. □

EXAMPLE 2.10. We consider the elliptic points of order $7, 9$ that arise for $\mathbb{Q}(\zeta_7)^+$ and $\mathbb{Q}(\zeta_9)^+$. For the former, we have one point of each of the types $(1, \pm 3, \pm 2; 7)$ by [**24**, Proposition 2.10 (ii)]. The $(1, 4, 2; 7)$ point has all defects 0, while the other three are isomorphic. Indeed, given a $(1, 3, 2; 7)$-singularity, we may replace the generator g of C_7 by g^4 to see that it is also a $(4, 5, 1; 7)$-singularity, or by g^5 to see that it is a singularity of type $(5, 1, 3; 7)$. The order of the a_i is of no importance. For each, there is a single simplex of volume $1/36$, giving an asymptotic of $q^3/252$ for the qth defect. The first 10 defects are $0, 0, 0, 0, 0, 0, 1, 1, 2, 3$.

Similarly, for $\mathbb{Q}(\zeta_9)^+$, according to [**24**, Proposition 2.10 (iii)] the types are $(1, \pm 2, \pm 4; 9)$, each occurring once. All defects are 0 for the point of type $(1, 7, 4; 9)$ and again the other three are equivalent. For $(1, 2, 4; 9)$ and those isomorphic to it there are two simplices of volume $1/6, 1/60$, but the second is contained in the first so the volume of the union is just $1/6$. So the asymptotic is $q^3/54$; the initial defects come to $0, 0, 0, 0, 1, 2, 4, 6, 10, 14$.

We display the relevant simplices in Figure 1.

REMARK 2.11. As pointed out in [**13**, Satz 2.0], the defects of cyclic quotient singularities of order $2, 3$ on Hilbert modular threefolds are 0; this is immediate from Theorem 2.8 as well. Thus, in light of the classification of cyclic quotient singularities of Hilbert modular varieties of cubic fields [**25**, Theorem 2.12] (less accessibly, though earlier, in [**11**]), the cyclic quotient singularities make no contribution to the defect for fields other than $\mathbb{Q}(\zeta_7)^+, \mathbb{Q}(\zeta_9)^+$.

I thank the referee for pointing out that this is analogous to the situation for quadratic fields, for which the fields of smallest discriminant $\mathbb{Q}(\sqrt{5}), \mathbb{Q}(\sqrt{2}), \mathbb{Q}(\sqrt{3})$ admit special quotient singularities that need to be considered separately (these are described in [**28**, Table 1.III]). In general [**27**, (7)] gives all totally real fields for which there are elliptic points with nonvanishing defects. Excluding quadratic fields and quartic fields containing $\sqrt{3}$, the list is finite.

In the next two sections of this paper we describe methods for calculating the sets $\Lambda_q(M)/V$ and estimating their asymptotic growth that can be applied to general totally real fields. The most interesting case for us is $M = \mathcal{O}_K, V = \mathcal{O}_{K,+}^\times$,

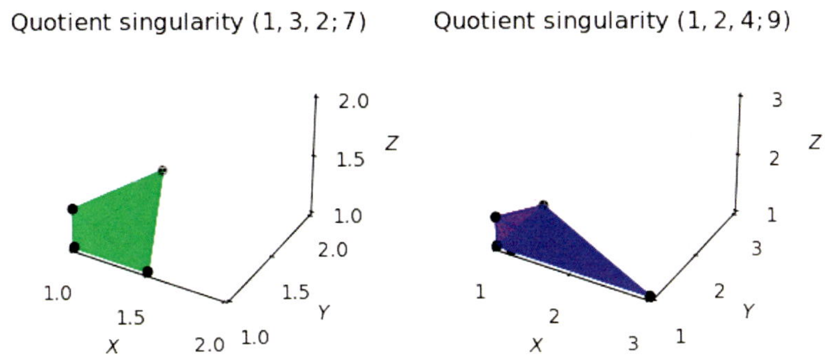

FIGURE 1. Simplices that describe the defects of $(1, 3, 2; 7)$ and $(1, 2, 4; 9)$-singularities. All simplices are defined by the condition that all coordinates are at least 1 and by one additional inequality. On the left, the choice $k = 1$ gives a single simplex defined by the inequality $x + 3y + 2z \leq 7$. On the right, the blue simplex is defined by $x + 2y + 4z \leq 9$ and the red simplex (contained in it) by $5x + y + 2z \leq 9$. Note that the scales are different.

which is associated to the cusp ∞, but imposing this restriction does not simplify the method, and we will consider more general singularities in Sections 6,7.

3. Basic algorithms

Here we describe the algorithms that underlie our computations. We assume the ring of integers, different, class group, and unit group of a number field K can be calculated, and that the prime factorization of a fractional ideal can be determined. Such calculations can be performed by various symbolic computation systems, such as Magma [**4**] and PARI/GP [**22**]. In addition to these, we need algorithms to find the elements of an ideal I of $\mathcal{O}_K$ whose real embeddings satisfy certain inequalities and to find the volume of a rational polytope.

3.1. Rational points in a polytope.

DEFINITION 3.1. A *rational polytope* P in $\mathbb{R}^n$ is an intersection of half-spaces defined by linear inequalities $\sum_{i=1}^n a_i x_i \leq c$ with $a_i, c \in \mathbb{Q}$ whose n-dimensional Lebesgue measure is positive and finite.

REMARK 3.2. We will assume that the volume of a rational polytope can be computed given the facets. For fixed dimension, this can be done in polynomial time [**5**, p. 4033]; in this paper we are always in $\mathbb{R}^3$. By [**27**, Theorem 2], as extended in Theorem 6.11, all components of Hilbert modular varieties are of general type for fields of degree greater than 6, so in the application to Hilbert modular varieties we never need to go beyond $\mathbb{R}^6$ and the volume can be computed in polynomial time.

In order to find the integral points in a general (not necessarily rational) polytope, we begin by finding a rational polytope that contains it. This can be done from crude bounds on the individual coordinates of every point of the polytope. We use a simple method to find the integral points in a rational polytope:

ALGORITHM 3.3. *Given a rational polytope P, determine its integral points.*

(1) *If* $\dim P = 1$ *the problem is trivial.*
(2) *For each coordinate* x_i*, determine lower and upper bounds* m_i, M_i *for* x_i*. If for any* i *there is only one integer in* $[m_i, M_i]$*, then set* x_i *to that integer and reduce the dimension by* 1.
(3) *Choose the* i *that maximizes* $M_i - m_i$ *and* $c_i \in (m_i, M_i)\setminus\mathbb{Z}$ *(in practice one chooses* c_i *near* $(m_i + M_i)/2$*). Apply the algorithm to the two polytopes* $P \cap (x_i \geq \lceil c_i \rceil), P \cap (x_i \geq \lfloor c_i \rfloor)$.

REMARK 3.4. The same algorithm can be used to determine the P-points of any rational lattice: such a lattice can be taken to $\mathbb{Z}^n$ by a rational change of coordinates, which takes a rational polytope to a rational polytope.

REMARK 3.5. Magma provides a built-in function to enumerate the integral points in a polytope; however, it is often slow. Empirically we have found that combining Algorithm 3.3 with calls to the built-in function when the polytope is small enough gives better performance than either one does individually.

3.2. Union of polytopes. A second important problem is to compute the volume of a union of rational polytopes $\cup_{i=1}^k P_i$. We use the following algorithm.

ALGORITHM 3.6. *Given a finite set of polytopes* $P_1, \dots, P_k$*, find* $v = \mathrm{vol}(\cup_{i=1}^k P_i)$.

(1) *Sort the* P_i *in order of increasing volume to obtain a list* $\mathcal{P}_0 = \mathcal{P}$*. For each pair* (i, j) *with* $i < j$*, determine whether* $P_i \subseteq P_j$*, and if so remove* P_i *from* $\mathcal{P}$*. For each pair* (i, j) *with* $i < j$*, determine whether* $\mathrm{vol}(P_i \cap P_j) > 0$*, and if so add it to the list* $\mathcal{L}$ *of pairs.*
(2) *While* $\mathcal{L}$ *is nonempty, repeat the following steps:*
   (a) *Choose an element* (i, j).
   (b) *Choose a face* F *of* P_i *that contains points of the interior of* P_j *(if there is no such face then* $(i, j) \notin \mathcal{L}$*). Let* P_+, P_- *be the intersections of* P_j *with the half-spaces defined by* F.
   (c) *Remove* P_j *from* $\mathcal{P}$ *and all pairs containing* j *from* $\mathcal{L}$.
   (d) *Determine whether* P_+, P_- *are contained in any element of* $\mathcal{P}$.
   (e) *For each* $P \in \{P_+, P_-\}$ *not contained in any element of* $\mathcal{P}$*, add* P *to* $\mathcal{P}$*, and add all pairs* $(i, \pm)$ *such that* $P_i \cap P$ *has positive volume to* $\mathcal{L}$.
(3) *Now that* $\mathcal{L}$ *is empty, return* $\sum_{P_i \in \mathcal{P}} \mathrm{vol}\, P_i$.

REMARK 3.7. To determine whether $P_i \subseteq P_j$ in Step 1 (resp. $P_\pm \subseteq P$ in Step 2d), it suffices to check whether $\mathrm{vol}\, P_i \cap P_j = \mathrm{vol}\, P_i$ (resp. $\mathrm{vol}\, P = \mathrm{vol}(P_\pm \cap P))$. Since our polytopes are given by linear inequalities, we can intersect them by concatenating the lists of inequalities. Magma easily handles these issues.

THEOREM 3.8. *Algorithm* 3.6 *terminates and returns the volume of* $\cup_{i=1}^k P_i$

PROOF. To prove that the algorithm terminates, let Π be the set of polytopes obtained by intersecting an element of $\mathcal{P}_0$ with a collection of half-spaces determined by the facets of elements of $\mathcal{P}_0$. The set Π is finite and every polyhedron considered in the course of the algorithm belongs to Π. Further, every pass through the while-loop (Steps 2a–2e) either increases $\#\mathcal{P}$ (if neither P_+ nor P_- is contained in any other element of $\mathcal{P}$) or decreases the sum of volumes of pairwise overlaps (otherwise). Again, we have $\#\mathcal{P} \leq \#\Pi$ and the set of possible sums of volumes of pairwise overlaps is finite, being a subset of the set of sums of subsets of the volumes of intersections of pairs of elements of Π. Thus the algorithm must terminate.

We now show that the algorithm is correct. It is clear that a pass through the loop cannot alter $\cup_{P_i \in \mathcal{P}} P_i$, so the volume does not change, and that when $\mathcal{L}$ is empty the intersection of any two elements of $\mathcal{P}$ has volume 0. Thus by inclusion-exclusion the volume is $\sum_{P_i \in \mathcal{P}} \operatorname{vol} P_i$ at that point. □

REMARK 3.9. The upper bound on running time implied by an explicit version of the argument for termination given just above is horrifyingly large. However, in practice the running time seems quite reasonable: see Section 6 for some examples.

4. Trace-minimal elements and reducers

In this section we present the geometric and combinatorial results that underlie our estimates for the defect of the cusp singularities of a Hilbert modular variety, in particular a threefold. We use three basic concepts in the geometry of the ring of integers in a totally real number field, which we introduce here.

DEFINITION 4.1. Let K be a totally real number field and let $d = [K : \mathbb{Q}]$; we write $x \gg 0$ to mean that x is totally positive and we write K_+ for $\{x \in K : x \gg 0\}$. Let $U_+ = \{x \in \mathcal{O}_K^\times : x \gg 0\}$ be the group of totally positive units of K. Given $x \in K$, we say that x is *trace-minimal* if $\operatorname{tr} ux \geq \operatorname{tr} x$ for all $u \in U_+$.

Let $b \in \mathbb{R}^+$, and for $x \in K_+$, let $x_1, \ldots, x_d$ be the real embeddings of x. If $0 < \max_i x_i / \min_i x_i \leq b$ we say that x is *b-balanced*.

If I is a fractional ideal and $x \gg 0 \in I$ has the property that $\operatorname{tr} xy < \min(\mathbb{Q}^+ \cap I) \operatorname{tr} y$ for some trace-minimal $y \in K$, then x is an *I-reducer*. We abbreviate $\min(\mathbb{Q}^+ \cap I)$ to $\min I$. When $I = \mathcal{O}_K$ we refer simply to a *reducer*. The set of I-reducers will be denoted $\mathcal{R}_I$. We also denote $\mathcal{R}_I \cup \{\min I\}$ by $\mathcal{R}'_I$.

EXAMPLE 4.2. In $\mathbb{Q}(\sqrt{7})$, the equality $(3 - \sqrt{7})(3 + \sqrt{7}) = 2$ shows that $3 \pm \sqrt{7}$ are reducers. In Remark 2.5 we showed that $\frac{9 \pm \sqrt{65}}{2}$ are reducers in $\mathbb{Q}(\sqrt{65})$.

REMARK 4.3. Not all ideals have reducers. In particular, the key property of the fields considered by Thomas-Vasquez and Grundman in [**8**, **9**, **23**] is that (1) has no reducers.

4.1. Trace-minimal and balanced elements. We now prove some basic properties of these definitions. Some of these results, in particular Lemma 4.6, Lemma 4.9, Corollary 4.10, are essentially contained in [**21**, Lemma 3], and our proofs are closely related as well.

DEFINITION 4.4. For this entire section, let us fix a set $S_\pm$ of units of $\mathcal{O}_K$ in bijection with the set of proper nonempty subsets of the set of real places of K such that the unit u_R corresponding to a subset R is totally positive and greater than 1 exactly at the places in R. In particular, for $1 \leq i \leq d$ let ${}_iu \in S_\pm$ be such that ${}_iu_j > 1$ if and only if $j = i$ (recall that ${}_iu_j$ refers to the jth real embedding of ${}_iu$). In addition, let $S_d = \{v_1, \ldots, v_d\}$, where v_i is a unit that is totally positive and greater than d in the ith real embedding.

REMARK 4.5. The set $S_\pm$ can be found by standard lattice techniques. In the log embedding of the units, we are looking for units with given negative coordinates. One way to find such a unit is simply to enumerate units up to a given Euclidean norm in this embedding until the desired one is found. The argument does not depend on the choice, but the bounds will be better if the ${}_iu_i$ are as small as possible. Constructing S_d is easy.

LEMMA 4.6. *A trace-minimal element is totally positive or* 0.

PROOF. Let $x \in K$. If x is totally negative, choose i such that $|x_i|$ is maximal; then $\operatorname{tr}(v_i x) < dx_i < \operatorname{tr} x$. If x is positive in some real embeddings and negative in others, let u_N be as in Definition 4.4, where N is the set of real embeddings where x is negative. Then $\operatorname{tr} u_N x < \operatorname{tr} x$. □

NOTATION 4.7. Let $b_K = \max_{i \neq j} \frac{{}_i u_i - 1}{1 - {}_i u_j}$.

LEMMA 4.8. *Every totally positive element of K that is not b_K-balanced is reduced by one of the ${}_i u$.*

PROOF. Let $x \gg 0$ be trace-minimal. Suppose that x_i is the smallest embedding of x and x_j is the largest. Then

$$\operatorname{tr} x - \operatorname{tr}({}_i u x) = \sum_{k=1}^{d} (x - {}_i u x)_k \geq (x - {}_i u x)_i + (x - {}_i u x)_j = x_i(1 - {}_i u_i) + x_j(1 - {}_i u_j).$$

Now $b_K \geq \frac{{}_i u_i - 1}{1 - {}_i u_j}$, so if $\frac{x_j}{x_i} > b_K$ this is positive, and so x is reduced by ${}_i u$. □

We now describe a finite set S_K of units such that every totally positive element that is not trace-minimal is reduced by one of them.

LEMMA 4.9. *Let b_K be the bound of Notation 4.7. Let S_K consist of the units ${}_i u$ of Definition 4.4, together with all totally positive units all of whose real embeddings are at most db_K. If $x \gg 0 \in K$ satisfies $\operatorname{tr} ux \geq \operatorname{tr} x$ for all $u \in S_K$, then x is trace-minimal.*

PROOF. If x is not b_K-balanced, then by Lemma 4.8 it is reduced by one of the ${}_i u$, which belongs to S_K. Let u be a totally positive unit with $u_i > db_K$ for some i. If x is b_K-balanced, we then have $\operatorname{tr} ux \geq db_K x_i \geq db_K x_{\min} \geq dx_{\max} \geq \operatorname{tr} x$, where $x_{\min}, x_{\max}$ are the smallest and largest real embeddings of x, respectively. It follows that if x is b_K-balanced and not trace-minimal, then x is reduced by a unit whose real embeddings are at most db_K and which therefore belongs to S_K. □

We can find the finitely many totally positive units all of whose real embeddings are at most db_K: indeed, in the Minkowski embedding $x \to (\log |x_i|)$, these correspond to lattice points within a compact subset of the hyperplane $\sum_i r_i = 0$. Thus Algorithm 3.3 applies.

COROLLARY 4.10. *There is a finite set S'_K of totally positive units such that every element $x \in K$ (totally positive or not) that is not trace-minimal satisfies $\operatorname{tr} ux < \operatorname{tr} x$ for some $u \in S'_K$.*

PROOF. The sets S_K, $S_\pm$, and S_d are all finite. Let S'_K be their union. □

We extend the concept of trace-minimality to $K \otimes_{\mathbb{Q}} \mathbb{R}$. If we use a basis for $K \otimes_{\mathbb{Q}} \mathbb{R}$ given by elements of K, then all of the inequalities defining our polyhedra have coefficients in $\mathbb{Q}$, so Algorithm 3.3 applies and we can always determine whether an integral point in the enlarged rational polytope belongs to the original without the possibility of error due to round-off.

COROLLARY 4.11. *The set of trace-minimal elements of $K \otimes \mathbb{R}$ is a polyhedral cone with finitely many rational faces. Thus, for $q \in \mathbb{Q}^+$, the set of such elements with $\operatorname{tr} x \leq q$ is a rational polytope.*

PROOF. Every element $u \in S'_K$ (Corollary 4.10) defines a half-space by $\operatorname{tr} ux \geq \operatorname{tr} x$, and the intersection of these is the trace-minimal cone. □

We summarize the discussion of this section in an algorithm.

ALGORITHM 4.12. *Let $V \subseteq \mathcal{O}_{K,+}^\times$ be a subgroup of finite index, where K is a totally real number field. Determine the V-trace-minimal cone.*

(1) *For each real place R_i of K, determine a unit ${}_iv \in V$ greater than 1 at R_i and less than 1 at all other real places (see Remark 4.5). Let U_V be the set of these units. Calculate a constant $b_{K,V}$ as in Notation 4.7.*
(2) *Let S_V be the subset of V consisting of units less than or equal to $db_{K,V}$ at all real places. Again, this is a standard lattice calculation.*
(3) *Return $\cap_{u \in S_V \cup U_V} H_u$, where H_u is the half-space defined by $\operatorname{tr} xu \geq \operatorname{tr} x$.*

4.2. Reducers. We now consider the problem of determining the complete set of I-reducers in a totally real number field.

LEMMA 4.13. *Let b_K be the bound of Notation 4.7. Every I-reducer is at most $(\min I)db_K$ at all real places.*

PROOF. This follows from the same argument that we used in Lemma 4.9. □

Lemma 4.13 allows one to compute a finite subset of K that contains all the reducers via Algorithm 3.3, but it does not give a good bound, nor does it allow us to determine whether an element is in fact a reducer. Thus we analyze the situation more closely.

LEMMA 4.14. *Let M_K be the trace-minimal cone inside $K \otimes \mathbb{R}$, and let its extremal rays be $\mathbb{R}^+ v_i$ for $1 \leq i \leq m$, where $v_i \in K$. An element $x \in I$ is an I-reducer if and only if $\operatorname{tr} xv_i < \min I \operatorname{tr} v_i$ for some i.*

PROOF. The v_i are totally positive, so "if" holds by definition. For "only if", suppose that $\operatorname{tr} xy < \min I \operatorname{tr} y$ with y trace-minimal; by definition $y = \sum_i c_i v_i$ with $c_i \geq 0$. Thus $\operatorname{tr} xy = \sum_i c_i \operatorname{tr} xv_i$. If $\operatorname{tr} xv_i \geq \min I \operatorname{tr} v_i$ for all i, then

$$\min I \operatorname{tr} y > \operatorname{tr} xy = \sum_i c_i \operatorname{tr} xv_i \geq \min I \sum_c c_i \operatorname{tr} v_i = \min I \operatorname{tr} y.$$

This contradiction establishes "only if". □

LEMMA 4.15. *Given $v \gg 0$, the elements $x \gg 0 \in I$ with $\operatorname{tr} xv < \min I \operatorname{tr} v$ are naturally in correspondence with the integral points of a simplex.*

PROOF. The set of such x is the set of integral points of the region in $\mathbb{R}^d$ defined by $x_i \geq 0$ for all i and $\sum_{i=1}^d (x_i - \min I)v_i = 0$ that belong to the sublattice of $\mathbb{Z}^d$ defined by I. A coordinate change converts this sublattice to the standard one. □

Thus we may use Algorithm 3.3 to list the reducers efficiently. Again we summarize the discussion in an algorithm.

ALGORITHM 4.16. *Given a fractional ideal M, determine the M-reducers.*

(1) *Determine the trace-minimal cone (Algorithm 4.12).*
(2) *For each extremal ray $v\mathbb{R}^+$ of the trace-minimal cone, choose a representative $v \in K$ and determine the set $R_v = \{x \in M_+ : \operatorname{tr} xv < \min M \operatorname{tr} v\}$ as in Lemma 4.15.*
(3) *The answer is $\cup_v R_v$.*

5. Algorithms for defects

We now consider two closely related problems concerned with counting elements with multiples of bounded trace. Let M be a lattice in K and $V \subseteq \mathcal{O}_{K,+}^{\times}$ be a subgroup of finite index preserving M (by Lemma 2.2 such a subgroup exists). We recall the notation from Definition 2.1 and the fact [**13**, Satz 2.4] that the qth defect of a cusp of type (M, V) is $\delta_{M,V}(q) = \#\Lambda_q(M)/V$. Our problems are as follows:

PROBLEM 5.1. Given M, V, q, compute $\delta_{M,V}(q)$.

PROBLEM 5.2. Given M, V, give an asymptotic formula for $\delta_{M,V}(q)$.

For the first of these problems, most of the ideas of Section 4 are not necessary. By rescaling we may assume that $M \subseteq \mathcal{O}_K$. For simplicity, and because it is the only case needed in this paper by [**2**, Proposition 3.3.8], we assume that M is an ideal of $\mathcal{O}_K$.

LEMMA 5.3. *Let $\mathfrak{d}_K$ be the different ideal of $\mathcal{O}_K$. Let $t \in \mathfrak{d}_K^{-1}$ and let $x \neq 0 \in \hat{M}$. Define I_1, I_2 by $(x) = \mathfrak{d}_K^{-1}M^{-1}I_1$, $(t) = \mathfrak{d}_K^{-1}I_2$. Then there exists $m \in M$ with $xm = t$ if and only if $I_1|I_2$.*

PROOF. Since $m \in K$ with $xm = t$ is unique, we need only determine whether $x^{-1}t \in M$. Since I_1, I_2 are integral and $(x^{-1}t) = MI_2I_1^{-1}$, the result follows. □

LEMMA 5.4. *For all ideals M, all $q > 0$, and all V of finite index in $\mathcal{O}_{K,+}^{\times}$, we have $\delta_{M,V}(q) - 1 = [\mathcal{O}_{K,+}^{\times} : V]\left(\delta_{M,\mathcal{O}_{K,+}^{\times}} - 1\right)$.*

PROOF. The quotient map $(\Lambda_q(M) \setminus \{0\})/V \to (\Lambda_q(M) \setminus \{0\})/\mathcal{O}_{K,+}^{\times}$ is $[\mathcal{O}_{K,+}^{\times} : V]$-to-1. □

Thus we obtain an algorithm to calculate the defects as follows:

ALGORITHM 5.5. *Given M, V where M is an ideal, compute the first q defects $(\delta_{M,V}(i))_{i=1}^{q}$.*

(1) *List all totally positive elements of $\mathfrak{d}_K^{-1}$ with trace less than q, using Algorithm 3.3. (We proved in Corollary 4.11 that these are the points of a lattice inside a rational polytope, so the algorithm is applicable.)*
(2) *Determine the sets of ideals $\mathcal{I}_i = \{(t\mathfrak{d}_K) : t \in \mathfrak{d}_K^{-1}, \operatorname{tr} t < i\}$ for $1 \leq i \leq q$.*
(3) *For each $I \in \mathcal{I}_q$, determine the set of divisors D_I of I, and let $D_{I,+} = \{J \in D_I : IJ^{-1}M$ is narrowly principal$\}$. For each i let $\mathcal{D}_i = \cup_{I \in \mathcal{I}_i} D_{I,+}$.*
(4) *The answer is $([\mathcal{O}_{K,+}^{\times} : V]\#\mathcal{D}_i + 1)_{i=1}^{q}$.*

PROPOSITION 5.6. *Algorithm 5.5 terminates and is correct.*

PROOF. Termination is immediate, since this algorithm has no loops and all of the steps are effective. For correctness, first we assume that $V = \mathcal{O}_{K,+}^{\times}$. Let $t \in \mathfrak{d}_{K,+}^{-1}$ with $\operatorname{tr} m < q$ and let $x \in \hat{M}_+$; define I_1, I_2 as in Lemma 5.3. If $tx^{-1} \in M$ then, from the above, $I_1I_2^{-1} = t^{-1}xM^{-1}$ and $I_1I_2^{-1}M$ is narrowly principal, being generated by $t^{-1}x$. The converse follows similarly. Counting narrowly principal ideals is the same as counting totally positive generators of those ideals up to totally positive units, so the result follows. The last step, giving the answer, is justified by Lemma 5.4. □

We now consider the problem of determining an asymptotic.

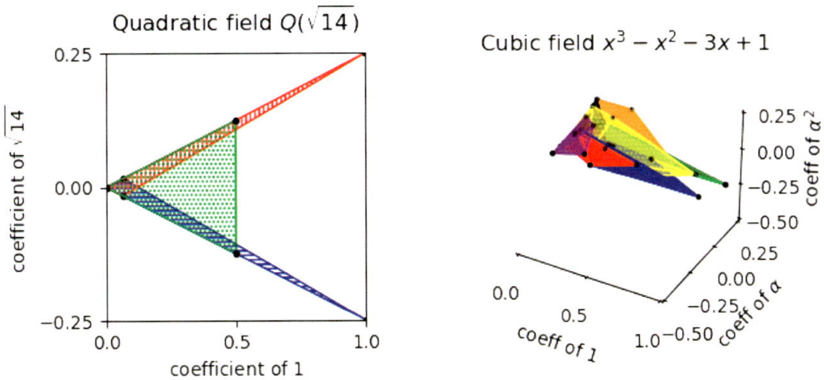

FIGURE 2. Regions describing elements of the quadratic and cubic fields of discriminant 56 and 148 respectively having an integral multiple with bounded trace. For the quadratic field we have a union of 3 triangles T_1, T_2, T_3, corresponding to the reducers $4 - \sqrt{14}, 4 + \sqrt{14}$ and 1, and shown in the figure in red, blue, and green respectively. The vertices of T_1 are $(0,0), (1,-1/4), (1/15, 1/60)$, and those of T_3 are $(0,0), (1/2, -1/8), (1/2, 1/8)$. We obtain T_2 by reflecting T_1 in the y-axis. For the cubic field, the region is a union of 8 polyhedra, each with 7 vertices.

DEFINITION 5.7. For $r \gg 0 \in K$, let $T_{M,V,r}(q)$ be the intersection of the V-trace-minimal cone with the half-space $\operatorname{tr} rx \leq q$. Let $T_{M,V}(q) = \cup_{r \in \mathcal{R}'_M} T_{M,V,r}(q)$.

Equivalently we could take the union over elements $x \in \mathcal{R}'_M$ such that $x - y$ is not totally positive for any $y \in \mathcal{R}'_M$. Indeed, letting T be the totally positive cone, if $x - y$ is totally positive then $T \cap (\operatorname{tr} ry \leq q) \supset T \cap (\operatorname{tr} rx \leq q)$.

EXAMPLE 5.8. We illustrate this definition for a quadratic and a cubic field. First let $K = \mathbb{Q}(\sqrt{14})$. The trace-minimal cone is bounded by rays through $4 \pm \sqrt{14}$, and the reducers are $4 \pm \sqrt{14}$. Thus the set $\mathcal{R}'_{\mathcal{O}_K}$ consists of 3 elements, the corresponding triangles being shown in the first plot of Figure 2. Next we consider the smallest cubic field that has reducers, namely $K = 3.3.148.1$, obtained by adjoining a root α of $x^3 - x^2 - 3x + 1$. The maximal order is $\mathbb{Z}[\alpha]$ and so we specify an element of K as a triple (a, b, c) representing $a + b\alpha + c\alpha^2$. The trace-minimal cone is bounded by rays through

$$(4, 26, 21), (15, -32, 14), (17, 18, 6), (53, -44, 10), (68, -2, -13), (139, 34, -38)$$

(for each x in this list one checks the necessary condition that there are at least two totally positive units $u_x \neq 1$ with $\operatorname{tr} x = \operatorname{tr} xu_x$); we find 7 reducers, namely

$$(0, 1, 1), (1, -2, 1), (1, 2, 1), (2, -7, 3), (2, -3, 1), (4, 1, -1), (5, 0, -1).$$

Thus $T_{\mathcal{O}_K, \mathcal{O}_{K,+}^\times}(1)$ is a union of 8 polyhedra, shown in the second plot of Figure 2. The largest polyhedron, containing a factor of 256/275 of the volume of the union, corresponds to $T_{\mathcal{O}_K, \mathcal{O}_{K,+}^\times, 1}(1)$. A Jupyter notebook in [**15**] contains an interactive version of this plot.

LEMMA 5.9. *$T_{M,V}(1)$ is a finite union of rational polytopes and $T_{M,V}(q)$ is obtained by scaling $T_{M,V}(1)$ by q.* □

LEMMA 5.10. *Let $x \in \hat{M}_+$ and suppose that $y \in M_+$ is such that $\operatorname{tr} xy < q$. Then there is $r \in \mathcal{R}'_M$ such that $\operatorname{tr} xr < q$.*

PROOF. Suppose that $\operatorname{tr} xy < q$. If $y \in \mathcal{R}_M$ we take $r = y$. If not, then $\operatorname{tr} x < q/\min M$ and we take $r = \min M$. □

LEMMA 5.11. *There is a surjective map from the set of $\hat{M}$-points of $T_{M,V}(q)$ to $\Lambda_{q+1}(M)/V$, taking $x \in \hat{M}$ to xV. If $x \in \hat{M}$ is in the interior of $T_{M,V}(q)$, then no other element of the set has the same image as x.*

PROOF. First, this is indeed a map of the given sets: for $x \in \hat{M}$, being in $T_{M,V}(q)$ implies that there exists y with $\operatorname{tr} xy \leq q$ which is either a reducer or $\min M$, so $xV \subset \Lambda_{q+1}(M)$. For surjectivity, fix $xV \in \Lambda_{q+1}(M)/V$ and choose x to be a trace-minimal element of xV. By definition we have $\operatorname{tr} xy \leq q$ for some $y \in M$, and by Lemma 5.10 we may take $y \in \mathcal{R}'_M$. Thus $x \in T_{M,V,y}(q) \subseteq T_{M,V}(q)$.

For the injectivity statement, suppose that $x, x' \in \hat{M} \cap T_{M,V}(q)$ have the same image in $\Lambda_q(M)/V$. Since $x^{-1}x' \in V$ and x, x' are trace-minimal, we must have $\operatorname{tr} x = \operatorname{tr} x'$, which means that x, x' are on the boundary of the trace-minimal cone and hence of $T_{M,V}(q)$. □

ALGORITHM 5.12. *Given M, V, where M is a fractional ideal of K and V is a subgroup of finite index in $\mathcal{O}_{K,+}^{\times}$, compute the rational constant $c_{M,V}$ such that $\delta_{M,V}(q) \sim c_{M,V} q^d$.*

(1) *Determine the trace-minimal cone by means of Algorithm 4.12.*
(2) *Determine the reducers $\mathcal{R}_M$ using Algorithm 4.16.*
(3) *Let $\mathcal{P}$ be the set of polyhedra given by intersecting the trace-minimal cone with the half-spaces $\operatorname{tr} rx \leq \min M \operatorname{tr} x$ for each element of $\mathcal{R}'_M$. Use Algorithm 3.6 to determine the volume E of $\cup_{P \in \mathcal{P}} P$.*
(4) *Return $E[\mathcal{O}_{K,+}^{\times} : V]/N_{K/\mathbb{Q}}(\hat{M})$, where $N_{K/\mathbb{Q}}$ is the norm map on fractional ideals.*

PROPOSITION 5.13. *Algorithm 5.12 terminates and is correct.*

PROOF. As before, termination is immediate from the termination of each step, since there are no loops. For correctness, by combining Lemmas 5.9, 5.11 we see that asymptotically the elements of $\Lambda_q(M)/\mathcal{O}_{K,+}^{\times}$ are in bijection with the points of $\hat{M}$ lying in $\cup_{P \in \mathcal{P}} P$ scaled by q. (The failure of injectivity only affects the points on polytopes of lower dimension, which are asymptotically negligible.) The correctness of the algorithm follows for $V = \mathcal{O}_{K,+}^{\times}$, since $\operatorname{covol} \hat{M}/\operatorname{covol} \mathcal{O}_K = N(\hat{M})$. The more general result follows from Lemma 5.4. The desired asymptotic is then a standard fact [**3**, Lemma 3.19], and since $\delta_{M,V}(q) = \#\Lambda_q(M)/V$ the result follows. □

COROLLARY 5.14. *The constant $c_{M,V}$ depends only on the narrow ideal class of M, and $c_{M,V} = [\mathcal{O}_{K,+}^{\times} : V] c_{M,\mathcal{O}_{K,+}^{\times}}$.*

PROOF. Let I be narrowly principal and generated by p. There is an obvious bijection $\Lambda_q(MI)/V \leftrightarrow \Lambda_q(M)/V$ given by multiplication by p. For the second statement, if we pass to a subgroup of V of index n, the volume of the fundamental domain is multiplied by n. The result follows. □

REMARK 5.15. Although $c_{pM,V} = c_{M,V}$ for $p \gg 0$, the time required to determine it using Algorithm 5.12 seems to grow rapidly with $[pM : (\min(pM))]$, and it is wise to choose p so as to minimize this quantity.

We have now solved Problems 5.1, 5.2.

REMARK 5.16. Even in dimension 2, it is not true that $c_{M,V}$ is independent of M. For example, take $K = \mathbb{Q}(\sqrt{3})$ and consider the two ideals $(1), (\sqrt{3})$ that represent the narrow class group. If $M = (1)$ we find that $\delta_M(q) = 1+q(q-1)$, while for $M' = (\sqrt{3})$ it turns out that $\delta_{M'}(q) = 1 + q(q-1)/2$. By [**1**] or [**28**, Example II.5.1] the self-intersections of the curves in the cusp resolutions are respectively -4 and $-3, -2$, so this is in accordance with [**28**, Proposition III.3.6]. (Note that [**28**] refers to the obstruction to extending forms that vanish at the cusp across a resolution, not all forms, so the quantity considered by van der Geer is 1 less than our defects.) In general it appears that, as I ranges over genus representatives, the largest $c_{I,\mathcal{O}_{K,+}^\times}$ occurs for the principal genus. For examples in dimension 3, compare Tables 2 and 4.

REMARK 5.17. By a standard result on rational polytopes, first proved by Ehrhart in 1962 but more easily accessible as [**3**, Theorem 3.23], the $\delta_{M,V}(q)$ for a given cusp are polynomial on residue classes. However, computations of the $\delta_{M,V}(q)$ indicate that they satisfy a simpler formula than might be expected, especially in the case $M = \mathcal{O}_K, V = \mathcal{O}_{K,+}^\times$. See Remark 6.9. We therefore suspect that there is some further structure to the $\delta(q)$ that remains to be elucidated.

6. Results at level 1

In this section we will describe the application of the algorithms presented here to Hilbert modular threefolds of level 1. To do so we need asymptotic formulas for both the dimension of the space of modular forms of weight $2k$ and the defects. For the first of these, we already have the result in Proposition 1.14. We easily derive a useful consequence:

PROPOSITION 6.1. *Suppose that $(-1)^d \cdot 2\zeta_K(-1) > \sum_i c_i$, where the sum ranges over the cusps of H_K and the qth defect of the ith cusp is asymptotic to $c_i q^d$. Then H_K is of general type.*

PROOF. We substitute the asymptotics for $\dim M_{2q}$ and $\sum_P \delta_P(q)$ into (1.1). The assumptions imply that $\mathrm{h}^0(qK)$ is not $o(q^d)$ and so H_K is of general type. □

Thus we will apply Algorithm 5.12 to determine the asymptotic rate of growth $c_2 k^3$ of the kth defect and compare it to $-2\zeta_K(-1)$. In particular, we recall that if $-2\zeta_K(-1) > c_2$ then H_K is of general type (see Remark 1.9). We begin by reexamining one of the threefolds of arithmetic genus 1 proved by Grundman [**9**] to have positive Kodaira dimension.

EXAMPLE 6.2. Let $K = K_{2,5}$ be the field obtained by adjoining a root α of $x(x-2)(x-5)-1$. As in [**9**], this field has discriminant 761, the ring of integers is generated by α, and the unit group is generated by $t, t-2, -1$, with the totally positive units generated by $t, (t-2)^2$. The class number is 1, so there is only one cusp up to equivalence, and the narrow class number is 2. We consider the variety H_K corresponding to the principal genus.

The bound b_K of Lemma 4.8 can be taken to be 76. Every totally positive element that is not trace-minimal is reduced by $(t-2)^i t^j$ for some $(i,j) \in S$, where

$$S = \{(-2,-3),(-2,-2),(-2,-1),(-2,0),(-2,1),(-2,2),(0,-2),\\ (0,-1),(0,1),(0,2),(0,3),(2,-1),(2,0),(2,1),(4,0)\}.$$

There are no reducers for (1), so determining the trace-minimal cone is enough to calculate the defects. The rays defining this cone are spanned by

$$(23,29,51), \quad (350,-55,-18), \quad (682,132,-109),\\ (-35,-375,154), \quad (84,139,87), \quad (1179,130,-165),$$

where (a,b,c) abbreviates $a+b\alpha+c\alpha^2$.

Cutting the cone by the hyperplane $\operatorname{tr} x = 1$, we obtain a polyhedron whose volume is 13/4. This matches Grundman's formula [**9**, Theorem 1], from which it follows that the qth defect of the cusp defined by the group of totally positive units is asymptotic to $13q^3/4$. Thus the defect for the group of squares of units is asymptotic to $13q^3/2$. Since $-2\zeta_K(-1) = 20/3 > 13/2$, Proposition 6.1 implies that principal component of the Hilbert modular threefold for K is of general type. On the author's laptop (a modest computational resource by the standards of the year 2025), this computation takes only 0.5 seconds.

REMARK 6.3. Similar considerations apply to $K_{3,5} = 3.3.985.1$, giving the result that the principal component of H_K is of general type. Though Grundman only states that at least one plurigenus of each of these varieties is positive, we consider the statement that they are of general type to be implicit in her work.

EXAMPLE 6.4. We now consider a more involved example: the cubic field 3.3.473.1. It is generated by a root α of t^3-5t-1. Then $\mathcal{O}_K = \mathbb{Z}[\alpha]$, while $\mathcal{O}_K^\times = \langle -1, \alpha, \alpha+2\rangle$ and $\mathcal{O}_{K,+}^\times = \mathcal{O}_K^{\times 2}$. We have $h_K = h_K^+ = 1$. Every non-trace-minimal element is reduced by $\alpha^i(\alpha+2)^j$ for some $(i,j) \in S$, where

$$S = \{(-4,-2),(-2,-2),(-2,0),(-2,2),(0,-2),(0,2),(2,-2),(2,0),(2,2),(4,0)\}.$$

It turns out that there are 19 reducers, whose norms range from 3 to 15. We find that 8 of the 20 polyhedra that correspond to the elements of $\mathcal{R}'_{\mathcal{O}_K}$ are redundant, so we need only find the volume of a union of 12 polyhedra. This is small enough that we can check the result of Algorithm 3.6 by inclusion-exclusion, finding by both methods that the volume is $79/24D_K$. On the other hand, we have $-2\zeta_K(-1) = 10/3$. Since $10/3 - 79/24 > 0$, this proves that H_K is of general type. This example takes under 4 seconds, most of which is used for the volume computation, which passes through the main loop of Algorithm 3.6 (step (2)) 22 times.

We now survey the fields for which $p_g(H_K) \le 1$ [**10**, Table I], beginning with those for which we cannot prove $H_{K;I}$ to be of general type. (Although we do not have a real result for these, the information shown here will be useful in Section 7.) In the tables, we use the following notation. The columns labeled h^+, r, n, t, t' refer to the narrow class number, the number of reducers, the number of passes through the main loop of Algorithm 3.6, the time for the whole calculation, and the time taken by one particular run of Algorithm 3.6. We omit the class number and the cusp because $h_K = 1$ for all cubic fields K with $p_g(H_K) \le 1$, and we omit h^+ in tables for the nonprincipal genus, since it is always 2 there. The exceptional speed of the examples of discriminant 49, 81, 169, 229, 257, 361, 697, 761, 985, already considered by Thomas-Vasquez or Grundman [**8**, **9**, **23**], reflects that these are fields with no reducers and a very simple cusp resolution. The field 3.3.1489.1 is also $K_{1,8}$. Likewise, in the genus-1 case, 3.3.1369.1 is the cubic subfield of $\mathbb{Q}(\zeta_{37})$, which is K_7 (Definition 1.6) and has no reducers. On the other hand, although 3.3.1765.1, for example, can be defined by the special polynomial $p(t) = t(t-2)(t-42)-1$, the maximal order is not generated by a root of p and the unit group is not generated by $t, t-2$, so this does not imply the existence of a particularly simple resolution.

PROPOSITION 6.5. *For the first* 11 *cubic fields as ordered by discriminant, the zeta values and asymptotic growth of defects coming from the cusp resolution are as shown in Table* 1.

PROOF. We simply apply Algorithm 3.6 to compute the growth of defects. □

TABLE 1. Hilbert modular threefolds for fields of discriminant at most 469, not shown to be of general type.

K	$-2\zeta_K(-1)$	$c_{\mathcal{O}_K,\mathcal{O}_{K,+}^\times}$	h^+	r	n	t	t'
3.3.49.1	2/21	5/12	1	0	0	0.250	0.010
3.3.81.1	2/9	3/4	1	0	0	0.130	0.000
3.3.148.1	2/3	55/36	1	7	8	0.950	0.700
3.3.169.1	2/3	17/12	1	0	0	0.190	0.000
3.3.229.1	4/3	3	2	0	0	0.240	0.000
3.3.257.1	4/3	26/9	2	0	0	0.200	0.000
3.3.316.1	8/3	137/36	1	60	35	7.160	5.500
3.3.321.1	2	8/3	1	17	21	3.050	2.510
3.3.361.1	2	29/12	1	0	0	0.390	0.000
3.3.404.1	10/3	143/36	1	94	55	8.640	6.920
3.3.469.1	4	49/12	1	58	25	6.330	3.990

THEOREM 6.6. *Let K be a cubic field of discriminant at least* 473 *and class number* 1 *such that the geometric genus of H_K is at most* 1. *Then the principal component of H_K and $\hat{H}_K$ is of general type, unless possibly $D(K) = 697$ or* 788.

PROOF. This is computed by the method of Example 6.4. When $h^+ = 1$, there is no difference between H_K and $\hat{H}_K$. When $h^+ = 2$, the dimension of the space of cusp forms of weight $2k$ for PSL_2 is asymptotically twice that for PGL_2^+, but the stabilizer of the cusp for PSL_2 is of index 2 in that for PGL_2^+, so the defect is also asymptotically twice as large. The elliptic points do not contribute by Remark 2.11. Thus the calculation is the same for the two groups. Tables 2, 3 show the results of our computations for the $22 + 14 = 36$ real cubic fields satisfying the hypotheses of the theorem. The general type result follows by noting that the second column, the constant in the asymptotic for the dimension of the space of modular forms, is greater than the third, the constant in the asymptotic for the defect. □

TABLE 2. Hilbert modular threefolds of general type, except for discriminant 697 and 788 and geometric genus 0 (principal genus).

K	$-2\zeta_K(-1)$	$c_{\mathcal{O}_K,\mathcal{O}_{K,+}^\times}$	h^+	r	n	t	t'
3.3.473.1	10/3	79/24	1	19	22	3.750	3.020
3.3.564.1	6	1021/180	1	227	84	25.230	20.530
3.3.568.1	20/3	1141/180	1	477	60	28.130	17.700
3.3.621.1	20/3	413/72	1	219	95	34.110	28.080
3.3.697.1*	16/3	67/12	2	0	0	0.330	0.010
3.3.733.1	8	221/36	1	251	75	30.510	22.020
3.3.756.1	26/3	469/72	1	297	102	55.440	32.830
3.3.761.1	20/3	13/2	2	0	0	0.510	0.010
3.3.785.1	22/3	29/6	1	71	76	19.080	16.820
3.3.788.1*	28/3	113/12	2	50	16	7.850	2.930
3.3.837.1	32/3	1381/180	1	1154	95	75.750	41.770
3.3.892.1	40/3	1091/90	2	141	54	20.220	13.610
3.3.940.1	44/3	703/72	1	1209	194	154.230	117.120
3.3.985.1	28/3	70/9	2	0	0	0.660	0.010
3.3.993.1	34/3	2377/360	1	120	68	35.030	27.330
3.3.1076.1	44/3	719/60	2	28	11	5.590	3.580
3.3.1257.1	16	101/9	2	15	15	5.880	4.540
3.3.1300.1	18	629/72	1	742	94	102.790	59.860
3.3.1345.1	46/3	779/120	1	95	143	55.150	50.250
3.3.1396.1	64/3	535/36	2	45	19	9.950	5.140
3.3.1489.1	16	134/15	2	0	0	0.920	0.000
3.3.1593.1	64/3	253/20	2	7	8	3.100	1.520

TABLE 3. Hilbert modular threefolds and geometric genus 1 (principal genus), all of general type.

K	$-2\zeta_K(-1)$	$c_{\mathcal{O}_K,\mathcal{O}_{K,+}^\times}$	h^+	r	n	t	t'
3.3.1016.1	52/3	124/9	2	842	70	85.090	29.070
3.3.1101.1	52/3	247/24	1	3198	163	278.630	125.560
3.3.1129.1	44/3	109/10	2	41	27	12.080	9.490
3.3.1229.1	56/3	37/3	2	64	43	20.580	17.330
3.3.1369.1	14	65/12	1	0	0	1.300	0.000
3.3.1373.1	68/3	1861/168	1	1351	171	173.030	109.780
3.3.1425.1	58/3	1519/180	1	481	145	97.420	72.400
3.3.1492.1	68/3	667/45	2	28	17	7.980	5.880
3.3.1573.1	76/3	2671/252	1	1088	230	234.060	179.500
3.3.1620.1	86/3	37/3	1	4708	274	623.380	300.940
3.3.1765.1	92/3	658/45	2	908	68	238.690	36.970
3.3.1825.1	68/3	401/36	2	3	4	2.810	1.230
3.3.1929.1	92/3	667/45	2	192	57	84.140	26.730
3.3.1937.1	28	397/30	2	18	26	12.890	10.270

THEOREM 6.7. *Let K be a cubic field of discriminant greater than 229 with $h_K^+ = 2$ such that the geometric genus of H_K is at most 1. Then the component of H_K or $\hat{H}_K$ corresponding to the nonprincipal genus is of general type.*

PROOF. Again, this is computed as above, with the same argument showing that only one of $H_K, \hat{H}_K$ need be considered. See Tables 4, 5. These tables were computed by choosing the ideal I representing the nonprincipal genus to be of minimal norm among integral ideals that are not narrowly principal. □

TABLE 4. Hilbert modular threefolds for the nonprincipal genus for fields K with $h^+ = 2$ for which $p_g(H_K) = 0$, all but the first shown to be of general type.

K	$-2\zeta_K(-1)$	$c_{\mathcal{O}_K,\mathcal{O}_{K,+}^\times}$	r	n	t	t'
3.3.229.1*	4/3	14/9	8	9	1.360	0.850
3.3.257.1	4/3	1	20	16	1.960	1.580
3.3.697.1	16/3	7/6	77	27	5.020	3.900
3.3.761.1	20/3	9/4	37	12	2.590	1.610
3.3.788.1	28/3	32/9	561	61	28.220	18.210
3.3.892.1	40/3	113/18	547	66	34.480	22.740
3.3.985.1	28/3	16/9	109	47	12.220	10.000
3.3.1076.1	44/3	127/30	281	67	25.830	20.500
3.3.1257.1	16	53/12	141	74	23.270	19.960
3.3.1396.1	64/3	40/9	1274	121	82.220	57.000
3.3.1489.1	16	4/3	182	40	17.560	12.700
3.3.1593.1	64/3	311/90	417	104	49.450	37.640

TABLE 5. Hilbert modular threefolds for the nonprincipal genus for fields K with $h^+ = 2$ for which $p_g(H_K) = 1$, all of general type.

K	$-2\zeta_K(-1)$	$c_{\mathcal{O}_K,\mathcal{O}_{K,+}^\times}$	r	n	t	t'
3.3.1016.1	52/3	73/9	3414	117	220.450	96.780
3.3.1129.1	44/3	173/36	472	39	33.160	18.250
3.3.1229.1	56/3	313/45	255	97	48.970	40.720
3.3.1492.1	68/3	35/9	850	146	90.950	67.550
3.3.1765.1	92/3	382/45	3649	100	310.810	75.830
3.3.1825.1	68/3	35/18	342	60	52.250	36.420
3.3.1929.1	92/3	53/9	1789	112	190.440	75.420
3.3.1937.1	28	21/4	134	101	73.210	64.580

We now discuss the three cubic fields for which we can prove that $\kappa_{H_K} > 0$ but not that H_K is of general type, aside from the field of discriminant 697 for which this is already known [**8**]. These correspond to the last two rows of Table 1 and the tenth row of Table 2, where $-2\zeta_K(-1)$ is less than the scaled volume of the union of polyhedra, so we cannot conclude that H_K is of general type.

PROPOSITION 6.8. (1) *Let* K_1 *be the cubic field* 3.3.404.1 $= \mathbb{Q}(\alpha)$, *where* α *is a root of* $x^3 - x^2 - 5x - 1$. *Then* $\kappa_{H_{K_1}} \geq 0$.
(2) *Let* K_2 *be the cubic field* 3.3.469.1 $= \mathbb{Q}(\alpha)$, *where* α *is a root of* $x^3 - x^2 - 5x + 4$. *Then* $\kappa_{H_{K_2}} > 0$.
(3) *Let* K_3 *be the cubic field* 3.3.788.1 $= \mathbb{Q}(\alpha)$, *where* α *is a root of* $x^3 - x^2 - 7x - 3$. *Then* $\kappa_{H_{K_3}} > 0$.

PROOF. In each case we compare the dimension of the space of modular forms to the defect for small q. For K_1 we use [**25**, Theorem 3.10] to find that $\dim M_4 = 13$, while Algorithm 5.5 shows that the defect for forms of weight 4 is 12. Thus $\mathrm{h}^0(K^{\otimes 2}) \geq 1$ and $\kappa_{H_{K_1}} \geq 0$. Similarly, for K_2 the dimension and defect are 15 and 12 respectively. Thus $\mathrm{h}^0(K^{\otimes 2}) \geq 3$ and $\kappa_{H_{K_2}} > 0$. □

For K_2, we compute the dimensions of M_{2k} as $1, 1, 15, 64, 172, 365, 668, 1098$ for $0 \leq k \leq 7$. The first two defects are as always $0, 1$, but the next few are $12, 60, 170, 365, 670, 1111$, and from then on they are presumably always greater than the dimension of M_{2k}, so nothing further is learned. We see that $\mathrm{h}^0(K^{\otimes 3}) \geq 4$ and $\mathrm{h}^0(K^{\otimes 4}) \geq 2$, which would also suffice to prove that $\kappa_{H_{K_2}} \geq 1$. Similarly, for K_3, the dimensions of M_{2k} are $1, 1, 34, 146, 402, 850, 1556, 2562$ for $0 \leq k \leq 7$, while the defects are $0, 1, 27, 139, 391, 843, 1547, 2563$. Considering $\mathrm{h}^0(K^{\otimes i})$ for any $2 \leq i \leq 6$, we see that $\kappa_{H_{K_3}} \geq 1$. In contrast, for K_1 the dimensions are $1, 1, 13, 53, 144, 304, 557, 915$, while the initial defects are $0, 1, 12, 59, 166, 356, 653, 1082$, and only $K^{\otimes 2}$ is seen in this way to have nonzero sections. Of course, if $K^{\otimes 2}$ has nonzero sections then so does $K^{\otimes 2n}$ for all $n \geq 0$.

REMARK 6.9. As alluded to in Remark 5.17, it appears that the defect series $\sum_{q=0}^{\infty} \delta(q) t^q$ is the rational function

$$\frac{-x^6 + 12x^5 + 38x^4 + 51x^3 + 36x^2 + 10x + 1}{(1-x)^2(1-x^2)(1-x^3)}.$$

This could be proved by a calculation like that used to prove [**8**, Theorem 1], but with much greater effort because we end up with a union of 9 convex polyhedra rather than a single one as in [**8**]. For other cubic fields, we find a similar formula, given by a polynomial of degree 6 divided by $(1-x)^2(1-x^2)(1-x^3)$. When the inverse different is replaced by a fractional ideal containing it with index d, it appears that the denominator changes to $(1-x)(1-x^2)(1-x^3)(1-x^d)$.

We have a similar result for the smallest cubic field of narrow class number 2 and the nonprincipal genus.

PROPOSITION 6.10. *Let* K_3 *be the cubic field* 3.3.229.1 $= \mathbb{Q}(\alpha)$, *where* α *is a root of* $x^3 - 4x - 1$, *and let* A *represent the nonprincipal genus. Then* $\kappa_{H_{K;A}} \geq 0$.

PROOF. The argument is essentially identical to the above; we have $\dim M_4 = 6$, while the defect for the cusp relative to the full group of totally positive units is 3. By Lemma 5.4, the defect for the group of totally positive units is $2(3-1)+1 = 5$, and we have $\mathrm{h}^0(K^{\otimes 2}) \geq 1$. □

6.1. Fields for which the geometric genus is greater than 1. To close this section, we describe what would be required to prove that $H_{K;A}$ is of general type for all K with $p_g(H_K) > 1$. We begin by showing that the criterion of [**27**, Theorem 1] applies to all genera.

THEOREM 6.11. *Let K be a totally real field of degree $d > 2$ with class number h, regulator R, and discriminant $\mathcal{D}_K$. Suppose that K is not $\mathbb{Q}(\zeta_n)^+$ for $n \in 7, 9, 15, 20$, and that if $d = 4$ then 3 is not a square in K. If*

$$2^{-2d+2}\pi^{-2d}d^d\frac{d_K\zeta_K(2)}{hR} > 1$$

then all components of H_K are of general type.

PROOF. The statement is the same as that of [**27**, Theorem 1] except that we do not restrict to the principal genus and (for simplicity) do not allow the quotient by a nontrivial subgroup of $\operatorname{Aut} K$. The proof is a straightforward adaptation of Tsuyumine's to this slightly more general situation.

In our notation Tsuyumine defines $\theta(I, q) = \{\nu \in I\mathcal{D}_K^{-1} : \nu \gg 0, \operatorname{tr}(\nu\beta \leq q)$ for some $\beta \gg 0 \in I^{-1}$ (cf. Definition 2.1), according to which this would be $\Lambda_{q+1}(I^{-1}))$ and $u(I, q) = \#(\theta(I, q)/\mathcal{O}_K^{\times 2})$. Then [**27**, Lemma 4] (again in our notation, and simplifying by ignoring the possibility of a subgroup of $\mathcal{S}_n$ acting on the coordinates) shows that $\dim S_{2k}^{(m)}(\Gamma) \geq \dim S_{2k}(\Gamma) - \sum_{\mathfrak{a}_\lambda} u(\mathfrak{a}_\lambda^2, \frac{1}{2}k + m - 1)$. Here $\mathfrak{a}_\lambda$ runs over a set of ideals corresponding to the cusps and $S^{(m)}$ denotes a certain space of cusp forms [**27**, p. 271] such that if $S^{(1)} \neq 0$ and K is not one of the exceptional fields in the statement of the theorem then the Hilbert modular variety is of general type.

Let g be the number of components of the Hilbert modular variety. Asymptotically the dimension of the space of weight-$2k$ cusp forms on each component is equal by Proposition 1.14, so on each component we obtain

$$2^{-2+1}\,\pi^{-2d}\,D_K^{3/2}\,\zeta_K(2)\,(2k)^d + O(k^{d-1})$$

([**27**, p. 274]), while the inequality [**27**, Lemma 5] $u(I, q) \leq (2^{d-1}\,d^{-d}\,D_K^{1/2}\,|R|)q^d + O(q^{d-1})$ is independent of I. Now apply [**27**, Lemma 4] with $\mathfrak{a}_\lambda$ running only over ideals representing the cusps of one particular component. As in [**27**, (9)] we find that for each component

$$\dim S_{2k}^{(m)}\Gamma \geq \left(2^{-2d+1}\pi^{-2d}D_K^{3/2}\zeta_K(2) - 2^{-1}d^{-d}D_K^{1/2}hR\right)k^d + O(k^{d-1}),$$

whence as in [**27**, Theorem 1] if $2^{-2d+2}\pi^{-2d}d^d\frac{D_K\zeta_K(2)}{hR} > 1$ then the component is of general type. In this context Tsuyumine's $\hat{h}$ is equal to our h since there is no action by a nontrivial subgroup of $\mathcal{S}_n$. □

COROLLARY 6.12. *Let K be a totally real cubic field of discriminant greater than $2.77 \cdot 10^8$ and A an arbitrary genus of K. Then $H_{K;A}$ is of general type.*

PROOF. In view of Theorem 6.11, the calculation of [**7**, Theorem 2] applies equally to all genera. □

Now we would like to determine all K for which [**27**, p. 276] does not immediately show this to be the case.

HYPOTHESIS 6.13. *For all but 421 totally real cubic fields, the largest of their discriminants being 26601, the Tsuyumine-Grundman criterion $\frac{D_K\zeta_K(2)}{hR} \geq \frac{16\pi^6}{27}$ (*[**27**, Theorem 1]*;* [**7**, Corollary 10]*) is satisfied.*

In light of Grundman's observation that this holds for all K with $D_K > 2.77 \cdot 10^8$, this could be proved by a finite calculation (though one should note that the tables of cubic fields in the LMFDB [**14**] are not complete this far out). If cubic fields are ordered by discriminant, the field of discriminant 26601 is 1133rd, and we have checked that there are no further counterexamples among the first 50000 fields. We thus expect no more.

Let us assume Hypothesis 6.13. Of the 421 fields, there are 47 with $p_g(H_K) \leq 1$, and these have been studied earlier in this section. Of the remaining 374, there are 25 with class number greater than 1, so we need a lemma to describe the cusps.

LEMMA 6.14. *Let K be a totally real field and fix a genus $A \in \mathrm{Cl}^+(K)/2\,\mathrm{Cl}^+(K)$. The $h(K)$ cusps of $H_{K;A}$ are of type $(I^2A, \mathcal{O}_{K,+}^\times)$ as I runs over the ideal classes of $\mathcal{O}_K$.*

PROOF. Immediate from [**2**, Proposition 3.3.8 (a)]. □

Now we need only apply the existing code to the cusps as described above and compare their defect contribution to $-2\zeta_K(-1)$ to prove that the Hilbert modular varieties are of general type. For example, let us consider the first cubic field of class number 2, which does not satisfy the Tsuyumine-Grundman criterion.

EXAMPLE 6.15. The field $K = 3.3.1957.1$ is $\mathbb{Q}(\alpha)$, where $\alpha^3 - \alpha^2 - 9\alpha + 10 = 0$. We have $h_K = 2$, but Cl^+ is cyclic of order 4, so there are 2 genera. Let Cl^+ be represented by I_0, I_1, I_2, I_3, where I_0, I_2 are principal and I_0 is narrowly principal. We may choose the representatives to have norm $1, 2, 4, 4$ respectively (this specifies them uniquely). For the principal genus and PGL_2^+, the two cusps are of type $(I_0, \mathcal{O}_{K,+}^\times), (I_2, \mathcal{O}_{K,+}^\times)$, and for the nonprincipal genus, they are of type $(I_1, \mathcal{O}_{K,+}^\times), (I_3, \mathcal{O}_{K,+}^\times)$. If we used PSL_2 instead of PGL_2^+, the groups of units would be $\mathcal{O}_K^{\times 2}$ instead of $\mathcal{O}_{K,+}^\times$. We calculate that $-2\zeta_K(-1) = 104/3$, and that the volume of the unions of polyhedra for cusps of type $(I_j, \mathcal{O}_K^{\times 2})$ scaled by the covolume of $\hat{I}_j$ are respectively $34/3, 71/18, 44/15, 103/18$. Since $104/3 > 34/3 + 44/15, 71/18 + 103/18$, this shows that both $H_{K;A}$ are of general type.

HYPOTHESIS 6.16. *For all K with $p_g(H_K) > 1$ and all genus representatives A, the Hilbert modular threefold $H_{K;A}$ is of general type.*

In [**15**] there is a script that is expected to verify that Hypothesis 6.13 implies Hypothesis 6.16 if run long enough, as well as code that will check that there are no counterexamples to Hypothesis 6.13 among cubic fields in the LMFDB. Note that the analogous statement is false for quadratic fields; in the honestly elliptic case of [**28**, Theorem VII.3.3] there are many counterexamples.

7. Results at higher level

In this section we describe our results on the Kodaira dimension of Hilbert modular threefolds of the form $H_{K,I;A}$. As before, we can try to prove that such a surface is of general type by showing that the dimension of the space of modular forms grows faster than the defects, and if this fails we can still hope to prove that $\kappa_{H_{K,I;A}} > 0$ by finding q such that $\dim |qK| \geq 2$. If $p_g(H_K) = 1$ then we have already shown that H_K is of general type, so the same follows for its covers and so the most interesting cases are those with $p_g(H_K) = 0$.

We now begin to list the pairs K, I systematically for which the geometric genus of $H_{K,I}$ is at most 1. If $J|I$ and $p_g(H_{K,J}) > 0$, then by the theory of oldforms $p_g(H_{K,I}) > 2p_g(H_{K,J}) \geq 2$, so such cases can be ignored. We will use the trace formula [**2**, (5.1.2)] to bound the levels for which $p_g(H_{K,I}) \leq 1$. Let us begin by stating the formula, first introducing the notation of [**2**, (5.1.3), (5.2.2)].

NOTATION 7.1. Let K be a totally real field of degree d and let I be an ideal of $\mathcal{O}_K$. Let $\mathcal{S}$ be the set of orders containing one of the $\mathcal{O}_K[x]/(x^2 - tx + u)$ where u ranges over totally positive units modulo squares and t over elements of $\mathcal{O}_K$ such that $t^2 - 4u \gg 0$. For each order $S \in \mathcal{S}$, let $c_S = h(S)/2[S^\times : \mathcal{O}_K^\times]$ as in [**2**, (5.1.3)]. Let $A = (-1)^{d-1}h^+(\mathcal{O}_K)$, let $B = \frac{1}{2^{d-1}} \cdot |\zeta_K(-1)| \cdot N(I) \prod_{\mathfrak{p}|I} (1 + N(\mathfrak{p})^{-1})$, let $C(u,t) = \frac{1}{2} \sum_{S \in \mathcal{S}} \frac{h(S)}{[S^\times : \mathcal{O}_K^\times]} m(\hat{S}, \hat{\mathcal{O}}_K; \hat{\mathcal{O}}_K^\times)$, where the m are certain embedding numbers that are products over primes dividing the level, defined in [**29**, Section 30.6], and let $D_k(u,t) = \frac{\alpha(u,t)^{k+1} - \beta(u,t)^{k+1}}{\alpha(u,t) - \beta(u,t)}$, where $\alpha(u,t), \beta(u,t)$ are the roots of $T^2 - tT + u$.

THEOREM 7.2 ([**2**, (5.1.2)]). *The dimension of the space of Hilbert cusp forms for $\Gamma_0(I)$ of weight $2k$ is the coefficient of T^{2k} in*

$$AT^2 + BT \left(T \frac{d}{dT}\right)^n \left(\frac{T}{1 - T^2}\right) + (-1)^n \sum_{(u,t)} C(u,t) \sum_{m \geq 1} N_{K/\mathbb{Q}}(D_{2m-2}(u,t)) T^{2m}.$$

REMARK 7.3. We have corrected a typographical error in [**2**, (5.1.2)], where the sign before the term beginning with $\sum_{(u,t)} C(u,t)$ is omitted.

THEOREM 7.4. *Let K be a totally real cubic field satisfying $p_g(H_K) = 0$ and let $\mathfrak{p}$ be a prime ideal of $\mathcal{O}_K$. If $p_g(H_{K,\mathfrak{p}}) \leq 1$ then either $\mathfrak{p}$ divides the conductor of some $S \in \mathcal{S}$ or $h^+(\mathcal{O}_K) + (N(\mathfrak{p}) + 1) |\zeta_K(-1)| / 4 - 2 \sum_{S \in \mathcal{S}} c_S \leq 1$.*

REMARK 7.5. Of course this implies an effective upper bound for $N(\mathfrak{p})$.

PROOF. This is a consequence of [**2**, (5.1.2),(5.1.3)]. Indeed, the geometric genus is the dimension of the space of weight-2 cusp forms by Theorem 1.18. We estimate the constants A, B, C, D. The coefficient of T^2 in $T \left(T \frac{d}{dT}\right)^n \left(\frac{T}{1-T^2}\right)$ is always 1, and $(N(\mathfrak{p}) + 1) |\zeta_K(-1)| / 4$ is the value of B in this context (recall that $p_g = 0$ implies $h_K = 1$ for cubic fields). The coefficient D_0 is always 1 so that factor may be ignored in computing the dimension of the space of weight-2 cusp forms, while [**29**, Lemma 30.6.17] implies that the local embedding number $m(\hat{S}, \hat{\mathcal{O}}; \hat{\mathcal{O}}^\times)$ is at most 2 when the level is an unramified prime. The result follows. □

We now present the calculation of levels at which the geometric genus is 0 or 1 in detail for the smallest cubic field.

THEOREM 7.6. *Let $K = \mathbb{Q}(\zeta_7)^+$. The Hilbert modular variety $H_{K,I}$ is of geometric genus 0 if and only if $I \in \{(1), \mathfrak{p}_7, (2), \mathfrak{p}_{13}, \mathfrak{p}_{29}, \mathfrak{p}_{43}\}$, and 1 if and only if $I \in \{(3), \mathfrak{p}_{41}, \mathfrak{p}_7^2, 2\mathfrak{p}_7, (4), \mathfrak{p}_{71}, \mathfrak{p}_7\mathfrak{p}_{13}, \mathfrak{p}_{97}, \mathfrak{p}_{113}, \mathfrak{p}_{127}, \mathfrak{p}_{13}^2\}$.*

PROOF. We apply Theorem 7.4. We have $h^+(\mathcal{O}_K) = 1$, while $\zeta_K(-1) = -1/21$ and $\sum_{S \in \mathcal{S}} c_s \leq 1/2(1/2 + 2(1/3) + 6(1/7)) = 85/84$, where each term $1/2, 2(1/3), 6(1/7)$ appears in the sum if and only if $\mathfrak{p}$ splits in $K(i), K(\zeta_3), K(\zeta_7)$ respectively (we assume that $\mathfrak{p}$ is not one of the primes $\mathfrak{p}_7, (2), (3)$ dividing one of the conductors). It follows that if $N(\mathfrak{p}) > 169$ then $p_g(H_{K,\mathfrak{p}}) > 1$, and one checks cases up to there.

Next we allow the level to be the product of two primes. If the primes are distinct then the local embedding number is $\prod_{i=1}^{2} 1 + \left(\frac{N(\mathfrak{p}_i)}{d}\right)$, and so the contribution cannot exceed $2(85/42)$. So again if $N(\mathfrak{p}_1\mathfrak{p}_2) > 340$ we have $p_g > 1$, and it is not difficult to check up to this point. At level $\mathfrak{p}^2$ the local embedding numbers are the same as for p except for primes above $2, 3, 7$, and again it is straightforward to verify that $p_g > 1$ for primes of norm 27 or greater.

Finally, if the level is a product of three or more primes, then p_g can be less than 2 only if $p_g = 0$ for all proper divisors. However, it has already been verified that if $p_g = 0$ then I is prime, so this is not possible. □

We now study the Kodaira dimension of these covers.

DEFINITION 7.7. Let $\mathfrak{p}$ be a prime of $\mathcal{O}_K$, where K is a totally real cubic field. If $K = \mathbb{Q}(\zeta_7)^+$, then define $c_\mathfrak{p}$ to be $2, 1, 0$ according as $N(\mathfrak{p}) \equiv 1, 0, -1 \bmod 7$. If $K = \mathbb{Q}(\zeta_9)^+$, then let $c_\mathfrak{p} = 2, 1, 0$ according as $N(\mathfrak{p}) \equiv 1, 0, -1 \bmod 3$. For a general ideal I of $\mathbb{Q}(\zeta_7)^+$ or $\mathbb{Q}(\zeta_9)^+$, let $e_I = 0$ if I is divisible by the square of the ramified prime and otherwise $\frac{\prod_{\mathfrak{p}|I} c_\mathfrak{p}}{c_K}$, where $c_K = 84$ for $\mathbb{Q}(\zeta_7)^+$ and 18 for $\mathbb{Q}(\zeta_9)^+$. For all other fields define $c_\mathfrak{p} = e_I = 0$.

THEOREM 7.8. *Let K be a totally real cubic field with $h(K) = 1$, let $z = -2\zeta_K(-1)$, let A be a genus representative for K, and let v be the volume of the cusp for A as computed in Algorithm* 3.6. *Let $\mathfrak{p}$ be a prime ideal of $\mathcal{O}_K$ such that $(N(\mathfrak{p}) + 1)z > 2v + e_\mathfrak{p}$. Then $H_{K,\mathfrak{p};A}$ is of general type.*

PROOF. Just as with the modular curve $X_0(p)$, the Hilbert modular variety $H_{K,\mathfrak{p};A}$ has two cusp orbits when p is prime, corresponding to $\infty, 0$. The stabilizers are respectively the groups of upper and lower triangular matrices contained in $\Gamma_0(\mathfrak{p}; A)$; in particular, if we conjugate in $\mathrm{GL}_2(K)$ to an upper triangular subgroup of SL_2, the group of units in the stabilizer is the full group of squares of units. (The classification of cusps given here is a special case of the results of [**2**, Sections 3.1–3.3].) Since the class number is 1, it follows that the defects of both cusps are equal to the defects of the cusp for $H_{K;A}$. In view of Example 2.10, the leading coefficient in the asymptotic to the contribution of elliptic points to the defect of $H_{K,\mathfrak{p}}$ is given by $e_\mathfrak{p}$, since the number of quotient singularities is multiplied by $1 + \left(\frac{\mathfrak{p}}{K(\alpha)}\right)$, where $K(\alpha)$ is the quadratic field corresponding to the elliptic point. On the other hand, the index of $\Gamma_0(\mathfrak{p})$ in $\mathrm{SL}_2(\mathcal{O}_K)$ is $N(\mathfrak{p}) + 1$, and the asymptotic dimension of the space of modular forms of weight k is multiplied by this factor. □

NOTATION 7.9. We use $\mathfrak{p}_n$ to denote a prime of norm n. If q is a rational prime that factors as $\mathfrak{P}_1\mathfrak{P}_2^2$ in $\mathcal{O}_K$, we denote $\mathfrak{P}_1, \mathfrak{P}_2$ by $\mathfrak{p}_q, \mathfrak{r}_q$ respectively.

It happens not to be necessary to distinguish the factors of primes that split completely. In the case of a Galois extension it is understood that the Galois images of all ideals mentioned have the same property. By [**17**, Theorem 1] it is unnecessary to distinguish between different genera, and so the rows referring to 3.3.229.1, 3.3.257.1 describe both genera.

TABLE 6. Fields and primes for which $p_g(H_{K,\mathfrak{p}}) \leq 1$.

K	$\mathfrak{p} : p_g = 0$	$\mathfrak{p} : p_g = 1$
3.3.49.1	$\mathfrak{p}_7, (2), \mathfrak{p}_{13}, \mathfrak{p}_{29}, \mathfrak{p}_{43}$	$(3), \mathfrak{p}_{41}, \mathfrak{p}_{71}, \mathfrak{p}_{97}, \mathfrak{p}_{113}, \mathfrak{p}_{127}$
3.3.81.1	$\mathfrak{p}_3, \mathfrak{p}_{19}, \mathfrak{p}_{37}$	$(2), \mathfrak{p}_{17}, \mathfrak{p}_{73}$
3.3.148.1	$\mathfrak{p}_2, \mathfrak{p}_5, \mathfrak{p}_{13}$	$\mathfrak{p}_{17}, \mathfrak{p}_{25}$
3.3.169.1	$\mathfrak{p}_5, \mathfrak{p}_{13}$	(2)
3.3.229.1	$\mathfrak{p}_2, \mathfrak{p}_4, \mathfrak{p}_7$	
3.3.257.1	$\mathfrak{p}_3, \mathfrak{p}_5, \mathfrak{p}_7$	
3.3.316.1	$\mathfrak{r}_2$	$\mathfrak{p}_2$
3.3.321.1	$\mathfrak{r}_3$	$\mathfrak{p}_3, \mathfrak{p}_7$
3.3.361.1		$\mathfrak{p}_7$
3.3.404.1		$\mathfrak{p}_2$
3.3.469.1		$\mathfrak{p}_4$
3.3.568.1		$\mathfrak{r}_2$

PROPOSITION 7.10. *Table 6 gives the pairs consisting of a field and a prime* $\mathfrak{p}$ *for which* $p_g(H_{K,\mathfrak{p}}) \leq 1$.

PROOF. See [**15**]. Our implementation is not very efficient, but it is fast for all but the two smallest real cubic fields, so we did not feel a need to improve it. □

PROPOSITION 7.11. *The Hilbert modular varieties* $H_{K,\mathfrak{p}}$ *for the primes in Table* 6 *are of general type, except possibly as shown in Table* 7.

TABLE 7. Fields, genera, and primes for which we cannot show that $H_{K,\mathfrak{p},I}$ is of general type. The g column is $+$ for the narrowly principal genus when $h^+ = 2$; otherwise it is left blank. There are no examples with the nonprincipal genus.

K	g	$\mathfrak{p}$
3.3.49.1		$\mathfrak{p}_7, (2)$
3.3.81.1		$\mathfrak{p}_3$
3.3.148.1		$\mathfrak{p}_2$
3.3.229.1	$+$	$\mathfrak{p}_2$
3.3.257.1	$+$	$\mathfrak{p}_3$

PROOF. Use [**15**] to check the condition of Theorem 7.8. □

TABLE 8. Fields and nonprime ideals which $p_g(H_{K,I}) \leq 1$.

K	$I : p_g = 0$	$I : p_g = 1$
3.3.49.1		$\mathfrak{p}_7^2, 2\mathfrak{p}_7, (4), \mathfrak{p}_7\mathfrak{p}_{13}, \mathfrak{p}_{13}\mathfrak{p}_{13}'$
3.3.81.1	$\mathfrak{p}_3^2$	$(3), \mathfrak{p}_3\mathfrak{p}_{19}$
3.3.148.1	$\mathfrak{p}_2^2, \mathfrak{p}_2\mathfrak{p}_5$	$(2), \mathfrak{p}_2^2\mathfrak{p}_5$
3.3.169.1		$\mathfrak{p}_5\mathfrak{p}_5'$
3.3.229.1	$\mathfrak{p}_2^2, \mathfrak{p}_2^3$	
3.3.316.1	$\mathfrak{r}_2^2$	$\mathfrak{r}_2^3$
3.3.321.1	$\mathfrak{r}_3^2$	

PROPOSITION 7.12. *Table 8 gives the pairs consisting of a field and a nonprime ideal I for which $p_g(H_{K,I;A}) \leq 1$.*

PROOF. (Sketch.) Such an ideal can only be divisible by primes listed in Table 6 and all of its divisors other than (1) must be in one of the two tables. We verify using our implementation of [**2**, (5.1.2),(5.1.3)] that if I is in Table 8 and $\mathfrak{p}$ is in Table 6 but $I\mathfrak{p}$ is not in Table 8 then $p_g(H_{K,I}) > 1$. As in Corollary 7.10, it is unnecessary to distinguish between different genera here. Again, see [**15**]. □

PROPOSITION 7.13. *Let K be a cubic field for which $p_g(H_K) \leq 1$. For all proper nonprime ideals $I \subset \mathcal{O}_K$ and genus representatives A, the corresponding Hilbert modular variety is of general type, except possibly for $\mathfrak{p}_3^2$ in $\mathbb{Q}(\zeta_9)^+$, $\mathfrak{p}_2^2$ in 3.3.148.1, and $\mathfrak{p}_2^2$ for 3.3.229.1 (principal genus).*

PROOF. If I has a prime divisor $\mathfrak{p}$ for which $H_{K,\mathfrak{p}}$ is of general type, then $H_{K,I}$ too is of general type, so we need only consider powers of the ideals from Proposition 7.11 and $2\mathfrak{p}_7$ in $\mathbb{Q}(\zeta_7)^+$. We start with $I = \mathfrak{p}_7^2$ for $\mathbb{Q}(\zeta_7)^+$. The index of $\Gamma_0(I)$ is $N(\mathfrak{p}_7)^2 + N(\mathfrak{p}_7) = 56$, just as for the index of $\Gamma_0(p^2)$ in $\mathrm{SL}_2(\mathbb{Z})$. The ray class group mod $\mathfrak{p}_7\infty_1\infty_2\infty_3$ is of order 2, so by [**2**, Corollary 3.1.18] there are 4 cusps. These are represented by $\infty, 0, 1/\pi_7, 3/\pi_7$, where $(\pi_7) = \mathfrak{p}_7$. For the first two of these the group of units is the full group of totally positive units as before, but for the other two it is the subgroup of totally positive units congruent to 1 mod $\mathfrak{p}_7$ and the defects are multiplied by the index of this subgroup, which is 3. Thus the dimension of the space of modular forms of weight $2n$ is asymptotically 56 times that for level 1, which makes it $16n^3/3$, while the defects are asymptotic to $(8 \cdot 5/12)n^3$, which is smaller. There are no elliptic points of order 7. Similarly, taking $I = (2)^2$ for $\mathbb{Q}(\zeta_7)^+$, the index is 72, the ray class group mod $2\infty_1\infty_2\infty_3$ is trivial, and so there are 3 cusps, represented by $\infty, 0, 1/2$. Again, the groups of units for the first two are $\mathcal{O}_K^{\times 2}$, but for the last we get the totally positive units congruent to 1 mod 2, a subgroup of index 7. In this case we have 8 elliptic points of order 7, of which 6 contribute to the defect. So the dimension is asymptotically $2 \cdot 72n^3/21$, while the defect is asymptotically $(9 \cdot 5/12 + 6/252)n^3$ and we have general type. The reader may check that $H_{\mathbb{Q}(\zeta_7)^+,2\pi_7}$ is of general type, the dimension of the space of cusp forms of weight $2n$ being asymptotic to $48n^3/7$ and the defect to $71n^3/42$.

For (3) in $\mathbb{Q}(\zeta_9)^+$, the index is 36 and there are 6 cusps. The stabilizers of these have index $27, 3, 3, 1, 1, 1$ in the SL_2-stabilizer, and the unit indices are respectively $1, 3, 3, 1, 1, 1$ (the 3 arises as in [**1**, Proposition 3.3.8 (a)] because the totally positive units congruent to $1 \bmod \mathfrak{p}^2$ are of index 3). Since $\zeta_{\mathbb{Q}(\zeta_9)^+}(-1) = -1/9$, the dimension of the space of forms is asymptotic to $(36 \cdot 2/9)n^3$, while the defect is 10 times that for level 1 since there are no elliptic points of order 9. We compute that the defect for the cusp at level 1 is asymptotic to $3n^3/4$. Since $36 \cdot 2/9 > 10 \cdot 3/4$, the result follows.

The arguments for π_2^3 in 3.3.148.1, 3.3.229.1 are similar and we will only sketch them; likewise for π_3^2 in 3.3.257.1. The index of π_2^3 in both cases is 12 and there are 4 cusps. For 3.3.148.1, all totally positive units are $1 \bmod \pi_2^2$ and so the unit indices are $1, 2, 1, 1$; thus the cusps create a defect asymptotic to $5 \cdot 55/36n^3$. On the other hand, the dimension of the space of modular forms grows like $12(-2\zeta_K(-1))n^3 = 8n^3$, which is larger.

In 3.3.229.1, not all totally positive units are 1 mod $\mathfrak{p}_2^2$ and the unit indices are all 1, so the dimension grows like $12(4/3)n^3$ and the defect like $4(3)n^3$. Again, in 3.3.257.1, the index of $\Gamma_0(\mathfrak{p}_3^2)$ is 12, not all totally positive units are 1 mod $\mathfrak{p}_3$, and there are 3 cusps with unit indices $1, 2, 1$. So the dimension of the space of modular forms and the defect grow like $12(4/3)n^3$ and $4(26/9)n^3$ respectively. □

REMARK 7.14. The argument does not apply to $\mathfrak{p}_3^2$ in $\mathbb{Q}(\zeta_9)^+$. The index is 12 and the ray class group has order 2, so again there are 4 cusps. Again the intermediate cusps $1/\pi_3, 2/\pi_3$ are stabilized by totally positive units congruent to 1 mod $\mathfrak{p}_3$, but this time that is all of them, so the defects are the same for all cusps. The dimension of the space of modular forms of weight $2n$ is 12 times that for level 1, so $8n^3/3$, while the defect is 4 times that for level 1 and is asymptotic to $3n^3$. Similarly for $\mathfrak{p}_2^2$ in 3.3.148.1; the dimension of the space of modular forms is asymptotic to $4n^3$ and the defect sum to $55n^3/12$. Again, for $\mathfrak{p}_2^2$ in 3.3.229.1, the asymptotics are $8n^3$ and $9n^3$ respectively, and we do not have a general type result.

REMARK 7.15. For reasons discussed at the end of Section 6, we certainly expect that the hypothesis that $p_g(H_K) < 1$ is unnecessary.

PROPOSITION 7.16. *In Table 9, the varieties $H_{K,I}$ have at least the indicated Kodaira dimension.*

PROOF. Table 9 also gives an n such that the dimension of the space of Hilbert cusp forms of level I and weight n is equal to resp. greater than the sum of the defects for $\kappa = 0, 1$ respectively. See [**15**] for details. □

TABLE 9. Fields and levels for which $H_{K,I}$ can be shown to be of nonnegative Kodaira dimension, but not of general type.

K	I	type	κ	n	$\dim M_n$	$\sum_i \delta_i(n)$
3.3.49.1	(2)		1	4	6	4
3.3.81.1	$\mathfrak{p}_3^2$		0	4	13	12
3.3.148.1	$\mathfrak{p}_2^2$		1	4	18	15
3.3.229.1	$\mathfrak{p}_2^2$	GL_2^+	1	4	33	27
3.3.229.1	$\mathfrak{p}_2^2$	SL_2	1	4	30	27
3.3.257.1	$\mathfrak{p}_3$	GL_2^+	1	4	22	18
3.3.257.1	$\mathfrak{p}_3$	SL_2	1	4	20	18

Acknowledgments

This paper is inspired by the work done on Hilbert modular forms in the context of the Simons Collaboration on Algebraic Geometry, Number Theory, and Computation. Some of the code from [**1**] was used in this work. I would like to thank all of the authors of [**1, 2**] and especially Eran Assaf, Abhijit Mudigonda, and John Voight for their patient replies to my numerous questions, Colin Ingalls for a helpful conversation on quotient singularities, and Leland McInnes for advice on interactive three-dimensional figures. I also thank the referees for their detailed and valuable criticism which led to substantial improvements in the exposition and the technical reviewer for pointing out some bugs and flaws in the code. In addition, I thank the Department of Pure Mathematics at the University of Waterloo, where this work was begun, for its hospitality, and the Tutte Institute for Mathematics and Computation for its support of my external research and for the opportunity to take leave from my employment there.

References

[1] Eran Assaf, Angelica Babei, Ben Breen, Sara Chari, Edgar Costa, Juanita Duque-Rosero, Aleksander Horawa, Jean Kieffer, Avinash Kulkarni, Grant Molnar, Abhijit Mudigonda, Michael Musty, Sam Schiavone, Shikhin Sethi, Samuel Tripp, and John Voight, *Software for computing with Hilbert modular forms*, 2024, `https://github.com/edgarcosta/hilbertmodularforms` (retrieved June 12, 2025).

[2] Eran Assaf, Angelica Babei, Ben Breen, Edgar Costa, Juanita Duque-Rosero, Aleksander Horawa, Jean Kieffer, Avinash Kulkarni, Grant Molnar, Sam Schiavone, and John Voight, *A database of basic numerical invariants of Hilbert modular surfaces*, LuCaNT: LMFDB, computation, and number theory, Contemp. Math., vol. 796, Amer. Math. Soc., [Providence], RI, [2024] ©2024, pp. 285–312, DOI 10.1090/conm/796/16006. MR4732692

[3] Matthias Beck and Sinai Robins, *Computing the continuous discretely: Integer-point enumeration in polyhedra*, 2nd ed., Undergraduate Texts in Mathematics, Springer, New York, 2015. With illustrations by David Austin, DOI 10.1007/978-1-4939-2969-6. MR3410115

[4] Wieb Bosma, John Cannon, and Catherine Playoust, *The Magma algebra system. I. The user language*, J. Symbolic Comput. **24** (1997), no. 3-4, 235–265, DOI 10.1006/jsco.1996.0125. Computational algebra and number theory (London, 1993). MR1484478

[5] C. A. Floudas and P. M. Pardalos (eds.), *Encyclopedia of optimization*, Springer, New York, 2009. Second edition [of MR1865755], DOI 10.1016/j.tcs.2009.07.038. MR3593766

[6] Eberhard Freitag, *Hilbert modular forms*, Springer-Verlag, Berlin, 1990, DOI 10.1007/978-3-662-02638-0. MR1050763

[7] H. G. Grundman, *On the classification of Hilbert modular threefolds*, Manuscripta Math. **72** (1991), no. 3, 297–305, DOI 10.1007/BF02568281. MR1118548

[8] H. G. Grundman, *Defects of cusp singularities and the classification of Hilbert modular threefolds*, Math. Ann. **292** (1992), no. 1, 1–12, DOI 10.1007/BF01444605. MR1141781

[9] H. G. Grundman, *Defect series and nonrational Hilbert modular threefolds*, Math. Ann. **300** (1994), no. 1, 77–88, DOI 10.1007/BF01450476. MR1289831

[10] H. G. Grundman and L. E. Lippincott, *Hilbert modular threefolds of arithmetic genus one*, J. Number Theory **95** (2002), no. 1, 72–76, DOI 10.1006/jnth.2001.2747. MR1916080

[11] Karl-Bernhard Gundlach, *Some new results in the theory of Hilbert's modular group*, Contributions to function theory (Internat. Colloq. Function Theory, Bombay, 1960), Tata Inst. Fund. Res., Bombay, 1960, pp. 165–180. MR125986

[12] Friedrich E. P. Hirzebruch, *Hilbert modular surfaces*, Enseign. Math. (2) **19** (1973), 183–281. MR393045

[13] Friedrich Wilhelm Knöller, *Beispiele dreidimensionaler Hilbertscher Modulmannigfaltigkeiten von allgemeinem Typ* (German, with English summary), Manuscripta Math. **37** (1982), no. 2, 135–161, DOI 10.1007/BF01168507. MR651131

[14] The LMFDB Collaboration, *The L-functions and modular forms database*, `https://www.lmfdb.org`, 2025, [Online; accessed 22 January 2025].

[15] Adam Logan, *Magma scripts and code supporting the claims of this paper*, Github repository, `https://github.com/adammlogan/hilbert-modular-threefolds`, 2025.

[16] Adam Logan, *Rings of Hilbert modular forms, computations on Hilbert modular surfaces, and the Oda-Hamahata conjecture*, Res. Number Theory **11** (2025), no. 2, Paper No. 57, 38, DOI 10.1007/s40993-025-00636-8. MR4908187

[17] Adam Logan and John Voight, *Online appendix to this paper*, 2025, `arXiv:2501.15719`.

[18] Kenji Matsuki, *Introduction to the Mori program*, Universitext, Springer-Verlag, New York, 2002, DOI 10.1007/978-1-4757-5602-9. MR1875410

[19] David R. Morrison and Glenn Stevens, *Terminal quotient singularities in dimensions three and four*, Proc. Amer. Math. Soc. **90** (1984), no. 1, 15–20, DOI 10.2307/2044659. MR722406

[20] Hideo Shimizu, *On discontinuous groups operating on the product of the upper half planes*, Ann. of Math. (2) **77** (1963), 33–71, DOI 10.2307/1970201. MR145106

[21] Takuro Shintani, *On evaluation of zeta functions of totally real algebraic number fields at non-positive integers*, J. Fac. Sci. Univ. Tokyo Sect. IA Math. **23** (1976), no. 2, 393–417. MR427231

[22] The PARI Group, Univ. Bordeaux, *PARI/GP version 2.15.4*, 2023, available from `http://pari.math.u-bordeaux.fr/`.

[23] E. Thomas and A. T. Vasquez, *On the resolution of cusp singularities and the Shintani decomposition in totally real cubic number fields*, Math. Ann. **247** (1980), no. 1, 1–20, DOI 10.1007/BF01359864. MR565136

[24] E. Thomas and A. T. Vasquez, *Chern numbers of Hilbert modular varieties*, J. Reine Angew. Math. **324** (1981), 192–210, DOI 10.1515/crll.1981.324.192. MR614525

[25] E. Thomas and A. T. Vasquez, *Rings of Hilbert modular forms*, Compositio Math. **48** (1983), no. 2, 139–165. MR700001

[26] Emery Thomas, *Defects of cusp singularities*, Math. Ann. **264** (1983), no. 3, 397–411, DOI 10.1007/BF01459133. MR714112

[27] Shigeaki Tsuyumine, *On the Kodaira dimensions of Hilbert modular varieties*, Invent. Math. **80** (1985), no. 2, 269–281, DOI 10.1007/BF01388606. MR788410

[28] Gerard van der Geer, *Hilbert modular surfaces*, Ergebnisse der Mathematik und ihrer Grenzgebiete (3) [Results in Mathematics and Related Areas (3)], vol. 16, Springer-Verlag, Berlin, 1988, DOI 10.1007/978-3-642-61553-5. MR930101

[29] John Voight, *Quaternion algebras*, Graduate Texts in Mathematics, vol. 288, Springer, Cham, [2021] ©2021, DOI 10.1007/978-3-030-56694-4. MR4279905

The Tutte Institute for Mathematics and Computation, P.O. Box 9703, Terminal, Ottawa, ON K1G 3Z4, Canada

Current address: School of Mathematics and Statistics, 4302 Herzberg Laboratories, 1125 Colonel By Drive, Carleton University, Ottawa, ON K1S 5B6, Canada AND McGill University, Department of Mathematics and Statistics, Burnside Hall, 805 Sherbrooke Street West, Montreal, QC H3A 0B9, Canada

Email address: `adam.m.logan@gmail.com`

Contemporary Mathematics
Volume **840**, 2026
https://doi.org/10.1090/conm/840/16818

Computing the mod-3 Galois image of a principally polarized abelian surface over the rationals

Shiva Chidambaram

Abstract. A lot of work has gone into computing images of Galois representations coming from elliptic curves. This article presents an algorithm to determine the image of the mod-3 Galois representation associated to a principally polarized abelian surface over $\mathbb{Q}$. Conjugacy class distribution of subgroups of $\mathrm{GSp}(4, \mathbb{F}_3)$ is a key ingredient. While this ingredient is feasible to compute for $\mathrm{GSp}(4, \mathbb{F}_\ell)$ for any small prime ℓ, distinguishing Gassmann-equivalent subgroups is a delicate problem. We accomplish it for $\ell = 3$ using several techniques. The algorithm does not require the knowledge of endomorphisms.

1. Introduction

Given an elliptic curve $E/\mathbb{Q}$, it is a fundamental problem to understand the Galois action on its ℓ-torsion points for a prime ℓ. This leads to the concrete question of determining the image of the associated mod-ℓ Galois representation $\overline{\rho}_{E,\ell} : G_\mathbb{Q} \to \mathrm{GL}(2, \mathbb{F}_\ell)$ up to conjugacy in $\mathrm{GL}(2, \mathbb{F}_\ell)$. There are efficient algorithms [**14**] accomplishing this task. Moreover, when E does not have complex multiplication, the work [**13**] determines the ℓ-adic Galois image, and Zywina goes further and computes the full adelic Galois image [**17**].

A natural next step is to tackle this problem for a principally polarized abelian surface $A/\mathbb{Q}$. For any prime ℓ, let $G_\ell = \mathrm{GSp}(4, \mathbb{F}_\ell)$. The ℓ-torsion subgroup $A[\ell]$ is a 4-dimensional vector space over $\mathbb{F}_\ell$, with a non-degenerate, Galois covariant, alternating pairing – the Weil pairing. Thus the Galois action leads to a mod-ℓ Galois representation $\overline{\rho} := \overline{\rho}_{A,\ell} : G_\mathbb{Q} \to G_\ell$, such that composing with the similitude character $\chi_{\mathrm{sim}} : G_\ell \to \mathbb{F}_\ell^\times$ gives the mod-ℓ cyclotomic character χ_ℓ. Given $A/\mathbb{Q}$ and ℓ, the task is to determine the mod-ℓ Galois image $\mathrm{im}(\overline{\rho}_{A,\ell}) \subseteq G_\ell$ up to conjugacy.

Remark 1.1. When $A_{\overline{\mathbb{Q}}}$ has no extra endomorphisms, Serre's open image theorem implies that $\mathrm{im}(\overline{\rho}_{A,\ell}) \neq G_\ell$ for finitely many primes ℓ. Dieulefait gave an algorithm to determine a finite set containing the non-surjective primes [**2**, **10**]. Building on this, [**16**] gives a complete algorithm for computing the rational isogeny class of A.

2020 *Mathematics Subject Classification.* Primary 11F80; 11G10.
The author was supported by Simons Foundation grant 550033.

Since $\chi_\ell = \chi_{\text{sim}} \circ \overline{\rho}$ is surjective, and complex conjugation is an order 2 element $c \in G_\mathbb{Q}$ with $\chi_\ell(c) = -1$, there are constraints on a subgroup $H \subseteq G_\ell$ to be the mod-ℓ Galois image. To be precise, H must be an eligible subgroup (see Definition 1.2).

Definition 1.2. A subgroup H of G_ℓ is said to be eligible if the restriction of χ_{sim} to H is surjective, and there exists $x \in H$ of order 2 such that $\chi_{\text{sim}}(x) = -1$.

We develop a method to completely accomplish this task when $\ell = 3$. It works for all principally polarized abelian surfaces without any restriction on endomorphism type. It consists of a probabilistic Monte Carlo Algorithm 2.3 that determines the conjugacy class distribution of the image of $\overline{\rho}_{A,3}$, followed by a deterministic Algorithm 2.10 that determines $\text{im}(\overline{\rho}_{A,3})$ up to G_3-conjugacy. We present the algorithms for Jacobians of genus 2 curves, but they also apply for Weil restrictions of an elliptic curve over a quadratic field. They also apply for curves over any number field K not containing ζ_3. If $\zeta_3 \in K$, the eligible subgroups are now those contained in $\text{Sp}(4, \mathbb{F}_3)$ and Algorithm 2.10 will then require an analysis of these subgroups.

Magma [**3**] code implementing the algorithm is available at the GitHub repository [**9**]. The main intrinsic is called `mod3Galoisimage`, which takes a genus 2 curve as input, and optionally a lower bound for probability ϵ, and two prime bounds B_1, B_2 (see Algorithm 2.3). Worked out examples can be found in the file `examples.m`. We have used it to compute the mod-3 Galois images for all 66 158 genus 2 curves in the L-functions and Modular Forms Database LMFDB [**12**], and for a bigger dataset consisting of 487 493 curves with 5-smooth conductors. Section 2 presents our main algorithms and Section 3 discusses the challenges for $\ell > 3$.

Any representation $\overline{\rho} : G_\mathbb{Q} \to G_3$ with $\chi_{\text{sim}} \circ \overline{\rho} = \chi_3$ is known to arise as the mod-3 Galois representation of infinitely many prinicipally polarized abelian surfaces over $\mathbb{Q}$ [**4**]. This is because the corresponding twist of the Siegel modular variety $\mathcal{A}_2(3)$, which may not be rational over $\mathbb{Q}$ [**6**], is nevertheless unirational over $\mathbb{Q}$. Consequently any eligible subgroup of G_3 arises as the mod-3 Galois image. Up to conjugacy G_3 has 492 subgroups; 330 of them are not contained in $\text{Sp}(4, \mathbb{F}_3)$ and 280 of them are eligible subgroups. This leads to the following question.

Question 1.3. Can we *explicitly* realize each of the 280 eligible subgroups of G_3 as the mod-3 Galois image of some principally polarized abelian surface over $\mathbb{Q}$?

The work [**5**] outlines a method to construct a genus 2 curve starting from a mod-3 Galois representation $\overline{\rho}$. It can be applied to realize eligible subgroups of G_3 of small order as mod-3 Galois images. Section 4 discusses progress towards this.

Notation: Throughout this article, we use the LMFDB labeling scheme [**1**] for referring to subgroups of G_ℓ. The label is a string ℓ`.i.n`, where `i` is the index in G_ℓ, and `n` is a counter giving a canonical ordering of all index-i subgroups of G_ℓ.

Acknowledgement: The author was supported by Simons Foundation grant 550033. I thank Andrew Sutherland, David Roberts and Nils Bruin for valuable discussions, Noam Elkies for Example 2.6, and the referees whose inputs greatly improved the article, and simplified Example 4.1.

2. Computing the mod-3 Galois image

Suppose A is given as the Jacobian of a genus 2 curve $C/\mathbb{Q} : y^2 = f(x)$ with $\deg(f) = 5$ or 6. Let S denote the set of primes of bad reduction for A. We want to determine $\operatorname{im}(\overline{\rho}_{A,3}) \subseteq G_3$ up to conjugacy. This group is abstractly isomorphic to the Galois group of the 3-torsion field. Although the latter can be computed from a 3-division polynomial, it is not enough to determine $\operatorname{im}(\overline{\rho}_{A,3})$ up to G_3-conjugacy. The difficulty comes from the presence of outer automorphisms. Remark 2.1 discusses this issue in the simpler case $\ell = 2$.

REMARK 2.1. For $\ell = 2$, since the 2-torsion field of A is the splitting field of $f(x)$, the mod-2 Galois image is abstractly isomorphic to the Galois group $\operatorname{Gal}(f) \subseteq S_6 \simeq G_2$. But the identification $S_6 \simeq G_2$ must be chosen carefully since S_6 has an outer automorphism. This is done, for instance, by noting that pairs of Weierstrass points on C correspond to points in $A[2]$, and matching the two conjugacy classes of order-48 subgroups correctly: the subgroup of S_6 that has orbits of size 2 and 4 must match with the subgroup of G_2 having a fixed point, while the other transitive subgroup of S_6 must match with the one having no fixed points.

2.1. Computing the Gassmann-equivalence class.

When computing any Galois group, the Chebotarev density theorem is a very useful tool. Suppose the Galois group is a subgroup of a group G. Then one computes the conjugacy class of Frob_p for all unramified primes $p \leq X$. This sampled frequency is compared against the conjugacy class distributions of all subgroups of G, and ideally this pins down the Galois group with high certainty, since Chebotarev guarantees that the distribution of Frob_p for $p \leq X$ matches that of the Galois group as $X \to \infty$.

This general approach yields the probabilistic method of Algorithm 2.3, wherein we prune out subgroups whose conjugacy class distribution is unlikely to yield the sampled Frobenius distribution. Since computing the Frobenius conjugacy class in G_3 is intensive, we do it only when needed. Instead we focus on two essential (and efficiently computable) features, together called the signature of a conjugacy class in G_3: the characteristic polynomial, and the dimension of the 1-eigenspace.

DEFINITION 2.2. For a prime p of good reduction for A, the Frobenius signature of A at p is defined to be the tuple $\langle L_p(A,t) \pmod 3, \dim_{\mathbb{F}_p} A[3](\mathbb{F}_p)\rangle$, where $L_p(A,t)$ is the Euler factor of A at p. Note that $\det(I - \overline{\rho}_{A,3}(\operatorname{Frob}_p)t) \equiv L_p(A,t)$ $\pmod 3$.

Although Algorithm 2.3 goes far, it cannot always determine the mod-3 Galois image due to the existence of non-conjugate but Gassmann-equivalent [**15**, Def 2.8] subgroups, i.e., subgroups giving rise to the same conjugacy class distribution. The 280 eligible subgroups of G_3 give rise to 230 distinct conjugacy class distributions. These are 38 pairs, 3 triples and 2 quadruples of Gassmann-equivalent subgroups.

ALGORITHM 2.3. *Input:*

- a genus 2 curve C over $\mathbb{Q}$ such that $A = \operatorname{Jac}(C)$
- a positive real number ϵ close to 0; two integers $B_1 \geq B_2 \geq 1$.
- a pre-computed list L of the 280 eligible subgroups of G_3 along with their signature distributions and conjugacy class distributions.

Output: the Gassmann-equivalence class of subgroups of G_3 containing $\operatorname{im}(\overline{\rho}_{A,3})$.

Step 1 For each prime $p \notin S \cup \{3\}, p \leq B_1$, compute the Frobenius signature.

Step 2 Using the pre-computed list of signature distributions, apply Bayes' rule to calculate for each $H \in L$ the probablity that $\mathrm{im}(\overline{\rho}_{A,3}) = H$ given the sampled Frobenius sign distribution from **Step 1**.

Step 3 Remove those $H \in L$ whose probability is smaller than ϵ. If the remaining subgroups form a unique Gassmann-equivalence class, return it.

Step 4 Otherwise, for each prime $p \notin S \cup \{3\}, p \leq B_2$, compute the conjugacy class of $\overline{\rho}_{A,3}(\mathrm{Frob}_p) \in G_3$ by computing $A[3](\overline{\mathbb{F}}_p)$ as a subgroup of $A(\mathbb{F}_{p^k})$ for some $k \geq 1$, and constructing a symplectic basis of $A[3]$.

Step 5 Using the pre-computed list of conjugacy class distributions, apply Bayes' rule to calculate for each remaining $H \in L$ the probablity that $\mathrm{im}(\overline{\rho}_{A,3}) = H$ given the sampled Frobenius conjugacy class distribution from **Step 4**.

Step 6 Remove those H whose probability is smaller than ϵ. If the remaining subgroups form a unique Gassmann-equivalence class, return it.

PROPOSITION 2.4. For any $0 < \epsilon < \frac{1}{4}$, there exists B such that Algorithm 2.3 with $B_i \geq B$ returns the Gassmann-equivalence class containing $\mathrm{im}(\overline{\rho}_{A,3})$.

PROOF. By the Chebotarev density theorem, the sampled Frobenius distribution converges to the conjugacy class distribution of the Gassmann-equivalence class containing $\mathrm{im}(\overline{\rho}_{A,3})$, as $B_2 \to \infty$. The proposition follows immediately. □

EXAMPLE 2.5. The largest subgroup of G_3 that fails the local-global principle for stabilizing a maximal isotropic subspace of $\mathbb{F}_3^4$ is $H = 3.1080.4$, i.e., every element of H stabilizes some 2-dimensional isotropic plane, but H does not. Consider the genus 2 curve with LMFDB label 25600.f.512000.1. Algorithm 2.3 returns that the mod-3 Galois image for this curve is H. So the Jacobian of C presents an example where the local-global principle for the existence of isogenies fails.

EXAMPLE 2.6. Let H denote the stabilizer in G_3 of an isotropic plane in $\mathbb{F}_3^4$. It is the group 3.40.2, and has three index-2 subgroups: $H \cap \mathrm{Sp}(4, \mathbb{F}_3)$ and the two eligible subgroups 3.80.3 and 3.80.4. We note that 3.80.3 is the largest subgroup of G_3 that does not occur as the mod-3 Galois image for any genus 2 curve in the LMFDB dataset. Consider the moduli space $X/\mathbb{Q}$ parametrizing $(3,3)$-isogenies of principally polarized abelian surfaces (analog of the modular curve $X_0(3)$), i.e., the moduli space of abelian surfaces A with $\mathrm{im}(\overline{\rho}_{A,3}) \subseteq 3.40.2$. The three index-2 subgroups correspond to three degree-2 covers. If $\mathbb{Q}(X)$ denotes the function field of X, the function fields of the three covers are respectively $\mathbb{Q}(\sqrt{-3})(X)$, $\mathbb{Q}(X)(\sqrt{-3f})$ and $\mathbb{Q}(X)(\sqrt{f})$ for some $f \in \mathbb{Q}(X)$. Using a birational model of X and the curves in LMFDB whose mod-3 Galois image is 3.80.4, Noam Elkies guessed the function f. Then a search for rational points on X where $-3f$ is a square, yields the curve $C : y^2 = -27x^6 + 54x^5 - 693x^4 + 1278x^3 - 543x^2 - 60x \quad 16$ with conductor $3^2\ 7^4\ 13\ 43^2$. Algorithm 2.3 verifies that its mod-3 Galois image is indeed 3.80.3.

REMARK 2.7. The output of Algorithm 2.3 is rigorous only when it is the whole group G_3. An effective Chebotarev density theorem [**11**] can be used to make Algorithm 2.3 completely rigorous, if we sample a very large number of primes. Let δ be the minimum of the L^∞ distance between any two conjugacy class distributions of eligible subgroups of G_3. Bounding the error term in Chebotarev density by $\delta/2$ pins down a unique Gassmann-equivalence class. By [**11**, Thm 1.1], which assumes GRH for zeta functions of number fields, this can be accomplished by sampling primes up to x, if $x > O(\exp(k\delta^{-1}))$ for some explicit constant k and x is greater than the root discriminant of $\mathbb{Q}(A[3])|\mathbb{Q}$. Note that the local root discriminant of $\mathbb{Q}(A[3])|\mathbb{Q}$ at a bad prime $p \neq 2, 3$ is less than or equal to p.

REMARK 2.8. If the abstract isomorphism class of $\mathrm{Gal}(\mathbb{Q}(A[3])|\mathbb{Q})$ is known, the list L in Algorithm 2.3 has to contain only those eligible subgroups of G_3 in this isomorphism class. This makes the output of Algorithm 2.3 rigorous in more cases. But whenever this class contains non-conjugate subgroups of G_3 meeting the same set of G_3-conjugacy classes, the probabilistic nature of Algorithm 2.3 is unavoidable.

2.2. Distinguishing Gassmann-equivalent subgroups. We begin by noting a canonical example of non-conjugate Gassmann-equivalent subgroups of G_ℓ.

EXAMPLE 2.9. Let $J_2 = \begin{bmatrix} & 1 \\ -1 & \end{bmatrix}$, $\mathcal{J}_2 = \begin{bmatrix} & 1 \\ 1 & \end{bmatrix}$. Suppose $\mathcal{J}_2' = \begin{bmatrix} & \mathcal{J}_2 \\ -\mathcal{J}_2 & \end{bmatrix}$ represents the standard alternating form on the column space $\mathbb{F}_\ell^4$, that is preserved by G_ℓ up to scalars. Let $H_1 \subseteq G_\ell$ be the stabilizer of a vector in $\mathbb{F}_\ell^4$, so $H_1 = \left\{ \begin{bmatrix} 1 & B & c \\ & A & D \\ & & x \end{bmatrix} : D = AJ_2B^t, \det(A) = x \right\}$. Let $H_2 \subseteq G_\ell$ be the transpose of H_1. Then H_1 and H_2 are clearly non-conjugate, but they are Gassmann-equivalent [**15**, Prop 2.6] since transposing gives a G_ℓ-conjugacy preserving bijection $H_1 \leftrightarrow H_2$.

In [**14**], while computing the mod-ℓ Galois image $\mathrm{im}(\overline{\rho}_{E,\ell})$ of an elliptic curve $E/\mathbb{Q}$, Sutherland tackles the problem of distinguishing Gassmann-equivalent subgroups of $\mathrm{GL}(2, \mathbb{F}_\ell)$ by computing the degree of the minimal number field over which E acquires ℓ-torsion points. For example, if $\mathrm{im}(\overline{\rho}_{E,\ell}) = \begin{bmatrix} 1 & * \\ 0 & * \end{bmatrix}$, then $E(\mathbb{Q})$ must have an ℓ-torsion point, whereas if $\mathrm{im}(\overline{\rho}_{E,\ell}) = \begin{bmatrix} * & * \\ 0 & 1 \end{bmatrix}$, then the minimal degree of a number field K such that $E(K)$ has an ℓ-torsion point is $\ell - 1$.

We develop this idea further in Algorithm 2.10 to distinguish any two Gassmann-equivalent subgroups of G_3 that are not conjugate in $\mathrm{GL}(4, \mathbb{F}_3)$. There are exactly 5 Gassmann-equivalence classes of size 2, listed in Table 1, containing $\mathrm{GL}(4, \mathbb{F}_3)$-conjugate subgroups. These are indistinguishable by such methods. Since these groups are small (order ≤ 32), we distinguish them in Algorithm 2.10 by literally constructing a symplectic basis of $A[3]$ over the 3-torsion field K.

We set up some notation before describing the algorithm. Let $\mathcal{K} = A/\{\pm\}$ denote the Kummer surface of A, and let $\mathcal{X}$ denote the image of $A[3]$ under the quotient map. Explicit formulas are known for doubling on $\mathcal{K}$ [**8**, Thm 3.4.1]. Using these, equations for the 3-torsion locus $\mathcal{X} \subset \mathcal{K}$ can be computed, as in [**7**], by writing $2P = -P$ in terms of the coordinates of an arbitrary point $P \in \mathcal{K}$. Although taking successive resultants of these equations can yield a 3-division polynomial, knowing a general 3-division polynomial at the outset makes **Step 1** and **Step 3** of Algorithm 2.10 faster. We use such a polynomial given to us by David Roberts.

ALGORITHM 2.10. *Input:*

- a genus 2 curve C over $\mathbb{Q}$ such that $A = \mathrm{Jac}(C)$
- a Gassmann-equivalence class L of subgroups of G_3 containing $\mathrm{im}(\overline{\rho}_{A,3})$.

Output: the image of $\overline{\rho}_{A,3}$ up to conjugacy in G_3.

TABLE 1. Non-conjugate Gassmann-equivalent subgroups of $\mathrm{GSp}(4, \mathbb{F}_3)$ that are conjugate in $\mathrm{GL}(4, \mathbb{F}_3)$.

$\|H\|$	Label	Entries in generators read horizontally
32	3.3240.6	[0, 0, 1, 2, 2, 2, 0, 1, 1, 1, 1, 0, 2, 1, 1, 0]
		[2, 1, 0, 1, 2, 2, 0, 1, 1, 1, 2, 1, 2, 1, 1, 0]
		[1, 1, 2, 1, 2, 0, 0, 2, 1, 2, 0, 2, 2, 1, 1, 2]
32	3.3240.7	[2, 0, 0, 0, 1, 0, 2, 0, 2, 2, 0, 0, 1, 2, 2, 1]
		[1, 0, 2, 1, 0, 1, 1, 2, 0, 0, 2, 0, 0, 0, 0, 2]
		[2, 1, 2, 0, 2, 2, 0, 2, 1, 1, 1, 2, 2, 1, 1, 1]
16	3.6480.16	[2, 0, 0, 1, 0, 1, 0, 0, 0, 1, 2, 0, 0, 0, 0, 1]
		[0, 2, 1, 0, 1, 2, 1, 1, 2, 0, 0, 1, 1, 2, 2, 2]
16	3.6480.3	[0, 2, 1, 0, 2, 1, 2, 1, 1, 1, 2, 1, 2, 1, 1, 0]
		[2, 2, 1, 0, 0, 1, 1, 1, 0, 0, 2, 1, 0, 0, 0, 1]
16	3.6480.13	[1, 2, 1, 0, 1, 1, 0, 1, 2, 2, 2, 1, 1, 2, 2, 2]
		[1, 0, 2, 1, 0, 1, 1, 2, 0, 0, 2, 0, 0, 0, 0, 2]
16	3.6480.17	[0, 1, 2, 2, 2, 2, 2, 2, 1, 1, 1, 2, 0, 1, 1, 0]
		[2, 0, 2, 1, 2, 1, 2, 2, 1, 1, 2, 0, 2, 1, 1, 1]
16	3.6480.14	[2, 1, 0, 1, 0, 2, 0, 0, 2, 2, 1, 2, 0, 2, 0, 1]
		[2, 0, 1, 2, 2, 1, 2, 1, 0, 0, 2, 0, 0, 0, 1, 1]
		[1, 1, 1, 0, 2, 2, 0, 2, 2, 0, 2, 1, 0, 1, 2, 1]
16	3.6480.15	[1, 2, 0, 2, 0, 1, 0, 0, 1, 1, 2, 1, 0, 1, 0, 2]
		[2, 0, 1, 2, 1, 0, 0, 1, 2, 2, 0, 0, 1, 2, 2, 1]
		[2, 0, 1, 1, 1, 1, 1, 1, 2, 1, 2, 0, 1, 2, 2, 1]
8	3.12960.5	[2, 0, 0, 1, 0, 1, 0, 0, 0, 1, 2, 0, 0, 0, 0, 1]
		[1, 0, 2, 1, 2, 0, 0, 2, 1, 1, 0, 0, 2, 1, 1, 2]
8	3.12960.11	[2, 2, 1, 0, 1, 1, 1, 1, 2, 0, 2, 1, 1, 2, 2, 1]
		[0, 2, 1, 1, 1, 1, 1, 1, 2, 2, 2, 1, 0, 2, 2, 0]

Step 1 If L is a class given in Table 1, compute the three-torsion field $K = \mathbb{Q}(A[3])$ and its automorphism group $\mathrm{Gal}(K|\mathbb{Q})$. Find all 41 geometric points on the three-torsion locus $\mathcal{X}$, i.e., compute $\mathcal{X}(\overline{\mathbb{Q}}) = \mathcal{X}(K)$, and lift points to find a basis of $A[3]$. Fix a symplectic basis by working in some residue field. Compute matrices with respect to this basis corresponding to the action of the generators of $\mathrm{Gal}(K|\mathbb{Q})$, thus determining $\mathrm{im}(\overline{\rho})$.

Step 2 Otherwise, for each $H \in L$, compute $\dim(\mathbb{F}_3^4)^{H_0}$ where $H_0 = H \cap \mathrm{Sp}(4, \mathbb{F}_3)$. Also compute $\max\limits_{[H:H_1]=d} \dim(\mathbb{F}_3^4)^{H_1}$ for each $d \in D = \{1, 2, 3, 6, 8, 12\}$.

Step 3 Compute $\mathcal{X}(\mathbb{Q}(\zeta_3))$, and the points in $\mathcal{X}$ whose degree belongs to the set D. Lift them to 3-torsion points on A, and thus compute $\dim_{\mathbb{F}_3}(A[3](\mathbb{Q}(\zeta_3)))$, and $\max\limits_{[K:\mathbb{Q}]=d} \dim_{\mathbb{F}_3}(A[3](K))$ for each $d \in D$.

Step 4 If there is a unique group $H \in L$, whose data computed in **Step 2** matches the data computed in **Step 3**, return it.

THEOREM 2.11. *Given any genus 2 curve $C/\mathbb{Q}$ with $A = \mathrm{Jac}(C)$, and the Gassmann-equivalence class L containing* $\mathrm{im}(\overline{\rho}_{A,3})$, *Algorithm* 2.10 *returns* $\mathrm{im}(\overline{\rho}_{A,3})$.

PROOF. Suppose L is a Gassmann-equivalence class of size > 1 not listed in Table 1. If L appears in Tables 2 or 3, the corresponding set of indices d shown in the table distinguish all subgroups appearing in L. Otherwise, the subgroups in L are distinguished by the dimension of their fixed spaces. □

EXAMPLE 2.12. Let H denote the stabilizer in G_3 of an line in $\mathbb{F}_3^4$. It is the group 3.40.1, and two of its index-2 subgroups $\{3.80.1, 3.80.2\}$ form a Gassmann-equivalence class. While 3.80.1 is the stabilizer of a point, 3.80.2 acts trivially on a 1-dimensional quotient of $\mathbb{F}_3^4$. Consider the genus 2 curve C with LMFDB label 277.a.277.2. Running Algorithm 2.3 on C returns the correct Gassmann-equivalence class. Further running Algorithm 2.10 returns that the mod-3 Galois image is the group 3.80.2. So the local-global principle for the existence of a 3-torsion point fails for $\mathrm{Jac}(C)$. This is consistent with Mordell-Weil group $\mathrm{Jac}(C)(\mathbb{Q}) \simeq \mathbb{Z}/5$.

3. Challenges in computing mod-ℓ Galois image for $\ell > 3$

For any prime $\ell > 3$, Algorithm 2.3 still works in principle to determine the Gassmann-equivalence class of $\mathrm{im}(\overline{\rho}_{A,\ell})$. The precomputation of all eligible subgroups of G_ℓ becomes hard for large ℓ, but for practical purposes, it is enough to compute only those subgroups whose index is less than a well-chosen bound.

But the problem of distinguishing Gassmann-equivalent subgroups gets considerably harder. Firstly, there are many more Gassmann-equivalent subgroups, for eg., G_5 has 1125 eligible subgroups, but they give rise to only 773 distinct conjugacy class distributions. Secondly, getting information about $A[\ell]$ as in **Step 3** of Algorithm 2.10 is a lot more challenging. For eg., computing equations for the ℓ-torsion locus on the Kummer surface seems hard already for $\ell = 7$. Thirdly, there are fairly big Gassmann-equivalent subgroups of G_ℓ that are conjugate inside $\mathrm{GL}(4, \mathbb{F}_\ell)$. For eg., there is such a pair of subgroups $\{5.48750.5, 5.48750.6\}$ of G_5 of order 768.

We can overcome some of these difficulties by taking endomorphisms into account. They constrain Galois action leading to fewer candidates for the Galois image. Also, for fixed endomorphism data, having a very small image is rare. An upcoming work of the author will fully address the case of no extra endomorphisms.

4. Towards explicitly realizing all mod-3 Galois images

As mentioned in the introduction, [**5**] yields in theory a way to explicitly realize any of the 280 eligible subgroups H of G_3 as the mod-3 Galois image of the Jacobian of a genus 2 curve over $\mathbb{Q}$. Given such a subgroup, one first obtains a number field K properly solving the embedding problem

$$\begin{array}{ccccccccc} & & & & & & G_{\mathbb{Q}} & & \\ & & & & & \overset{\overline{\rho}}{\swarrow} & \downarrow{\scriptstyle \chi_3} & & \\ 0 & \longrightarrow & H \cap \mathrm{Sp}(4, \mathbb{F}_3) & \longrightarrow & H & \longrightarrow & \mathbb{Z}/2 & \longrightarrow & 0 \end{array} \tag{4.1}$$

i.e., $\overline{\rho}$ is surjective and K is the fixed field of $\ker(\overline{\rho})$. Since G_3 is a split extension of $\mathbb{Z}/2$ by $\mathrm{Sp}(4, \mathbb{F}_3)$, this embedding problem is always solvable and such a number field K exists. Then [**5**] gives a recipe to construct the corresponding twisted Burkhardt quartic threefold B, and associate to the rational points on B certain genus 2 curves. Finally, we make a suitable quadratic twist to ensure that the mod-3 Galois image of the Jacobian is exactly equal to H. Here is an illustrative example to conclude.

TABLE 2. Gassmann-equivalent subgroups distinguished by $\dim(\mathbb{F}_3^4)^{H_0}$ and $\max_{[H:H_1]=d} \dim(\mathbb{F}_3^4)^{H_1}$ for $d \leq 3$.

$\|H\|$	Label	Entries in generators read horizontally	H_0	d=1	2	3
324	3.320.1	[2, 2, 2, 0, 2, 1, 1, 0, 2, 0, 1, 2, 2, 0, 2, 2]	1	1	1	1
		[1, 0, 2, 0, 2, 1, 1, 2, 2, 2, 0, 1, 2, 1, 0, 1]				
	3.320.2	[1, 1, 1, 2, 0, 0, 2, 1, 2, 0, 0, 1, 1, 1, 2, 2]	1	0	1	0
		[0, 1, 0, 1, 2, 1, 2, 1, 1, 2, 1, 0, 0, 2, 2, 1]				
	3.320.5	[1, 2, 1, 1, 0, 1, 2, 0, 0, 0, 2, 2, 0, 0, 0, 2]	0	0	1	1
		[2, 2, 0, 0, 2, 1, 2, 1, 0, 0, 1, 2, 1, 1, 2, 2]				
	3.320.6	[1, 2, 1, 1, 1, 0, 2, 0, 0, 2, 1, 0, 2, 2, 1, 1]	0	0	1	0
		[1, 0, 2, 1, 1, 1, 0, 0, 1, 0, 0, 1, 1, 1, 2, 1]				
162	3.640.1	[2, 2, 0, 0, 2, 0, 1, 2, 0, 2, 0, 0, 1, 2, 0, 2]	1	1	1	2
		[0, 0, 1, 2, 1, 1, 0, 1, 0, 0, 2, 0, 2, 0, 2, 0]				
	3.640.2	[0, 1, 0, 1, 0, 1, 0, 2, 2, 2, 2, 1, 1, 1, 2, 1]	1	1	1	1
		[1, 1, 1, 0, 0, 0, 2, 1, 0, 1, 2, 2, 0, 0, 0, 1]				
		[2, 2, 2, 1, 0, 0, 1, 2, 0, 1, 0, 1, 0, 0, 0, 1]				
	3.640.3	[2, 2, 0, 2, 1, 1, 1, 0, 1, 2, 1, 0, 1, 2, 0, 2]	1	0	1	0
		[2, 2, 0, 2, 0, 2, 1, 2, 1, 2, 1, 0, 2, 0, 2, 2]				
	3.640.4	[0, 0, 2, 2, 0, 0, 2, 1, 2, 1, 1, 2, 1, 1, 2, 2]	1	0	1	1
		[2, 1, 2, 1, 2, 2, 0, 0, 0, 1, 2, 0, 1, 1, 2, 1]				
		[2, 1, 2, 2, 2, 2, 0, 1, 0, 0, 1, 1, 1, 2, 0, 2]				
36	3.2880.13	[0, 0, 1, 2, 2, 2, 2, 1, 1, 1, 1, 0, 0, 1, 1, 0]	0	0	2	0
		[2, 0, 0, 0, 0, 2, 0, 0, 1, 1, 1, 0, 2, 1, 0, 1]				
		[2, 1, 1, 1, 2, 2, 2, 1, 2, 1, 2, 2, 2, 2, 1, 2]				
	3.2880.17	[1, 2, 1, 0, 1, 1, 0, 1, 0, 1, 0, 1, 2, 0, 2, 0]	0	0	1	0
		[0, 1, 0, 2, 1, 2, 2, 0, 0, 2, 0, 2, 2, 0, 2, 2]				
		[2, 2, 0, 2, 2, 1, 2, 0, 0, 1, 2, 1, 1, 0, 1, 1]				
18	3.5760.2	[1, 1, 1, 2, 0, 0, 2, 1, 0, 2, 0, 2, 0, 0, 0, 2]	2	1	2	2
		[2, 1, 2, 0, 0, 0, 2, 2, 1, 0, 2, 2, 2, 1, 0, 0]				
	3.5760.5	[2, 1, 2, 0, 0, 0, 2, 2, 1, 0, 2, 2, 2, 1, 0, 0]	2	1	2	1
		[1, 2, 1, 2, 1, 0, 2, 1, 0, 1, 0, 1, 2, 0, 2, 2]				
12	3.8640.2	[1, 0, 0, 0, 0, 1, 0, 0, 1, 1, 2, 0, 2, 1, 0, 2]	0	0	2	0
		[1, 2, 2, 0, 2, 2, 1, 2, 2, 1, 2, 1, 0, 2, 1, 0]				
	3.8640.4	[0, 2, 1, 1, 1, 0, 2, 1, 1, 2, 0, 1, 1, 1, 2, 0]	0	0	1	0
		[2, 1, 0, 0, 0, 1, 0, 0, 0, 0, 2, 2, 0, 0, 0, 1]				
12	3.8640.12	[1, 1, 0, 0, 0, 2, 0, 0, 2, 0, 1, 2, 1, 2, 0, 2]	0	0	2	0
		[2, 0, 1, 1, 0, 2, 1, 1, 2, 1, 1, 0, 1, 2, 0, 1]				
	3.8640.13	[2, 1, 1, 1, 0, 0, 1, 1, 1, 1, 1, 2, 2, 1, 0, 2]	0	0	1	0
		[2, 0, 2, 1, 2, 1, 0, 2, 0, 0, 2, 0, 0, 0, 1, 1]				

EXAMPLE 4.1. Consider the subgroup $H \subset G_3$ with label 3.12960.9. It is isomorphic to $(\mathbb{Z}/2)^3$. We use the `BurkhardtModel` intrinsic with input $K = \mathbb{Q}(\zeta_{24})$ and an isomorphism $\overline{\rho} : \mathrm{Gal}(K|\mathbb{Q}) \simeq H$ solving Eq. (4.1), to find a model for the corresponding twisted Burkhardt quartic threefold B. We then follow the recipe in [**5**] with a random rational point on B. This produces the quadratic twist by 2 of the curve $C : y^2 = x(x^4 - 6840x^2 + 456976)$, and Algorithm 2.3 verifies that its

TABLE 3. Gassmann-equivalent subgroups distinguished by $\max_{[H:H_1]=d} \dim(\mathbb{F}_3^4)^{H_1}$ for $d = 6, 8, 12$.

$\lvert H\rvert$	Label	Entries in generators read horizontally	d=6	8	12
432	3.240.6	[0, 0, 0, 2, 0, 1, 2, 2, 0, 0, 2, 0, 2, 0, 1, 0]	0	1	0
		[1, 0, 0, 2, 0, 2, 1, 0, 0, 0, 2, 0, 2, 0, 0, 2]			
	3.240.7	[2, 0, 1, 1, 0, 1, 1, 2, 0, 0, 2, 0, 2, 0, 1, 2]	0	1	1
		[0, 0, 2, 2, 1, 2, 0, 2, 0, 0, 1, 0, 2, 0, 2, 0]			
324	3.320.3	[1, 1, 0, 2, 1, 0, 2, 0, 0, 2, 1, 2, 2, 0, 2, 0]	2	-	2
		[2, 0, 0, 1, 2, 1, 1, 1, 0, 1, 0, 0, 1, 2, 0, 0]			
	3.320.4	[1, 2, 2, 1, 2, 1, 2, 0, 2, 0, 0, 2, 2, 0, 2, 2]	1	-	2
		[1, 1, 1, 2, 2, 1, 2, 0, 2, 0, 0, 1, 2, 1, 0, 1]			
		[2, 1, 1, 1, 1, 2, 1, 1, 1, 1, 1, 2, 1, 1, 2, 1]			
48	3.2160.9	[1, 1, 1, 1, 0, 2, 1, 1, 0, 0, 1, 2, 0, 0, 0, 2]	1	1	2
		[0, 2, 0, 1, 2, 1, 1, 0, 2, 0, 1, 0, 2, 1, 0, 2]			
	3.2160.10	[0, 2, 1, 0, 1, 0, 0, 1, 0, 1, 2, 2, 2, 1, 0, 2]	1	2	2
		[2, 2, 2, 0, 0, 2, 0, 2, 1, 0, 0, 0, 2, 0, 2, 0]			

mod-3 Galois image is indeed H. The Jacobian of C is isogenous to the product of the elliptic curve 14.a3 and its twist by -1. It has conductor $2^5 7^2 = 1568$, mod-3 image 3.25920.3 and 3-torsion field $\mathbb{Q}(\zeta_{12})$. The curve C is not currently in LMFDB.

References

[1] *Labelling subgroups of GSp(4,ℓ)*, `https://www.lmfdb.org/knowledge/show/g2c.galois_rep_modell_image`, 2025, [Online; accessed 18 June 2025].

[2] Barinder S. Banwait, Armand Brumer, Hyun Jong Kim, Zev Klagsbrun, Jacob Mayle, Padmavathi Srinivasan, and Isabel Vogt, *Computing nonsurjective primes associated to Galois representations of genus* 2 *curves*, LuCaNT: LMFDB, computation, and number theory, Contemp. Math., vol. 796, Amer. Math. Soc., [Providence], RI, [2024] ©2024, pp. 129–163, DOI 10.1090/conm/796/16000. MR4732686

[3] Wieb Bosma, John Cannon, and Catherine Playoust, *The Magma algebra system. I. The user language*, J. Symbolic Comput. **24** (1997), no. 3-4, 235–265, DOI 10.1006/jsco.1996.0125. Computational algebra and number theory (London, 1993). MR1484478

[4] George Boxer, Frank Calegari, Toby Gee, and Vincent Pilloni, *Abelian surfaces over totally real fields are potentially modular*, Publ. Math. Inst. Hautes Études Sci. **134** (2021), 153–501, DOI 10.1007/s10240-021-00128-2. MR4349242

[5] Nils Bruin and Eugene Filatov, *Twists of the Burkhardt quartic threefold*, Res. Number Theory **8** (2022), no. 4, Paper No. 73, 16, DOI 10.1007/s40993-022-00372-3. MR4487859

[6] Frank Calegari and Shiva Chidambaram, *Rationality of twists of the Siegel modular variety of genus 2 and level 3*, Proc. Amer. Math. Soc. **150** (2022), no. 5, 1975–1984, DOI 10.1090/proc/15854. MR4392333

[7] Frank Calegari, Shiva Chidambaram, and Alexandru Ghitza, *Some modular abelian surfaces*, Math. Comp. **89** (2020), no. 321, 387–394, DOI 10.1090/mcom/3434. MR4011548

[8] J. W. S. Cassels, *Arithmetic of curves of genus* 2, Number theory and applications (Banff, AB, 1988), NATO Adv. Sci. Inst. Ser. C: Math. Phys. Sci., vol. 265, Kluwer Acad. Publ., Dordrecht, 1989, pp. 27–35. MR1123068

[9] Shiva Chidambaram, Mod-3 Galois Image of Abelian Surfaces, `https://github.com/shiva-chid/mod3Galoisimage`, 2025

[10] Luis V. Dieulefait, *Explicit determination of the images of the Galois representations attached to abelian surfaces with* $\mathrm{End}(A) = \mathbb{Z}$, Experiment. Math. **11** (2002), no. 4, 503–512 (2003). MR1969642

[11] J. C. Lagarias and A. M. Odlyzko, *Effective versions of the Chebotarev density theorem*, Algebraic number fields: L-functions and Galois properties (Proc. Sympos., Univ. Durham, Durham, 1975), Academic Press, London-New York, 1977, pp. 409–464. MR447191

[12] The LMFDB Collaboration, *The L-functions and modular forms database*, `https://www.lmfdb.org`, 2024, [Online; accessed 18 June 2025].

[13] Jeremy Rouse, Andrew V. Sutherland, and David Zureick-Brown, *ℓ-adic images of Galois for elliptic curves over $\mathbb{Q}$ (and an appendix with John Voight)*, Forum Math. Sigma **10** (2022), Paper No. e62, 63, DOI 10.1017/fms.2022.38. With an appendix with John Voight. MR4468989

[14] Andrew V. Sutherland, *Computing images of Galois representations attached to elliptic curves*, Forum Math. Sigma **4** (2016), Paper No. e4, 79, DOI 10.1017/fms.2015.33. MR3482279

[15] ———, *Stronger arithmetic equivalence*, Discrete Anal. (2021), Paper No. 23, 23. MR4341956

[16] Raymond van Bommel, Shiva Chidambaram, Edgar Costa, and Jean Kieffer, *Computing isogeny classes of typical principally polarized abelian surfaces over the rationals*, LuCaNT: LMFDB, computation, and number theory, Contemp. Math., vol. 796, Amer. Math. Soc., [Providence], RI, [2024] ©2024, pp. 187–214, DOI 10.1090/conm/796/16002. MR4732688

[17] David Zywina, *Explicit open images for elliptic curves over $\mathbb{Q}$*, 2022. `https://arxiv.org/abs/2206.14959`

Department of Mathematics, University of Wisconsin-Madison, Madison WI 53706-1325, USA

Email address: `chidambaram3@wisc.edu`

Contemporary Mathematics
Volume **840**, 2026
https://doi.org/10.1090/conm/840/16819

Completely decomposable modular Jacobians

Jennifer Paulhus and Andrew V. Sutherland

Abstract. We use recently developed algorithms and a new database of modular curves constructed for the *L-functions and Modular Forms Database* to enumerate completely decomposable modular Jacobians of level $N < 240$. In particular, we find examples in 13 previously unknown genera of Jacobian varieties isogenous to a product of elliptic curves over $\mathbb{Q}$. The new genera are: 38, 68, 75, 76, 77, 78, 113, 135, 137, 157, 159, 169, and 409.

As abelian varieties, Jacobian varieties can be decomposed (up to isogeny) into a product of simple abelian varieties. Many interesting questions can be asked about these decompositions. In particular, Ekedahl and Serre [**ES93**] studied the question of when a Jacobian variety factors into a product of elliptic curves (abelian varieties of dimension one), called a *completely decomposable Jacobian*. They asked if the genera for which there is such a curve are bounded, and if not, whether there is an example in every genus of a curve whose Jacobian is completely decomposable. Toward the latter question, they produced curves of many genera (up to genus 1297) with completely decomposable Jacobians defined over number fields. Both questions remain open over fields of characteristic zero (in positive characteristic one can realize infinitely many genera; see [**Yui80**, **Gon99**, **DS07**]).

But there were still many genera below 1297 for which Ekedahl and Serre were unable to produce an example. In the intervening years some gaps have been filled in. Work of Yamauchi [**Yam07**] on decompositions of Jacobian varieties of modular curves found some new examples. The first author in collaboration with Rojas [**PR17**] used classifications of the automorphism groups of curves as well as a new technique involving intermediate quotients of curves with automorphisms to produce many more examples. Rojas and Rodríguez recently demonstrated an example in genus 101 [**RR24**]. See [**Ser20**, §3.11] for a brief summary of this problem and prior results.

In this short note we present results obtained from a new database of modular curves constructed for the L-functions and Modular Forms Database (LMFDB) [**LMF25**] to obtain completely decomposable Jacobians in 13 new genera, all of which are defined over $\mathbb{Q}$. In particular, we find an example in genus 38, which increases the least genus for which no completely decomposable Jacobian over a number field (or even over $\mathbb{C}$) is known to 56. The complete list of genera known to occur in characteristic zero now stands as follows (with the 13 new genera in bold):

2020 *Mathematics Subject Classification.* Primary 14G35; Secondary 11G05, 11G18, 14Q05.
Sutherland was supported by Simons Foundation grant 550033.

0–37, **38**, 39–55, 57–58, 61–65, 67, **68**, 69, 71–73, **75–78**, 79–82, 85, 89, 91, 93, 95, 97, 101, 103, 105–107, 109, **113**, 118, 121, 125, 129, **135**, **137**, 142, 145, 154, **157**, **159**, 161, 163, **169**, 193, 199, 205, 211, 213, 217, 244, 257, 325, **409**, 433, 649, 1297.

Many of the examples produced by Ekedahl and Serre in [**ES93**] are Jacobians $J_0(N)$ of classical modular curves $X_0(N)$ whose non-cuspidal points parameterize elliptic curves with a cyclic isogeny of degree N. They noted that $X_0(N)$ is completely decomposable whenever the eigenvalues of the Hecke operators T_p acting on weight-2 cusp forms of level N are rational for all $p \nmid N$, and used an early precursor to the LMFDB (unpublished tables of modular forms computed by Cohen, Skoruppa, and Zagier in the early 1990s), to determine the $N \leq 1000$ satisfying these conditions. Yamauchi later determined the entire list of N for which $J_0(N)$ is completely decomposable [**Yam07**, Thm 1.1], modulo errors at $N = 28$ (inadvertently omitted) and $N = 672$ (incorrectly included), as noted in [**EHR14**, p. 82]. The genera that arise are:

0–11, 13, 17, 19, 21, 25, 29, 33, 37, 43, 49, 53, 55, 57, 61, 73, 97, 121, 161, and 205.

Recently, a database of modular curves was added to the beta version of the LMFDB (`https://beta.lmfdb.org/ModularCurve/Q/`) [**LMF25**] as part of a project supported by the Simons Collaboration in Arithmetic Geometry, Number Theory, and Computation. This database includes all modular curves X_H associated with open $H \leq \mathrm{GL}_2(\widehat{\mathbb{Z}})$ of level $N \leq 70$ with $\det(H) = \widehat{\mathbb{Z}}^\times$. The *level* of H is the least integer N for which H is the inverse image of its projection to $\mathrm{GL}_2(\mathbb{Z}/N\mathbb{Z})$, and the constraint on $\det(H)$ ensures that $X_H/\mathbb{Q}$ is a *nice* (smooth, projective, geometrically integral) curve. The X_H are generalizations of the classical modular curves $X_0(N)$ and $X_1(N)$, and they parameterize elliptic curves with *H-level structure*; we refer the reader to [**RSZB22**, §2] for further details.

This database includes the decomposition of $J_H := \mathrm{Jac}(X_H)$ into simple modular abelian varieties A_f associated to Galois orbits of weight-2 newforms f. As proved in [**RSZB22**], all such f are (not necessarily new) normalized eigenforms in $S_2(\Gamma_1(N) \cap \Gamma_0(N^2))$. These decompositions were computed using the algorithm described in [**RSZB22**, §6], which we briefly recall. If $[f_1], \ldots, [f_m]$ is a complete list of Galois orbits of eigenforms in $S_2(\Gamma_1(N) \cap \Gamma_0(N^2))$, then there is a unique sequence of nonnegative integers $e_1, \ldots, e_m$ for which

$$L(J_H, s) = \prod_{i=1}^{m} L(A_{f_i}, s)^{e_i}, \tag{1}$$

which implies the integer linear constraint $a_p(J_H) = \sum_{i=1} e_i a_p(\mathrm{Tr}(f_i))$ for each prime p. Here $\mathrm{Tr}(f_i)$ denotes the *traceform* of f_i, the sum of the Galois conjugates of f, see [**BBB+21**, §4.5]). We also have the constraint $\sum_{i=1}^{m} e_i a_1(\mathrm{Tr}(f_i)) = g(X_H)$.

The integer q-expansions $\sum_{n=1} a_n q^n$ of the traceforms $\mathrm{Tr}(f)$ were computed using the algorithms described in [**BBB+21**] for $n \leq B$. By increasing the bound B as required, one eventually obtains a linear system determined by the constraints on $a_p(J_H)$ and $g(X_H)$ that has a unique solution in $\mathbb{Q}$ (which is necessarily integral). This reduces the problem of computing the isogeny decomposition of J_H to linear algebra, provided one can efficiently compute the integers $a_p(J_H)$ and the q-expansions of all the relevant traceforms $\mathrm{Tr}(f_i)$. An efficient algorithm for computing $a_p(J_H)$ is described in [**RSZB22**, §5], and a Magma implementation of this algorithm is available in the associated GitHub repository. The computations of the traceforms is more difficult (this requires decomposing spaces of modular forms of level N^2, including spaces with nontrivial character), but for H of level $N \leq 70$ most of the required traceforms were already present in the LMFDB thanks to [**BBB$^+$21**] (this work involved hundreds of CPU years of computations that we fortunately did not need to repeat). This made it a simple matter to search the database for examples where J_H is completely decomposable, leading to the discovery of 11 previously unknown genera.

The database also includes many X_H with H of level $N > 70$, but for $N > 70$ it does not include decompositions of the Jacobians J_H due to the difficulty of computing the modular forms in $S_2(\Gamma_1(N) \cap \Gamma_0(N^2))$. Indeed, $N = 71$ is already problematic in this regard, as the space of newforms with LMFDB label `5041.2.g` of dimension 9120 has resisted all efforts to decompose it to date. However, the modular abelian varieties A_f corresponding to Galois orbits of newforms f in this space all have dimension divisible by 24 (the degree of the field of values $\mathbb{Q}(\zeta_{35})$ of the Dirichlet character orbit `5041.g`), which implies that they cannot appear as isogeny factors of a completely decomposable J_H. Only weight-2 newforms with rational coefficients, corresponding to elliptic curves $E/\mathbb{Q}$, are relevant to our search, and these E are easy to determine. Indeed, the LMFDB contains all elliptic curves $E/\mathbb{Q}$ of conductor less than 500,000, which is enough to handle any $N \leq 707$.

It is not computationally feasible to enumerate all open $H \leq \mathrm{GL}_2(\widehat{\mathbb{Z}})$ of level $N \leq 707$; at level 80 there are already nearly 20 million H with $\det(H) = \widehat{\mathbb{Z}}^\times$. But it is not necessary to enumerate all such H. If J_H is not completely decomposable, then neither is J_K for any $K \leq H \leq \mathrm{GL}_2(\widehat{\mathbb{Z}})$, since the inclusion $K \leq H$ induces a morphism of curves $X_K \to X_H$ that makes J_H an isogeny factor of J_K. This suggests the following strategy: for a given level N, enumerate the subgroup lattice of $\mathrm{GL}_2(\mathbb{Z}/N\mathbb{Z})$ up to conjugacy from the top down by successively computing (conjugacy classes of) maximal subgroups (a task that MAGMA [**BCFSe25**] performs very efficiently), and do not proceed beyond any H for which J_H is not completely decomposable. An open subgroup $H \leq \mathrm{GL}_2(\widehat{\mathbb{Z}}) = \varprojlim_N \mathrm{GL}_2(\mathbb{Z}/N\mathbb{Z})$ of level N is uniquely determined by its projection to $\mathrm{GL}_2(\mathbb{Z}/N\mathbb{Z})$, and we may identify H with the conjugacy class of its image in $\mathrm{GL}_2(\mathbb{Z}/N\mathbb{Z})$, since conjugate subgroups give rise to isomorphic modular curves. Not every subgroup $K \leq \mathrm{GL}_2(\mathbb{Z}/N\mathbb{Z})$ has level N, it may have level M properly dividing N, but such K are still of interest because they may contain subgroups H of level N; as noted above, if J_K is not completely decomposable then neither is J_H.

As we enumerate the subgroup lattice of $\mathrm{GL}_2(\mathbb{Z}/N\mathbb{Z})$ we may restrict our attention to H for which $\det(H) = (\mathbb{Z}/N\mathbb{Z})^\times$, as we are interested in nice curves over $\mathbb{Q}$ (if $\det(H) \neq (\mathbb{Z}/N\mathbb{Z})^\times$ then X_H will not be geometrically irreducible), and we can also require H to contain $-I$, since if H does not contain $-I$ and we put $H' = \langle H, -I\rangle$, the curves X_H and X'_H will be isomorphic (even though they have different moduli interpretations); see [**RSZB22**, §2] for further details. Note that if H does not satisfy both these constraints than neither does any of its subgroups.

When J_H is not completely decomposable we can prove this using the L-series of the elliptic curves $E/\mathbb{Q}$ of conductor dividing N^2. Explicitly, let $E_1, \ldots, E_m$ be the list of all such elliptic curves; this is a finite list, and for $N \leq 707$ it can be easily extracted from the LMFDB. If J_H is completely decomposable there are nonnegative integers $e_1, \ldots, e_m$ for which

$$L(J_H, s) = \prod_{i=1}^{m} L(E, s)^{e_i}. \tag{2}$$

We may now assemble a list of m primes $p_1, \ldots, p_m$ for which the $m \times m$ integer matrix $A := [a_{p_i}(E_j)]_{ij}$ has nonzero determinant. Such a list necessarily exists (the newforms corresponding to the E_j are linearly independent), and in practice it does not take long to find. We now invert the matrix A (over $\mathbb{Q}$), which we note depends only on the level N, and can be computed before enumerating any subgroups. As we enumerate subgroups $H \leq \mathrm{GL}_2(\mathbb{Z}/N\mathbb{Z})$ of level N with $\det(H) = \widehat{\mathbb{Z}}^\times$ containing $-I$, for each H we compute the integer vector $v := (a_{p_1}(J_H), \ldots a_{p_m}(J_H))$ using the algorithm in [**RSZB22**, §5] and then consider $A^{-1}v$. If this vector does not lie in $\mathbb{Z}^m_{\geq 0}$ then J_H cannot be completely decomposable, since its L-function cannot satisfy (2). For $N < 240$ the number of $E/\mathbb{Q}$ we need to consider is less than 1000, and the primes $p \leq 8192$ are sufficient to obtain a matrix $[a_{p_i}(E_j)]$ of full rank from which we can extract a matrix with nonzero determinant by picking a subset of linearly independent rows.

If $A^{-1}v$ is a vector of nonnegative integers, this does not necessarily imply that J_H is completely decomposable (although in practice this is almost always true). If J_H is not completely decomposable there certainly exists a list of primes $p_1, \ldots, p_m$ we could use to prove this, but it might not be the list we chose, and we don't want to take the time to check all possibilities up to the Sturm bound for $S_2(\Gamma_1(N) \cap \Gamma_0(N^2))$. But while searching for unrealized genera we are happy to defer a rigorous verification to a later stage, since most of the X_H we encounter will have genera that have already been realized (and when we find one that has not been realized, there will often be many, and we only need one example).

We implemented this strategy in Magma, using code from [**RSZB22**]; see the file `gl2split.m` in the GitHub repository linked to below. We ran this algorithm on every level $N < 240$ (we included $N \leq 70$ as a sanity check). These computations were performed on a 256-core AMD EPYC 9754 2.25GHz CPU with 1.5TB RAM running Ubuntu 24.04 and took only a few days (less than a CPU-year of computation). This yielded a list of more than half a million modular Jacobians J_H that appear to be completely decomposable. For each genus g realized by one of these examples we determined the minimal H with $g(X_H) = g$, with the H ordered according to their LMFDB label $N.i.g.c.n$, whose first three parts encode the level N, index i, and genus g of H. These minimal examples are listed in Table 1, and the full dataset is available at

`https://github.com/AndrewVSutherland/CompletelyDecomposableModularJacobians/.`

All but five of the minimal examples we found have level $N \leq 70$, meaning the decomposition of J_H is already available in the LMFDB and no further verification was required. The exceptions all have level 120, and after computing all the newforms in $S_2(\Gamma_1(120) \cap \Gamma_0(120^2))$, we were able to use the algorithm in [**RSZB22**, §6] to rigorously determine the decomposition of $\mathrm{Jac}(X_H)$. While $\dim(S_2(\Gamma_1(120) \cap \Gamma_0(120^2)) = 12032$ is rather large, it is the direct product of 279 newspaces $S_2(M, \chi)$ with $M|N^2$ of dimension at most 360, so this was a feasible computation (the fact that $N = 120$ is highly composite helps us here). The examples at level 120 yielded two previously unknown genera: 113 and 169.

TABLE 1. Minimal modular curves of level $N < 240$ with completely decomposable Jacobians ordered by genus g. New genera are in bold, previously known modular genera are in italic. Unlinked labels are not yet in the LMFDB.

g	LMFDB label	g	LMFDB label	g	LMFDB label	g	LMFDB label
0	1.1.0.a.1	*21*	20.360.21.a.1	45	60.640.45.a.1	85	20.1440.85.b.1
1	6.6.1.a.1	22	20.360.22.a.1	*49*	30.720.49.a.1	89	120.1152.89.sn.1
2	10.30.2.a.1	23	30.360.23.a.1	50	120.640.50.a.1	*97*	60.1440.97.y.1
3	7.168.3.a.1	24	24.384.24.a.1	51	60.720.51.u.1	103	60.1440.103.a.1
4	10.90.4.a.1	*25*	24.384.25.a.1	52	60.720.52.i.1	105	60.1440.105.g.1
5	10.120.5.a.1	26	11.660.26.c.1	*53*	60.720.53.a.1	109	60.1440.109.a.1
6	10.180.6.b.1	27	20.480.27.a.1	*55*	36.864.55.a.1	**113**	120.1440.113.hflr.1
7	10.180.7.a.1	28	18.486.28.a.1	*57*	60.864.57.a.1	*121*	60.1728.121.a.1
8	11.220.8.a.1	*29*	20.480.29.a.1	*61*	60.864.61.c.1	129	120.1728.129.vwth.1
9	15.240.9.c.1	31	28.576.31.b.1	65	60.960.65.m.1	**135**	60.1920.135.a.1
10	15.180.10.a.1	32	30.540.32.a.1	**68**	60.960.68.a.1	**137**	60.1920.137.a.1
11	24.192.11.i.1	*33*	24.576.33.a.1	69	60.960.69.c.1	**157**	60.2160.157.a.1
12	11.330.12.a.1	34	30.540.34.a.1	72	60.960.72.a.1	**159**	60.2160.159.gu.1
13	10.360.13.b.1	35	60.480.35.a.1	*73*	24.1152.73.a.1	*161*	48.2304.161.cax.1
14	30.240.14.a.1	36	30.540.36.a.1	**75**	60.1080.75.gu.1	163	60.2160.163.c.1
15	20.240.15.a.1	*37*	20.720.37.i.1	**76**	60.1080.76.dc.1	**169**	120.2160.169.dcty.1
16	28.288.16.q.1	**38**	60.540.38.y.1	**77**	60.1080.77.ge.1	193	60.2880.193.dw.1
17	20.360.17.e.1	39	60.540.39.e.1	**78**	60.1080.78.a.1	*205*	60.2880.205.a.1
18	60.240.18.m.1	40	60.540.40.a.1	79	60.1080.79.bg.1	217	60.2880.217.c.1
19	18.324.19.c.1	41	20.720.41.i.1	81	48.1152.81.id.1	325	60.4320.325.a.1
20	20.360.20.a.1	*43*	20.720.43.a.1	82	60.1080.82.a.1	**409**	60.5760.409.c.1

REMARK 1. We believe that Table 1 includes every genus for which a completely decomposable modular Jacobian is known. It includes every genus arising for X_H of level $N < 240$, and every genus arising from $X_0(N)$ for some N, the largest of which is $N = 1200$ (we realize $g(X_0(1200)) = 205$ as $g(X_H)$ with H of level 60). Table 1 does not include 32 genera for which nonmodular completely decomposable Jacobians are known, but not all of these 32 are realized over $\mathbb{Q}$. This suggests two refinements of the question posed by Ekedahl and Serre: (1) in which genera are there completely decomposable modular Jacobians, and (2) in which genera are there nice curves $X/\mathbb{Q}$ with completely decomposable Jacobians? Question (1) is a subset of (2), if we restrict to open $H \leq \mathrm{GL}_2(\widehat{\mathbb{Z}})$ with $\det(H) = \widehat{\mathbb{Z}}^\times$ as we do here.

REMARK 2. Ekedahl and Serre also consider quotients of modular curves. This allows them to realize several genera not known to occur for completely decomposable modular Jacobians; the genus 47 quotient $X_0(600)/\langle w_{24} \rangle$ is one such example. We have not attempted to construct quotients of the completely decomposable modular Jacobians we found, but our search yielded more than 100,000 examples of genus $g > 100$ whose quotients might yield new examples. One could also consider modular Jacobians that are nearly but not completely decomposable, as these may admit quotients whose Jacobians are completely decomposable.

REMARK 3. The modular abelian variety A_f associated to the Galois orbit of a non-CM newform is geometrically simple, so in most cases the $\mathbb{Q}$-isogeny decomposition of J_H implied by (1) is a $\overline{\mathbb{Q}}$-isogeny decomposition (and in fact a decomposition over $\mathbb{C}$). But the presence of complex multiplication complicates matters. In our search we only checked for J_H that are completely decomposable over $\mathbb{Q}$ and might have missed examples that are completely decomposable over $\overline{\mathbb{Q}}$.

References

[BBB+21] Alex J. Best, Jonathan Bober, Andrew R. Booker, Edgar Costa, John E. Cremona, Maarten Derickx, Min Lee, David Lowry-Duda, David Roe, Andrew V. Sutherland, and John Voight, *Computing classical modular forms*, Arithmetic geometry, number theory, and computation, Simons Symp., Springer, Cham, [2021] ©2021, pp. 131–213, DOI 10.1007/978-3-030-80914-0_4. MR4427962

[BCFSe25] W. Bosma, J. J. Cannon, C. Fieker, and A. Steel (eds.). Handbook of Magma functions, version 2.28-17. https://magma.maths.usyd.edu.au/magma/handbook/, 2025.

[DS07] Claus Diem and Jasper Scholten, *Ordinary elliptic curves of high rank over $\overline{\mathbb{F}}_p(x)$ with constant j-invariant. II*, J. Number Theory **124** (2007), no. 1, 31–41, DOI 10.1016/j.jnt.2006.08.005. MR2320989

[EHR14] Noam D. Elkies, Everett W. Howe, and Christophe Ritzenthaler, *Genus bounds for curves with fixed Frobenius eigenvalues*, Proc. Amer. Math. Soc. **142** (2014), no. 1, 71–84, DOI 10.1090/S0002-9939-2013-11839-3. MR3119182

[ES93] Torsten Ekedahl and Jean-Pierre Serre, *Exemples de courbes algébriques à jacobienne complètement décomposable* (French, with English and French summaries), C. R. Acad. Sci. Paris Sér. I Math. **317** (1993), no. 5, 509–513. MR1239039

[Gon99] Josep González, *Fermat Jacobians of prime degree over finite fields*, Canad. Math. Bull. **42** (1999), no. 1, 78–86, DOI 10.4153/CMB-1999-009-7. MR1695866

[LMF25] The LMFDB Collaboration. The L-functions and modular forms database. `https://beta.lmfdb.org`, 2025. [Online; accessed 24 January 2025].

[PR17] Jennifer Paulhus and Anita M. Rojas, *Completely decomposable Jacobian varieties in new genera*, Exp. Math. **26** (2017), no. 4, 430–445, DOI 10.1080/10586458.2016.1199981. MR3684576

[RR24] Rubí E. Rodríguez and Anita M. Rojas. Decomposing abelian varieties into simple factors: algorithms and applications, 2024.

[RSZB22] Jeremy Rouse, Andrew V. Sutherland, and David Zureick-Brown, *ℓ-adic images of Galois for elliptic curves over $\mathbb{Q}$ (and an appendix with John Voight)*, Forum Math. Sigma **10** (2022), Paper No. e62, 63, DOI 10.1017/fms.2022.38. With an appendix with John Voight. MR4468989

[Ser20] Jean-Pierre Serre, *Rational points on curves over finite fields*, Documents Mathématiques (Paris) [Mathematical Documents (Paris)], vol. 18, Société Mathématique de France, Paris, [2020] ©2020. With contributions by Everett Howe, Joseph Oesterlé and Christophe Ritzenthaler; Edited by Alp Bassa, Elisa Lorenzo García, Christophe Ritzenthaler and René Schoof. MR4242817

[Yam07] Takuya Yamauchi, *On $\mathbb{Q}$-simple factors of Jacobian varieties of modular curves*, Yokohama Math. J. **53** (2007), no. 2, 149–160. MR2302608

[Yui80] Noriko Yui, *On the Jacobian variety of the Fermat curve*, J. Algebra **65** (1980), no. 1, 1–35, DOI 10.1016/0021-8693(80)90236-7. MR578793

Department of Mathematics and Statistics, Mount Holyoke College, South Hadley, MA, 01075

Email address: `jpaulhus@mtholyoke.edu`

Department of Mathematics, Massachusetts Institute of Technology, Cambridge, MA 02139

Email address: `drew@math.mit.edu`

Contemporary Mathematics
Volume **840**, 2026
https://doi.org/10.1090/conm/840/16820

Labeling abelian varieties over finite fields

Edgar Costa, Taylor Dupuy, Stefano Marseglia, David Roe, and Christelle Vincent

Abstract. We describe a deterministic process to associate a practical, permanent label to isomorphism classes of abelian varieties defined over finite fields with commutative endomorphism algebra as long as they are ordinary or defined over a prime field. In the ordinary case, we also provide labels for the polarizations they admit.

1. Introduction

Let k be a field of size $q = p^a$, where p is a prime. Building on the work of Deligne [**Del69**], Howe [**How95**] and Centelghe–Stix [**C15**] the third author developed in [**Mar20**, **Mar21**, **Mar24**, **Mar25**] tools that, when combined with the work of Hofmann–Sircana [**HS20**], allow the enumeration of k-isomorphism classes of abelian varieties of dimension g defined over k contained in isogeny classes satisfying (2.1) and (2.2), as well as their polarizations when the class is ordinary. Along with Mckenzie West, the authors have implemented these algorithms and enumerated some of these isomorphism classes and low-degree polarizations for forthcoming inclusion in the LMFDB [**LMFDBa**] to expand the data about isogeny classes (see [**DKRV21**]).

The purpose of this note is to describe a deterministic process to associate a **label**, by which we mean a unique permanent identifier, to each of these objects. We believe that useful, short labels that can gain widespread adoption in the community are crucial for the long-term usefulness and intelligibility of results in the field. Note that, unlike the labeling scheme for isogeny classes, the scheme we propose does not include enough data to directly recompute the abelian variety (without re-enumerating the isogeny class); it is also not stable under base change.

The third author's algorithms rely on a bijection between certain isomorphism classes of ideals and the isomorphism classes of abelian varieties to be enumerated (§2). Accordingly, the labeling scheme we propose ultimately labels these isomorphism classes of ideals. Furthermore, when the abelian variety is ordinary we also compute polarizations of low degree; this is done by defining a distinguished representative I in the isomorphism class of ideals and giving a polarization λ on the corresponding variety A as an element of the endomorphism algebra.

We close by describing our proposed labeling: a polarized abelian variety corresponding to a pair (I, λ) as above is labeled as

(1.1) `g.q.isog - N.i.w.j - d.k`,

where `g.q.isog` is the label of the isogeny class (`g` is the dimension, `q` is the size of the finite field, `isog` encodes the coefficients of the Weil polynomial; see [**LMFDBb**]), `N.i.w.j` is the label of the isomorphism class of I (§§3.3, 3.4 and 3.5 – `N.i` determines the endomorphism ring §3.1), `d` is the degree d of λ and `k` is determined by the sort key of λ (§3.6). When no polarization λ is given the last part is omitted.

2. Abelian varieties and ideal classes

Throughout, let $\mathcal{I}$ be a k-isogeny class of abelian varieties of dimension g satisfying:

(2.1) the k-endomorphism ring of any abelian variety in $\mathcal{I}$ is commutative and

(2.2) the isogeny class is ordinary or k is the prime field $\mathbb{F}_p$.

By [**Del69**] and [**C15**], there is an equivalence between the category of abelian varieties in $\mathcal{I}$ (with k-homomorphisms) and the category of fractional R-ideals (with R-linear morphisms), where R is the *Frobenius order*, a ring attached to $\mathcal{I}$. This equivalence induces a bijection between $\mathrm{ICM}(R)$, the *ideal class monoid* of R, and the isomorphism classes in $\mathcal{I}$, see [**Mar21**]. In this section we introduce the related notation, definitions, and results we need, and refer to [**Mar24, Mar25**] for details.

Let $h(x)$ be the characteristic polynomial of any abelian variety in $\mathcal{I}$. By [**Tat66**, Theorem 2.(c)] and assumpion (2.1), the $2g$ complex roots of $h(x)$ are distinct. We sort the distinct monic irreducible factors $h_1(x), \dots, h_n(x)$ over $\mathbb{Q}$ of $h(x)$ according to the lexicographical order of their coefficients, starting from the constant term.

Consider the étale $\mathbb{Q}$-algebra $K := \mathbb{Q}[x]/h(x)$; throughout we write

$$K = K_1 \times \cdots \times K_n, \tag{2.3}$$

where $K_i := \mathbb{Q}[x]/h_i(x)$ is a number field, and denote by F the class of the variable x in K and by $\mathcal{B}_K$ the ordered $\mathbb{Q}$-basis $(V^{g-1}, \dots, V, 1, F, \dots, F^g)$ of K, where $V = q/F$. Recall that an **order** S in K is a subring of K which is a full $\mathbb{Z}$-lattice, that is, the underlying additive group of S has rank $2g$. The **Frobenius order** is then $R := \mathbb{Z}[F, V]$, the $\mathbb{Z}$-span of $\mathcal{B}_K$, and an **overorder** (of R) is an order in K containing R. We denote the unique maximal order of K by $\mathcal{O}_K$; using (2.3) we have $\mathcal{O}_K = \mathcal{O}_{K_1} \oplus \cdots \oplus \mathcal{O}_{K_n}$, where $\mathcal{O}_{K_i}$ is the ring of integers of K_i.

For S an order in K, a **fractional** S**-ideal** is a full $\mathbb{Z}$-lattice in K which is closed under multiplication by elements of S. A fractional S-ideal I is **invertible** if for every maximal ideal $\mathfrak{p}$ of S there exists $a \in K^\times$ such that $I_\mathfrak{p} = aS_\mathfrak{p}$. If S is not maximal then not all fractional ideals are invertible, and ideal multiplication induces a commutative monoid structure on the set of all fractional ideals of S.

The **multiplicator ring** of a fractional R-ideal I is the overorder defined by $(I : I) := \{a \in K : aI \subseteq I\}$. Two fractional R-ideals I and J are **weakly equivalent** if $I_\mathfrak{p} \simeq J_\mathfrak{p}$ as $R_\mathfrak{p}$-modules for each maximal ideal $\mathfrak{p}$ of R. The multiplicator ring is an invariant of the weak equivalence class, and if S is an overorder, we denote by W_S the set of weak equivalence classes of fractional R-ideals with multiplicator ring S.

We need a finer equivalence relation on fractional R-ideals: two ideals I and J are **isomorphic** if there exists $a \in K^\times$ such that $I = aJ$. Multiplication is well defined on these classes, and the set of isomorphism classes $[I]$ of fractional R-ideals I is denoted $\mathrm{ICM}(R)$, the **ideal class monoid of** R. Within $\mathrm{ICM}(R)$, the set of isomorphism classes of invertible fractional S-ideals for S an overorder is denoted by $\mathrm{Pic}(S)$; ideal multiplication induces a group structure on $\mathrm{Pic}(S)$.

The **trace dual ideal** of a fractional R-ideal I is $I^t := \{a \in K : \mathrm{Tr}_{K/\mathbb{Q}}(aI) \subseteq \mathbb{Z}\}$. For S an overorder and $\mathfrak{p}$ a maximal ideal of S, we denote by $\mathrm{type}_{\mathfrak{p}}(S)$ the **Cohen-Macaulay type of S at $\mathfrak{p}$**, defined as $\dim_{(S/\mathfrak{p})}(S^t/\mathfrak{p}S^t)$. Note that $\mathrm{type}_{\mathfrak{p}}(S) = 1$ whenever $\mathfrak{p}$ is invertible. We then define the **Cohen-Macaulay type** of S to be $\mathrm{type}(S) := \max_{\mathfrak{p}}\{\mathrm{type}_{\mathfrak{p}}(S)\}$ as $\mathfrak{p}$ ranges over the maximal ideals of S.

When $\mathcal{I}$ is ordinary, we can compute the polarizations on an abelian variety A belonging to $\mathcal{I}$ from the data of the corresponding fractional R-ideal I in the following manner: let $\Phi := \{\varphi_1, \dots, \varphi_g\}$ be a CM-type of K which satisfies the Shimura-Taniyama condition; see for example [**CCO14**, § 2.1.4.1]. By [**How95**], the dual abelian variety $A^\vee$ corresponds to $\overline{I}^t$, where $\bar{\cdot}$ denotes complex conjugation on K, defined by $\overline{F} := V$, and a polarization corresponds to an element $\lambda \in K^\times$ such that $\lambda I \subseteq \overline{I}^t$, $\lambda = -\overline{\lambda}$ and $\Im(\varphi(\lambda)) > 0$ for every $\varphi \in \Phi$. Moreover, its degree is $|\overline{I}^t/\lambda I|$. Two polarizations λ and λ' of I are isomorphic if there exists $u \in S^\times$ such that $\lambda = u\overline{u}\lambda'$.

3. Labeling ideal classes and polarizations

The goal of this section is to define a deterministic procedure to attach a label and a distinguished representative to each ideal class in $\mathrm{ICM}(R)$, where R is the Frobenius order defined in §2, and to the polarizations they admit, when applicable.

3.1. Labeling overorders. An overorder S is labeled `N.i` where $N = [\mathcal{O}_K : S]$ and i is the index of S when all overorders of index N in $\mathcal{O}_K$ are sorted in lexicographic order according to the sort key $s(S) := [d, n_1, \dots, n_{g(2g+1)}]$, defined as follows. Let H be the matrix whose rows are the coefficients of any $\mathbb{Z}$-basis of S written with respect to $\mathcal{B}_K$. Then d is the least common multiple of the denominators of the entries of H and $n_1, \dots, n_{g(2g+1)}$ are the entries of the upper triangular part of the (upper triangular) Hermite Normal Form of dH.

3.2. Sorting maximal ideals. In what follows we will need ordered sets of maximal ideals for a fixed overorder S. We do this by defining sort keys and using the lexicographical order; while our process only sorts maximal ideals within the same order, this is sufficient for our purposes. Let $p_i \colon K \to K_i$ be the natural projection onto the i-th component in (2.3). For $\mathfrak{P}$ a maximal ideal of $\mathcal{O}_K$, there exists a unique index $1 \leq j \leq n$ such that $p_j(\mathfrak{P})$ is a maximal ideal of $\mathcal{O}_{K_j}$ and $p_l(\mathfrak{P}) = \mathcal{O}_{K_l}$ for $l \neq j$. We define then the sort key $s(\mathfrak{P})$ of $\mathfrak{P}$ to be $[j, m, n]$ where `m.n` is the LMFDB label of $p_j(\mathfrak{P})$; here m is the norm of the ideal and n is a tiebreaker [**CPS20**]. If $\mathfrak{p}$ is a maximal ideal of a non-maximal order S, then the sort key of $\mathfrak{p}$ is defined to be the lexicographically smallest sort key among those of the finitely many maximal ideals $\mathfrak{P}$ of $\mathcal{O}_K$ above $\mathfrak{p}$.

3.3. Labeling and distinguished representatives of ideal classes. Let J be a fixed fractional R-ideal with multiplicator ring S. For any fractional R-ideal I which is weakly equivalent to J we have $I = J(I : J)$ where $(I : J) := \{a \in K : aJ \subseteq I\}$ is an invertible fractional S-ideal. In fact, by [**Mar20**, Corollary 4.5, Theorem 4.6], we have a bijection

$$\mathrm{ICM}(R) \longleftrightarrow \bigsqcup_{R \subseteq S \subseteq \mathcal{O}_K} (W_S \times \mathrm{Pic}(S)), \qquad [I] \mapsto (\omega_J, [(I : J)])$$

where S runs over the finitely many overorders of R.

We define the distinguished representative of an arbitrary ideal class $[I]$ as the multiplication of the distinguished representative J of the weak equivalence class ω_I of I (see §3.4) with the distinguished representative of $[(I : J)]$ (see §3.5). We define the label `N.i.w.j` of an arbitrary ideal class $[I]$ as the concatenation of the label `N.i.w` of ω_I defined in §3.4 together with the index `j` of $[(I : J)] \in \mathrm{Pic}(S)$ when enumerated using the sort-key defined in §3.5.

3.4. Labeling and distinguished representatives of weak equivalence classes. We now construct a label and a distinguished representative J for each weak equivalence class ω of W_S for a fixed overorder S. Throughout, let `N.i` be the label of S and $\mathcal{S} := (\mathfrak{p}_1, \ldots, \mathfrak{p}_u)$ be the ordered set of non-invertible maximal ideals of S. Define the label (resp. distinguished representative) of the invertible class of W_S as `N.i.1` (resp. $1 \cdot S$). For each class $\omega \in W_S$ and any $I \in \omega$, set $s(\omega) := [1]$ if $\mathcal{S} = \emptyset$ and

$$s(\omega) := \left[\dim_{(S/\mathfrak{p}_i)}(I/\mathfrak{p}_i I) : 1 \le i \le u\right],$$

otherwise. If $\mathrm{type}(S) \le 2$, the string $s(\omega)$ is a complete invariant: if $\mathrm{type}_{\mathfrak{p}_i}(S) = 1$ then $I_{\mathfrak{p}_i} \simeq S_{\mathfrak{p}_i}$ by, for example, [**Mar24**, Proposition 3.4]; if $\mathrm{type}_{\mathfrak{p}_i}(S) = 2$ then either $I_{\mathfrak{p}_i} \simeq S_{\mathfrak{p}_i}$ or $I_{\mathfrak{p}_i} \simeq S^t_{\mathfrak{p}_i}$ by [**Mar24**, Theorem 6.2]. We lexicographically sort W_S accordingly and define the label of the wth class as `N.i.w`.

If $\mathrm{type}(S) = 1$, then we are done, since W_S consists only of the invertible class. If $\mathrm{type}(S) = 2$, it remains to define the distinguished representative of each non-invertible class $\omega \in W_S$: Let d be the smallest positive integer such that $dS^t \subseteq S$, $\mathcal{S}_0 := (\mathfrak{q}_1, \ldots, \mathfrak{q}_r) \subseteq \mathcal{S}$ be the ordered set of maximal ideals of S at which $\mathrm{type}_{\mathfrak{q}_i}(S) = 2$, and $m_1, \ldots, m_r$ be positive integers such that $\mathfrak{q}_i^{m_i} S_{\mathfrak{q}_i} \subseteq (dS^t)_{\mathfrak{q}_i}$. Then for I any member of ω, we define the distinguished representative J of ω to be

$$J := \sum_{i=1}^{r} \Big((I_i + \mathfrak{q}_i^{m_i}) \prod_{j \neq i} \mathfrak{q}_j^{m_j}\Big),$$

where $I_i := S$ if $I_{\mathfrak{q}_i} \simeq S_{\mathfrak{q}_i}$ and $I_i := dS^t$ if $I_{\mathfrak{q}_i} \simeq (dS^t)_{\mathfrak{q}_i}$; see [**Mar24**, Lemma 6.4].

Now, we consider the case $\mathrm{type}(S) > 2$. The following procedure is more time consuming than the previous one. For each $\mathfrak{p}_i \in \mathcal{S}$, put $T_i := (\mathfrak{p}_i : \mathfrak{p}_i)$; $S \subsetneq T_i$ yields a surjective group homomorphism $\mathrm{Pic}(S) \to \mathrm{Pic}(T_i)$ induced by the extension map. Let $\mathfrak{K}_i$ be its kernel and set $G_i := (T_i^\times / S^\times) \times \mathfrak{K}_i$. Sort the orders T_i by the size of G_i from smallest to largest, breaking ties using the ordering on the ideals $\mathfrak{p}_i$. Let T be first among the sorted orders T_i. Let U be a transversal of $T^\times / S^\times$ and $\mathcal{K}$ be a set of representatives L of the corresponding $\mathfrak{K}_i$ satisfying $LT = T$. By recursion, we assume that we have already computed distinguished representatives for the elements of W_T, which we denote by $J_1, \ldots, J_t$.

Fix now a non-invertible $\omega \in W_S$. By [**Mar25**, Proposition 6.2], the class ω admits a representative I_0 such that $I_0 T = J_i$ for a unique index i. Every fractional S-ideal I_1 weakly equivalent to I_0 satisfying $I_1 T = J_i$ is of the form $I_1 = u \cdot L \cdot I_0$ for unique $u \in U$ and $L \in \mathcal{K}$. Since U and $\mathcal{K}$ are finite sets, we list all such ideals I_1 and sort them according to their sort key $s(I_1)$, defined in the same way as if I_1 were an order (§3.1). Finally, define the distinguished representative J of ω to be the first ideal of the sorted list and the sort key $\tilde{s}(\omega)$ of ω as $s(\omega)$ concatenated with $s(J)$. Sort W_S accordingly and define the label of the wth class as `N.i.w`.

3.5. Labeling and distinguished representatives of invertible ideal classes. We start by computing some description of $\mathrm{Pic}(R)$ using [**KP05**]; we then need to fix an ordering of the elements and ideals of R representing each. We choose a set $\mathcal{P}$ of generators of $\mathrm{Pic}(R)$ by iterating over the maximal ideals of R, sorted first by norm and then by the sort-key defined in §3.2, keeping only the ones that enlarge the group generated, until we generate all of $\mathrm{Pic}(R)$. Now, consider an overorder S, and write $\mathrm{Pic}(S) \simeq \mathbb{Z}/m_1\mathbb{Z} \times \cdots \times \mathbb{Z}/m_k\mathbb{Z}$ with $m_i \mid m_{i+1}$. The ordered set $\mathcal{P}_S := (\mathfrak{p}S : \mathfrak{p} \in \mathcal{P})$ generates $\mathrm{Pic}(S)$, and we seek to construct a basis from it by iteratively choosing elements $g_k, \ldots, g_1 \in \mathrm{Pic}(S)$ of order $m_k, \ldots, m_1$. At stage i, let $H_i := \langle g_{i+1}, \ldots, g_k \rangle$ and choose L_i with class $g_i \in \mathrm{Pic}(S)$ to be the first product of elements of $\mathcal{P}_S$ (by lexicographic order on exponent vectors) with order m_i both in $\mathrm{Pic}(S)$ and $\mathrm{Pic}(S)/H_i$. We sort the elements $g \in \mathrm{Pic}(S)$ by writing them as $g = g_1^{e_1} \cdots g_k^{e_k}$ with $0 \leq e_i \leq m_i - 1$ and we define the distinguished representative of g as $L_1^{e_1} \cdots L_k^{e_k}$.

3.6. Labeling and distinguished representative of polarizations. Assume now that the isogeny class $\mathcal{I}$ is ordinary, and that I is the distinguished representative of its isomorphism class. Set $S := (I : I)$. For each isomorphism class of polarizations of I, we define a distinguished representative as follows. We will sort elements $oa \in K^\times$ by lexicographically ordering the sequence $(e, n_1, \ldots, n_{2g})$ where $(\frac{n_1}{e}, \ldots, \frac{n_{2g}}{e})$ are the coefficients of a written with respect to the basis $\mathcal{B}_K$. The image of the multiplicative group $U := \langle u\overline{u} \ : \ u \in S^\times \rangle$ under the map

$$\mathrm{Log}_\Phi : K \longrightarrow \mathbb{R}^g \quad a \mapsto (\log|\varphi_1(a)|, \ldots, \log|\varphi_g(a)|),$$

is a lattice $\mathcal{L}$ in $\mathbb{R}^g$. Fix a representative λ of a fixed isomorphism class of polarizations of I. Consider the set L of elements $u \in S^\times$ giving the points $\mathrm{Log}_\Phi(u\overline{u}) \in \mathcal{L}$ which are the closest to $\mathrm{Log}_\Phi(\lambda) \in \mathbb{R}^g$. For each $u \in L$, add uv to L where v runs over the units in $S^\times$ such that $|\mathrm{Log}_\Phi(u\overline{u}\lambda)| = |\mathrm{Log}_\Phi(uv\overline{uv}\lambda)|$, so that L is independent of the choice of the initial representative λ. Let u_0 be the first element of L according to sorting scheme described above. Then we set the distinguished representative of the isomorphism class of λ to be $\lambda_0 := \lambda u_0 \overline{u}_0$, and sort the isomorphism classes of polarizations of I of the same degree d by sorting their distinguished representatives using the procedure described above.

3.7. Examples. A Magma implementation of the procedures described above is available at [**LMFDBx**]. We list some interesting examples.

(i) For $g = 2$, the smallest q so that there exists an isogeny class with an endomorphism ring with Cohen Macaulay type 3 (largest possible for $g = 2$ by [**Mar24**, Proposition 4.9]) has label 2.5.a_g. The particular overorder S is the unique order with $[\mathcal{O}_K : S] = 8$, has 5 weak equivalence classes and trivial Picard group. This gives 5 isomorphism classes, with labels `2.5.a_g-8.1.w.1` for $1 \leq w \leq 5$.

(ii) For $g = 2$ and $q = 5$, there are two isogeny classes (twists of each other) with maximum size of Picard group. In both cases the Picard groups is C_{12}. See 2.5.b_ac and 2.5.ab_ac.

(iii) For $g = 2$ and $q = 5$, whenever the Picard group has size 8, it is isomorphic to $C_4 \times C_2$, and there are three such examples. In one case, 2.5.a_e, the Frobenius order is maximal; the other two examples (2.5.b_e and 2.5.ab_e) have Frobenius order of index 50.

Acknowledgements

We thank the anonymous referees for corrections and suggestions. Costa was supported by Simons Foundation grant SFI-MPS-Infrastructure-00008651. Costa and Roe were supported by Simons Foundation grant 550033. Dupuy is supported by National Science Foundation grant DMS-2401570. Marseglia was supported by NWO grant VI.Veni.202.107, and by Marie Skłodowska-Curie Actions - Postdoctoral Fellowships 2023 (project 101149209 - AbVarFq). Vincent is supported by a Simons Foundation Travel Support for Mathematicians grant.

References

[C15] Tommaso Giorgio Centeleghe and Jakob Stix, *Categories of abelian varieties over finite fields, I: Abelian varieties over* $\mathbb{F}_p$, Algebra Number Theory **9** (2015), no. 1, 225–265, DOI 10.2140/ant.2015.9.225. MR3317765

[CCO14] Ching-Li Chai, Brian Conrad, and Frans Oort, *Complex multiplication and lifting problems*, Mathematical Surveys and Monographs, vol. 195, American Mathematical Society, Providence, RI, 2014, DOI 10.1090/surv/195. MR3137398

[CPS20] John Cremona, Aurel Page, and Andrew V. Sutherland, *Sorting and labelling integral ideals in a number field*, 2020. `arXiv:2005.09491`.

[Del69] Pierre Deligne, *Variétés abéliennes ordinaires sur un corps fini* (French), Invent. Math. **8** (1969), 238–243, DOI 10.1007/BF01406076. MR254059

[DKRV21] Taylor Dupuy, Kiran Kedlaya, David Roe, and Christelle Vincent, *Isogeny classes of abelian varieties over finite fields in the LMFDB*, Arithmetic geometry, number theory, and computation, Simons Symp., Springer, Cham, 2021, pp. 375–448, DOI 10.1007/978-3-030-80914-0_13. MR4427971

[HS20] Tommy Hofmann and Carlo Sircana, *On the computation of overorders*, Int. J. Number Theory **16** (2020), no. 4, 857–879, DOI 10.1142/S179304212050044X. MR4093387

[How95] Everett W. Howe, *Principally polarized ordinary abelian varieties over finite fields*, Trans. Amer. Math. Soc. **347** (1995), no. 7, 2361–2401, DOI 10.2307/2154828. MR1297531

[KP05] Jürgen Klüners and Sebastian Pauli, *Computing residue class rings and Picard groups of orders*, J. Algebra **292** (2005), no. 1, 47–64, DOI 10.1016/j.jalgebra.2005.04.013. MR2166795

[LMFDBa] The LMFDB Collaboration, *The L-functions and modular forms database*, 2024. [Online; accessed 20 November 2024], `http://www.lmfdb.org`.

[LMFDBb] The LMFDB Collaboration, *Labels for isogeny classes of abelian varieties over finite fields*, `https://www.lmfdb.org/knowledge/show/av.fq.lmfdb_label?timestamp=1477792334876000`.

[LMFDBx] The LMFDB Collaboration, `https://github.com/stmar89/AbVarFq_LMFDBLabels`.

[Mar20] Stefano Marseglia, *Computing the ideal class monoid of an order*, J. Lond. Math. Soc. (2) **101** (2020), no. 3, 984–1007, DOI 10.1112/jlms.12294. MR4111932

[Mar21] Stefano Marseglia, *Computing square-free polarized abelian varieties over finite fields*, Math. Comp. **90** (2021), no. 328, 953–971, DOI 10.1090/mcom/3594. MR4194169

[Mar24] Stefano Marseglia, *Cohen-Macaulay type of orders, generators and ideal classes*, J. Algebra **658** (2024), 247–276, DOI 10.1016/j.jalgebra.2024.05.051. MR4767439

[Mar25] Stefano Marseglia, *Local isomorphism classes of fractional ideals of orders in étale algebras*, J. Algebra **673** (2025), 77–102, DOI 10.1016/j.jalgebra.2025.02.030. MR4878799

[Tat66] John Tate, *Endomorphisms of abelian varieties over finite fields*, Invent. Math. **2** (1966), 134–144, `https://doi.org/10.1007/BF01404549`. MR206004

Edgar Costa, Department of Mathematics, Massachusetts Institute of Technology, Cambridge, MA 02139-4307, USA
Email address: edgarc@mit.edu
URL: https://edgarcosta.org

Taylor Dupuy, Department of Mathematics and Statistics, Innovation Hall E220, 82 University Place, Burlington, VT, 05405, USA
Email address: taylor.dupuy@gmail.com
URL: http://tdupu.github.io

Stefano Marseglia, Laboratoire Jean Alexandre Dieudonné, Université Côte Azur, 06108 Nice Cedex 2, France
Email address: stefano.marseglia@univ-cotedazur.fr
URL: https://stmar89.github.io

David Roe, Department of Mathematics, Massachusetts Institute of Technology, Cambridge, MA 02139-4307, USA
Email address: roed@mit.edu
URL: https://math.mit.edu/~roed

Christelle Vincent, Department of Mathematics and Statistics, Innovation Hall E220, 82 University Place, Burlington, VT, 05405, USA
Email address: christelle.vincent@uvm.edu
URL: https://www.uvm.edu/~cvincen1

Contemporary Mathematics
Volume **840**, 2026
https://doi.org/10.1090/conm/840/16821

On pairs of primes with small order reciprocity

Craig Costello and Gaurish Korpal

ABSTRACT. We give a sieving algorithm for finding pairs of primes with small multiplicative orders modulo each other. This problem is a necessary condition for obtaining constructions of 2-cycles of pairing-friendly curves, which have found use in cryptographic applications. Our database of examples suggests that, except for a well-known infinite family of such primes, instances become increasingly rare as the size of the primes increase. This leads to some interesting open questions for which we hope our database prompts further investigation.

1. The problem

This paper is concerned with finding pairs of primes that satisfy certain instances of the following definition.

DEFINITION 1.1 (Primes with (k,k')-order reciprocity). The prime numbers p and q have (k,k')-*order reciprocity* if $\mathrm{ord}_q(p) = k$ and $\mathrm{ord}_p(q) = k'$. That is, k and k' are the smallest natural numbers such that $p^k \equiv 1 \pmod q$ and $q^{k'} \equiv 1 \pmod p$.

Herein we are interested in instances of Definition 1.1 where p and q are *large* and where k and k' are *small*. To be more concrete about what we mean by large and small, we must first discuss why this problem finds practical relevance in cryptography. In 2014, Ben-Sasson, Chiesa, Tromer, and Virza [**BSCTV14**] showed that certain types of pairing-based zero-knowledge proof systems can be instantiated in a very efficient and scalable way if we have a 2-*cycle* of *pairing-friendly* elliptic curves. Let p and q be large primes. We say that the elliptic curves $E/\mathbb{F}_p$ and $E'/\mathbb{F}_q$ are a 2-cycle if and only if $\#E(\mathbb{F}_p) = q$ and $\#E'(\mathbb{F}_q) = p$. Furthermore, we say that $E/\mathbb{F}_p$ is pairing-friendly if its *embedding degree* $k = \mathrm{ord}_q(p)$ is small enough for the order-q Weil and/or Tate pairings to be efficiently computable. Similarly, we say that $E'/\mathbb{F}_q$ is pairing-friendly p if its embedding degree $k' = \mathrm{ord}_p(q)$ is suitably small. These embedding degrees need to be small enough that the pairings in $\mathbb{F}_{p^k}^\times$ and $\mathbb{F}_{q^{k'}}^\times$ can be computed efficiently. In practice, and indeed in this paper, we follow the standard rule-of-thumb and impose that both k and k' are no larger than 50 [**FST10**].

2020 *Mathematics Subject Classification.* Primary 11A07, 11T22, 11T71, 11A15, 11Y16.

Part of this work was done while Gaurish was an intern at Microsoft Research.

On the other hand, for the associated proof system to be cryptographically secure, the elliptic curve discrete logarithm problems (ECDLPs) and the finite field discrete logarithm problems (DLPs) all need to be computationally infeasible. In practice, this means p and q must both be at least 2^{160} for ECDLP security, and k and k' must be large enough for the fields $\mathbb{F}_{p^k}$ and $\mathbb{F}_{q^{k'}}$ to offer comparable DLP security.

1.1. A necessary condition for 2-cycles. The only known instances of 2-cycles of ordinary pairing-friendly elliptic curves arise from the Miyaji-Nakabayashi-Takano (MNT) family [**MNT01**], which is parameterised as

$$p = x^2 - x + 1 \qquad \text{and} \qquad q = x^2 + 1, \tag{1.1}$$

from which it follows that $(k, k') = (\text{ord}_q(p), \text{ord}_p(q))$ are always such that $k \mid 6$ and $k' \mid 4$.

It turns out that the MNT 2-cycle is suboptimal in practical applications, because the embedding degrees are too small to optimally balance the ECDLP and DLP securities. With $(k, k') = (4, 6)$, meeting the requisite DLP security requires that $\mathbb{F}_{p^4}$ and $\mathbb{F}_{q^6}$ are at least a few thousand bits [**SG18**, Table 1], which forces the sizes of p and q to be orders of magnitude larger than they would be in the optimal scenario where k and k' are larger. The subsequent hunt for 2-cycles of ordinary elliptic curves with larger embedding degrees has only produced negative results [**CCW19, BMUS23**]. For example, Chiesa, Chua and Weidner [**CCW19**] proved that 2-cycles of ordinary elliptic curves $E/\mathbb{F}_p$ and $E'/\mathbb{F}_q$ with $(k, k') \in \{(5, 10), (8, 8), (12, 12)\}$ do not exist. In what follows, we use a toy example to motivate the work in this paper by showing the existence of a 2-cycle with $(k, k') = (12, 12)$ under relaxed conditions, in light of recent work in [**CCN24**].

Example 1.2. The prime pair $(p, q) = (620461, 15493)$ is such that $(k, k') = (\text{ord}_q(p), \text{ord}_p(q)) = (12, 12)$. Proposition 3 of [**CCW19**] rules out the existence of a 2-cycle with $q = \#E(\mathbb{F}_p)$ and $p = \#E'(\mathbb{F}_q)$. Indeed, the Hasse bound for $E'/\mathbb{F}_q$ in this instance is $15246 \leq \#E'(\mathbb{F}_q) \leq 15742$, ruling out $p = \#E'$. If we relax the 2-cycle condition to allow cofactors as in [**CCW19**, §7] and to allow extension fields as in [**CCN24**], we might instead seek to find two elliptic curves $E/\mathbb{F}_p$ and $E'/\mathbb{F}_{q^2}$ such that $\#E(\mathbb{F}_p) = hq$ and $\#E'(\mathbb{F}_{q^2}) = h'p$. While such an E does exist (with $h = 40$), there is no multiple of p inside the Hasse interval for $E'/\mathbb{F}_{q^2}$; $h' = 386$ is too small for $h'p$ to be inside the interval, while $h' = 387$ is too large. If, however, we instead consider a genus 2 curve $C/\mathbb{F}_q$, close examination of the possible group orders of its Jacobian (see [**GKS11**, §2.1]) reveals that $h' \in \{374, \ldots, 399\}$ are all plausible cofactors. Indeed, we readily find one such example as $C/\mathbb{F}_q \colon y^2 = x^6 + 6611x^5 + 13858x^4 + 6818x^3 + 5652x^2 + 10423x + 1795$, which is such that $\#\mathcal{J}_C(\mathbb{F}_q) = 383 \cdot p$, forming a 2-cycle with $E/\mathbb{F}_p \colon y^2 = x^3 + 30984x + 426966$, which is such that $\#E(\mathbb{F}_p) = 40 \cdot q$. Both $\mathcal{J}_C$ and E are ordinary and have embedding degree 12.

The above example shows that it is possible to construct ordinary 2-cycles under relaxed conditions, even when their existence has been ruled out in the case where both curves are elliptic curves of prime order, so long as we have primes with (k, k')-order reciprocity[1]. In our view, the existence of large primes with small order

[1]The notion and terminology *order reciprocity* is found in a 2019 post by Lee [**Lee19**].

reciprocity is *the* fundamental and necessary requirement for determining whether relaxed 2-cycles can exist in the interesting cryptographic ranges.

1.2. Open questions. One can show that there are no prime pairs with $(2,2)$-order reciprocity, but infinitely many prime pairs with $(p-1, q-1)$-order reciprocity[2]. However, in between these two extremes there are a number of questions that are of potential relevance to cryptography. For example, we saw above that the pair $(p,q) = (620461, 15493)$ has $(12,12)$-order reciprocity, but this is the only such example we found. We state some open questions that warrant further investigation:

(1) *Is* $(620461, 15493)$ *the only prime pair with* $(12,12)$*-order reciprocity?*
(2) *Are there any fixed values of* (k,k') *with* $\min(k,k) > 4$ *for which there are an infinite number of primes with* (k,k')*-order reciprocity?*
(3) *Are there any fixed values of* (k,k') *with* $\min(k,k) > 2$ *for which there are no pairs of primes with* (k,k')*-order reciprocity?*

Finally, we pose the question of interest that underlies this work:

(4) *Are there* **any** *large pairs of primes with small* (k,k')*-order reciprocity?*

In the search for answers to the above questions, we derived an algorithm for finding primes with (k,k')-order reciprocity – this is presented in Section 2. We used this algorithm to exhaustively search up to the 200 millionth prime for pairs with small order reciprocity – the database of examples we found is summarised in Section 3.

We reiterate that we are not interested in the known examples which come from the MNT parameterisation in (1.1), and that in practice we are therefore interested in examples with, say, $5 \leq k, k' \leq 50$. By *large* primes, we mean cryptographically large, i.e. at least 2^{160}.

2. The algorithms

Let p be a prime number and let $\alpha_p \in \mathbb{F}_p$ be such that $\mathbb{F}_p^\times = \langle \alpha_p \rangle$, where elements in $\mathbb{F}_p^\times$ are represented by the corresponding least positive remainder in $(\mathbb{Z}/p\mathbb{Z})^\times$. Then Algorithm 1 returns all $q < p$ such that $\mathrm{ord}_q(p) = k$ and $\mathrm{ord}_p(q) = k'$. Furthermore, let P be a list of prime numbers and A be a list such that $A[i]$ is a primitive element modulo $P[i]$. Algorithm 2 finds all the primes pairs (p,q) with $p \in P$ and $q < p$ such that $\mathrm{ord}_q(p) = k$ and $\mathrm{ord}_p(q) = k'$ or $\mathrm{ord}_q(p) = k'$ and $\mathrm{ord}_p(q) = k$.

2.1. Proof of correctness. We start by showing that Algorithm 1 returns all primes $q < p$ such that $\mathrm{ord}_q(p) = k$ and $\mathrm{ord}_p(q) = k'$. Any such q must be the least positive remainder of some power of α_p^t with $t = (p-1)/k'$ in $(\mathbb{Z}/p\mathbb{Z})^\times$; Lines 4–5 exhaust these possibilities, while Line 6 avoids testing those q which cannot have exact order k'. Line 13 only keeps those q which are prime and for which elements of order k exist in $(\mathbb{Z}/q\mathbb{Z})^\times$. The remainder of Algorithm 1 simply keeps those p which have exact order k in $(\mathbb{Z}/q\mathbb{Z})^\times$. Algorithm 2 loops over all combinations of $p, q \in P$ and, for $k \neq k'$, makes an additional call to Algorithm 1 to reverse the order of k and k'.

[2]For the former, $p < q$ implies $p = q - 1$, the only option for which is $(p,q) = (2,3)$, which has $\mathrm{ord}_p(q) = 1$. For the latter, see [**Fre19**].

Algorithm 1 `reciprocity`(p, α_p, k, k')

```
Q ← [] // empty list
if p ≡ 1 (mod k') then
    t ← (p−1)/k'
    for s = 1 to k' − 1 do
        m ← s · t
        if gcd(m, p − 1) = t then
            q ← α_p^m
            if q ≡ 1 (mod k) and q is prime then
                // we have found q such that ord_p(q) = k'.
                if p^k ≡ 1 (mod q) then
                    b ← 1
                    for d = 1 to k − 1 do
                        if k ≡ 0 (mod d) and p^d ≡ 1 (mod q) then
                            b ← 0
                            break // ord_q(p) ≠ k
                    if b = 1 then
                        Q.append(q) // ord_q(p) = k
return Q
```

Algorithm 2 `primepairs`(P, A, k, k')

```
L ← [] // empty list
for p in P do
    Q ← reciprocity(p, α_p, k, k')
    if #Q ≠ 0 then
        for q ∈ Q do
            L.append((p, q))
    if k ≠ k' then
        Q' ← reciprocity(p, α_p, k', k)
        if #Q' ≠ 0 then
            for q' ∈ Q' do
                L.append((q', p))
return L
```

2.2. Complexity analysis. Let p be the largest prime in the list P, whose cardinality is N. We will analyse the time complexity of Algorithm 2 as $p \to \infty$, assuming that k and k' remain fixed, small constants. We will also assume that, as $p \to \infty$, the average size of the primes in P is in $O(p)$; this assumption holds if P contains all primes within a given interval and, in particular, if P contains all primes up to p. Algorithm 2 makes either one or two calls to Algorithm 1 for each prime in P, so it follows that the time complexity of Algorithm 2 is in $O(N \cdot R)$, where R is the complexity of Algorithm 1 on input of a prime of size in $O(p)$.

We will now proceed step by step through Algorithm 1, stating the time complexity of each step together with the probability of advancing through the control statements. These can be accumulated in a straightforward way to obtain a precise big-O complexity of Algorithm 1, but this becomes a rather messy and nonillustrative expression, so we will eventually ignore all logarithmic factors to instead state a final big-$\tilde{O}$ time complexity.

For small, fixed k and k', calls to Algorithm 1 advance to Line 6 $O(p)$ times on average. The time complexity of the GCD algorithm here is in $O(\log p)$ and, heuristically, $O(1)$ of these tests advance to Line 7. The asymptotic complexity of this exponentiation is in $O(\log p \log^2(\log p) \log\log\log p)$, using the Schönhage–Strassen algorithm [**SS71**]. In Line 8, an optimised probabilistic primality test (like Miller-Rabin [**Rab80**]) has time complexity $O(\log^3 p)$, while the deterministic AKS primality test [**AKS04**] has time complexity $O(\log^6 p)$. Since the size of the prime q is in $O(p)$, the probability of advancing to the remaining steps is in $O(1/\log p)$ by the Prime Number Theorem.

It follows that Algorithm 1 has time complexity $\tilde{O}(p)$, and thus that Algorithm 2 has complexity $\tilde{O}(N \cdot p)$. In particular, if P contains all the primes up to p, then Algorithm 2 has complexity $\tilde{O}(p^2)$ by the Prime Number Theorem.

3. The database

We used the `parallel GP` interface of `PARI/GP` [**The24**] running on AMD Ryzen Threadripper PRO 3995WX with 128 GB memory to perform all the computations. The `PARI/GP` implementation of the algorithms and the data are available in

`https://github.com/gkorpal/order-reciprocity`.

Let P_1 be the list of first 100 million primes and P_2 be the list of the next 100 million primes. We first performed the primitive element pre-computation for P_1 and P_2 using the in-built `znprimroot` function [**Coh93**, Algorithm 1.4.4].

We then ran Algorithm 2 to find prime pairs in P_1 and for prime pairs (where the largest prime is) in P_2 with (k, k')-order reciprocity, for $2 \le k \le k' \le 51$. We could not parallelise the outermost for loop of this algorithm because of the memory sharing restrictions. However, we utilized the `parallel GP` interface to simultaneously run the computations for $(k, k') \in \{\{m, m+1\} \times \{n, n+1\} \mid m, n \in 2\mathbb{Z} \text{ and } 2 \le m \le n \le 50\}$ such that $k \le k'$.

Tables 1 and 2 present sample data, while the repository offers access to the full dataset. Table 1 highlights that $(4, 6)$-order reciprocity remains the most prevalent, even as prime size increases. Meanwhile, Table 2 shows that larger prime sizes result in most (k, k')-order reciprocities yielding no prime pairs.

TABLE 1. Counts of prime pairs with selected (k, k')-order reciprocity. The bottom row counts the number of pairs when the larger prime is in P_2.

(k, k')		$(4,6)$	$(3,10)$	$(3,14)$	$(4,46)$	$(2,35)$	$(11,49)$	$(15,45)$	$(38,45)$
#prime	P_1	738	20	14	8	5	4	3	2
pairs	P_2	258	0	0	2	2	3	2	2

TABLE 2. Counts of (k, k')-order reciprocities with selected number of prime pairs. The bottom row counts the number of (k, k')-order reciprocities when the larger prime is in P_2.

#prime pairs		0	1	2	3	4 to 12	14	20	258	738	Total
# (k, k')	P_1	97	147	209	231	588	1	1	0	1	1275
	P_2	1108	159	6	1	0	0	0	1	0	1275

References

[AKS04] Manindra Agrawal, Neeraj Kayal, and Nitin Saxena, *PRIMES is in P*, Ann. of Math. (2) **160** (2004), no. 2, 781–793, DOI 10.4007/annals.2004.160.781. MR2123939

[BMUS23] Marta Bellés-Muñoz, Jorge Jiménez Urroz, and Javier Silva, *Revisiting cycles of pairing-friendly elliptic curves*, Advances in cryptology—CRYPTO 2023. Part II, Lecture Notes in Comput. Sci., vol. 14082, Springer, Cham, 2023, pp. 3–37, DOI 10.1007/978-3-031-38545-2_1. MR4656175

[BSCTV14] Eli Ben-Sasson, Alessandro Chiesa, Eran Tromer, and Madars Virza, *Scalable zero knowledge via cycles of elliptic curves*, Advances in cryptology—CRYPTO 2014. Part II, Lecture Notes in Comput. Sci., vol. 8617, Springer, Heidelberg, 2014, pp. 276–294, DOI 10.1007/978-3-662-44381-1_16. MR3239473

[CCN24] Maria Corte-Real Santos, Craig Costello, and Michael Naehrig, *On cycles of pairing-friendly abelian varieties*, Advances in cryptology—CRYPTO 2024. Part IX, Lecture Notes in Comput. Sci., vol. 14928, Springer, Cham, 2024, pp. 221–253, DOI 10.1007/978-3-031-68400-5_7. MR4826122

[CCW19] Alessandro Chiesa, Lynn Chua, and Matthew Weidner, *On cycles of pairing-friendly elliptic curves*, SIAM J. Appl. Algebra Geom. **3** (2019), no. 2, 175–192, DOI 10.1137/18M1173708. MR3934100

[Coh93] Henri Cohen, *A course in computational algebraic number theory*, Graduate Texts in Mathematics, vol. 138, Springer-Verlag, Berlin, 1993, DOI 10.1007/978-3-662-02945-9. MR1228206

[Fre19] FredH, *Order reciprocity*, Mathematics Stack Exchange, 2019, `https://math.stackexchange.com/q/3118669` (version: 2019-02-19).

[FST10] David Freeman, Michael Scott, and Edlyn Teske, *A taxonomy of pairing-friendly elliptic curves*, J. Cryptology **23** (2010), no. 2, 224–280, DOI 10.1007/s00145-009-9048-z. MR2578668

[GKS11] Pierrick Gaudry, David Kohel, and Benjamin Smith, *Counting points on genus 2 curves with real multiplication*, Advances in cryptology—ASIACRYPT 2011, Lecture Notes in Comput. Sci., vol. 7073, Springer, Heidelberg, 2011, pp. 504–519, DOI 10.1007/978-3-642-25385-0_27. MR2935020

[Lee19] S. Lee, *Order reciprocity*, Mathematics Stack Exchange, 2019, `https://math.stackexchange.com/q/3118453` (version: 2019-02-19).

[MNT01] Atsuko Miyaji, Masaki Nakabayashi, and Shunzo Takano, *Characterization of elliptic curve traces under FR-reduction*, Information security and cryptology—ICISC 2000 (Seoul), Lecture Notes in Comput. Sci., vol. 2015, Springer, Berlin, 2001, pp. 90–108, DOI 10.1007/3-540-45247-8_8. MR1897375

[Rab80] Michael O. Rabin, *Probabilistic algorithm for testing primality*, J. Number Theory **12** (1980), no. 1, 128–138, DOI 10.1016/0022-314X(80)90084-0. MR566880

[SG18] Michael Scott and Aurore Guillevic, *A new family of pairing-friendly elliptic curves*, Arithmetic of finite fields, Lecture Notes in Comput. Sci., vol. 11321, Springer, Cham, 2018, pp. 43–57, DOI 10.1007/978-3-030-05153-2_2. MR3905068

[SS71] A. Schönhage and V. Strassen, *Schnelle Multiplikation grosser Zahlen* (German, with English summary), Computing (Arch. Elektron. Rechnen) **7** (1971), 281–292, DOI 10.1007/bf02242355. MR292344

[The24] The PARI Group, Univ. Bordeaux, *PARI/GP version 2.15.5*, 2024, available from `http://pari.math.u-bordeaux.fr/`.

Queensland University of Technology, Brisbane, Australia
Email address: `craig.costello@qut.edu.au`

University of Arizona, Tucson, United States
Email address: `gkorpal@arizona.edu`

Contemporary Mathematics
Volume **840**, 2026
https://doi.org/10.1090/conm/840/16822

Genus formulas for families of modular curves

Asimina S. Hamakiotes and Jun Bo Lau

Abstract. For each open subgroup $H \leq \mathrm{GL}_2(\widehat{\mathbb{Z}})$, there is a modular curve X_H, defined as a quotient of the full modular curve $X(N)$, where N is the level of H. A formula for the genus of a modular curve is well known for the modular curves $X_0(N)$, $X_1(N)$, $X(N)$, $X_{\mathrm{sp}}(N)$, $X^*_{\mathrm{sp}}(N)$, $X_{\mathrm{ns}}(N)$, $X^*_{\mathrm{ns}}(N)$, and $X_{S_4}(p)$ for p prime. We explicitly work out the invariants of the genus formulas for $X^+_{\mathrm{sp}}(N)$, $X^+_{\mathrm{ns}}(N)$, and $X_{\mathrm{arith},1}(M, MN)$. We provide the invariants of the genus formulas for the aforementioned modular curves in a table at the end of the article.

1. Introduction

Let E be an elliptic curve defined over a number field K, with algebraic closure $\overline{K}$. For an integer $N \geq 2$, let $E[N]$ be the N-torsion subgroup of $E(\overline{K})$. Note that $E[N]$ is a free $\mathbb{Z}/N\mathbb{Z}$-module of rank 2. The absolute Galois group $\mathrm{Gal}_K := \mathrm{Gal}(\overline{K}/K)$ acts on $E[N]$ via the natural action on the coordinates of the points in $E[N]$, which induces the following Galois representation after fixing a basis of $E[N]$:

$$\rho_{E,N} : \mathrm{Gal}_K \to \mathrm{Aut}(E[N]) \cong \mathrm{GL}_2(\mathbb{Z}/N\mathbb{Z}).$$

For a subgroup $H \leq \mathrm{GL}_2(\mathbb{Z}/N\mathbb{Z})$, the K-rational points of the modular curve $X_H(N)$ parametrize isomorphism classes of elliptic curves E/K with $\rho_{E,N}(\mathrm{Gal}_K) \leq H$. Classifying rational points on modular curves has been a subject of study for many years, e.g., Serre's Uniformity Problem [**Ser72**], Mazur's "Program B" [**Maz77**]. Faltings' Theorem states that for a smooth curve C with genus $g \geq 2$, there are finitely many rational points on C. The genus of a curve C provides information on the arithmetic of the rational points on C and suggests methods that could be used to compute rational points.

In the classification of possible 2-adic and ℓ-adic images of Galois representations attached to elliptic curves over $\mathbb{Q}$, Rouse, Sutherland, and Zureick-Brown developed computational tools to determine the genus of modular curves [**RZB15**, **RSZB22**]. More precisely, given an integer $N > 1$ and a subgroup type $H \leq$

2020 *Mathematics Subject Classification.* Primary 14G35, 11G18, 11G30.

Lau was supported partly by the Simons Foundation grant #550023 for the Simons Collaboration on Arithmetic Geometry, Number Theory, and Computation, and in part by the European Research Council (ERC) under the European Union's Horizon 2020 research and innovation programme (grant agreement ISOCRYPT - No. 101020788), by the Research Council KU Leuven grant C14/24/099 and by CyberSecurity Research Flanders with reference number VR20192203.

$\mathrm{GL}_2(\mathbb{Z}/N\mathbb{Z})$, one can compute the generators of H, the index of the subgroup, number of elliptic points of order 2 and 3, and the number of cusps.

In this paper, we determine genus formulas of modular curves associated to subgroups of $\mathrm{GL}_2(\mathbb{Z}/N\mathbb{Z})$ which admit a model over $\mathbb{Q}$, using only congruence conditions. The list includes various families of interest which can be found in the L-functions and Modular Forms Database (LMFDB), see [**LMF24**], and contains the list from [**Ser72**]. Below is a summary of the families of modular curves for which the genus formula is known, with references to where one can find the genus formulas.

(1) Let $N \geq 1$ be an integer. For the families $X_0(N), X_1(N)$, and $X(N)$, the genus formulas are worked out in [**DS05**, **Shi71**].
(2) Let p be a prime and $r \geq 1$ an integer.The invariants of the genus formulas for $X_{\mathrm{sp}}(p^r)$, $X_{\mathrm{sp}}^*(p^r)$, $X_{\mathrm{ns}}(p^r)$, and $X_{\mathrm{ns}}^*(p^r)$ are in [**DLM22**, Table 3.1]. Since the genus formula invariants are multiplicative when the group is the full preimage under the Chinese Remainder map, one can determine the invariants of the genus formula for composite levels N, see [**DLM22**, Equation 3.14]. Also, see [**DMS19**, Remark 4.3] and [**Bar10**, Proposition 7.10, Remark 3.3].
(3) Let p be a prime such that $p \equiv \pm 3 \bmod 8$. Ligozat determined the genus formula of $X_{S_4}(p)$ [**Lig77**]. Also, see [**Ser72**, **SD73**].

We prove genus formulas for the following families of modular curves: $X_{\mathrm{sp}}^+(N)$, $X_{\mathrm{ns}}^+(N)$, $X_{\mathrm{arith},1}(M, MN)$ and $X_{\mathrm{arith},\pm 1}(M, MN)$. The results of this paper are summarized in Table 1, where we list the invariants of the genus formulas of the modular curves. In addition, the results of this work are published on the LMFDB in the "Modular curves over $\mathbb{Q}$" section [**LMF24**]. There are pages for each family of modular curves in the database, where the invariants of the genus formula for that family are displayed.

1.1. Structure of the paper. In Section 2, we review the background for modular curves, the genus formula and Cartan subgroups. For composite numbers M, N, we determine the genus formulas of $X_{\mathrm{sp}}^+(N)$ in Section 3; $X_{\mathrm{ns}}^+(N)$ in Section 4; and lastly, $X_{\mathrm{arith},1}(M, MN)$ and $X_{\mathrm{arith},\pm 1}(M, MN)$ in Section 5.

2. Background

Let N be a positive integer. There is a functor sending a $\mathbb{Z}[1/N]$-algebra R to the set of (isomorphism classes of) pairs (E, ϕ) where E is an elliptic curve over R and $\phi : E[N] \to (\mathbb{Z}/N\mathbb{Z})^2$ is an isomorphism of R-group schemes. If $N \geq 3$, then the functor is representable by a smooth affine $\mathbb{Z}[1/N]$-scheme, denoted by $Y(N)$. If $N < 3$, we take the coarse moduli space to get a scheme. We denote by $X(N)$ the compactification of $Y(N)$ and call this the full modular curve of level N.

Every matrix $\gamma \in \mathrm{GL}_2(\mathbb{Z}/N\mathbb{Z})$ is an automorphism of $(\mathbb{Z}/N\mathbb{Z})^2$, with γ acting on (E, ϕ) by sending it to $(E, \phi\circ\gamma)$. For a subgroup $H \leq \mathrm{GL}_2(\mathbb{Z}/N\mathbb{Z})$, we denote by X_H the quotient $X(N)/H$. As a coarse moduli space, $Y_H := Y(N)/H$ parametrizes isomorphism classes of elliptic curves with H-structure. More precisely, the equivalence class of elliptic curves with H-structure is given by (E, ϕ) where E is an elliptic curve over a $\mathbb{Z}[1/N]$-scheme R and $\phi : E[N] \to (\mathbb{Z}/N\mathbb{Z})^2$ is an isomorphism of R-group schemes and $(E, \phi) \sim_H (E', \phi')$ if and only if $(\phi')^{-1} \circ \iota|_{E[N]} \circ \phi = h$ for some $h \in H$ and $\iota : E \xrightarrow{\sim} E'$.

On the other end, it is well known that for a subgroup $\Gamma \leq \mathrm{SL}_2(\mathbb{Z})$ with finite index, one can define a complex structure on $\mathbb{H}^*/\Gamma$, where $\mathbb{H}^*$ is the union of the upper half plane, $\mathbb{Q}$ and ∞. In particular, X_H defined above can be realized as a Riemann surface $\mathbb{H}^*/\Gamma_H$, where $\Gamma_H \leq \mathrm{SL}_2(\mathbb{Z})$ is defined as the lift of $H \cap \mathrm{SL}_2(\mathbb{Z}/N\mathbb{Z})$.

Before stating the genus formula of a modular curve, we will define elliptic points and cusps of a congruence subgroup $\Gamma \leq \mathrm{SL}_2(\mathbb{Z})$. For each point $\tau \in \mathbb{H}^*$, let Γ_τ be the subgroup of Γ that fixes τ, i.e., $\Gamma_\tau = \{\gamma \in \Gamma : \gamma(\tau) = \tau\}$. A point $\tau \in \mathbb{H}^*$ is an *elliptic point* of Γ if Γ_τ is nontrivial as a group of transformations, i.e., if $\{\pm I\}\Gamma_\tau \supset \{\pm I\}$. The *cusps* of Γ are the Γ-equivalence classes of $\mathbb{Q} \cup \{\infty\}$.

The following invariants have a contribution to the genus formula of a modular curve X_H:

- The PSL_2*-index* of X_H, denoted $i(\Gamma_H)$, is the index of $H \cap \mathrm{PSL}_2(\mathbb{Z}/N\mathbb{Z})$ in $\mathrm{PSL}_2(\mathbb{Z}/N\mathbb{Z})$.
- The *number of elliptic points of order* 2 on X_H, denoted $\varepsilon_2(\Gamma_H)$, is the number of right cosets of Γ_H in $\mathrm{SL}_2(\mathbb{Z}/N\mathbb{Z})$ fixed by the right action of $\left(\begin{smallmatrix} 0 & -1 \\ 1 & 0 \end{smallmatrix}\right)$.
- The *number of elliptic points of order* 3 on X_H, denoted $\varepsilon_3(\Gamma_H)$, is the number of right cosets of Γ_H in $\mathrm{SL}_2(\mathbb{Z}/N\mathbb{Z})$ fixed by the right action of $\left(\begin{smallmatrix} 0 & -1 \\ 1 & -1 \end{smallmatrix}\right)$.
- The *number of cusps* on X_H, denoted $\varepsilon_\infty(\Gamma_H)$, is the number of orbits of the right coset space $\Gamma_H \backslash \mathrm{SL}_2(\mathbb{Z}/N\mathbb{Z})$ under the right action of $\left(\begin{smallmatrix} 1 & 1 \\ 0 & 1 \end{smallmatrix}\right)$.

That the number of elliptic points and cusps are equal to the number of cosets mentioned above is discussed in [**DS05**, Sections 2.3, 2.4]. It follows from the Riemann-Hurwitz formula that the genus of the modular curve X_H is given by:

$$g(\Gamma_H) = 1 + \frac{i(\Gamma_H)}{12} - \frac{\varepsilon_2(\Gamma_H)}{4} - \frac{\varepsilon_3(\Gamma_H)}{3} - \frac{\varepsilon_\infty(\Gamma_H)}{2}.$$

Common notation used in various genus formulas is: $\varphi(N)$ which is the usual Euler totient function, $\omega(N)$ which is the number of primes dividing N, and $\left(\frac{\cdot}{\cdot}\right)$ which is the usual Legendre symbol.

REMARK 2.1. Let $H \leq \mathrm{GL}_2(\mathbb{Z}/N\mathbb{Z})$. An index 2 or *quadratic refinement* of H is a subgroup $H' \leq H$ such that $H = \pm H'$. The modular curves X_H and $X_{H'}$ are isomorphic as curves, but the moduli problem is refined from H to H'. Naturally, the genus of X_H and $X_{H'}$ are equal. The genus formulas for $X_{\pm 1}(N)$, $X_{\mathrm{arith},\pm 1}(M, MN)$, and their quadratic refinements are included in Table 1.

2.1. Cartan Subgroups. Let R be a commutative ring. Given a free rank 2 étale R-algebra A, for $a \in A^\times$, the multiplication-by-a map defines an injective homomorphism $A^\times \hookrightarrow \mathrm{GL}_2(R)$. The image of this homomorphism is called a *Cartan subgroup* of $\mathrm{GL}_2(R)$. There is a canonical involution of A which gives another element of $\mathrm{GL}_2(R)$. The group generated by the image of $A^\times$ and the involution is called the *extended Cartan subgroup*. When $R = \mathbb{Z}/p^e\mathbb{Z}$, where p^e is some prime power, A has two possibilities: $A = R \times R$ (*split* Cartan) or $A = \mathcal{O}/p^e\mathcal{O}$ (*nonsplit* Cartan), where $\mathcal{O}$ is either a degree 2 unramified extension of $\mathbb{Z}_p$ or a quadratic order where p is inert. In general, for $R = \mathbb{Z}/N\mathbb{Z}$, the extended Cartan subgroup is called the *normalizer of Cartan subgroup* and it is denoted by either $C^+_{\mathrm{sp}}(N)$ or $C^+_{\mathrm{ns}}(N)$, depending on whether A is split or nonsplit, respectively, at each prime dividing N. The Cartan subgroup has index 2 in the extended Cartan subgroup. In this paper, the normalizer of the Cartan subgroup will mean the extended Cartan subgroup.

REMARK 2.2. In the literature, sometimes the normalizer of the Cartan subgroup is defined as the group generated by the Cartan subgroup along with *an involution for each prime dividing* N, denoted by $C^*_{\rm sp}$ or $C^*_{\rm ns}$, depending on whether A is split or nonsplit, respectively, at each prime dividing N. In contrast with the above, the Cartan subgroup here has index $2^{\omega(N)}$ in the extended Cartan subgroup and is generated by the Cartan subgroup and involutions from each prime dividing N. Geometrically, $X^*_{\rm sp}$ (resp. $X^*_{\rm ns}$) is isomorphic to the fiber product of split (resp. nonsplit) Cartan curves from lower levels.

EXAMPLE 2.3. Consider the modular curve associated to the normalizer of a split Cartan subgroup of level 21, namely $X^+_{\rm sp}(21)$. Following our notation, $X^+_{\rm sp}(21)$ correponds to the LMFDB label `21.336.17.d.1` with genus 17. In [**Bar10**], the curve with the same notation is defined as in Remark 2.2 and has LMFDB label `21.168.9.f.1` with genus 9.

3. Split Cartan and normalizer of the split Cartan

The split Cartan modular curve, denoted $X_{\rm sp}(N)$, is the modular curve X_H for the subgroup $H \leq \mathrm{GL}_2(\widehat{\mathbb{Z}})$ given by the inverse image of a Cartan subgroup $\{\left(\begin{smallmatrix} * & 0 \\ 0 & * \end{smallmatrix}\right)\}$ that is split at every prime dividing N. As a moduli space its K-points parametrize isomorphism classes of triples (E, C, D), where E is an elliptic curve over K, and C and D are Gal_K-stable cyclic subgroups such that $E[N](\overline{K}) \simeq C \oplus D$.

The normalizer of the split Cartan, denoted $X^+_{\rm sp}(N)$, is the modular curve X_H for the subgroup $H \leq \mathrm{GL}_2(\widehat{\mathbb{Z}})$ given by the inverse image of an extended Cartan subgroup $\{\left(\begin{smallmatrix} * & 0 \\ 0 & * \end{smallmatrix}\right)\} \cup \{\left(\begin{smallmatrix} 0 & * \\ * & 0 \end{smallmatrix}\right)\} \leq \mathrm{GL}_2(\mathbb{Z}/N\mathbb{Z})$ that is split at every prime dividing N. As a moduli space it parametrizes isomorphism classes of pairs $(E, \{C, D\})$, where E is an elliptic curve over K, and $\{C, D\}$ is a Gal_K-stable pair of cyclic subgroups such that $E[N](\overline{K}) \simeq C \oplus D$. Note that neither C nor D need be Gal_K-stable and the split Cartan subgroup is an index 2 subgroup in the normalizer.

In this section, we describe the genus formulas for the normalizer of the split Cartan modular curve $X^+_{\rm sp}(N)$. We derive the invariants of their genus formulas for composite level N, using methods communicated by Pietro Mercuri for prime power levels p^r in [**DMS19**, Remark 4.3].

From the definition above, the orders of the split Cartan subgroup $C_{\rm sp}(N)$ and its normalizer $C^+_{\rm sp}(N)$ are $\varphi(N)^2$ and $2\varphi(N)^2$, respectively. Therefore, the indices are given by

$$[\mathrm{GL}_2(\mathbb{Z}/N\mathbb{Z}) : C_{\rm sp}(N)] = N^2 \prod_{p \mid N} \left(1 + \frac{1}{p}\right),$$

$$[\mathrm{GL}_2(\mathbb{Z}/N\mathbb{Z}) : C^+_{\rm sp}(N)] = \frac{N^2}{2} \prod_{p \mid N} \left(1 + \frac{1}{p}\right).$$

Let $N = \prod_i p_i^{e_i}$ be the prime factorization of N. The Chinese Remainder Theorem implies that there is a bijection:

$$\mathrm{GL}_2(\mathbb{Z}/N\mathbb{Z})/C_{\rm sp}(N) \to \prod_i \mathrm{GL}_2(\mathbb{Z}/p_i^{e_i}\mathbb{Z})/C_{\rm sp}(p_i^{e_i}).$$

Let $\omega := \left(\begin{smallmatrix} 0 & 1 \\ 1 & 0 \end{smallmatrix}\right) \in \mathrm{GL}_2(\mathbb{Z}/N\mathbb{Z})$ be an involution for the split Cartan subgroup. The subgroup generated by ω acts on $\mathrm{GL}_2(\mathbb{Z}/N\mathbb{Z})/C_{\mathrm{sp}}(N)$ by left multiplication, which induces a natural action of $\vec{\omega} := (\omega, \omega, \dots, \omega)$ on $\prod_p \mathrm{GL}_2(\mathbb{Z}/p_i^{e_i}\mathbb{Z})/C_{\mathrm{sp}}(p_i^{e_i})$, giving the following bijection:

$$\Big(\mathrm{GL}_2(\mathbb{Z}/N\mathbb{Z})/C_{\mathrm{sp}}(N)\Big)/\langle\omega\rangle \to \Big(\prod_i \mathrm{GL}_2(\mathbb{Z}/p_i^{e_i}\mathbb{Z})/C_{\mathrm{sp}}(p_i^{e_i})\Big)/\langle\vec{\omega}\rangle.$$

Note that the left hand side is equal to $\mathrm{GL}_2(\mathbb{Z}/N\mathbb{Z})/C_{\mathrm{sp}}^+(N)$. We first compute a set of coset representatives for prime power levels and lift them to coset representatives on the left hand side in the bijection above.

LEMMA 3.1. *Let p be a prime, a set of coset representatives $\mathrm{GL}_2(\mathbb{Z}/p^r\mathbb{Z})/C_{\mathrm{sp}}(p^r)$ have the form:*

$$\alpha(u,v) = \begin{pmatrix} 1+uv & u \\ v & 1 \end{pmatrix}, \quad \beta(u,v) = \begin{pmatrix} u & -1 \\ 1 & pv \end{pmatrix},$$

where $u, v \in \mathbb{Z}/p^r\mathbb{Z}$. A set of coset representatives $\mathrm{GL}_2(\mathbb{Z}/p^r\mathbb{Z})/C_{\mathrm{sp}}^+(p^r)$ are the same as the ones above under the following identifications:

$$\alpha(u,v) \sim \begin{cases} \alpha((1+uv)v^{-1}, -v) & \text{if } v \text{ is invertible mod } p^r, \\ \beta(u, -\frac{v}{p}(1+uv)^{-1}) & \text{if } v \text{ is not invertible mod } p^r. \end{cases}$$

In particular, the identification extends to the above bijection, thereby identifying pairs of tuples consisting of α's and β's for each prime dividing the level.

PROOF. The number of coset representatives is equal to $[\mathrm{GL}_2(\mathbb{Z}/p^r\mathbb{Z}) : C_{\mathrm{sp}}(p^r)]$ and that $\alpha(u',v')^{-1}\alpha(u,v), \beta(u',v')^{-1}\beta(u,v), \alpha(u',v')^{-1}\beta(u,v) \in C_{\mathrm{sp}}(p^r)$ or $C_{\mathrm{sp}}^+(p^r)$ if and only if $u' \equiv u \pmod{p^r}$ and $v' \equiv v \pmod{p^r}$. The identification agrees with the involution defined above. □

With this setup, we calculate the quantities $\varepsilon_2, \varepsilon_3$, and ε_∞ in the genus formula.

PROPOSITION 3.2. *Let ε_∞ and ε_∞^+ denote the number of cusps in $X_{\mathrm{sp}}(N)$ and $X_{\mathrm{sp}}^+(N)$, respectively. Then we have*

$$\varepsilon_\infty = N \prod_{p|N}\left(1 + \frac{1}{p}\right),$$
$$\varepsilon_\infty^+ = \varepsilon_\infty/2.$$

PROOF. Note that $\mathrm{SL}_2(\mathbb{Z})_\infty = \langle\left(\begin{smallmatrix} 1 & 1 \\ 0 & 1 \end{smallmatrix}\right)\rangle$ and denote $\Gamma_{\mathrm{sp}}(N)$ to be the lift of $C_{\mathrm{sp}}(N)$ in $\mathrm{SL}_2(\mathbb{Z})$. Let $g \in \mathrm{SL}_2(\mathbb{Z})/\Gamma_{\mathrm{sp}}(N)$ be such that $\left(\begin{smallmatrix} 1 & a \\ 0 & 1 \end{smallmatrix}\right) \in g\Gamma_{\mathrm{sp}}(N)g^{-1}$, for some $a \in \mathbb{Z}_{>0}$. It follows that $a = N$ and by [**Shi71**, Prop 1.37], the ramification index for each cusp is N. Since the map $X_{\mathrm{sp}}(N) \to X(1)$ has degree $N^2\prod_p(1+1/p)$, we have $\varepsilon_\infty = N\prod_p(1+1/p)$. Furthermore, the morphism $X_{\mathrm{sp}}(N) \to X_{\mathrm{sp}}^+(N)$ is of degree 2 and unramified over the cusps. It follows that $\varepsilon_\infty^+ = \varepsilon_\infty/2$. For $N = 2$, one can work directly with the definition of ε_2^+ as the number of cosets in $\mathrm{SL}_2(\mathbb{Z}/2\mathbb{Z})$ fixed by the right action of $\left(\begin{smallmatrix} 0 & -1 \\ 1 & 0 \end{smallmatrix}\right)$ and find that there are 2 cusps. □

PROPOSITION 3.3. *Let* ε_3 *and* ε_3^+ *denote the number of elliptic points of order 3 in* $X_{\mathrm{sp}}(N)$ *and* $X_{\mathrm{sp}}^+(N)$*, respectively. Then,*

$$\varepsilon_3 = \begin{cases} 0 & \text{if } 2 \mid N \text{ or } 3 \mid N, \\ \prod_{p \mid N} \left(1 + \left(\frac{-3}{p}\right)\right) & \text{otherwise.} \end{cases}$$
$$\varepsilon_3^+ = \varepsilon_3/2.$$

PROOF. Let $\rho = e^{2\pi i/3}$ be a third root of unity. Note that $\mathrm{SL}_2(\mathbb{Z})_\rho = \langle \left(\begin{smallmatrix} 0 & -1 \\ 1 & -1 \end{smallmatrix}\right) \rangle$. By [**Shi71**, Prop 1.37], the elliptic points of order 3 on $X_{\mathrm{sp}}(N)$ are the points in the inverse image $f^{-1}(\rho)$, where $f : \mathbb{H}^*/\Gamma_{\mathrm{sp}}(N) \to \mathbb{H}^*/\mathrm{SL}_2(\mathbb{Z})$, with ramification index 1. By the same proposition, there exists $g \in \mathrm{SL}_2(\mathbb{Z})/\Gamma_{\mathrm{sp}}(N)$ such that $\mathrm{SL}_2(\mathbb{Z})_\rho \le g\Gamma_{\mathrm{sp}}(N)g^{-1}$. Therefore, it is sufficient to find the number of coset representatives γ such that $\gamma^{-1}\left(\begin{smallmatrix} 0 & -1 \\ 1 & -1 \end{smallmatrix}\right)\gamma \in C_{\mathrm{sp}}(N)$ and $C_{\mathrm{sp}}^+(N)$, respectively.

We begin with prime power levels $N = p^r$. We have the following matrices:

(1)
$$\alpha(u,v)^{-1}\left(\begin{smallmatrix} 0 & -1 \\ 1 & -1 \end{smallmatrix}\right)\alpha(u,v) = \begin{pmatrix} -u^2v + uv - u - v & -u^2 + u - 1 \\ u^2v^2 - uv^2 + 2uv + v^2 - v + 1 & u^2 - uv + u + v - 1 \end{pmatrix},$$
(2)
$$\beta(u,v)^{-1}\left(\begin{smallmatrix} 0 & -1 \\ 1 & -1 \end{smallmatrix}\right)\beta(u,v) = (1 + puv)^{-1}\begin{pmatrix} u - pv - 1 & -p^2v^2 - pv - 1 \\ u^2 - u + 1 & -puv - u + pv \end{pmatrix}.$$

For Equation (1) to be an element of $C_{\mathrm{sp}}(N)$, we require that the off-diagonal entries to be congruent to 0 modulo N:

$$-u^2 + u - 1 \equiv 0 \pmod{N}, \tag{3}$$
$$-v^2(-u^2 + u - 1) + 2uv - v + 1 \equiv 0 \pmod{N}. \tag{4}$$

Equation (3) has two solutions if $p \equiv 1 \pmod 3$, one solution if $N = 3$, and zero solutions otherwise. There exists a unique v to Equation (4) if $(2u - 1) \not\equiv 0 \pmod{N}$. For a matrix from Equation (2) to be an element of $C_{\mathrm{sp}}(N)$, we require that:

$$-p^2v^2 - pv - 1 \equiv 0 \pmod{N}, \tag{5}$$
$$u^2 - u + 1 \equiv 0 \pmod{N}. \tag{6}$$

However, this does not yield any solution since p is not invertible modulo N in Equation (5).

Writing the level N as a product of primes, we have two solutions for each prime p dividing N satisfying $p \equiv 1 \pmod 3$ and this gives ε_3. By the same argument in calculating ε_∞^+, the morphism $X_{\mathrm{sp}}(N) \to X_{\mathrm{sp}}^+(N)$ is of degree 2 and unramified over the elliptic points of order 3 and therefore, we have $\varepsilon_3^+ = \varepsilon_3/2$. ∎

PROPOSITION 3.4. *Let ε_2 and ε_2^+ denote the number of elliptic points of order 2 in $X_{\mathrm{sp}}(N)$ and $X_{\mathrm{sp}}^+(N)$, respectively. Then we have*

$$\varepsilon_2 = \begin{cases} 0 & \text{if } 2 \mid N, \\ \prod_{p \mid N} \left(1 + \left(\frac{-1}{p}\right)\right) & \text{otherwise.} \end{cases}$$

$$\varepsilon_2^+ = \frac{\varepsilon_2}{2} + \left(\frac{N}{2} \prod_{\substack{p \mid N, \\ p \equiv 1 \bmod 4}} \left(1 - \frac{1}{p}\right) \prod_{\substack{p \mid N, \\ p \equiv 3 \bmod 4}} \left(1 + \frac{1}{p}\right)\right),$$

PROOF. Note that $\mathrm{SL}_2(\mathbb{Z})_i = \langle \left(\begin{smallmatrix} 0 & -1 \\ 1 & 0 \end{smallmatrix}\right) \rangle$. By [**Shi71**, Prop 1.37], the elliptic points of order 2 on $X_{\mathrm{sp}}(N)$ are the points in the inverse image $f^{-1}(i)$, where $f : \mathbb{H}^*/\Gamma_{\mathrm{sp}}(N) \to \mathbb{H}^*/\,\mathrm{SL}_2(\mathbb{Z})$, with ramification index 1. By the same proposition, there exists $g \in \mathrm{SL}_2(\mathbb{Z})/\Gamma_{\mathrm{sp}}(N)$ such that $\mathrm{SL}_2(\mathbb{Z})_i \leq g\Gamma_{\mathrm{sp}}(N)g^{-1}$. Therefore, it is sufficient to find the number of coset representatives γ such that $\gamma^{-1}\left(\begin{smallmatrix} 0 & -1 \\ 1 & 0 \end{smallmatrix}\right)\gamma \in C_{\mathrm{sp}}(N)$ and $C_{\mathrm{sp}}^+(N)$, respectively.

We begin with prime power levels $N = p^r$. We have the following matrices:

$$\alpha(u,v)^{-1}\left(\begin{smallmatrix} 0 & -1 \\ 1 & 0 \end{smallmatrix}\right)\alpha(u,v) = \begin{pmatrix} -u^2v - u - v & -u^2 - 1 \\ u^2v^2 + 2uv + v^2 + 1 & u^2v + u + v \end{pmatrix}, \tag{7}$$

$$\beta(u,v)^{-1}\left(\begin{smallmatrix} 0 & -1 \\ 1 & 0 \end{smallmatrix}\right)\beta(u,v) = (1 + puv)^{-1} \begin{pmatrix} u - pv & -p^2v^2 - 1 \\ u^2 + 1 & pv - u \end{pmatrix}. \tag{8}$$

Following a similar argument in Proposition 3.3, we establish congruence equations from the off-diagonal entries and count the number of solutions. Writing the composite level N as a product of prime powers, we obtain ε_2.

For ε_2^+, we begin with a prime power level $N = p^r$. For matrices from Equations ((7), (8)) to be elements of $C_{\mathrm{sp}}^+(N)$, in addition to congruence equations obtained from the off-diagonal entries, we also require congruence conditions from the diagonal entries, giving the following equations:

$$u^2v + u + v \equiv 0 \pmod{N}, \tag{9}$$

$$u - pv \equiv 0 \pmod{N}. \tag{10}$$

In Equation (9), note that $v \equiv -u(u^2+1)^{-1} \pmod{N}$. We observe that $u^2 + 1$ is not invertible when $p \equiv 1 \pmod 4$ and is always invertible when $p \equiv 3 \pmod 4$. It follows that there are p^r solutions when $p \equiv 3 \pmod 4$ and $p^r - 2p^{r-1}$ solutions when $p \equiv 1 \pmod 4$. When $p = 2$, $u^2 + 1$ is not invertible modulo N half of the time, so there are $2^r/2 = 2^{r-1}$ solutions. In Equation (10), there are p^{r-1} solutions, with no restrictions on p.

The identification from Lemma 3.1 pairs the two solutions from ε_2 via

$$\alpha(u, -u(u^2+1)^{-1}) \sim \begin{cases} \alpha(-u^{-1}, u(u^2+1)^{-1}) & \text{if } u \text{ is invertible,} \\ \beta(u, u/p) & \text{if } u \text{ is not invertible.} \end{cases}$$

This recovers the formula in [**DLM22**] for prime power levels $N = p^r$:

$$\varepsilon_2^+ = \begin{cases} 2^{r-1} & \text{if } p = 2, \\ 1 + (p^r - p^{r-1})/2 & \text{if } p \equiv 1 \pmod 4, \\ (p^r + p^{r-1})/2 & \text{if } p \equiv 3 \pmod 4. \end{cases}$$

The involution ω acts on $\mathrm{GL}_2(\mathbb{Z}/N\mathbb{Z})/C_{\mathrm{sp}}(N)$ and $\prod_i \mathrm{GL}_2(\mathbb{Z}/p_i^{e_i}\mathbb{Z})/C_{\mathrm{sp}}(p_i^{e_i})$, thus identifying pairs of solutions in $\prod_i \mathrm{GL}_2(\mathbb{Z}/p_i^{e_i}\mathbb{Z})/C_{\mathrm{sp}}(p_i^{e_i})$. Dividing the above analysis into two parts, one coming from $C_{\mathrm{sp}}(N)$ and the other from $C_{\mathrm{sp}}^+(N)$, we obtain the formula for ε_2^+. □

Using the multiplicative relations from [**DLM22**], one can produce formulas for ε_2^*, ε_3^*, and ε_∞^* of $X_{\mathrm{sp}}^*(N)$.

COROLLARY 3.5. *Let ε_2^*, ε_3^*, ε_∞^* denote the number of elliptic points of orders 2, 3, and cusps in $X_{\mathrm{sp}}^*(N)$, respectively. Then we have*

$$\varepsilon_\infty^* = \begin{cases} \frac{N}{3\cdot 2^{\omega(N)-2}} \prod_{p\mid N} (1+1/p) & \text{if } 2\| N, \\ \frac{N}{2^{\omega(N)}} \prod_{p\mid N} (1+1/p) & \text{otherwise,} \end{cases}$$

$$\varepsilon_3^* = \begin{cases} 1 & p \equiv 1 \pmod 3 \text{ for all } p \mid N, \\ 0 & \text{otherwise,} \end{cases}$$

$$\varepsilon_2^* = 2^{\nu_2(N)-1} \prod_{\substack{p\mid N,\\ p\equiv 1 \bmod 4}} \left(1 + \frac{p^{\nu_p(N)-1}(p-1)}{2}\right) \prod_{\substack{p\mid N,\\ p\equiv 3 \bmod 4}} \frac{p^{\nu_p(N)-1}(p+1)}{2}.$$

4. Non-split Cartan and normalizer of the non-split Cartan

The non-split Cartan subgroup can also be defined explicitly. Let $R = \mathbb{Z}[\alpha]$ be a quadratic order such that α satisfies an irreducible monic polynomial $X^2 - uX + v \in \mathbb{Z}[X]$. Suppose that the discriminant of R is coprime to N and that every prime dividing N is inert in R. Then $A = R/NR$ is free $\mathbb{Z}/N\mathbb{Z}$-module of rank 2. For $x \in A$, the multiplication-by-x map induces a map $A^\times \hookrightarrow \mathrm{GL}_2(\mathbb{Z}/N\mathbb{Z})$ and the image is a non-split Cartan subgroup, denoted $C_{\mathrm{ns}}(N)$.

For the normalizer of $C_{\mathrm{ns}}(N)$ in $\mathrm{GL}_2(\mathbb{Z}/N\mathbb{Z})$, there is an induced ring automorphism of order 2 from R in $\mathrm{GL}_2(\mathbb{Z}/N\mathbb{Z})$, denoted by S_N. Viewing $C_{\mathrm{ns}}(N)$ as $(\mathbb{Z}/N\mathbb{Z})[\alpha]^\times$, since α is a root of $X^2 - uX + v$, $S_N(\alpha)$ also satisfies this polynomial and is therefore equal to $u - \alpha$. This gives an action of S_N on $(\mathbb{Z}/N\mathbb{Z})[\alpha]$ in the expected manner:

$$\begin{aligned} 1 &\mapsto 1 \pmod N, \\ \alpha &\mapsto u - \alpha \pmod N. \end{aligned}$$

The normalizer of the non-split Cartan subgroup, denoted $C_{\mathrm{ns}}^+(N)$, is the group generated by the non-split Cartan subgroup and this involution $\langle C_{\mathrm{ns}}(N), S_N\rangle$.

We denote the modular curves associated to $C_{\mathrm{ns}}(N)$ and $C_{\mathrm{ns}}^+(N)$ by $X_{\mathrm{ns}}(N)$ and $X_{\mathrm{ns}}^+(N)$, respectively. There are natural finite morphisms ϕ_1 and ϕ_2 such that

$$X_{\mathrm{ns}}(N) \xrightarrow{\phi_1} X_{\mathrm{ns}}^+(N) \xrightarrow{\phi_2} X(1),$$

where $\deg(\phi_1) = 2$ and $\deg(\phi_2) = N\varphi(N)/2$. Note that $\deg(\phi_2) = [\mathrm{GL}_2(\mathbb{Z}/N\mathbb{Z}) : C_{\mathrm{ns}}^+(N)]$.

Let $C'_{\rm ns}(N) := C_{\rm ns}(N) \cap \mathrm{SL}_2(\mathbb{Z}/N\mathbb{Z})$ and let $C^{+\prime}_{\rm ns}(N) := C^+_{\rm ns}(N) \cap \mathrm{SL}_2(\mathbb{Z}/N\mathbb{Z})$, where $C'_{\rm ns}(N)$ and $C^{+\prime}_{\rm ns}(N)$ can be identified with the following groups:

$$\begin{aligned} C'_{\rm ns}(N) =& \{M_y : y \in (\mathbb{Z}/N\mathbb{Z})[\alpha]^\times, N(\alpha) = 1\}, \\ C^{+\prime}_{\rm ns}(N) =& \{M_y : y \in (\mathbb{Z}/N\mathbb{Z})[\alpha]^\times, N(\alpha) = 1\} \cup \\ & \{M_y \circ S_N : y \in (\mathbb{Z}/N\mathbb{Z})[\alpha]^\times, N(\alpha) = -1\}, \end{aligned}$$

where M_y is the multiplication-by-y map on $(\mathbb{Z}/N\mathbb{Z})[\alpha]$, $N(x) = x\bar{x}$ is the norm map, and $\overline{a + b\alpha} = a + b(u - \alpha)$. For every element $a \in (\mathbb{Z}/N\mathbb{Z})^\times$ or $a \in (\mathbb{Z}/N\mathbb{Z})^\times/\pm 1$, choose $y_a \in (\mathbb{Z}/N\mathbb{Z})[\alpha]^\times$ such that $N(y_a) = a$. We define the sets $\mathcal{Y}(N)$ and $\mathcal{Y}_\pm(N)$ as follows:

$$\begin{aligned} \mathcal{Y}(N) &= \{y_a \in (\mathbb{Z}/N\mathbb{Z})[\alpha]^\times : N(y_a) = a, a \in (\mathbb{Z}/N\mathbb{Z})^\times\}, \\ \mathcal{Y}_\pm(N) &= \{y_a \in (\mathbb{Z}/N\mathbb{Z})[\alpha]^\times : N(y_a) = a, a \in (\mathbb{Z}/N\mathbb{Z})^\times / \pm 1\}. \end{aligned}$$

The following proposition provides a matrix representation for a set of coset representatives of $\Gamma_{\rm ns}$ and $\Gamma^+_{\rm ns}$ in $\mathrm{SL}_2(\mathbb{Z})$.

PROPOSITION 4.1 (Proposition 6.2 and 6.3, [**Bar10**]). *A set of coset representatives of $C'_{\rm ns}(N)$ (resp. $C^{+\prime}_{\rm ns}(N)$) in $SL_2(\mathbb{Z}/N\mathbb{Z})$ can be represented as linear maps that transform the basis $\{1, \alpha\}$ as*

$$\begin{aligned} 1 &\mapsto y^{-1}, \\ \alpha &\mapsto \bar{y}(\alpha + x), \end{aligned}$$

where $x \in \mathbb{Z}/N\mathbb{Z}$ and $y \in \mathcal{Y}(N)$ (resp. $y \in \mathcal{Y}_\pm(N)$).

PROOF. We have the following equality

$$[\mathrm{SL}_2(\mathbb{Z}) : \Gamma_{\rm ns}(N)] = [\mathrm{SL}_2(\mathbb{Z}/N\mathbb{Z}) : C'_{\rm ns}(N)] = [\mathrm{GL}_2(\mathbb{Z}/N\mathbb{Z}) : C_{\rm ns}(N)]$$

and a similar statement holds for $C^+_{\rm ns}(N)$. The number of cosets of $C_{\rm ns}(N)$ and $C^+_{\rm ns}(N)$ in $\mathrm{GL}_2(\mathbb{Z}/N\mathbb{Z})$ are $N\varphi(N)$ and $N\varphi(N)/2$, respectively, and $|\mathcal{Y}(N)| = \varphi(N)$ and $|\mathcal{Y}_\pm(N)| = \varphi(N)/2$. It remains to show that, under the identification above, the linear maps represent different cosets.

Suppose we have two elements in the same coset of $\mathrm{SL}_2(\mathbb{Z}/N\mathbb{Z})/C'_{\rm ns}(N)$:

$$\begin{aligned} 1 &\mapsto y^{-1}, & 1 &\mapsto y'^{-1}, \\ \alpha &\mapsto \bar{y}(\alpha + x), & \alpha &\mapsto \overline{y'}(\alpha + x'). \end{aligned}$$

Then we have $N(y) = N(y')$ and since $y, y' \in \mathcal{Y}(N)$, it follows that $y = y'$, which implies $x = x'$.

In the case of $C^+_{\rm ns}(N)$, observe that the linear maps that transform the basis $\{1, \alpha\}$ have two possibilities:

$$\begin{aligned} 1 &\mapsto y & 1 &\mapsto \bar{y} \\ \alpha &\mapsto y\alpha & \alpha &\mapsto \bar{y}\bar{\alpha}. \end{aligned}$$

where $y \in (\mathbb{Z}/N\mathbb{Z})[\alpha]^\times$ with $N(y) = 1$ and $N(y) = -1$, respectively. Suppose we have two elements in the same coset of $\mathrm{SL}_2(\mathbb{Z}/NZ)/C^+_{\rm ns}(N)$:

$$\begin{aligned} 1 &\mapsto y^{-1}, & 1 &\mapsto y'^{-1}, \\ \alpha &\mapsto \bar{y}(\alpha + x), & \alpha &\mapsto \overline{y'}(\alpha + x'). \end{aligned}$$

The same argument shows that we have $N(y) = N(y')$ and since $y, y' \in \mathcal{Y}_\pm(N)$, it follows that $y = y'$, which implies $x = x'$. □

With this setup, we begin calculating the invariants $\varepsilon_2, \varepsilon_3$, and ε_∞ in the genus formula.

PROPOSITION 4.2. *Let ε_∞ and ε_∞^+ denote the number of cusps in $X_{\rm ns}(N)$ and $X_{\rm ns}^+(N)$, respectively. Then we have*

$$\varepsilon_\infty = \varphi(N),$$
$$\varepsilon_\infty^+ = \begin{cases} 1 & \text{if } N = 2, \\ \varphi(N)/2 & \text{otherwise.} \end{cases}$$

PROOF. Note that $\mathrm{SL}_2(\mathbb{Z})_\infty = \langle \left(\begin{smallmatrix} 1 & 1 \\ 0 & 1 \end{smallmatrix}\right) \rangle$. Let $g \in \mathrm{SL}_2(\mathbb{Z})/\Gamma_{\rm ns}(N)$ such that $\left(\begin{smallmatrix} 1 & a \\ 0 & 1 \end{smallmatrix}\right) \in g\Gamma_{\rm ns}(N)g^{-1}$, for some $a \in \mathbb{Z}_{>0}$, which stabilizes ∞. The element $g^{-1}\left(\begin{smallmatrix} 1 & a \\ 0 & 1 \end{smallmatrix}\right)g \in \Gamma_{\rm ns}(N)$ can be represented in two ways: one in the manner of Proposition 4.1, and the other as a multiplication-by-k matrix, with $k \in (\mathbb{Z}/N\mathbb{Z})[\alpha]^\times$, giving the following system of equations:

$$y^{-1} \equiv y^{-1}k \pmod N,$$
$$\bar{y}(\alpha + x) \equiv \bar{y}(\alpha + x)k \pmod N,$$

where $x \in \mathbb{Z}/N\mathbb{Z}, y \in \mathcal{Y}(N), k \in (\mathbb{Z}/N\mathbb{Z})[\alpha]^\times$, and $N(k) = 1$. It follows that $k = 1$ and therefore $a = 0 \pmod N$. By [**Shi71**, Prop 1.37], the ramification index for each cusp of $X_{\rm ns}(N)$ is N and since the degree of the morphism $X_{\rm ns}(N) \to X(1)$ is $N\varphi(N)$, we have $\varepsilon_\infty = \varphi(N)$.

The same argument for $X_{\rm ns}^+(N)$ shows that the ramification index for each cusp is N and since the degree of the morphism $X_{\rm ns}^+(N) \to X(1)$ is $N\varphi(N)/2$, we have $\varepsilon_\infty^+ = \varphi(N)/2$. When $N = 2$, $\varepsilon_\infty^+ = 1$. $\square$

PROPOSITION 4.3. *Let ε_3 and ε_3^+ denote the number of elliptic points of order 3 in $X_{\rm ns}(N)$ and $X_{\rm ns}^+(N)$, respectively. Then we have*

$$\varepsilon_3 = \begin{cases} 2^{\omega(N)} & p \equiv 2 \pmod 3 \text{ for all } p \mid N, \\ 0 & \text{otherwise,} \end{cases}$$
$$\varepsilon_3^+ = \varepsilon_3/2.$$

PROOF. Let $\rho = e^{2\pi i/3}$ be a third root of unity. Note that $\mathrm{SL}_2(\mathbb{Z})_\rho = \langle \left(\begin{smallmatrix} 0 & -1 \\ 1 & 1 \end{smallmatrix}\right) \rangle$. By [**Shi71**, Prop 1.37], the elliptic points of order 3 on $X_{\rm ns}(N)$ are the points in the inverse image $f^{-1}(\rho)$, where $f : \mathbb{H}^*/\Gamma_{\rm ns}(N) \to \mathbb{H}^*/\mathrm{SL}_2(\mathbb{Z})$, with ramification index 1. By the same proposition, there exists $g \in \mathrm{SL}_2(\mathbb{Z})/\Gamma_{\rm ns}(N)$ such that $\mathrm{SL}_2(\mathbb{Z})_\rho \leq g^{-1}\Gamma_{\rm ns}(N)g$. Therefore, ε_3 is equal to the number of coset representatives g such that

$$g \begin{pmatrix} 0 & -1 \\ 1 & 1 \end{pmatrix} g^{-1} = \gamma,$$

for some $\gamma \in \Gamma_{\rm ns}(N)$. By Proposition 4.1, we can represent γ as a multiplication-by-k matrix for some $k \in (\mathbb{Z}/N\mathbb{Z})[\alpha]^\times$, and we obtain the following system of equations:

$$\bar{y}(\alpha + x) \equiv y^{-1}k \pmod N, \tag{11}$$
$$\bar{y}(\alpha + x) - y^{-1} \equiv \bar{y}(\alpha + x)k \pmod N, \tag{12}$$

where $x \in \mathbb{Z}/N\mathbb{Z}, y \in \mathcal{Y}(N), k \in (\mathbb{Z}/N\mathbb{Z})[\alpha]^\times$, and $N(k) = 1$.

We compute the number of solutions to the above equations, following [**Bar10**, Lem 7.7]. Suppose that x, y, k satisfy the equations above. Equation (11) can be rearranged as

$$N(y)\alpha + N(y)x \equiv k \pmod{N}.$$

Since $N(y) \in (\mathbb{Z}/N\mathbb{Z})^\times$ and $k \notin (\mathbb{Z}/N\mathbb{Z})^\times$, equation (11) can be substituted into equation (12) as follows,

$$y^{-1}k - y^{-1} \equiv y^{-1}k^2 \pmod{N},$$

which implies that

$$k^2 - k + 1 \equiv 0 \pmod{N}. \tag{13}$$

We can reduce equation (13) modulo the primes p dividing N. Hensel's lemma and the Chinese Remainder Theorem imply that there are two nontrivial solutions for each prime dividing N. Since $\{1, \alpha\}$ is a basis for $(\mathbb{Z}/N\mathbb{Z})[\alpha]$, there are two unique solutions $(x, y) \in \mathbb{Z}/p^r\mathbb{Z} \times \mathcal{Y}(p^r)$ for the equation

$$N(y)\alpha + N(y)x \equiv k \pmod{p^r},$$

for each prime $p \equiv 2 \pmod 3$ dividing N and $p^r \| N$. If there is a prime $p \not\equiv 2 \pmod 3$, then there will be no solutions. This yields the formula for ε_3. Since the morphism $X_{\mathrm{ns}}(N) \to X_{\mathrm{ns}}^+(N)$ is of degree 2 and unramified over the elliptic points of order 3, we obtain $\varepsilon_3^+ = \varepsilon_3/2$. □

PROPOSITION 4.4. *Let ε_2 and ε_2^+ denote the number of elliptic points of order 2 in $X_{\mathrm{ns}}(N)$ and $X_{\mathrm{ns}}^+(N)$, respectively. Then we have*

$$\varepsilon_2 = \prod_{p|N}\left(1 - \left(\frac{-1}{p}\right)\right),$$

$$\varepsilon_2^+ = \frac{1}{2}\sum_{p|N}\left(1 - \left(\frac{-1}{p}\right)\right) + \left(\frac{N}{2}\prod_{p|N}\left(1 + \frac{1}{p}\right) - \#S\right),$$

where $S = \{a + b\alpha \in (\mathbb{Z}/N\mathbb{Z})[\alpha]^\times/\pm 1 : N(a + b\alpha) = -1, gcd(b, N) > 1\}$.

PROOF. Note that $\mathrm{SL}_2(\mathbb{Z})_i = \left\langle \left(\begin{smallmatrix} 0 & -1 \\ 1 & 0 \end{smallmatrix}\right)\right\rangle$ and the elliptic points of order 2 on $X_{\mathrm{ns}}(N)$ are the points in the inverse image $f^{-1}(i)$, where $f : \mathbb{H}^*/\Gamma_{\mathrm{ns}}(N) \to \mathbb{H}^*/\mathrm{SL}_2(\mathbb{Z})$, with ramification index 1. Therefore, by [**Shi71**, Prop 1.37], ε_2 is equal to the number of coset representatives $g \in \mathrm{SL}_2(\mathbb{Z})/\Gamma_{\mathrm{ns}}(N)$ such that the equality

$$g^{-1}\begin{pmatrix} 0 & -1 \\ 1 & 0 \end{pmatrix} g = \gamma,$$

holds for some $\gamma \in \Gamma_{\mathrm{ns}}(N)$. Considering the above equality as linear maps using Proposition 4.1 we have the equations:

$$\bar{y}(\alpha + x) \equiv y^{-1}k \pmod{N}, \tag{14}$$

$$-y^{-1} \equiv \bar{y}(\alpha + x)k \pmod{N}, \tag{15}$$

where $x \in \mathbb{Z}/N\mathbb{Z}, y \in \mathcal{Y}(N), k \in (\mathbb{Z}/N\mathbb{Z})[\alpha]^\times$, and $N(k) = 1$.

We compute the number of solutions to the above equations, following [**Bar10**, Lem 7.8]. Suppose that x, y, k satisfy the above equations. Equation (14) can be rearranged and also substituted into equation (15) to get the following:

$$N(y)\alpha + N(y)x \equiv k \pmod{N}, \tag{16}$$
$$-y^{-1} \equiv y^{-1}k^2 \pmod{N}. \tag{17}$$

Since $N(y) \in (\mathbb{Z}/N\mathbb{Z})^\times$, equation (16) implies that $k \notin \mathbb{Z}/N\mathbb{Z}$. From equation (17), we obtain

$$-1 \equiv k^2 \pmod{N}.$$

By the Chinese Remainder Theorem and Hensel's lemma, since k is not an element of $(\mathbb{Z}/N\mathbb{Z})^\times$ and hence of $(\mathbb{Z}/p\mathbb{Z})^\times$ for any prime p dividing N, we have $p \equiv 3 \pmod{4}$ for all primes p dividing N, and there are two solutions for the system of equations. This proves the ε_2 part of the proposition.

Recall that,

$$\begin{aligned} C_{\text{ns}}^{+\prime}(N) =& \{M_y : y \in (\mathbb{Z}/N\mathbb{Z})[\alpha]^\times, N(\alpha) = 1\} \cup \\ & \{M_y \circ S_N : y \in (\mathbb{Z}/N\mathbb{Z})[\alpha]^\times, N(\alpha) = -1\}. \end{aligned}$$

Therefore, for the normalizer of a non-split Cartan subgroup, the same argument as above gives rise to two systems of equations, corresponding to the norm $+1$ and -1 elements:

$$\begin{aligned} \bar{y}(\alpha + x) &\equiv y^{-1}k \pmod{N}, & \bar{y}(\alpha + x) &\equiv \bar{y}^{-1}k \pmod{N}, \\ -y^{-1} &\equiv \bar{y}(\alpha + x)k \pmod{N}, & -y^{-1} &\equiv y(\bar{\alpha} + x)k \pmod{N} \end{aligned}$$

where $x \in \mathbb{Z}/N\mathbb{Z}, y \in \mathcal{Y}_\pm(N), k \in (\mathbb{Z}/N\mathbb{Z})[\alpha]^\times$, and $N(k) = +1$ and -1, respectively. Notice that the system of equations on the left are the same as the systems of equations in the ε_2 case. Since the covering map $X_{\text{ns}}^+(N) \to X_{\text{ns}}(N)$ is of degree 2 and is unramified over these points, the norm $+1$ elements contribute $\sum_{p|N} \frac{1-\left(\frac{-1}{p}\right)}{2}$ to ε_2^+.

For the norm -1 elements, since $N(k) = -1$, the equations above on the right hand side are conjugate. Therefore, we want to know the number of solutions $(x, y) \in \mathbb{Z}/N\mathbb{Z} \times \mathcal{Y}_\pm(N)$ for which there exists a $k \in (\mathbb{Z}/N\mathbb{Z})[\alpha]^\times$ such that the following equation holds

$$\bar{y}(\alpha + x) \equiv \bar{y}^{-2}k \pmod{N}.$$

Taking norms and rearranging terms, we obtain

$$N(\alpha + x) \equiv -N(y)^{-2} \pmod{N}.$$

Since the norm map $N : \mathcal{Y}_\pm(N) \to (\mathbb{Z}/N\mathbb{Z})^\times / \pm 1$ is an isomorphism, we need to count the number of $x \in \mathbb{Z}/N\mathbb{Z}$ and $v \in (\mathbb{Z}/N\mathbb{Z})^\times / \pm 1$ such that the following equality holds

$$N(\alpha + x) \equiv -v^2 \pmod{N}.$$

Now, we write $h = \frac{\alpha + x}{v} \in (\mathbb{Z}/N\mathbb{Z})[\alpha]^\times / \pm 1$, with $N(h) \equiv -1 \pmod{N}$. The problem at hand is to find how many such h's exist. Note that the norm map

$$N : (\mathbb{Z}/N\mathbb{Z})[\alpha]^\times / \pm 1 \to (\mathbb{Z}/N\mathbb{Z})^\times$$

is surjective and its kernel has order

$$\left(\frac{1}{2}N^2\prod_{p|N}\left(1-\frac{1}{p^2}\right)\right)/\varphi(N)=\frac{1}{2}\prod_{p|N}p^{r-1}(p+1).$$

Therefore, the number of elements in $(\mathbb{Z}/N\mathbb{Z})[\alpha]^\times/\pm 1$ of norm -1 is equal to $\frac{1}{2}\prod_{p|N}p^{r-1}(p+1)$. Note that this count also includes elements whose α-coefficient is not coprime to N. We exclude these elements in the next few paragraphs.

Let $\mathcal{P}\in\mathbb{P}(N):=\text{PowerSet}(\{p: p \text{ prime}, p\mid N\})$ and define

$$\begin{aligned}\Theta_{\mathcal{P}}:(\mathbb{Z}/N\mathbb{Z})[\alpha]^\times/\pm 1 &\to (\mathbb{Z}/N\mathbb{Z})^\times\times\prod_{p\in\mathcal{P}}\frac{(\mathbb{Z}/p\mathbb{Z})[\alpha]^\times}{(\mathbb{Z}/p\mathbb{Z})^\times},\\ a+b\alpha &\mapsto (N(a+b\alpha),(a_p+b_p\alpha)_p),\end{aligned}$$

where a_p,b_p are the representatives of a,b under the reduction modulo p and projection maps.

Then the image of $\Theta_{\mathcal{P}}$ is given by

$$\begin{aligned}&\operatorname{Im}(\Theta_{\mathcal{P}})\\ &\quad=\{(b,(c_p)_p): b\equiv N(c_p)v_p^2 \pmod p \text{ for some } v_p\in(\mathbb{Z}/p\mathbb{Z})^\times,\ \text{for all } p\in\mathcal{P}\},\end{aligned}$$

which has cardinality $\varphi(N)\prod_{p\in\mathcal{P}}\frac{p+1}{2}\cdot 2^\delta$, where $\delta=1$ when $2\in\mathcal{P}$ and $\delta=0$ otherwise. The cardinality of the kernel can be calculated as

$$\begin{aligned}\#\ker(\Theta_{\mathcal{P}})&=\#C_{\text{ns}}^+(N)/\#\operatorname{Im}(\Theta_{\mathcal{P}}),\\ &=\frac{1}{2}N^2\left(\prod_{p|N}\left(1-\frac{1}{p^2}\right)\right)/\left(\varphi(N)\prod_{p\in\mathcal{P}}\frac{p+1}{2}\cdot 2^\delta\right),\\ &=2^{\#\mathcal{P}-1-\delta}N\prod_{p|N}\left(1+\frac{1}{p}\right)/\prod_{p\in\mathcal{P}}(p+1).\end{aligned}$$

We are interested in

$$\Theta_{\mathcal{P}}^{-1}(-1,(1,\ldots,1))=\{a+b\alpha: N(a+b\alpha)=-1,\ \gcd(b,p)>1,\ \text{for all } p\in\mathcal{P}\}.$$

Note that this set has the same cardinality as the kernel, and if there exists $p\equiv 3 \pmod 4\in\mathcal{P}$, then the preimage will be empty since the congruence condition

$$-1\equiv 1\cdot v_p^2 \pmod p$$

does not have a solution when $p\equiv 3 \pmod 4$. To summarize,

$$\begin{aligned}&\#\Theta_{\mathcal{P}}^{-1}(-1,(1,\ldots,1))\\ &\quad=\begin{cases}0 & \exists p\in\mathcal{P}, p\equiv 3 \pmod 4,\\ 2^{\#\mathcal{P}-1-\delta}N\prod_{p|N}\left(1+\frac{1}{p}\right)/\prod_{p\in\mathcal{P}}(p+1) & \text{otherwise.}\end{cases}\end{aligned}$$

By an inclusion-exclusion argument, one can compute the cardinality of the set

$$S:=\{a+b\alpha: N(a+b\alpha)=-1,\ \gcd(b,N)>1\}.$$

Let $\mathbb{P}(N)_k := \{\mathcal{P} \in \mathbb{P}(N) : \#\mathcal{P} = k\}$. Then,

$$\#S = \sum_{k=1}^{\omega(N)} (-1)^{k+1} \left(\sum_{\mathcal{P} \in \mathbb{P}(N)_k} \#\Theta_{\mathcal{P}}^{-1}(-1, \underbrace{(1, \ldots, 1)}_{k}) \right)$$

and we obtain

$$\varepsilon_2^+ = \underbrace{\sum_{p|N} \frac{1 - \left(\frac{-1}{p}\right)}{2}}_{\text{norm}=+1} + \underbrace{\frac{N}{2} \prod_{p|N} \left(1 + \frac{1}{p}\right) - \#S}_{\text{norm}=-1}.$$

□

REMARK 4.5. Propositions 4.2, 4.3, and 4.4 recover the formulas from [**Bar10**] when $N = p^r$.

Since the index, number of elliptic points and cusps are multiplicative in the level, one can produce formulas for ε_2^*, ε_3^*, and ε_∞^* of $X_{\text{ns}}^*(N)$ from prime power levels (see [**DLM22**]).

COROLLARY 4.6. *Let ε_2^*, ε_3^*, ε_∞^* denote the number of elliptic points of orders* 2, 3, *and cusps in $X_{\text{ns}}^*(N)$, respectively. Then we have*

$$\varepsilon_\infty^* = \begin{cases} \varphi(N)/2^{\omega(N)-1} & \text{if } 2 \| N, \\ \varphi(N)/2^{\omega(N)} & \text{otherwise,} \end{cases}$$

$$\varepsilon_3^* = \begin{cases} 1 & p \equiv 2 \pmod 3 \text{ for all } p \mid N, \\ 0 & \text{otherwise,} \end{cases}$$

$$\varepsilon_2^* = 2^{\nu_2(N)-1} \prod_{\substack{p|N, \\ p\equiv 1 \bmod 4}} \frac{p^{\nu_p(N)-1}(p-1)}{2} \prod_{\substack{p|N, \\ p\equiv 3 \bmod 4}} \left(1 + \frac{p^{\nu_p(N)-1}(p+1)}{2}\right).$$

REMARK 4.7. One can consider a different proof of the genus formula that follows from the ideas of Chen and de Smit-Edixhoven (see [**DLM22**, Theorem 3.8] and the references therein). The genus of the modular curve is equal to the dimension of its Jacobian; there is an isogeny between the Jacobian of the Cartan modular curve and the Jacobian of the Borel modular curve. Since the Borel case is well-known, this yields another proof for the genus formula of X_{sp}, X_{sp}^*, X_{ns}, and X_{ns}^*.

5. Modular Covers

For $N \geq 1$ an integer, $X_{\text{arith}}(N)$ is the modular curve X_H for $H \leq \mathrm{GL}_2(\widehat{\mathbb{Z}})$ given by the inverse image of $\{\left(\begin{smallmatrix} 1 & 0 \\ 0 & * \end{smallmatrix}\right)\} \leq \mathrm{GL}_2(\mathbb{Z}/N\mathbb{Z})$. As a moduli space, X_{arith} parametrizes isomorphism classes of triples (E, ϕ, P), where E is a generalized elliptic curve, P is a point of order N, and $\phi\colon E \to E'$ is a cyclic N-isogeny such that $E[N]$ is generated by P and $\ker \phi$. Alternatively, it parametrizes isomorphism classes of pairs (E, ψ), where E is a generalized elliptic curve and $\psi\colon \mu_N \times \mathbb{Z}/N\mathbb{Z} \to E[N]$ is a symplectic isomorphism. The curve $X_{\text{arith}}(N)$ corresponds to the subgroup of matrices H such that $H \cap \mathrm{SL}_2(\mathbb{Z}/N\mathbb{Z})$ is trivial and is a connected component of the Riemann surface $X(N)$. Therefore, the genus formula for $X_{\text{arith}}(N)$ is the same as the genus formula for $X(N)$.

In this section, we are interested in modular curves parametrizing full arithmetic level M-arithmetic structure with a compatible torsion point of level MN. Let E be an elliptic curve defined over a number field K of degree d. It is known that the torsion subgroup $\#E_{\text{tors}}(K)$ can be generated by two elements and is uniformly bounded in the degree of K [**Mer96**]. The modular curve $X_{\text{arith},1}(M, MN)$ is defined by the inverse image of $\{(\begin{smallmatrix} 1 & M* \\ 0 & * \end{smallmatrix})\} \subset \mathrm{GL}_2(\mathbb{Z}/MN\mathbb{Z})$. As a moduli space it parametrizes isomorphism classes of triples (E, P, C), where E is an elliptic curve over K, $P \in E[MN](K)$ is a point of order MN, and $C \leq E[M](K)$ is a Gal_K-stable cyclic subgroup of order M such that $E[M] = \langle NP \rangle + C$. The modular curve $X_{\text{arith},1}(M, MN)$ is one of the quadratic refinements of $X_{\text{arith},\pm 1}(M, MN)$, which is defined by the inverse image of $\{\pm(\begin{smallmatrix} 1 & M* \\ 0 & * \end{smallmatrix})\} \subset \mathrm{GL}_2(\mathbb{Z}/MN\mathbb{Z})$, see Remark 2.1 and [**DS17**].

For $M, N > 1$, the index of $\{(\begin{smallmatrix} 1 & M* \\ 0 & * \end{smallmatrix})\} \leq \mathrm{GL}_2(\mathbb{Z}/MN\mathbb{Z})$ is $M^3N^2 \prod_{p|MN}(1 - \frac{1}{p^2})$ and since $-I \notin \{(\begin{smallmatrix} 1 & M* \\ 0 & * \end{smallmatrix})\}$, the PSL_2-index is $\frac{1}{2}M^3N^2 \prod_{p|MN}(1 - \frac{1}{p^2})$. Note that $X_{\text{arith},1}(M, MN) \to X_1(MN)$ is a covering map of degree M. In particular, when $M = 1$, the corresponding modular curve $X_{\text{arith},1}(M, MN)$ is equal to $X_1(N)$, and when $N = 1$, the corresponding modular curve is equal to $X_{\text{arith}}(M)$. The reader can verify that the genus formulas of $X_1(N)$ and $X_{\text{arith}}(M)$ follow from $X_{\text{arith},1}(M, MN)$ after Propositions 5.1, 5.2.

PROPOSITION 5.1. *For $M, N > 1$, there are no elliptic points of orders 2 and 3 for $\Gamma_{\text{arith},1}(M, MN)$.*

PROOF. The elliptic points of orders 2 and 3 for $\mathrm{SL}_2(\mathbb{Z})$ are $\mathrm{SL}_2(\mathbb{Z})i$ and $\mathrm{SL}_2(\mathbb{Z})e^{2\pi i/3}$, respectively. Any elliptic points of order $h \in \{2, 3\}$ of a congruence subgroup $\Gamma \leq \mathrm{SL}_2(\mathbb{Z})$ must map to one of the above two points. Since $X_{\text{arith},1}(M, MN) \to X_1(MN)$ is a covering map, and there are no elliptic points of order h for $\Gamma_1(MN)$ when $M, N > 1$, there are no elliptic points of order h for $\Gamma_{\text{arith},1}(M, MN)$. ☐

PROPOSITION 5.2. *For $M, N > 1$, let ε_∞ denote the number of cusps in $X_{\text{arith},1}(M, MN)$. Then we have:*

$$\varepsilon_\infty = \begin{cases} 1 & \text{if } M, N = 1, \\ 2 & \text{if } M = 1,\ N = 2, \\ 3 & \text{if } M = 1,\ N = 4, \\ \frac{1}{2} \sum\limits_{d|MN} \varphi\left(\frac{MN}{d}\right) \varphi(d) \gcd\left(M, \frac{MN}{d}\right) & \text{otherwise.} \end{cases}$$

PROOF. We follow a similar approach to that in [**DS05**, Section 3.8]. Let $s' = a'/c'$, $s = a/c \in \mathbb{Q} \cup \infty$ and $\Gamma_{\text{arith},1} := \Gamma_{\text{arith},1}(M, MN)$. The coset decomposition $\Gamma_{\text{arith},1} = \bigcup_{j=0}^{N-1} \Gamma(MN)(\begin{smallmatrix} 1 & Mj \\ 0 & 1 \end{smallmatrix})$ gives the second equivalence:

$$\begin{aligned} \Gamma_{\text{arith},1}s' = \Gamma_{\text{arith},1}s &\iff s' \in \Gamma_{\text{arith},1}s, \\ &\iff s' \in \Gamma(MN)\left(\begin{smallmatrix} 1 & Mj \\ 0 & 1 \end{smallmatrix}\right) \text{ for some } j, \\ &\iff \left(\begin{smallmatrix} a' \\ c' \end{smallmatrix}\right) \equiv \pm\left(\begin{smallmatrix} a+Mcj \\ c \end{smallmatrix}\right) \pmod{MN} \text{ for some } j. \end{aligned}$$

This implies that the top row a is determined modulo $\gcd(Mc, MN)$ and since this is a cusp, we also have $\gcd(a, c, MN) = 1$. Let $d := \gcd(c, MN)$, there are

$\varphi(MN/d)$ elements c such that $0 \le c \le MN-1$ and $\gcd(c, MN) = d$. The set

$$\begin{aligned}&\{a : 0 \le a \le \gcd(Mc, MN), \gcd(a, c, MN) = 1\}\\ &= \{a : 0 \le a \le \gcd(Md, MN), \gcd(a, d) = 1\}\end{aligned}$$

has size $\varphi(d)\gcd(M, MN/d)$. Summing over divisors of MN yields the result. □

TABLE 1. Invariants of genus formula

X_H	i	ε_2
$X_0(N)$	$N\prod_{p\mid N}\left(1+\frac{1}{p}\right)$	0 if $N \equiv 0 \bmod 4$ $\prod_{p\mid N}\left(1+\left(\frac{-1}{p}\right)\right)$ otherwise
$X_1(N)$, $X_{\pm1}(N)$	1 if $N=1$ 3 if $N=2$ $\frac{N^2}{2}\prod_{p\mid N}\left(1-\frac{1}{p^2}\right)$ otherwise	1 if $N=1,2$ 0 otherwise
$X(N)$, $X_{\mathrm{arith}}(N)$	6 if $N=2$ $\frac{N^3}{2}\prod_{p\mid N}\left(1-\frac{1}{p^2}\right)$ if $N>2$	1 if $N=1$ 0 if $N>1$
$X_{\mathrm{arith},1}(M, MN)$, $X_{\mathrm{arith},\pm1}(M, MN)$	1 if $M, N=1$ 3 if $M=1, N=2$ 6 if $M=2, N=1$ $\frac{M^3N^2}{2}\prod_{p\mid MN}\left(1-\frac{1}{p^2}\right)$ otherwise	1 if $M, N=1$ 1 if $M=2, N=1$ 0 otherwise
$X_{\mathrm{sp}}(N)$	$N^2\prod_{p\mid N}\left(1+\frac{1}{p}\right)$	0 if $2\mid N$, $\prod_{p\mid N}\left(1+\left(\frac{-1}{p}\right)\right)$ otherwise
$X_{\mathrm{sp}}^+(N)$	$\frac{N^2}{2}\prod_{p\mid N}\left(1+\frac{1}{p}\right)$	See Proposition 3.4
$X_{\mathrm{sp}}^*(N)$	$\frac{N^2}{2^{\omega(N)}}\prod_{p\mid N}\left(1+\frac{1}{p}\right)$	See Corollary 3.5
$X_{\mathrm{ns}}(N)$	$N\varphi(N)$	$\prod_{p\mid N}\left(1-\left(\frac{-1}{p}\right)\right)$
$X_{\mathrm{ns}}^+(N)$	$\frac{1}{2}N\varphi(N)$	See Proposition 4.4
$X_{\mathrm{ns}}^*(N)$	$\frac{N}{2^{\omega(N)}}\varphi(N)$	See Corollary 4.6
$X_{S_4}(p)$	$\frac{1}{24}p(p^2-1)$	$\frac{1}{4}\left(p-\left(\frac{-1}{p}\right)\right)$

Table 1 continued on the next page ...

Table 1 continued ...

X_H	ε_3	ε_∞
$X_0(N)$	0 if $N \equiv 0 \bmod 9$ $\prod_{p\mid N}\left(1+\left(\frac{-3}{p}\right)\right)$ otherwise	$\sum_{\substack{d\mid N,\\ d>0}} \varphi\left((d, \frac{N}{d})\right)$
$X_1(N)$, $X_{\pm1}(N)$	1 if $N = 1, 3$ 0 otherwise	1 if $N = 1$ 2 if $N = 2$ 3 if $N = 4$ $\frac{1}{2}\sum_{d\mid N} \varphi(d)\varphi\left(\frac{N}{d}\right)$ otherwise
$X(N)$, $X_{\text{arith}}(N)$	1 if $N = 1$ 0 if $N > 1$	1 if $N = 1$ $\frac{i}{N}$ if $N > 1$
$X_{\text{arith},1}(M, MN)$, $X_{\text{arith},\pm1}(M, MN)$	1 if $M = 1$ and $N = 1, 3$ 0 otherwise	See Proposition 5.2
$X_{\text{sp}}(N)$	0 if $2 \mid N$ or $3 \mid N$, $\prod_{p\mid N}\left(1+\left(\frac{-3}{p}\right)\right)$ otherwise	$N\prod_{p\mid N}\left(1+\frac{1}{p}\right)$
$X_{\text{sp}}^+(N)$	0 if $2 \mid N$ or $3 \mid N$, $\frac{1}{2}\prod_{p\mid N}\left(1+\left(\frac{-3}{p}\right)\right)$ otherwise	$\frac{N}{2}\prod_{p\mid N}\left(1+\frac{1}{p}\right)$
$X_{\text{sp}}^*(N)$	See Corollary 3.5	See Corollary 3.5
$X_{\text{ns}}(N)$	$2^{\omega(N)}$ if $p \equiv 2 \pmod 3$ for all $p \mid N$, 0 otherwise,	$\varphi(N)$
$X_{\text{ns}}^+(N)$	$2^{\omega(N)-1}$ if $p \equiv 2 \pmod 3$ for all $p \mid N$, 0 otherwise,	1 if $N = 2$, $\frac{1}{2}\varphi(N)$ otherwise
$X_{\text{ns}}^*(N)$	See Corollary 4.6	See Corollary 4.6
$X_{S_4}(p)$	$\frac{1}{3}\left(p-\left(\frac{-3}{p}\right)\right)$	$\frac{1}{24}(p^2-1)$

Acknowledgments

The authors would like to thank Jennifer Balakrishnan, Harris Daniels, Jeremy Rouse, Andrew Sutherland, John Voight, and David Zywina for help with references. They would like to thank Pietro Mercuri for sharing his work on split Cartan subgroups, and the Simons Collaboration on Arithmetic Geometry, Number Theory, and Computation for funding the Modular Curves 3 Workshop hosted by MIT, where this work began. Lastly, they would like to thank the anonymous referees who reviewed this paper and gave thoughtful and constructive feedback.

References

[Bar10] Burcu Baran, *Normalizers of non-split Cartan subgroups, modular curves, and the class number one problem*, J. Number Theory **130** (2010), no. 12, 2753–2772, DOI 10.1016/j.jnt.2010.06.005. MR2684496 ↑294, 296, 301, 303, 304, 306

[DLM22] Valerio Dose, Guido Lido, and Pietro Mercuri, *Automorphisms of Cartan modular curves of prime and composite level*, Algebra Number Theory **16** (2022), no. 6, 1423–1461, DOI 10.2140/ant.2022.16.1423. MR4488580 ↑294, 299, 300, 306

[DMS19] Valerio Dose, Pietro Mercuri, and Claudio Stirpe, *Double covers of Cartan modular curves*, J. Number Theory **195** (2019), 96–114, DOI 10.1016/j.jnt.2018.05.021. MR3867436 ↑294, 296

[DS05] Fred Diamond and Jerry Shurman, *A first course in modular forms*, Graduate Texts in Mathematics, vol. 228, Springer-Verlag, New York, 2005. MR2112196 ↑294, 295, 307

[DS17] Maarten Derickx and Andrew V. Sutherland, *Torsion subgroups of elliptic curves over quintic and sextic number fields*, Proc. Amer. Math. Soc. **145** (2017), no. 10, 4233–4245, DOI 10.1090/proc/13605. MR3690609 ↑307

[Lig77] Gérard Ligozat, *Courbes modulaires de niveau* 11 (French), Modular functions of one variable, V (Proc. Second Internat. Conf., Univ. Bonn, Bonn, 1976), Lecture Notes in Math., Vol. 601, Springer, Berlin-New York, 1977, pp. 149–237. MR463118 ↑294

[LMF24] The LMFDB Collaboration, *The L-functions and modular forms database*, `https://www.lmfdb.org`, 2024, [Online; accessed 12 June 2024]. 294

[Maz77] B. Mazur, *Rational points on modular curves*, Modular functions of one variable, V (Proc. Second Internat. Conf., Univ. Bonn, Bonn, 1976), Lecture Notes in Math., Vol. 601, Springer, Berlin-New York, 1977, pp. 107–148. MR450283 ↑293

[Mer96] Loïc Merel, *Bornes pour la torsion des courbes elliptiques sur les corps de nombres* (French), Invent. Math. **124** (1996), no. 1-3, 437–449, DOI 10.1007/s002220050059. MR1369424 ↑307

[RSZB22] Jeremy Rouse, Andrew V. Sutherland, and David Zureick-Brown, *ℓ-adic images of Galois for elliptic curves over $\mathbb{Q}$ (and an appendix with John Voight)*, Forum Math. Sigma **10** (2022), Paper No. e62, 63, DOI 10.1017/fms.2022.38. With an appendix with John Voight. MR4468989 ↑293

[RZB15] Jeremy Rouse and David Zureick-Brown, *Elliptic curves over $\mathbb{Q}$ and 2-adic images of Galois*, Res. Number Theory **1** (2015), Paper No. 12, 34, DOI 10.1007/s40993-015-0013-7. MR3500996 ↑293

[SD73] H. P. F. Swinnerton-Dyer, *On l-adic representations and congruences for coefficients of modular forms*, Modular functions of one variable, III (Proc. Internat. Summer School, Univ. Antwerp, Antwerp, 1972), Lecture Notes in Math., Vol. 350, Springer, Berlin-New York, 1973, pp. 1–55. MR406931 ↑294

[Ser72] Jean-Pierre Serre, *Propriétés galoisiennes des points d'ordre fini des courbes elliptiques* (French), Invent. Math. **15** (1972), no. 4, 259–331, DOI 10.1007/BF01405086. MR387283 ↑293, 294

[Shi71] Goro Shimura, *Introduction to the arithmetic theory of automorphic functions*, Kanô Memorial Lectures, No. 1, Iwanami Shoten Publishers, Tokyo; Princeton University Press, Princeton, NJ, 1971. Publications of the Mathematical Society of Japan, No. 11. MR314766 ↑294, 297, 298, 299, 302, 303

Department of Mathematics, University of Connecticut, Storrs, CT 06269, USA
Email address: `asimina.hamakiotes@uconn.edu`
URL: `https://asiminah.github.io/`

Department of Electrical Engineering, KU Leuven, Kasteelpark Arenberg 10/2452, 3001 Leuven (Heverlee), Belgium
Email address: `junbo.lau@kuleuven.be`
URL: `https://junbolau.github.io/`

Contemporary Mathematics
Volume **840**, 2026
https://doi.org/10.1090/conm/840/16823

Evaluation of modular polynomial from supersingular elliptic curves

Maria Corte-Real Santos, Jonathan Komada Eriksen, Antonin Leroux, Michael Meyer, and Lorenz Panny

Abstract. We present several new algorithms to evaluate modular polynomials of level ℓ modulo a prime p on an input j. More precisely, we introduce two new generic algorithms, sharing the following similarities: they are based on a CRT approach; they make use of supersingular curves and the Deuring correspondence; and, their memory requirements are optimal.

The first algorithm combines the ideas behind a hybrid algorithm of Sutherland in 2013 with a recent algorithm to compute modular polynomials using supersingular curves introduced in 2023 by Leroux. The complexity (holding around several plausible heuristic assumptions) of the resulting algorithm matches the $\tilde{O}(\ell^3 \log^3 \ell + \ell \log p)$ time complexity of the best known algorithm by Sutherland, but has an optimal memory requirement.

Our second algorithm is based on a sub-algorithm that can evaluate modular polynomials efficiently on supersingular j-invariants defined over $\mathbb{F}_p$, and achieves heuristic complexity quadratic in both ℓ and $\log j$, and linear in $\log p$. In particular, it is the first generic algorithm with optimal memory requirement to obtain a quadratic complexity in ℓ.

Additionally, we show how to adapt our method to the computation of other types of modular polynomials such as the one stemming from Weber's function.

Finally, we provide an optimised implementation of the two algorithms detailed in this paper, though we emphasise that various modules in our codebase may find applications outside their use in this paper.

1. Introduction

The evaluation of modular polynomials is one of the main subroutines involved in the SEA point counting algorithm [**19**]. The main application of point counting is to find elliptic curves suitable for cryptography. The asymptotic bottleneck of the computation relating to "Elkies primes" in the SEA algorithm is modular evaluation, and it is thus important to find improvements to the theoretical and practical efficiency of modular evaluation.

More generally, modular polynomials play an important role in the theory of elliptic curves, and so the computation and evaluation of modular polynomials is a central task in algorithmic number theory. Aside from point counting, modular polynomials are related to isogeny computations. While most applications tend to

2020 *Mathematics Subject Classification.* 11Y40.

use the more efficient Vélu formulas [**5**, **44**], we still observe few instances where modular polynomials have been considered. For example, it is used in the CRS key exchange [**15**, **36**], the first isogeny-based protocol, the OSIDH construction [**12**], the reduction introduced in [**2**], or the fast implementation of Delfs–Galbraith proposed in [**37**].

In this work, we introduce several new algorithms for efficiently computing modular polynomials Φ_ℓ of level ℓ, evaluated in one of the variables.

Related work. Almost all known methods to efficiently evaluate modular polynomials first requires their computation, which is the primary reason most literature on the topic focuses on the computation of modular polynomials.

The historical approach to computing modular polynomials is based on the computation of the coefficients of the Fourier expansion of the modular j-function [**6**, **19**, **29**, **33**]. As this computation works over the integers, it can be applied to compute $\Phi_\ell \bmod p$.

An alternative approach based on the CRT method, which uses supersingular curves and ℓ-isogeny computations, was introduced by Charles and Lauter [**11**] to work entirely over $\mathbb{F}_p$.

Enge [**20**] uses interpolation and fast floating-point interpolation to obtain a quasi-linear algorithm over the integers under some heuristics.

One of the main problems behind the computation of modular polynomials over the integers is the large size of their coefficients. In theory, the size of coefficients is less of an issue when dealing with Φ_ℓ as the coefficients are reduced modulo p. However, this is only the case if we can avoid the computation over $\mathbb{Z}$ entirely, which is not easy to obtain in practice.

Bröker, Lauter, and Sutherland (BLS hereafter) [**8**] obtained the first quasi-linear complexity in both time and space with a careful application of the CRT method using ordinary curves.

More recently, Leroux [**31**] revisited the CRT approach from supersingular curves with new algorithmic results on the Deuring correspondence to obtain an algorithm with the same complexity as BLS. Leroux claims a better practical efficiency, but provided no experimental proof of this claim.

Robert [**35**] outlined another CRT method based on supersingular curves by using a p-adic approach in conjunction with the high-dimensional isogeny technique introduced in the context of the cryptanalysis of the SIDH key exchange. This method was later refined and implemented by Kunzweiler and Robert [**27**]. However, the main advantage of their method is mostly theoretical as it does not require any assumptions (not even GRH!), but it is unlikely to be practical due to a non-optimal high memory requirement. Furthermore, the implementation they provide is in SageMath [**43**], rather than a low-level implementation.

In 2013, Sutherland [**41**] provided the first evidence that a method tailored to the evaluation could do better than a computation algorithm. Sutherland showed how to adapt the CRT approach from BLS to obtain two evaluation algorithms with excellent memory requirements. However, these algorithms have essentially the same complexity as the BLS computation algorithm [**8**].

Contributions. Our main contribution is the introduction of three new algorithms targeted at the task of evaluating modular polynomials. Our work builds upon the previous idea of Leroux [**31**] to use supersingular curves with the Deuring correspondence for efficient modular polynomial computation. The fact that his algorithms could be improved to produce a more efficient evaluation algorithm was already mentioned by Leroux, but no precise method was described or even outlined. In this work, we show that there are several interesting approaches to consider in this setting. For a fixed value of ℓ, each of our new algorithms will achieve the best known complexity for some range of primes p (relative to the size of ℓ). Hereafter, we assume that the evaluation is done on an input $j \in \mathbb{F}_p$ that can be seen as an integer $0 \leq j \leq p-1$. All the stated complexities hold under several plausible heuristic assumptions.

(1) ModularEvaluationBigCharacteristic is a generic CRT evaluation algorithm built on top of the OrdersTojInvariantBigSet algorithm from [**31**] which combines [**41**, Algorithm 2.2] and ModularComputation from [**31**]. The complexity is $O\left(\ell^2(\ell\log\ell + \log p)\log^{2+\varepsilon}(\ell\log\ell+\log p)\right)$ and the memory requirement is $O\left((\ell\log\ell+\log p)\log(\ell\log\ell+\log p)+\ell\log p\right)$. The asymptotic complexity is the same as [**41**, Algorithm 1], but the space requirement is better. This algorithm will achieve the best known space/time complexity in cases where p (i.e., the characteristic) is large.

(2) SupersingularEvaluation works on supersingular j-invariants and has a complexity of $O\left(\ell(\log p^{4+\varepsilon} + \log\ell^{2+\varepsilon}\log^{1+\varepsilon}p) + p^{1/4}\log^{3+\varepsilon}p\right)$, using $O(\ell\log p + p^{1/4}\log^{1+\varepsilon}p)$ space. Despite a limited range of applications (due to the supersingularity constraint), this algorithm achieves the best known complexity when the prime p is small, and it is the only algorithm that is linear in ℓ.

(3) ModularEvaluationBigLevel is a generic CRT evaluation algorithm built on top of SupersingularEvaluation. It has two main steps: a CRT prime collection of complexity $O\left(\ell^2\log^2 j(\log^{2+\varepsilon}(\ell\log j))\right)$, and a CRT computation of complexity $O(\ell^2\log j\log^{3+\varepsilon}\ell + \ell^{3/2}\log^{3/2}j\log\ell^{3+\varepsilon} + \log^{2+\varepsilon}\ell\log^{1+\varepsilon}p)$. The global space requirement is $O(\ell\log(pj))$. This algorithm achieves the best known complexity of all generic evaluation algorithm if either one of j or p is small (with respect to ℓ). In the generic case, we will have $j = \Theta(p)$, and the complexity is quadratic in $\log p$, which is worse than the other algorithms. However, note that the quadratic component of the complexity comes from the CRT prime collection, and we note that this step only depends on the input j and can be shared among evaluations of different modular polynomials on the same j. This could prove interesting for applications such as point counting.

Outside of these algorithmic contributions, we also show how to adapt our methods to work for other modular polynomials, such as the one based on the Weber function, by exploiting a modular parametrization of elliptic curves with order 48 level structure underlying the Weber function. This is very convenient in practice as the sizes of the coefficients of modular polynomials associated to the Weber function are smaller by a large constant, which makes them approximately 1728 times faster to compute than standard modular polynomials of the same level.

Finally, in Section 4, we provide an efficient implementation of all our algorithms written in C++ and NTL [**38**]. Unfortunately, at their current state, the implementations of our two generic algorithms do not seem to outperform the state-of-the-art implementation from [**41**] for parameters that are within reasonable computational reach. A more detailed analysis of the concrete performance of our implementation can be found in Section 4.2. We emphasise, however, that many of the algorithms we implement are useful outside their application in this paper. We highlight in particular the following:

(1) An optimised implementation of polynomial interpolation, which outperforms NTL's in-built function for polynomials of large degree.
(2) An implementation of the Deuring correspondence for generic primes, thus providing a low-level optimised implementation of algorithms from [**22**], which outperforms that implementation by a large constant factor.
(3) An optimised implementation of the computation of the endomorphism ring of a supersingular elliptic curve defined over $\mathbb{F}_p$ for generic primes.

The last two routines are especially useful for the cryptanalysis and construction of isogeny-based cryptography.

Acknowledgements. We thank Drew Sutherland and Sam Frengley for helping us finding the maps that are needed to compute the modular polynomials associated to the Weber function.

This work was supported in part by the European Research Council (ERC) under the European Union's Horizon 2020 research and innovation programme (grant agreement ISOCRYPT – No. 101020788), by the Research Council KU Leuven grant C14/24/099, and by CyberSecurity Research Flanders with reference number VR20192203. This work was also supported by the European Research Council under grant No. 101116169 (AGATHA CRYPTY) and by the Deutsche Forschungsgemeinschaft (DFG, German Research Foundation) under SFB 1119 – 236615297.

1.1. Technical overview. Let p, ℓ be two distinct primes, and let $j \in \mathbb{F}_p$. Hereafter, we abuse notation and also write j for the unique integer between 0 and $p-1$ in the class of j modulo p. The goal of all the algorithms introduced in this work is to compute the polynomial $\Phi_\ell(X, j) \in \mathbb{F}_p[X]$.

Our work can be seen as an extension of the ideas introduced by Leroux in [**31**] to the setting of modular polynomial evaluation, using some ideas outlined by Sutherland in [**41**]. Note that the complexities of our algorithms rely on heuristic assumptions because the complexity of Leroux's algorithm are only proven assuming various heuristics.

All our algorithms rely on the Deuring correspondence and the OrdersTojInvariantSmallSet and OrdersTojInvariantBigSet algorithms from [**31**] whose goal is to compute the set of j-invariants corresponding (under the Deuring correspondence) to a set of maximal order types given as input. OrdersTojInvariantSmallSet is generally more efficient, unless the set of j-invariants to be computed is somewhat close to the entire set of supersingular j-invariants, where OrdersTojInvariantBigSet is tailored to be more efficient. These two algorithms were used by Leroux to build two new efficient algorithms to compute modular polynomials modulo p:

(1) SupersingularModularComputation directly applies OrdersTojInvariantSmallSet (or OrdersTojInvariantBigSet, if it is faster) to compute the necessary j-invariants to interpolate $\Phi_\ell(X, Y)$ over $\mathbb{F}_p$.
(2) ModularComputation is a CRT algorithm that applies SupersingularModularComputation on a set of small CRT primes p_i before reconstructing the result mod p. Here, SupersingularModularComputation is always used with OrdersTojInvariantBigSet because the p_i are small compared to ℓ.

The approach by Leroux is amenable to ideas introduced by Sutherland [**41**] to adapt the BLS modular polynomial computation algorithm to the evaluation setting. Sutherland observed that there are essentially two ways to evaluate Φ_ℓ on $j \in \mathbb{F}_p$. The first and most direct way is to consider j as an integer, compute $\Phi_\ell(X, j)$ over the integers and finally reduce the result modulo p. This leads to [**41**, Algorithm 2] which has optimal space complexity, but can be quite inefficient when $\log p$ grows and $j = \Theta(p)$. The main problem with that method is that the powers of j are much bigger than j over $\mathbb{Z}$. This increases the size of the coefficients of $\Phi_\ell(X, j) \in \mathbb{Z}[X]$ which has a negative impacts on the performances of this approach.

Sutherland's trick is to realize that we can avoid exponentiating over $\mathbb{Z}$ by exponentiating over $\mathbb{F}_p$ where the powers of j are all in $O(p)$. Thus, he proposes to lift each element $j^k \in \mathbb{F}_p$ to an integer $\widehat{j_k}$ for $1 \leq k \leq \ell + 1$, see $\Phi_\ell(X, j)$ as a multivariate polynomial $P_\ell(X, j_1, \ldots, j_{\ell+1})$ by replacing each j^k by a variable j_k and evaluate this polynomial on the $\widehat{j_k}$ before reducing the result modulo p. This gives [**41**, Algorithm 1], which obtains the best known complexity but requires more space than the more naive approach. This method has the same complexity as computing the whole modular polynomial, but requires less memory as the full polynomial is never stored in memory.

Sutherland also presents a "hybrid" version of [**41**, Algorithm 1] (see [**41**, Section 3.4]). This hybrid algorithm optimises the amount of data stored throughout the computation, and so it is slower but uses an optimal amount of space.

Our first algorithm, ModularEvaluationBigCharacteristic can be seen as an adaptation of Leroux's ModularComputation with the ideas behind Sutherland's hybrid algorithm. Similarly to Sutherland's idea, we lift powers of j to integers $\widehat{j}_1, \ldots, \widehat{j}_{\ell+1}$ before evaluating the polynomial $P_\ell(X, \widehat{j}_1, \ldots, \widehat{j}_{\ell+1})$ with a CRT method. The computation of $P_\ell(X, \widehat{j}_1, \ldots, \widehat{j}_{\ell+1})$ modulo the small CRT primes p_i mainly relies on OrdersTojInvariantBigSet in a manner similar to SupersingularModularComputation. It turns out that this method requires less storage than the version using ordinary curves, and so we can execute the analogue of Sutherland's hybrid version without any negative impact on the complexity.

Thus, just as ModularComputation matches the complexity of BLS, ModularEvaluationBigCharacteristic essentially matches the complexity of [**41**, Algorithm 1] taking $O\left(\ell^2(\ell \log \ell + \log p) \log^{2+\varepsilon}(\ell \log \ell + \log p)\right)$, but with a better space complexity of $O\left((\ell \log \ell + \log p) \log(\ell \log \ell + \log p) + \ell \log p\right)$ that is quasi-linear in ℓ.

Our second generic evaluation algorithm ModularEvaluationBigLevel applies the naive approach of computing $\Phi_\ell(X, j)$ over $\mathbb{Z}[X]$, but compensates the efficiency loss caused by the large size of the powers of j in $\mathbb{Z}$ by noticing that modulo some well-chosen CRT primes p_i, the computation of $\Phi_\ell(X, j) \bmod p_i$ can be made much more efficient than the computation of $P_\ell(X, \widehat{j}_1, \ldots, \widehat{j}_{\ell+1}) \bmod p_i$ (which is essentially equivalent to the computation of the full $\Phi_\ell(X, Y) \bmod p_i$). The well-chosen primes are those where $j \bmod p_i$ is the j-invariant of a supersingular curve and this efficient algorithm is our third algorithmic contribution: SupersingularEvaluation, whose complexity is linear in ℓ. The main downside of this idea is that the probability that $j \bmod p_i$ is supersingular is in $O(1/\sqrt{p_i})$ which makes the computation of the CRT primes quite costly with a complexity of $O\left(\ell^2 \log^2 j \log^{2+\varepsilon}(\ell \log j)\right)$. In the generic case where $\log j = \theta(\log p)$, this step will be the bottleneck.

Once the CRT primes have been computed, the rest of the computation takes $O\left(\ell^2 \log j \log \ell^{3+\varepsilon} + \ell^{3/2} \log^{3/2} j \log^{2+\varepsilon} \ell + \log^{2+\varepsilon} \log^{1+\varepsilon} p\right)$, and the overall space requirement is $O(\ell \log(pj))$ which is optimal given the size of the output.

SupersingularEvaluation consists of a rather straightforward application of OrdersTojInvariantSmallSet to compute the j-invariants ℓ-isogenous to the supersingular j-invariant given as input. This gives a linear complexity in ℓ, but comes at the cost of computing the endomorphism ring of the input curve, which takes $O(p^{1/4} \log p^{3+\varepsilon})$. The final complexity is

$$O\left(\ell(\log \ell^{2+\varepsilon} \log p^{1+\varepsilon} + \log p^{4+\varepsilon}) + p^{1/4} \log p^{3+\varepsilon}\right).$$

We note that SupersingularEvaluation is interesting in its own right, as it achieves the best known complexity to evaluate Φ_ℓ when the prime p is small. However, the constraint of having a supersingular j-invariant as input is quite limiting for practical applications.

Other modular functions. We also show how to adapt our method to compute modular polynomials associated to other modular functions such as the Weber function. It was already suggested by Leroux in [**31**] that, similarly to the BLS algorithm, the supersingular approach could be adapted to work for other modular functions. However, unlike the prediction by Leroux, who qualified the task as "not too daunting", concretely implementing this comes with a few technical obstacles. In particular, the approach taken in BLS cannot be made to work here. Our solution uses a level structure parametrization of Weber invariants and elimination theory to solve the issue.

Organisation of the article. The rest of this paper is organised as follows: in Section 2, we introduce some background on isogenies, quaternion algebras and the Deuring correspondence. The main technical contributions are introduced in Section 3 where we present our new algorithms. Finally, in Section 4, we give some details on our implementation.

2. Background material

Throughout the paper, we consider logarithm in base 2 that we simply write $\log(\cdot)$, and we use $\text{llog}(\cdot)$ for $\log(\log(\cdot))$.

Background on the main topics covered in this article can be found in:

(1) Silverman [**24**] for elliptic curves and isogenies;
(2) Voight [**45**] for quaternion algebras
(3) the thesis of Leroux [**30**] for the algorithmic aspects of the Deuring correspondence.

In short, the Deuring corrrespondence makes a link between elliptic curves and isogenies over $\mathbb{F}_p$ on one side, and orders and ideals of the quaternion algebra ramified at p and ∞ on the other side. This result thus provides a way to navigate efficiently the supersingular isogeny graph by using some simple operations over some lattices of dimension 4. This is what we call the effective Deuring correspondence.

Effective Deuring correspondence. More precisely, one important algorithm for us is a heuristic polynomial-time algorithm introduced in [**18**] which computes the j-invariant corresponding to a maximal order type given as input.

It makes use of a smooth isogeny connecting the desired j-invariant to a special j-invariant (for instance $j = 1728$ or $j = 0$). This smooth isogeny is found using the Deuring correspondence with the KLPT algorithm [**25**]. The smooth isogeny found with KLPT is usually quite large (the heuristic estimates suggest that one should be able to find an isogeny of degree $\tilde{O}(p^3)$ with that algorithm), and so the concrete computation of this isogeny, for a generic prime p, can be quite slow. The asymptotic cost of this computation is $O(\log^{4+\varepsilon} p)$, but the constants hidden in that complexity are dauntingly large for generic primes. Nevertheless, in [**22**] it was shown that this algorithm can be made practical, even for generic primes, by applying several practical improvements to the original algorithm from [**18**] Henceforth, we call this algorithm SingleOrderTojInvariant().

REMARK 2.1. There seems to be a typo in [**31**] where the cost of this algorithm is estimated to be $O(\log p^{5+\varepsilon})$. However, if we look at [**23**, Lemma 4] we see that the cost should be of $O(\log^{4+\varepsilon} p)$ (after a precomputation of $\log^{6+\varepsilon} p$ that can be amortized across all isogeny computations).

Efficient Deuring correspondence for several orders. In [**31**], Leroux looked at the problem of the effective Deuring correspondence for a set of order types. For a small number of orders (relative to the prime characteristic p), the best method seems to be simply to apply SingleOrderTojInvariant() to each order in the set. Leroux called this algorithm OrdersTojInvariantSmallSet, and we will keep that notation throughout the paper. If S is the size of the set of orders to be computed, the complexity of this algorithm is $O(S\log^{4+\varepsilon} \log p)$.

However, when the set contains a lot of order types, Leroux introduced a much faster algorithm called OrdersTojInvariantBigSet (the algorithm really becomes interesting for sets of size $O(p)$). The idea of this algorithm is to go through the entire set of supersingular curves through quaternions and collect the desired j-invariants along the way in order to minimize the number of isogeny computations required.

The asymptotic complexity of this algorithm is $O(S\log p^{2+\varepsilon} + p\log p^{1+\varepsilon})$, and it also has better hidden constants than OrdersTojInvariantSmallSet due to the fact that it involves the computation of isogenies of degree $O(p)$ instead of $O(p^3)$.

Application of the Deuring correspondence to the computation of modular polynomials. The principle that "quaternion operations" are generally more efficient than "elliptic curve operations" is at the heart of the recent algorithm ModularComputation of Leroux [**31**] for computing modular polynomials. Indeed, the main part of this computation is to collect pairs of ℓ-isogenous j-invariants. By working with supersingular j-invariants, Leroux showed how to exploit the effective Deuring correspondence to realize that collection with a minimal amount of "elliptic curve" operations by working mostly with quaternions. The idea is that the set of maximal order types corresponding to the j-invariants to be collected can be generated purely from quaternion operations. Then, it suffices to apply the algorithms we mention above to collect all the necessary j-invariants.

3. Evaluation of modular polynomials

In this section, we introduce our main theoretical contributions, namely the three new algorithms ModularEvaluationBigCharacteristic, SupersingularEvaluation, and ModularEvaluationBigLevel (respectively introduced in Sections 3.1 to 3.3). In Section 3.4, we provide a concrete comparison with several other algorithms from the literature. Finally, in Section 3.5, we explain how to adapt our method to evaluate modular polynomials associated with the Weber invariant.

3.1. A first CRT approach for big characteristic. Our first CRT approach is inspired by the "hybrid" algorithm described by Sutherland but where [**41**, Algorithm 2.2] is replaced with an algorithm SpecialSupersingularEvaluation described below. This algorithm combines the ideas from SupersingularModularComputation from [**31**], with the trick introduced by Sutherland to perform the evaluation with a minimal space requirement. The part of our algorithm that constructs the final output by CRT is identical to the one in [**41**, Algorithm 2]. We therefore do not describe these steps precisely, and instead refer the reader to [**40**, Section 6] for more details.

Algorithm 1 SpecialSupersingularEvaluation$(p, \ell, x_0, \dots, x_\ell)$

Input: A prime p, a prime ℓ with $\lceil p/12 \rceil + 1 < \ell$ and $\ell + 1$ values $x_0, \dots, x_\ell \in \mathbb{F}_p$.
Output: $P(Y) = \sum_{i,j} a_{i,j} x_i Y^j$ where $\Phi_\ell(X, Y) = \sum_{i,j} a_{i,j} X^i Y^j$.

Compute the set of maximal order types $\mathcal{O}_1, \dots, \mathcal{O}_m \subset \mathcal{B}_{p,\infty}$ and set $\mathfrak{S}$ as the list of these maximal order types
Compute $J =$ OrdersTojInvariantBigSet$(p, \mathfrak{S})$
Select a set $\mathcal{O}_{1,0}, \dots, \mathcal{O}_{\ell+2,0}$ in $\mathfrak{S}$
for $i = 1$ to $\ell + 2$ **do**
  Find $j_{i,0}$ as the j-invariant in J corresponding to $\mathcal{O}_{i,0}$
  Compute $I_{i,1}, \dots, I_{i,\ell+1}$, the $\ell + 1$ $\mathcal{O}_{i,0}$-ideals of norm ℓ
  for $k = 1$ to $\ell + 1$ **do**
    Select $j_{i,k}$ as the j-invariant in J corresponding to $\mathcal{O}_R(I_{i,k})$
  end for
  $P(j_{i,0}, Y) \leftarrow \prod_{k=1}^{\ell+1} (Y - j_{i,k})$
  Write $P(j_{i,0}, X) = \sum_{k=1}^{\ell+1} c_{i,k-1} X^{k-1}$
  $y_i \leftarrow \sum_{k=1}^{\ell+1} c_{i,k-1} x_{k-1}$
end for
Interpolate $P(Y) \bmod p$ as the polynomial of degree $\ell + 1$ with $P(j_{i,0}) = y_i$.
return $P(Y)$.

PROPOSITION 3.1. SpecialSupersingularEvaluation *is correct, and can be executed in* $O(\ell^2 \log^{2+\varepsilon} \ell \log^{1+\varepsilon} p + p \log^{2+\varepsilon} p)$, *and requires* $O(\ell \log p + p \log p)$ *space.*

PROOF. All maximal orders in $\mathcal{B}_{p,\infty}$ can be enumerated in $O(p \log p)$ with the techniques described for SupersingularModularComputation from [**31**]. By [**31**, Theorem 1], the running time of OrdersTojInvariantBigSet is $O(p \log^{2+\varepsilon} p)$. Thus, all the steps before the main loop can be run in $O(p \log^{2+\varepsilon} p)$.

Computing an ideal of norm ℓ takes $O(\log(\ell p))$. Thus, the asympotic cost of the loop is dominated by the cost of the polynomial reconstruction at the end which takes $O(\ell \log^{2+\varepsilon} \ell \log^{1+\varepsilon} p)$ with the fast interpolation algorithms. Since the loop is repeated ℓ times, the final cost is $O(\ell^2 \log^{2+\varepsilon} \ell \log^{1+\varepsilon} p)$.

The sets $\mathfrak{S}$ and J use $O(p \log p)$ space to store, and the space requirement of OrdersTojInvariantBigSet is $O(p \log p)$. The amount of space for each iteration of the loop is $O(\ell \log p)$ to store the j-invariants and the polynomials $P(j_{i,0}, Y)$. After each iteration of the loop, we need to store one value in $\mathbb{F}_p$, so the total is $O(\ell \log p)$. □

REMARK 3.2. The algorithm outlined as Algorithm 1 is correct up to some small detail: order types correspond to j-invariants up to Galois conjugacy. Since we work over $\mathbb{F}_{p^2}$, this means that there can be up to two distinct j-invariants with the same endomorphism ring (j and j^p). In practice, this means that our algorithm needs a small modification to know which one of the two is the correct among $j_{i,k}$ and $j_{i,k}^p$. This can be done by modifying OrdersTojInvariantBigSet to store not only couples of order types and j-invariants, but also ideals whose corresponding isogeny's codomain is the computed j-invariant. Then, we can check if the j-invariant corresponding to $I_{i,k}$ is $j_{i,k}$ or $j_{i,k}^p$ by checking if $I_{i,k}$ is equivalent or not to the ideal stored along with the type of $\mathcal{O}_R(I_{i,k})$.

Algorithm 2 ModularEvaluationBigCharacteristic(p, j, ℓ)

Input: A prime p, $j \in \mathbb{F}_p$, a prime ℓ.
Output: $\Phi_\ell(j, Y)$.

Let $(\overline{j_i})$ be the integers in $[0, p-1]$ equal to $j^i \bmod p$ for $1 \le i \le \ell + 1$
Set $B = 2^{6\ell \log \ell + 18\ell + \log p + \log(\ell+2)}$.
Set $\mathcal{P}_\ell = \{\}$, $P = 1$, $q = 12(\ell+2)+1$.
while $P < B$ **do**
  if q is prime **then**
    $P \leftarrow q \cdot P, \quad \mathcal{P}_\ell \leftarrow \mathcal{P}_\ell \cup \{q\}$
  end if
  $q \leftarrow q + 2$
end while
Perform the pre-computations for the explicit CRT mod p using $\mathcal{P}_\ell$
for $q \in \mathcal{P}_\ell$ **do**
  Set j_q to be the vector $(\overline{j_i})$ reduced mod$q$.
  $P_q(Y) \leftarrow$ SpecialSupersingularEvaluation(q, ℓ, j_q)
  Update the CRT sums for each coefficient of $P_q(Y)$
end for
Perform the post-computation for the explicit CRT to obtain $P(Y) \in \mathbb{F}_p[X]$
return $P(Y)$

PROPOSITION 3.3. *The expected running time of* ModularEvaluationBigCharacteristic *is* $O(\ell^2(\ell \log \ell + \log p) \log^{2+\varepsilon}(\ell \log \ell + \log p))$ *and requires* $O((\ell \log \ell + \log p) \log(\ell \log \ell + \log p) + \ell \log p)$ *memory.*

PROOF. With the same reasoning as in the proof of the expected running time of ModularComputation in [**31**], we can take $\#\mathcal{P}_\ell = O(\log B / \mathrm{llog} B)$ and $\max_{q \in \mathcal{P}_\ell} q = O(\log B)$. We obtain the result by combining $\log B = O(\ell \log \ell + \log p)$ with Proposition 3.1, and the cost of the CRT operations given in [**41**]. □

3.2. The supersingular case. Our algorithm to evaluate the modular polynomial on the j-invariant j of a supersingular curve is simple to describe. It mainly consists in the application of the OrdersTojInvariantSmallSet (or OrdersTojInvariantBigSet) algorithms from [**31**] to collect the j-invariants required to reconstruct $\Phi_\ell(j, X)$ from its roots. However, the input of these algorithms are the maximal orders isomorphic to the endomorphism ring of the curve ℓ-isogenous to j. The simplest way to compute these maximal orders is to know the endomorphism ring associated to j. Hence, the first building block we need is an algorithm to compute the endomorphism ring of a supersingular curve defined over $\mathbb{F}_p$.

Computing the endomorphism ring of a supersingular elliptic curve over $\mathbb{F}_p$. We introduce an algorithm EndomorphismRing of heuristic complexity $\widetilde{O}(p^{1/4})$. This algorithm relies on a subroutine that computes isogenies between two given supersingular curves over $\mathbb{F}_p$, which can morally be traced back to [**17**, Algorithm 1]. In this paper, it is simply refined to be efficient in practice and allow the computation of the ideal corresponding to the computed isogeny (via the Deuring correspondence). We then simply apply this algorithm between the target curve E and some starting curve E_0 of known endomorphism ring.

The main property behind this algorithm is that the set of supersingular curves E defined over $\mathbb{F}_p$ with a given $\mathbb{F}_p$-endomorphism ring $R := \mathrm{End}(E) \cap \mathbb{Q}(\pi)$, where $\pi\colon E \to E$ is the p-power Frobenius endomorphism, admits (very few exceptional cases exempt) a free and transitive group action by $\mathrm{Cl}(R)$. In the supersingular setting, the only two choices are $R = \mathbb{Z}[\pi] \cong \mathbb{Z}[\sqrt{-p}]$ and $R = \mathbb{Z}[\frac{1+\pi}{2}] \cong \mathbb{Z}[\frac{1+\sqrt{-p}}{2}]$. Thus, the set of supersingular curves over $\mathbb{F}_p$ with a given $\mathbb{F}_p$-endomorphism ring can be obtained by applying the action of a set of elements spanning $\mathrm{Cl}(R)$ on any given starting curve $E_0/\mathbb{F}_p$ with the correct choice of $\mathbb{F}_p$-endomorphism ring ($\mathbb{Z}[\pi]$ or $\mathbb{Z}[\frac{1+\pi}{2}]$, corresponding to the surface or floor of the 2-isogeny volcano). Heuristically, we can obtain such a set by considering the combination of small powers of $O(\log p)$ ideal classes in $\mathrm{Cl}(R)$. This is formalized as Heuristic 1.

This heuristic result is convenient in practice because computing a policyclic representation as in [**40**, Algorithm 2.2] requires a number of operations linear in the class number, whereas constructing the set used in the heuristic method can be done in polylogarithmic time.

HEURISTIC 1. *Write* $R \in \{\mathbb{Z}[\pi], \mathbb{Z}[\frac{1+\pi}{2}]\}$. *There exists a constant* C *such that for any prime* p*, any pair of elements* $[\mathfrak{g}_1], [\mathfrak{g}_2]$ *in* $\mathrm{Cl}(R)$*, any set of ideals* $\mathfrak{l}_1, \ldots, \mathfrak{l}_n$ *with pairwise distinct odd prime norms in* R*, and set of exponents* $e_1, \ldots, e_n$ *such that* $\prod_{i=1}(4e_i + 1) > Cp^{1/2}\log p$*, there exists* $\mathfrak{a} = \prod_{i=1}^n \mathfrak{l}_i^{b_i}$ *with each* $b_i \in [-2e_i, 2e_i]$ *such that* $[\mathfrak{g}_1] = [\mathfrak{a}][\mathfrak{g}_2]$ *in* $\mathrm{Cl}(R)$.

PROPOSITION 3.4. *Assuming Heuristic* 1, EndomorphismRing *always returns a maximal order isomorphic to the endomorphism ring of the input curve. The expected running time of* EndomorphismRing *is* $O(p^{1/4}\log p^{3+\varepsilon})$*, and the expected memory cost is* $O(p^{1/4}\log p^{2+\varepsilon})$*. The size of the output is* $O(\log p)$.

PROOF. Under *GRH*, the smallest value of q such that $\sqrt{-q}$ is contained in $\mathcal{B}_{p,\infty}$ is in $O(\log^2 p)$ by a result of Ankeny [**1**]. Thus, with the CM method, a curve E_0 can be found using $O(\mathsf{poly}(\log(p)))$ binary operations.

By Heuristic 1 and [**10**, Theorem 7], all supersingular j-invariants defined over $\mathbb{F}_p$ will be spanned by $\left(\prod_{i=1}^n \mathfrak{l}_i^{b_i}\right) * j_0$, for $b_i \in [-2e_i, 2e_i]$ with overwhelming probability. Thus, we know there will be a collision in J and J_0.

Under GRH, if we select the norms $\ell_1, \ldots, \ell_n$ of the ideals $\mathfrak{l}_1, \ldots, \mathfrak{l}_n$ as the smallest primes that are split in $\mathbb{Z}\sqrt{-n}$, then we have $\mathfrak{l}_i = O(\log p^{1+\varepsilon})$ for all i, and we have $n = O(\log p)$.

Algorithm 3 EndomorphismRing(p, j)

Input: A prime p, and the j-invariant of a supersingular curve over $\mathbb{F}_p$, denoted j.
Output: $\perp$, or maximal order $\mathcal{O} \in \mathcal{B}_{p,\infty}$ isomorphic to $\mathrm{End}(E)$ where $j(E) = j$.

Let $\mathcal{O}_0$ be one of the maximal orders of $\mathcal{B}_{p,\infty}$ given in [**25**, Lemma 2,3,4]
Compute E_0 a supersingular curve defined over $\mathbb{F}_p$ of known endomorphism ring whose endomorphism ring is isomorphic to $\mathcal{O}_0$
Compute a set of ideals $\mathfrak{l}_1, \ldots, \mathfrak{l}_n$ in $\mathfrak{O}$ of odd prime norm $\ell_1, \ldots, \ell_n$ in $\mathfrak{O}$ and exponents $e_1, \ldots, e_n < 10$ such that $\prod_{i=1}^n (4e_i + 1) > Cp^{1/2} \log p$
Set $J_0 = \{(j_0, \mathfrak{O})\}$
Set $J = \{(j, \mathfrak{O})\}$
for $i = 1$ to n **do**
    for $e \in [-e_i, e_i]$ and $(j_1, \mathfrak{a}_1), (j_2, \mathfrak{a}_2) \in J_0 \times J$ **do**
        $J_0 = J_0 \cup \{(\mathfrak{l}_i^b \star j_1, \mathfrak{l}_i^b \mathfrak{a}_1)\}$
        $J = J \cup \{(\mathfrak{l}_i^b \star j_2, \mathfrak{l}_i^b \mathfrak{a}_2)\}$
    end for
    if there is a collision $(j', \mathfrak{a}), (j', \mathfrak{b}) \in J_0 \times J$ **then**
        return $\mathcal{O} = \mathcal{O}_R(\mathcal{O}_0 \mathfrak{a}\mathfrak{b}^{-1})$
    end if
end for
return $\perp$

The computation of isogenies of degree ℓ_i can be done in $O(\ell_i M_{\mathbb{P}}(\ell_i) M_{\mathbb{Z}}(\log p))$, as Vélu's formulas can be computed in $O(\ell_i)$ operations over the field of definition of the kernel. The degree of the field of definition of the ℓ_i-torsion over $\mathbb{F}_p$ is $O(\ell_i)$ (see [**31**, Lemma 2] for instance). And so arithmetic operations over that field can be performed in $O(M_{\mathbb{P}}(\ell_i) M_{\mathbb{Z}}(\log p))$.

The kernel of the isogenies realizing the action of $\mathfrak{l}_i$ can be computed as the eigenvalue of the Frobenius morphism. These points can be computed by evaluating the Frobenius on a basis of the ℓ_i-torsion. The overall cost of this operation is $O(\sqrt{\ell_i} M_{\mathbb{P}}(\ell_i) M_{\mathbb{Z}}(\log p))$.

With $\ell_i = O(\log p)$, we get the final bound of $O(\log p^{3+\varepsilon})$ on the cost of each group action computation.

Under GRH, we have $h(\mathfrak{O}) = O(\sqrt{p}\,\mathrm{llog}(p))$, and so, by the birthday paradox, the number of j-invariants that needs to be computed in the sets J and J_0 before a collision can be found is in $O(p^{1/4} \log^{\varepsilon} p)$. Thus, the expected running time of EndomorphismRing is $O(p^{1/4} \log^{3+\varepsilon} p)$.

The order $\mathcal{O}_0$ can be computed with coefficients over $\mathbb{Q}$ of size $O(\log p)$ (see [**25**], for instance). The computation of $\mathcal{O}$ can be done in $O(\log p)$ as the coefficients of the ideal $\mathcal{O}\mathfrak{a}\mathfrak{b}^{-1}$ are in $O(p)$. The maximal order $\mathcal{O}$ can be given by 16 coefficients over $\mathbb{Q}$. Since $\mathcal{O}$ it is the right order of an ideal of norm in $O(p)$, the size of the coefficients is $O(\log p)$. The j-invariants and the ideals take $O(\log p)$ to store so the total memory cost is $O(p^{1/4} \log^{1+\varepsilon} p)$.

The output is the right order of the ideal $\mathcal{O}_0 \mathfrak{a}\mathfrak{b}^{-1}$. This ideal has norm $O(\log p)$ and so the coefficients of its right order over the basis $1, i, j, k$ of $\mathcal{B}_{p,\infty}$ can be upper-bounded by $O(\log p)$. Indeed, it can be easily verified that $\mathcal{O}n(\mathfrak{a}\mathfrak{b}) \subset \mathcal{O}_0$, and so we can express a basis of $\mathcal{O}$ as elements of $\mathcal{O}_0$ divided by $n(\mathfrak{a}\mathfrak{b})$. This gives the desired upper-bound on the size of $\mathcal{O}$. □

The supersingular evaluation algorithm. Computing $\Phi_\ell(j,X)$ mainly consists in computing the j-invariants that are ℓ-isogenous to j. When j is supersingular, we run this operation without any ℓ-isogeny computation by computing the endomorphism ring of j and using the Deuring correspondence to compute the ℓ-isogenous j-invariants.

Algorithm 4 SupersingularEvaluation(p,j,ℓ)

Input: A prime p, the j-invariant of a supersingular curve over $\mathbb{F}_p$, a prime ℓ.
Output: $\Phi_\ell(j,Y)$.
1: Compute E the curve whose j-invariant is j
2: Compute $\mathcal{O} = \mathsf{EndomorphismRing}(p,j)$
3: Compute $I_1,\cdots,I_{\ell+1}$ the $\ell+1$ $\mathcal{O}$-ideals of norm ℓ
4: Set $\mathfrak{S} = \{\mathcal{O}_R(I_i) | i \in \{1,\dots,\ell+1\}\}$
5: Compute $J = \mathsf{OrdersTojInvariantSmallSet}(p,\mathfrak{S})$ (or use OrdersTojInvariantBigSet$(p,\mathfrak{S})$ if this is more efficient)
6: **return** $\prod_{j_i\in J}(Y-j_i)$

PROPOSITION 3.5. SupersingularEvaluation *can be executed in*

$$O\left(\ell(\log \ell^{2+\varepsilon}\log p^{1+\varepsilon} + \log p^{2+\varepsilon}) + p\log p^{1+\varepsilon}\right)$$

binary operations and requires $O((\ell+p)\log p)$ *space when* OrdersTojInvariantBigSet *is used in Step* 5 *(assuming GRH,*[**31**, Claim 1] *and Heuristic* 1*). When using* OrdersTojInvariantSmallSet*, it can be executed in*

$$O\left(\ell(\log p^{4+\varepsilon} + \log \ell^{2+\varepsilon}\log^{1+\varepsilon} p) + p^{1/4}\log^{3+\varepsilon} p\right)$$

binary operations, requiring $O(\ell\log p + p^{1/4}\log^{1+\varepsilon} p)$ *space (under GRH, Heuristic* 1 *and the heuristics from* [**25**]*).*

PROOF. The correctness follows directly from the correctness of the sub-algorithms.

The computation of an $\mathcal{O}$-ideal of norm ℓ can be done in $O(C+\log\ell)$ where C is a bound on the size of the coefficients of $\mathcal{O}$. Thus, by Proposition 3.4, the complexity of computing the $O(\ell)$ ideals is $O(\ell(\log\ell+\log p))$. The computation of the rights orders has the same complexity.

Then, the result follows from [**31**, Theorem 1] for OrdersTojInvariantBigSet, the complexity of $O(\ell\log^{4+\varepsilon} p)$ for OrdersTojInvariantSmallSet for a set of size $O(\ell)$, Proposition 3.5 for EndomorphismRing and the complexity of the algorithm to compute a polynomial from its roots by building a product tree.

The memory requirements follow from the same results. ☐

3.3. A second CRT algorithm for big level. We obtain our second generic algorithm to evaluate modular polynomials by applying SupersingularEvaluation on a set of well-chosen small primes and then reconstructing the desired result using the CRT method. The only constraint on the choice of the CRT primes is that we need to find primes p_i where the reduction modulo p_i of the j-invariant to be evaluated is supersingular.

Algorithm 5 ModularEvaluationBigLevel(p, j, ℓ)

Input: A prime p, $j \in \mathbb{F}_p$, a prime ℓ.
Output: $\Phi_\ell(j, Y) \in \mathbb{F}_p[X]$.

Let $\overline{j}$ be the integer in $[0, p-1]$ equal to $j \mod p$
Set $B = 2^{6\ell \log \ell + 18\ell + (\ell+1)\log j + \log(\ell+2)}$
$\mathcal{P}_\ell(j) \leftarrow \{\}$, $P \leftarrow 1$
$\Delta \leftarrow \lceil \log B \rceil$
Compute $\mathcal{P}_\Delta$ as the set of primes smaller than Δ with Eratosthenes sieve
$q_\delta \leftarrow \max \mathcal{P}_\Delta$, $n \leftarrow q_\Delta + 2$
while $P < B$ **do**
  $S \leftarrow [n, n+1, \dots, n+\Delta]$
  Remove all multiples of the elements of $\mathcal{P}_\Delta$ from S
  for $q \in S$ **do**
    Let $j_q = \overline{j} \mod q$ and E_q be the elliptic curve over $\mathbb{F}_q$ of j-invariant equal to j_q
    if E_q is supersingular **then**
      $P \leftarrow q \cdot P$, $\mathcal{P}_\ell(j) \leftarrow \mathcal{P}_\ell(j) \cup \{q\}$
    end if
  end for
  $n \leftarrow n + \Delta + 1$
  if $q_\Delta^2 \leq n$ **then**
    $S = [q_\Delta + 2, q_\Delta + 3, \dots, 2q_\Delta + 1]$
    Remove all multiples of the elements of $\mathcal{P}_\Delta$ from S
    $\mathcal{P}_\Delta \leftarrow \mathcal{P}_\Delta \bigcup S$
    $q_\Delta \leftarrow \max \mathcal{P}_\Delta$
  end if
end while
Perform the pre-computations for the explicit CRT mod p using $\mathcal{P}_\ell$
for $q \in \mathcal{P}_\ell$ **do**
  $P_q(Y) \leftarrow$ SupersingularEvaluation(q, j_q, ℓ)
  Update the CRT sums for each coefficient of $P_q(Y)$
end for
Perform the post-computation for the explicit CRT to obtain $P(Y) \in \mathbb{F}_p[X]$
return $P(Y)$

PROPOSITION 3.6. ModularEvaluationBigLevel *is correct and, under the Lang–Trotter Conjecture* [**28**], *Heuristic* 1, *and the heuristic from* [**25**], *the computation of* $\mathcal{P}_\ell(j)$ *can be done in*

$$O\left(\ell^2 \log^2 j (\log^{1+\varepsilon}(\ell \log j))\right)$$

and the rest of the computation in

$$O\left(\ell^2 \log j \log^{3+\varepsilon} \ell + \ell^{3/2} \log^{3/2} j \log \ell^{3+\varepsilon} + \ell \log^{2+\varepsilon} \log^{1+\varepsilon} p\right).$$

The memory complexity is $O(\ell \log(pj))$ *for each step.*

PROOF. According to the Lang–Trotter conjecture, there are $O(\sqrt{x}/\log x)$ primes q smaller than x such that the reduction of any given $j \in \mathbb{Q}$ modulo

q is supersingular. To ensure that their product is bigger than B, we can take $x = O(\log^2 B)$, and there will be $O(\log B/\text{llog}B)$ different primes.

With $\log B = O(\ell \log j)$, we obtain $\#\mathcal{P}_\ell(j) = O(\ell \log j / \log(\ell \log j))$ and so $\max_{q \in \mathcal{P}_\ell(j)} q = O(\ell^2 \log j^2)$. The set $\mathcal{P}_\ell(j)$ takes $O(\ell \log j)$ memory to store.

The algorithm we propose to use to compute $\mathcal{P}_\ell(j)$ is a simple variation of the segmented sieve of Bays and Hudson [**4**] that can enumerate through all primes smaller than a given x in $O(x\text{llog}x)$ complexity while using $O(\sqrt{x})$ space when $\Delta = O(\sqrt{x})$, and combine it with supersingularity tests on the fly (to avoid storing a list of all the primes smaller than x). Supersingularity testing over $\mathbb{F}_q$ can be performed in $O(\log^{2+\varepsilon} q)$ [**42**], and it is only performed on prime numbers. Thus, we end up with a total complexity of $O\left(\ell^2 \log^2 j(\log^{1+\varepsilon}(\ell \log j))\right)$ and a space complexity of $O(\ell \log j)$ to compute $\mathcal{P}_\ell(j)$.

For each q, by Proposition 3.1 and the fact that $q = O(\ell^2 \log j)$, the complexity of the execution of SupersingularEvaluation with OrdersTojInvariantSmallSet (which will be faster than using OrdersTojInvariantSmallSetbecause $p = \Theta(\ell^2 \log^2 j)$) is

$$O\left(\ell(\log^{4+\varepsilon}(\ell \log j) + \log^{2+\varepsilon} \ell \log^{1+\varepsilon}(\ell \log j)) + \ell^{1/2} \log j^{1/2} \log(\ell \log j)^{3+\varepsilon}\right).$$

We deduce that the global cost of all the executions of SupersingularEvaluation is

$$O\left(\ell^2 \log j \log^{3+\varepsilon}(\ell \log j) + \ell^{3/2} \log^{3/2} j \log^{2+\varepsilon}(\ell \log p)\right).$$

Finally, the cost of the CRT computation is $O(\log^{2+\varepsilon} \log^{1+\varepsilon} p)$ as shown in [**8**]. This concludes the proof of the final complexity result.

The prime search can be done with $O(\ell \log j)$ memory (to store the set $\mathcal{P}_\ell$). Each execution of SupersingularEvaluation requires a memory of

$$O(\ell \log(\ell \log j)) + \ell^{1/2} \log^{1/2} j \log^{1+\varepsilon} \log(\ell \log j).$$

The only thing to store between each CRT prime computation are the CRT data whose size is $O(\ell \log p)$. This proves the result on the memory requirement. □

REMARK 3.7. We note that unlike most other CRT algorithms, in the generic case where $j = \Theta(p)$, the asymptotically dominant step of ModularEvaluationBigLevel is the selection of the primes.

Sharing the computation of the CRT primes. The set of primes $\mathcal{P}_\ell(j)$ only depends on the value of j. Thus, if we want to evaluate modular polynomials of different levels at the same j-invariant (like in the SEA algorithm), the cost of the selection of the primes can be done once and for all for the biggest levels, and then the computation for each level can simply use a subset of these primes. This means that for enough levels, the bottleneck will not necessarily be the computation of the primes anymore. In particular, in point counting for elliptic curves over a finite field of prime characteristic of bitsize n where one needs to evaluate j on $O(n)$ modular polynomials of level $O(n)$, the cost of the prime selection is amortized over the different levels.

3.4. Comparison between the existing methods. The best algorithms from the literature are Sutherland's algorithms from [**41**] and Robert's algorithm from [**35**]. We are going to compare our new algorithms with them. To establish what is the best asymptotic algorithm, we fix a value of ℓ and vary the value of p

(and j when it is relevant). We will see that the best algorithm will change as p grows.

In [**41**], Sutherland introduces three different CRT algorithms to compute modular polynomials. The main one has a complexity of

$$O\left(\ell^2(\ell \log \ell + \log p) \log^{2+\varepsilon}(\ell \log \ell + \log p)\right),$$

with a $O(\ell \log p + \ell^2 \log(\ell \log \ell + \log p))$ space requirement. The complexity is essentially the same as the cost of computing the entire modular polynomial with a CRT algorithm, but the memory requirement is smaller. Despite this, the space complexity is quadratic in ℓ, which is not optimal since the size of the output is $O(\ell \log p)$. The second algorithm due to Sutherland achieves a $O(\ell^{3+\varepsilon} \log p)$ complexity with an optimal space requirement. The third is a hybrid version of the first two and has a complexity of $O(\ell^3 \log^{6+\varepsilon} \ell)$. It also requires an optimal amount of space.

More recently in [**35**], Robert briefly outlined an algorithm with time complexity equal to $O(\ell^2 \log^{\alpha+\varepsilon} \ell \log p)$ where α is at best equal to $1 + 2u$ (according to [**35**, Proposition 3.1]) for $u \in \{1, 2, 4\}$ such that ℓ minus a product of small primes powers is equal to a sum of u squares. Heuristically, by adjusting the product of small primes, one can expect to find a case where $u = 2$. Thus, at best, the complexity of Robert's algorithm should be $O(\ell^2 \log^{5+\varepsilon} \ell \log p)$. However, note that Robert's paper only gives an outline of the algorithm and of its complexity analysis. Without clear statements, it is hard to assess the real complexity of the algorithm. Indeed, from the statements in [**35**], it is not completely clear to us if his asymptotic estimate can be used in the generic case. If not, then the correct value of α is actually $3 + 2u = 7$. The space complexity of Robert's algorithm is not clearly stated either, but it does not seem to be optimal.

Thus, for small values of p or j, our algorithm ModularEvaluationBigLevel appears to have the best known asymptotic complexity. In particular, if $j = O(p^\varepsilon)$ for $\varepsilon < 1$, then our algorithm has the best complexity of all existing algorithms.

If $j = \Theta(p)$, as p grows, the quadratic factor in $\log j$ in our complexity will mean that ModularEvaluationBigLevel will be outperformed by Robert's algorithm. The asymptotic complexity tells us that the breaking point should occur when $\log p \approx \log \ell^{1+2u}$. Although, note that Robert's algorithm probably requires more memory, therefore there may be some practical case where our algorithm would still outperform Robert's. Also note that if we remove the cost of the CRT prime computation (that can be shared among the computation for several levels as already argued), then we expect our algorithm to outperform Robert's for a much larger range of primes (up to $\log p \approx \ell$).

When $\ell = O(\log p)$, the best complexity will be obtained by Sutherland's first CRT algorithm and our ModularEvaluationBigCharacteristic algorithm, but ModularEvaluationBigCharacteristic has a better space complexity than Sutherland's algorithm (which is quadratic in ℓ).

Note that the proof of the complexity of Robert's algorithm does not require any assumptions, whereas our method and Sutherland's are only heuristic.

In Section 4.2, we will compare the practical performance of our C++ implementation with the one from Sutherland.

3.5. Other modular functions. Modular polynomials can be generalized to modular functions other than the j-function. Considering other kinds of modular polynomials is a well-known trick to make computations more efficient (see for instance [**8**]). Indeed, these alternate modular polynomials may have smaller coefficients. One of the most interesting examples appears to be the modular polynomial associated to the Weber function $\mathfrak{f}$. Over $\mathbb{C}(j)$, this function is a root of the polynomial

$$\Psi(X, j) = (X^{24} - 16)^3 - jX^{24} \tag{3.1}$$

and the logarithmic height bound of the associated modular polynomial Φ_ℓ^ℓ is expected to be 72 times smaller than the one of Φ_ℓ. Moreover, the coefficient of X^aY^b is non-zero if and only if $\ell a + b = \ell + 1 \mod 24$. These two facts imply that the complexity of computing $\Phi_\ell^{\mathfrak{f}}$ should be roughly $72 \times 24 = 1728$ faster than Φ_ℓ. By using the simple algebraic relation between j and $\mathfrak{f}$ given by Eq. (3.1), it quickly becomes more efficient to use $\Phi_\ell^{\mathfrak{f}}$ rather than Φ_ℓ.

Over $\mathbb{F}_q$, every j-invariant is associated to several $\mathfrak{f}$-invariants that are the roots of $\Psi(X, j)$ given in Eq. (3.1). In fact, this polynomial comes from the covering map of $X(1) \cong \mathbb{P}^1$ by X_H, a modular curve of level 48 isomorphic to the map labelled 48.72.0.d.1 in the (Beta version of the) LMFDB [**32**]. The polynomial $\Phi_\ell^{\mathfrak{f}}$ can be constructed from $\mathfrak{f}$-invariants in a manner analogous to Φ_ℓ from j-invariants. When ℓ is coprime to 48, it makes sense to talk about ℓ-isogenous $\mathfrak{f}$-invariants, and evaluating $\Phi_\ell^{\mathfrak{f}}(f_0, X)$ can be done by computing all the ℓ-isogenous $\mathfrak{f}$-invariants.

This idea was described in BLS [**8**], and Leroux already stated in [**31**] that the BLS method could be extended to this setting analogously. The goal of this section is to explain how this can be realized concretely. We find that the BLS method cannot be directly applied, and instead new ideas are needed. The main obstacle of using $\mathfrak{f}$-invariants instead of j-invariants is the multiplicity of the polynomial Ψ, which implies that there are several $\mathfrak{f}$-invariants corresponding to the same j-invariant (in the worst case, 72). Therefore, to find the ℓ-isogenous $\mathfrak{f}$-invariants, the usual method of projecting from X_H down to $X(1)$, finding an ℓ-isogenous j-invariant, and then lifting again to the correct $\mathfrak{f}$-invariant does not work. Indeed, it is hard to know which lift of j to consider.

The method used in [**8**]. Bröker, Lauter, and Sutherland overcome this obstacle by working purely over X_H. Indeed, by using the class group action of a quadratic imaginary order of conductor ℓ on $X(1)$, they enumerate the set of neighbours in the ℓ-isogeny graph. This action can then be extended to X_H. More explicitly, by using the modular polynomial of level ℓ_i, one can compute the action of ideals of small norm ℓ_i. When working over X_H, we can simply use the modular polynomial $\Phi_{\ell_i}^{\mathfrak{f}}$ instead, and the roots will directly give the ℓ_i-isogenous $\mathfrak{f}$-invariants. We furthermore remark that this method works as the BLS method only needs to consider points of $X_H(\mathbb{F}_p)$.

To adapt this idea to our setting of supersingular curves defined over $\mathbb{F}_{p^2}$, the main problem is that the required j-invariants are not computed with the help of modular polynomials, but rather as the codomain of certain isogenies. As there are no efficient isogeny formulæ for isogenies of degree ℓ between elements of $X_H(\mathbb{F}_{p^2})$, we cannot directly derive the ℓ-isogenous $\mathfrak{f}$-invariants. Our idea is to use a different interpretation of the curve X_H. Indeed, modular curves are known to parametrize elliptic curves enriched with level structure. Thus, if we have an explicit way to associate the elements of X_H with curves and associated level structure of order 48 (meaning that we can compute the value of the $\mathfrak{f}$-invariant only from the curve and the level structure), then it suffices to push the level structure through the isogenies of degree ℓ to be able to recover the ℓ-isogenous $\mathfrak{f}$-invariant. This will work for ℓ coprime to 48, hence for every prime ≥ 5.

The parametrization associated to X_H. The level of X_H is $48 = 3 \times 16$, so we instead consider the level structure of order 3 and 16 separately.

Let us first consider the level 3 part. For this, we look at Eq. (3.1) and observe that if f is a root of $\Psi(X, j)$ for a given j, then f^8 is a root of $\Psi'(X, j^{1/3}) = (X^3 - 16) - Xj^{1/3}$. This is convenient because the j-function is well-known to be the cube of another modular function of level 3, often denoted by γ_2. In terms of modular curves, this can be interpreted as the cover map

$$X_{ns}^+(3) \to X(1), \ t \mapsto t^3.$$

There is a classical formula to compute the three possible γ_2-invariants above a given j-invariant of a curve E and the x-coordinates of $E[3]$ (see for instance in [**9**, Section 6.6, Example 1]). If E is the curve $y^2 = x^3 + Ax + B$ and x_1, x_2, x_3, x_4 are the x-coordinates of the non-trivial points of $E[3]$, then the roots of $X^3 - j(E)$ are given by

$$\frac{-48A}{2A - 3(x_1x_2 + x_3x_4)}, \quad \frac{-48A}{2A - 3(x_1x_3 + x_2x_4)}, \quad \frac{-48A}{2A - 3(x_1x_4 + x_3x_2)}.$$

In summary, given an ordering for the x-coordinates of the non-trivial points of $E[3]$, we can compute the three possible γ_2-invariants associated to E. This covers the part of level 3.

We now consider the level 16 part. Here, we look for the solutions of $\Psi''(X, j) = (X^8 - 16)^3 - jX^8$, which will correspond to $\mathfrak{f}^3$. The map $t \mapsto \frac{(t^8-16)^3}{t^8}$ corresponds to the cover of $X(1)$ by the modular curve $X_{H'}$ of level 16 labelled as 16.24.0.p.1 in the new beta version of LMFDB. To the best of our knowledge, there are no known formulas to recover the $\mathfrak{f}^3$-invariant from a level structure of order 16. Such formulas certainly exist, but it is unclear to us how can they be computed efficiently. Instead, we propose to use the much better studied modular curve $X_s(16)$ above $X_{H'}$. In particular, $X_s(16)$ parametrizes curves together with two subgroups of order 16 which intersect trivially. In this way, given a curve $E(\mathbb{F}_q)$ and two subgroups $G_1, G_2 \subset E[16](\mathbb{F}_q)$, we can recover the corresponding element of $X_s(16)(\mathbb{F}_q)$ by looking at the common preimage of $j(E)$ through the two maps $X_s(16) \to X(1)$ factoring through the two covers

$$\begin{aligned} \psi_1 &\colon X_s(16) \to X_0(16), \quad E, G_1, G_2 \mapsto E, G_1, \\ \psi_2 &\colon X_s(16) \to X_0(16), \quad E, G_1, G_2 \mapsto E, G_2; \end{aligned}$$

see Figure 1.

Remark 3.8. The method that we outlined above works most of the time, but there are some problems when there exist 16-isogenies between the same pair of curves (as the map $X_0(16) \to X(1) \times X(1)$ is not injective anymore and so our GCD polynomial will have several roots). In that case, it is still possible to identify the element of $X_0(16)$ corresponding to the given subgroup of order 16 by decomposing the 16-isogeny associated to this subgroup as 4 isogenies of degree 2, finding the 4

$X_0(2)$ points corresponding to each of these 2-isogenies, and finally identifying the correct root of the GCD with the different existing maps from $X_0(16)$ to $X_0(2)$.

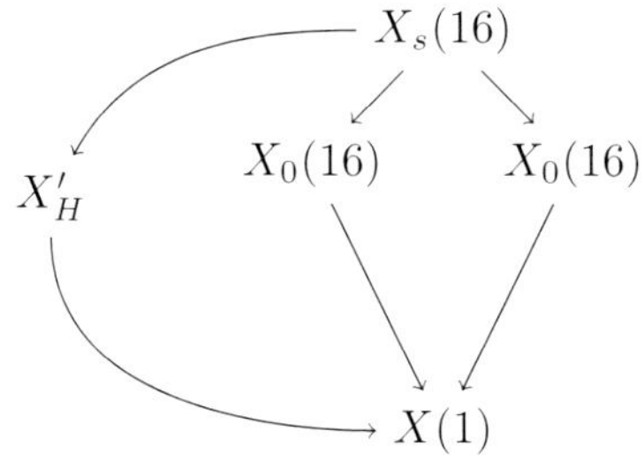

FIGURE 1. Modular covers of $X(1)$ by $X_s(16)$

We do this as follows:

(1) Compute the points of $X_0(16)$ lying above the two pairs (j, j_1) and (j, j_2) of j-invariants, i.e., for $k = 1, 2$, compute x_k such that $\rho(x_k) = (j, j_k)$ where $\rho : X_0(16) \to X(1) \times X(1)$ is given by

$$\rho(X) := \left(\frac{G_0(16)(X)}{H_0(16)(X)}, \frac{I_0(16)(X)}{J_0(16)(X)} \right),$$

for $G_0(16)(X), H_0(16)(X), I_0(16)(X), J_0(16)(X) \in \mathbb{Z}[X]$. If it fails because there are multiple solutions, use the method outlined in Remark 3.8

(2) Recover the point (x, y) on $X_s(16)$ above the two points of $X_0(16)$ by finding a solution to the system of equations consisting of

$$F_s(16)(X, Y) = 0 \qquad \frac{G_s(16)(X)}{H_s(16)(X, Y)} = (x_1) \qquad \frac{I_s(16)(X)}{J_s(16)(X, Y)} = (x_2),$$

for $F_s, G_s, H_s, I_s, J_s \in \mathbb{Z}[X, Y]$. Where F is the equation of $X_s(16)$, and $\frac{G_s}{H_s}, \frac{I_s}{J_s}$ give the two projection maps $X_s(16) \to X_0(16)$.

(3) Project this point (x, y) on $X_s(16)$ to X'_H via $\varphi \colon X_s(16) \to X'_H$, defined as

$$\varphi(X, Y) := \frac{-16X^6Y^{12} - X^6Y^4 + 64X^2Y^{16} + 20X^2Y^8 + X^2}{32Y^{15} + 4Y^7}.$$

Then, $\varphi(x, y) = \mathfrak{f}^3$.

For convenience, we recall the formulas for all the maps involved in the computation described above in **??**. Some of these maps can be found on the LMFDB, and the other maps were computed using MAGMA [**7**].

To develop an efficient algorithm that follows the procedure above, we use the method detailed in [**13**, §5]. We first define polynomials

$$\begin{aligned} f_1(X, Y) &:= F_s(16)(X, Y), \\ f_2(X, Y) &:= G_s(16)(X, Y) - \alpha_1(j, j_1)H_s(16)(X, Y), \\ f_3(X, Y) &:= I_s(16)(X, Y) - \alpha_2(j, j_2)J_s(16)(X, Y), \end{aligned}$$

where, for $k = 1, 2$, $\alpha_k \in \mathbb{Z}[j, j_k]$ is defined as

$$\alpha_k(j, j_k) := \gcd\big(G_0(16)(X) - jH_0(16)(X), I_0(16)(X) - j_kJ_0(16)(X)\big).$$

For $i, j \in \{1, 2, 3\}$, we also define polynomials

$$R_{i,j}(Y) := \mathrm{res}_X(f_i(X,Y), f_j(X,Y)) \in \mathbb{Z}[j, j_1, j_2][Y].$$

By the elimination property of the resultant (e.g., see [**16**, §3.6, Lemma 1]), the specialisations $(R_{i,j})_{[E,G_1,G_2]}(Y)$ given by evaluating the coefficients of each $R_{i,j}$ at $j := j(E)$, $j_1 := j(E/G_1)$ and $j_2 := j(E/G_2)$, vanish at the Y-coordinate of any common solution to the specialised polynomials $(f_j)_{[E,G_1,G_2]}(X,Y)$.

However, these resultants (generically) have factors which correspond to spurious (unwanted) solutions. Therefore, we instead consider polynomials where these spurious solutions have been removed, namely:

$$\begin{aligned} P_{1,2}(Y) &:= \frac{16}{Y^{108}(16Y^8+1)^7} \cdot R_{1,2}(Y), \text{ and} \\ P_{2,3}(Y) &:= \frac{16}{Y^{77}(16Y^8+1)^7} \cdot R_{2,3}(Y). \end{aligned}$$

Note that now $P_{1,2}$ and $P_{2,3}$ are coprime.

If there exist $x, y \in \mathbb{F}_p$ such that $(f_i)_{[E,G_1,G_2]}(x,y) = 0$ for each $i = 1, 2, 3$ then the degree of

$$g(Y) := \gcd\big((P_{1,2})_{[E,G_1,G_2]}(Y), (P_{2,3})_{[E,G_1,G_2]}(Y)\big)$$

is 1. Conversely, if $y \in \mathbb{F}_p$ is a root of $g(Y)$, then there exist $x, x' \in \mathbb{F}_p$ such that

$$\begin{aligned} (f_1)_{[E,G_1,G_2]}(x,y) &= (f_2)_{[E,G_1,G_2]}(x,y) = 0, \text{ and} \\ (f_2)_{[E,G_1,G_2]}(x',y) &= (f_3)_{[E,G_1,G_2]}(x',y) = 0. \end{aligned}$$

We assume $x = x'$ (otherwise, we throw an error and restart the procedure; note this will only happen with negligible probability).

Therefore, from a gcd computation, we can extract $y \in \mathbb{F}_p$, and then recover x via a square root computation. Then, (x, y) is a point on $X_s(16)$, and we have $\mathfrak{f}^3 = \varphi(x,y)$, where $\varphi : X_s(16) \to X'_H$.

In practice, rather than computing the polynomials $P_{1,2}$ and $P_{2,3}$ on the fly, we precompute and store them. Then, given E, G_1, G_2 defined over $\overline{\mathbb{F}}_p$, we evaluate the polynomials at corresponding j, j_1, j_2, reduce them modulo p, and compute their gcd. We summarise the discussion above in Algorithm 6. Here, Roots denotes a function that returns the roots of a polynomial over a finite field.

Algorithm 6 GetWeberCube(p, j, j_1, j_2)

Input: A prime p, and j-invariants j, j_1, j_2 of $E, E/G_1, E/G_2$ respectively.
Output: The cube $\mathfrak{f}^3$ of the $\mathfrak{f}$-invariant associated to E.

Evaluate the coefficients of $P_{1,2}(Y)$ at j, j_1, j_2 and reduce modulo p to obtain $h_1(Y) \in \mathbb{F}_p[Y]$
Evaluate the coefficients of $P_{2,3}$ at j, j_1, j_2 and reduce modulo p to obtain $h_2(Y) \in \mathbb{F}_p[Y]$
$c_0 Y + c_1 \leftarrow \gcd(h_1, h_2)$
$y \leftarrow -c_1 \cdot c_0^{-1}$
`rts` $\leftarrow$ Roots$(f_1(X, y))$
for r in `rts` **do**
  if $f_3(r, y) = 0$ **then**
    return $\varphi(r, y)$
  end if
end for
return $\perp$

In conclusion, an ordering x_1, x_2, x_3, x_4 of the x-coordinates of $E[3]$ and two subgroups of order 16 with trivial intersection in $E[16]$, we obtain two polynomials in X^8 and X^3. It then suffices to take the gcd to find the value of the Weber invariant associated to this level structure of order 48 on E: if t is the output of GetWeberCube(p, j, j_1, j_2), we compute

$$\gcd(X^3 - t, X^8 - \gamma_2)$$

and obtain $\mathfrak{f}$-invariant associated to E.

To find the ℓ-isogenous $\mathfrak{f}$-invariants corresponding to some ℓ-isogeny φ it suffices to evaluate the level structure in φ and repeat the procedure depicted above. Since our algorithms relies on computing alternative paths through the Deuring correspondence (in particular, we never explicitly compute any ℓ-isogenies), obtaining this evaluation requires the use of techniques which at this point are standard in isogeny-based cryptography. We refer to Section 4.1 for details.

4. Implementation results

We have implemented the algorithms ModularEvaluationBigLevel and ModularEvaluationBigCharacteristic in C++/NTL. The aim of this section is to give an overview of the code-base, before we present our implementation results.

4.1. Implementation details. We first describe the different parts of our library, which is available at:

`https://github.com/tonioecto/modular-polynomial-computation-and-evaluation-from-supersingular-curves`

NTL functionalities. Our library uses NTL to perform basic arithmetic over finite fields $\mathbb{F}_{p^k}$ (for $k \geq 1$), and can be compiled both using classes `NTL::zz_p` and `NTL::zz_pE` or `NTL::ZZ_p` and `NTL::ZZ_pE` if needed, integer arithmetic using class `NTL::ZZ`, and polynomial arithmetic using classes `NTL::zz_pX`, `NTL::zz_pEX` or `NTL::ZZ_pX`, `NTL::ZZ_pEX` if needed. We furthermore use NTL vectors and matrices constructed from these classes.

Finding CRT primes. We precomputed a list of all primes up to 2^{25}, and use a segmented version of the sieve of Eratosthenes [**4**] to find larger primes if required (with an upper bound of 2^{50}). Based on the implementation of [**14**], it takes an interval $[L, R]$ and finds all contained primes. Sieving different intervals is independent, so the segmented sieve parallelizes perfectly.

Instead of storing all primes and testing supersingularity of the corresponding elliptic curves later, we run supersingularity tests on the fly after identifying all primes in a sieving interval $[L, R]$. Since the overwhelming majority of curves we test is ordinary, our implementation benefits most from a test that discards those curves quickly. Following [**3**], the best choice in our context is Sutherland's supersingularity test based on determining whether the 2-isogeny graph has a volcano structure (ordinary case) or not (supersingular case) [**42**]. We implement Sutherland's test including the optimisations proposed in [**3**].

Field arithmetic. We require arithmetic in $\mathbb{F}_{p^{2k}}$ for small k. For all basic operations, we rely on NTL. Further, we precompute a matrix corresponding to the Frobenius action. Additionally, due to how the isogeny computations work, we also need a fixed, effective embedding $\iota_k : \mathbb{F}_{p^2} \hookrightarrow \mathbb{F}_{p^{2k}}$ for all k used. This embedding is given by a fixed image $\iota_k(\omega)$, where ω is a fixed generator of $\mathbb{F}_{p^2}$ of trace 0.

For an element $a + b\omega \in \mathbb{F}_{p^2}$ (represented in NTL as the tuple (a, b)), we can easily lift this to $\mathbb{F}_{p^{2k}}$ by computing $a + \iota_k(\omega)b$ in $\mathbb{F}_{p^{2k}}$. To coerce elements into $\mathbb{F}_{p^2}$, we proceed as follows. Given an element $\alpha \in \mathbb{F}_{p^{2k}}$ (represented in NTL as the tuple $(a_1, \ldots, a_{2k})$), which we know is in the image of ι_k, i.e., $\alpha = \iota_k(a + \omega b)$:

(1) Recover a from the trace as $a = \frac{\operatorname{tr} \alpha}{2k}$.
(2) Recover b as $a_j / \iota_k(\omega)_j$, where $j > 1$ is any non-zero index of the representation $(\iota_k(\omega)_1, \ldots, \iota_k(\omega)_{2k})$ of the element $\iota_k(\omega)$.

We extended these lifting and coersion operations to the polynomial rings $\mathbb{F}_{p^{2k}}[X]$, by lifting and coercing each coefficient, respectively.

Elliptic curves and isogenies. We implement standard elliptic curve operations using Weierstraßcurves. One of the main computational tasks required by the elliptic curve part of our library is the computation of smooth degree isogenies given by kernel generators of smooth order defined over small extensions $\mathbb{F}_{p^{2k}}$ of $\mathbb{F}_{p^2}$. This setting is similar to that in [**22**], and thus the strategy employed is the same. Namely, we use supersingular curves E with $\pi_E^2 = [-p]$, where π_E denotes the p-Frobenius endomorphism on E. As a result, all the isogenies will be defined over $\mathbb{F}_{p^2}$, even if the kernel generators are not. Due to this, we can compute our isogenies by recovering the kernel polynomial using [**22**, Algorithm 4], before employing Kohel's formula to recover the isogeny [**26**, Chapter 2.4].

Computing the endomorphism ring of a given supersingular elliptic curve E defined over $\mathbb{F}_p$ is done by brute-force searching an isogeny to a curve with known endomorphism ring. Concretely, we start by determining the $\mathbb{F}_p$-endomorphism ring R of E; the two cases $\mathbb{Z}[\pi] \cong \mathbb{Z}[\sqrt{-p}]$ and $\mathbb{Z}[\frac{1+\pi}{2}] \cong \mathbb{Z}[\frac{1+\sqrt{-p}}{2}]$ are readily distinguished by testing whether the rank of $E(\mathbb{F}_p)[2]$ equals 1 or 3 respectively. This boils down to checking whether the cubic polynomial $f(x)$ in the short Weierstraßequation $y^2 = f(x)$ defining E splits over $\mathbb{F}_p$ or not. Then, using a suitable starting curve E_0 with the same $\mathbb{F}_p$-endomorphism ring and known $\mathbb{F}_{p^2}$-endomorphism ring, there must exist an ideal class $[\mathfrak{a}] \in \mathrm{Cl}(R)$ such that $[\mathfrak{a}] * E_0 = E$, which we recover using a straightforward meet-in-the-middle brute-force search using a (heuristic) generating set of small generators for $\mathrm{Cl}(R)$. In practice, this means we do not set explicitly the constant C from Algorithm 3, we just increase the exponents until a solution is found.

Lattices in quaternion algebras. Our library identifies the quaternion algebra $B = H(-q, -p)$ simply by the pair $p, q \in \mathbb{Z}$, and the format of its elements consists of a reference to B together with a 5-tuple of integers $t, x, y, z, d \in \mathbb{Z}$ representing the quaternion $(t+xi+yj+zk)/d$. Arithmetic in this representation is straightforward. We further represent a quaternion lattice I by a basis matrix in $\mathbb{Q}^{4\times 4}$, which is in turn represented as a matrix in $\mathbb{Z}^{4\times 4}$ together with a denominator in $\mathbb{Z}$. All basic functionality for quaternion lattices, such as computing the intersection, product, or sum of two lattices, computing left or right orders, or finding canonical or reduced lattice bases, all essentially rely on the linear-algebra routines provided by NTL.

Our library also includes an implementation of a variant of the KLPT algorithm [**25**], which works for all primes where there exists an embedding $K \hookrightarrow B_{p,\infty}$, where K is an imaginary quadratic field of class number 1. Note that this is a very mild assumption on the prime p (a quick heuristic estimate is that only 1 in every 2^9 primes fail this requirement). Our implementation uses the standard improvement due to Petit and Smith [**34**] to reduce the output size. However, our implementation also differs in another way. The output size (with the improvement from [**34**]) of KLPT is typically stated as being in $\tilde{O}(p^3)$. However, this is based on the heuristic that the shortest prime-normed equivalent ideal is of norm $O(p^{1/2}\log(p))$. While this heuristic almost never fails in large characteristic, we are working in small enough characteristic to regularly encounter ideals where this fails. Hence, we change the size of the solution we search for on the fly to be in $\tilde{O}(p^2 N_I^2)$, where N_I denotes the norm of the shortest prime-normed equivalent ideal.[1]

Orders to j-invariants. In OrdersTojInvariantSmallSet, we repeatedly apply the same procedure as in [**22**] to translate orders to j-invariants. The only difference is that our implementation has precomputed elliptic curves $E/\mathbb{Q}$ with CM by an order of class number 1, together with the isogeny ι defining the complex multiplication. Thus, for the first step described in [**22**], we can take the starting curve $E_0/\mathbb{F}_p$, to be a supersingular reduction of one of our precomputed curves. Note that the requirement on p that one of the precomputed curves has supersingular reduction coincides with the requirement we described for KLPT (except for a constant number of primes p, where the precomputed curves might have bad reduction). Further, note that for $p \equiv 1 \pmod{12}$, the isomorphism $\mathcal{O}_0 \cong \mathrm{End}(E_0)$ cannot be determined solely by the (reduction of the) endomorphism ι; see [**22**, Section 3.1] for details, and how to resolve this ambiguity.

ModularEvaluationBigCharacteristic also requires the OrdersTojInvariantBigSet algorithm introduced in [**31**] to compute efficiently all the supersingular j-invariants and the corresponding maximal order. Our implementation follows quite closely the algorithm detailed by Leroux (see [**31**, Algorithm 1]). Maximal orders are represented by the 3 successive minima (that can be computed with LLL) of the trace-0 sublattice.

[1]Note that we do have the guarantee that $N_I \in \tilde{O}(p)$, hence the size of the output size is still upper bounded by $\tilde{O}(p^4)$.

Computing Weber invariants. To compute the Weber invariant of a curve with a given level structure, we implement the ideas discussed in Section 3.5. In particular, we implement subroutines to evaluate multivariate polynomials of degree 2 and 3 (NTL only supports univariate polynomials), and we use built-in NTL functions for gcd computations and root finding. Furthermore, the algorithm requires computing the gcd of two precomputed resultants, which are stored in `getresultants.cpp`. The memory cost of the hardcoded resultants is negligible in the context of the full algorithm, and the efficiency gain is significant.

To obtain the ℓ-isogenous Weber invariants in ModularEvaluationBigLevel, we also need to evaluate the level structure at the ℓ-isogenies. Since no ℓ-isogenies are ever explicitly computed, this is done by evaluating the level structure on a well-chosen endomorphism, before pushing it through the alternative isogeny path we have computed. Explicitly, this endomorphism equals the composition of the ℓ-isogeny with the dual of the alternative path, and can be found by multiplying the corresponding ideals.

Furthermore, OrdersTojInvariantBigSet needs to be adapted to work with Weber invariants instead of j-invariants. For that, we represent Weber invariants as a j-invariant together with some basis of the 48-torsion, and with some coefficients to represent the level structure in the basis. By exploiting the symmetries coming from the Weber function, we obtain an efficient algorithm to recover the full list of Weber invariants.

OrdersTojInvariantBigSet also needs to be adapted to work with Weber invariants instead of j-invariants, and so we need to modify OrdersTojInvariantBigSet to compute all supersingular $\mathfrak{f}$-invariants as well. For that, we will use our level structure parametrization of $\mathfrak{f}$-invariants. That way, by using a basis of the 48-torsion, each Weber invariant can be fully derived from a set of small integers (giving the coefficients of the points of the level structure in the basis). Then, it suffices to propagate the 48-torsion basis throughout the isogenies involved in OrdersTojInvariantBigSet. For each supersingular j-invariant, we will then need to compute all $\mathfrak{f}$-invariants above it from the basis. The symmetries coming from the Weber function allow us to be much more efficient to recover the full list of Weber invariants than one application of the algorithm we described in Section 3.5 for each of the 72 Weber invariants.

The CRT method. For the implementation of the CRT method in our library, we follow Sutherland's software `classpoly` [**39**] based on [**21**, **40**]. In particular, our CRT functions in `crt.cpp` take heavy inspiration from `crt.c` in Sutherland's software. The main difference is that our algorithms use C++/NTL rather than C/GMP. We therefore defer a detailed explanation on the implementation of the CRT method to these previous works.

Polynomial interpolation. In our library, we implement univariate polynomial interpolation following Algorithm 10.11 in [**46**, §10]. In particular, we implement two main functions:

- An algorithm to interpolate a univariate polynomial $f(X) \in \mathbb{F}_{p^{2k}}[X]$ of degree d at $d+1$ points $(x_0, f(x_0)), \ldots, (x_d, f(x_d))$.
- An algorithm to interpolate a univariate polynomial $f(X) \in \mathbb{F}_{p^{2k}}[X]$ of degree d from its roots $(x_1, \ldots, x_d)$. This is a special case of the first algorithm where $f(x_i) = 0$, but here we optimise for this case.

We remark that these algorithms outperform the in-built NTL interpolation function `interpolate` and the NTL function `BuildFromRoots`, respectively, for polynomials of relatively large degrees (which is our use case in this paper).

4.2. Results. In this section, we present the performance of our implementation of the algorithms ModularEvaluationBigLevel and ModularEvaluationBigCharacteristic, and compare it to the state-of-the-art [**41**].

Performance analysis. All code was compiled and run on a Linux server with two AMD EPYC 9754 CPUs running at 2.25 GHz (with dynamic frequency scaling and simultaneous multithreading disabled), 1 TB RAM, and using `g++` version 15.1.1 with flags `-std=c++17 -O3 -march=native -DNDEBUG`.

We ran experiments for the Weber variant of both algorithms ModularEvaluationBigLevel and ModularEvaluationBigCharacteristic (i.e., we compute the Weber modular polynomial evaluated at a Weber invariant). To understand the performance of these algorithms for varying level ℓ, we fix the characteristic $p = 2^{31} - 1$, the Weber invariant $w = 2$, and run both algorithms for the same set of growing levels ℓ. Since our implementation is heavily multithreaded, to enable a meaningful comparison with data collected on other hardware, we report the CPU *core* time (i.e., in single-core equivalents) rather than the wall-clock time consumed by each run of each algorithm. Our results are summarized in Table 1.

We see that despite a better asympotic complexity in ℓ, we were not able to reach a point where ModularEvaluationBigLevel is faster than ModularEvaluationBigCharacteristic. This is the due to the fact that the hidden constants in the asymptotic complexity of the SingleOrderTojInvariant algorithm are very large (due to the computation of a smooth isogeny of degree $\approx p^3$). Nonetheless, the current data is enough to witness that the scaling of ModularEvaluationBigLevel and ModularEvaluationBigCharacteristic are indeed quadratic and cubic in ℓ respectively.

Extrapolating the results presented in Table 1, we observe that our current implementation falls short of a record computation (using a similar computational power as in [**41**]). Take, for instance, the large characteristic setting. In the SEA point-counting record described in [**41**, Section 5] with $\log(p) \approx 16646$, the total time to compute evaluated modular polynomials using Weber invariants for level $\ell = 11681$ is reported to be around 2 CPU hours on a single thread.

For the same level ℓ, but a much smaller characteristic p, our implementation of ModularEvaluationBigCharacteristic took around 40 CPU hours. The dependency on $\log p$ in the asymptotic complexity comes from the $\log p$ term in the height bound. When taking $\log(p) \approx 16646$ for $\ell = 11681$, one sees that the height bound is essentially doubled (compared to $\log p \approx 30$). Thus, we can estimate that the same computation as Sutherland would have taken around 80 CPU hours with our implementation, which is 1 or 2 orders of magnitude slower. Since the computations were run on different hardware, it is difficult to be more precise than that.

Possible improvements. There are still some places where significant improvements could be made. We identify the following possible improvements to our software (in order of expected impact):

(1) Currently, the quaternion operations are the bottleneck in ModularEvaluationBigCharacteristic (taking up to 80% percent of the global

TABLE 1. Experimental data: CPU core time consumed by runs of ModularEvaluationBigLevel and ModularEvaluationBigCharacteristic for various levels ℓ, fixing the characteristic $p = 2^{31} - 1$ and the input Weber invariant $w = 2$.

Level ℓ	ModularEvaluationBigLevel	ModularEvaluationBigCharacteristic	Ratio
211	7.68 min	18.191 s	25.3
419	24.01 min	40.702 s	35.4
607	51.34 min	1.24 min	41.3
811	1.46 h	2.00 min	43.7
1019	2.28 h	3.07 min	44.5
2003	9.12 h	15.13 min	36.2
3011	21.54 h	46.53 min	27.8
4003	1.63 d	1.65 h	23.7
5003	2.66 d	3.31 h	19.3
6007	3.89 d	5.49 h	17.0
7019	5.32 d	8.42 h	15.2
8011	7.00 d	12.13 h	13.8
9007	9.13 d	17.93 h	12.2
10007	11.39 d	1.00 d	11.3
11681	15.78 d	1.56 d	10.1

run time) despite being asymptotically negligible compared to interpolation. One explanation is that the hidden constants in these operations can be quite big (one quaternion multiplication is already 17 integer multiplications), and most operations on orders and ideals require to apply operations on bases of four quaternion elements. Another reason is that our implementation uses `NTL::ZZ` which are integers of arbitrary size when the integers involved in all our quaternion computation can in fact be bounded. Thus, switching to fixed-sized integers should bring a noticeable improvement. Moreover, some of the operations on ideals use generic algorithms that could be improved with a tailored implementation.

(2) Using x-only arithmetic for elliptic curve operations, and in particular in isogeny computations which are by far the bottleneck in ModularEvaluationBigLevel due to the high constants involved in the complexity of computing the isogenies in OrdersTojInvariantSmallSet.

(3) Precomputing the action of each basis element of the endomorphism ring on the relevant ℓ^e-torsion bases of the starting curve would also greatly improve the cost of computing the isogeny kernels in OrdersTojInvariantSmallSet and thus positively impact the performances of ModularEvaluationBigLevel.

We are hopeful that the concrete efficiency gains from these improvements could allow to significantly improve our implementation to the point where ModularEvaluationBigCharacteristic could match or even beat the performances of the software from [**41**].

Despite the better asymptotic scaling, it seems unlikely that the improvements listed above would be enough for ModularEvaluationBigLevel to beat the other two algorithms for a realistic value of ℓ. To reach that point, one would likely need a deeper algorithmic improvement allowing us to decrease the degree of the isogenies used in SingleOrderTojInvariant.

We remark, however, that all the changes listed above require a significant implementation effort, and are thus out of scope for this paper.

5. Conclusion

We have introduced several new algorithms to evaluate modular polynomials using supersingular curves and the Deuring correspondence. These algorithms all improve the asymptotic time or space complexity of previously known algorithms for some range of inputs.

We implemented our algorithms in C++, exhibiting the practicality of the algorithms presented. We further suggest potential improvements that could lead to important practical speed-ups of our implementation, but they require an extensive optimization effort and are thus left for future work.

References

[1] N. C. Ankeny, *The least quadratic non residue*, Ann. of Math. (2) **55** (1952), 65–72, DOI 10.2307/1969420. MR45159

[2] Sarah Arpin, James Clements, Pierrick Dartois, Jonathan Komada Eriksen, Péter Kutas, and Benjamin Wesolowski, *Finding orientations of supersingular elliptic curves and quaternion orders*, Des. Codes Cryptogr. **92** (2024), no. 11, 3447–3493, DOI 10.1007/s10623-024-01435-5. MR4801719

[3] Gustavo Banegas, Valerie Gilchrist, and Benjamin Smith, *Efficient supersingularity testing over $\mathbb{F}_p$ and CSIDH key validation*, Mathematical Cryptology **2** (2022), no. 1, 21–35.

[4] Carter Bays and Richard H. Hudson, *The segmented sieve of Eratosthenes and primes in arithmetic progressions to* 10^{12}, Nordisk Tidskr. Informationsbehandling (BIT) **17** (1977), no. 2, 121–127, DOI 10.1007/bf01932283. MR447090

[5] Daniel J. Bernstein, Luca De Feo, Antonin Leroux, and Benjamin Smith, *Faster computation of isogenies of large prime degree*, ANTS XIV—Proceedings of the Fourteenth Algorithmic Number Theory Symposium, Open Book Ser., vol. 4, Math. Sci. Publ., Berkeley, CA, 2020, pp. 39–55, DOI 10.2140/obs.2020.4.39. MR4235105

[6] Ian F. Blake, János A Csirik, Michael Rubinstein, and Gadiel Seroussi, *On the computation of modular polynomials for elliptic curves*, HP Laboratories Technical Report (1999).

[7] Wieb Bosma, John Cannon, and Catherine Playoust, *The Magma algebra system. I. The user language*, J. Symbolic Comput. **24** (1997), no. 3-4, 235–265, DOI 10.1006/jsco.1996.0125. Computational algebra and number theory (London, 1993). MR1484478

[8] Reinier Bröker, Kristin Lauter, and Andrew V. Sutherland, *Modular polynomials via isogeny volcanoes*, Math. Comp. **81** (2012), no. 278, 1201–1231, DOI 10.1090/S0025-5718-2011-02508-1. MR2869057

[9] Reinier Martijn Bröker, *Constructing elliptic curves of prescribed order*, Ph.D. thesis, Leiden University, 2006.

[10] Wouter Castryck, Tanja Lange, Chloe Martindale, Lorenz Panny, and Joost Renes, *CSIDH: an efficient post-quantum commutative group action*, Advances in cryptology—ASIACRYPT 2018. Part III, Lecture Notes in Comput. Sci., vol. 11274, Springer, Cham, 2018, pp. 395–427, DOI 10.1007/978-3-030-03332-3_15. MR3897883

[11] Denis Charles and Kristin Lauter, *Computing modular polynomials*, LMS J. Comput. Math. **8** (2005), 195–204, DOI 10.1112/S1461157000000954. MR2166572

[12] Leonardo Colò and David Kohel, *Orienting supersingular isogeny graphs*, J. Math. Cryptol. **14** (2020), no. 1, 414–437, DOI 10.1515/jmc-2019-0034. MR4165916

[13] Maria Corte-Real Santos, Craig Costello, and Sam Frengley, *An algorithm for efficient detection of (N, N)-splittings and its application to the isogeny problem in dimension 2*, Public-Key Cryptography—PKC 2024. Part III, Lecture Notes in Comput. Sci., vol. 14603, Springer, Cham, [2024] ©2024, pp. 157–189, DOI 10.1007/978-3-031-57725-3_6. MR4763489

[14] Craig Costello, Michael Meyer, and Michael Naehrig, *Sieving for twin smooth integers with solutions to the Prouhet-Tarry-Escott problem*, Advances in cryptology—EUROCRYPT 2021. Part I, Lecture Notes in Comput. Sci., vol. 12696, Springer, Cham, [2021] ©2021, pp. 272–301, DOI 10.1007/978-3-030-77870-5_10. MR4284266

[15] Jean Marc Couveignes, *Hard homogeneous spaces*, IACR Cryptology ePrint Archive, 2006.

[16] David Cox, John Little, and Donal O'Shea, *Ideals, varieties, and algorithms*, 3rd ed., Undergraduate Texts in Mathematics, Springer, New York, 2007. An introduction to computational algebraic geometry and commutative algebra, DOI 10.1007/978-0-387-35651-8. MR2290010

[17] Christina Delfs and Steven D. Galbraith, *Computing isogenies between supersingular elliptic curves over $\mathbb{F}_p$*, Des. Codes Cryptogr. **78** (2016), no. 2, 425–440, DOI 10.1007/s10623-014-0010-1. MR3451433

[18] Kirsten Eisenträger, Sean Hallgren, Kristin Lauter, Travis Morrison, and Christophe Petit, *Supersingular isogeny graphs and endomorphism rings: reductions and solutions*, Advances in cryptology—EUROCRYPT 2018. Part III, Lecture Notes in Comput. Sci., vol. 10822, Springer, Cham, 2018, pp. 329–368, DOI 10.1007/978-3-319-78372-7_11. MR3794837

[19] Noam D. Elkies, *Elliptic and modular curves over finite fields and related computational issues*, Computational perspectives on number theory (Chicago, IL, 1995), AMS/IP Stud. Adv. Math., vol. 7, Amer. Math. Soc., Providence, RI, 1998, pp. 21–76, DOI 10.1090/amsip/007/03. MR1486831

[20] Andreas Enge, *The complexity of class polynomial computation via floating point approximations*, Math. Comp. **78** (2009), no. 266, 1089–1107, DOI 10.1090/S0025-5718-08-02200-X. MR2476572

[21] Andreas Enge and Andrew V. Sutherland, *Class invariants by the CRT method*, Algorithmic number theory, Lecture Notes in Comput. Sci., vol. 6197, Springer, Berlin, 2010, pp. 142–156, DOI 10.1007/978-3-642-14518-6_14. MR2721418

[22] Jonathan Komada Eriksen, Lorenz Panny, Jana Sotáková, and Mattia Veroni, *Deuring for the people: Supersingular elliptic curves with prescribed endomorphism ring in general characteristic*, LuCaNT 2023, 2023.

[23] Steven D. Galbraith, Christophe Petit, and Javier Silva, *Identification protocols and signature schemes based on supersingular isogeny problems*, Advances in cryptology—ASIACRYPT 2017. Part I, Lecture Notes in Comput. Sci., vol. 10624, Springer, Cham, 2017, pp. 3–33, DOI 10.1007/978-3-319-70694-8_1. MR3747691

[24] Joseph H. Silverman, *The arithmetic of elliptic curves*, 2nd ed., Graduate Texts in Mathematics, vol. 106, Springer, Dordrecht, 2009, DOI 10.1007/978-0-387-09494-6. MR2514094

[25] David Kohel, Kristin Lauter, Christophe Petit, and Jean-Pierre Tignol, *On the quaternion ℓ-isogeny path problem*, LMS J. Comput. Math. **17** (2014), no. suppl. A, 418–432, DOI 10.1112/S1461157014000151. MR3240818

[26] David Russell Kohel, *Endomorphism rings of elliptic curves over finite fields*, ProQuest LLC, Ann Arbor, MI, 1996. Thesis (Ph.D.)–University of California, Berkeley. MR2695524

[27] Sabrina Kunzweiler and Damien Robert, *Computing modular polynomials by deformation*, Res. Number Theory **11** (2025), no. 1, Paper No. 10, 28, DOI 10.1007/s40993-024-00596-5. MR4837977

[28] Serge Lang and Hale Trotter, *Frobenius distributions in GL_2-extensions*, Lecture Notes in Mathematics, Vol. 504, Springer-Verlag, Berlin-New York, 1976. Distribution of Frobenius automorphisms in GL_2-extensions of the rational numbers. MR568299

[29] Frank Lehmann, Markus Maurer, Volker Müller, and Victor Shoup, *Counting the number of points on elliptic curves over finite fields of characteristic greater than three*, Algorithmic number theory (Ithaca, NY, 1994), Lecture Notes in Comput. Sci., vol. 877, Springer, Berlin, 1994, pp. 60–70, DOI 10.1007/3-540-58691-1_44. MR1322712

[30] Antonin Leroux, *Quaternion algebra and isogeny-based cryptography*, Ph.D. thesis, Ecole doctorale de l'Institut Polytechnique de Paris, 2022.

[31] Antonin Leroux, *Computation of Hilbert class polynomials and modular polynomials from supersingular elliptic curves*, 2023.

[32] The LMFDB Collaboration, *The L-functions and modular forms database*, `https://www.lmfdb.org`, 2024, [Online; accessed 7 June 2024].

[33] François Morain, *Calcul du nombre de points sur une courbe elliptique dans un corps fini: aspects algorithmiques* (French, with English and French summaries), J. Théor. Nombres Bordeaux **7** (1995), no. 1, 255–282, DOI 10.5802/jtnb.143. Les Dix-huitièmes Journées Arithmétiques (Bordeaux, 1993). MR1413579

[34] Christophe Petit and Spike Smith, *An improvement to the quaternion analogue of the ℓ-isogeny problem*, Presentation at MathCrypt (2018).

[35] Damien Robert, *Some applications of higher dimensional isogenies to elliptic curves (overview of results)*, IACR Cryptology ePrint Archive, 2022.

[36] Alexander Rostovtsev and Anton Stolbunov, *Public-key cryptosystem based on isogenies*, IACR Cryptology ePrint Archive, 2006.

[37] Maria Corte-Real Santos, Craig Costello, and Jia Shi, *Accelerating the Delfs-Galbraith algorithm with fast subfield root detection*, Advances in cryptology—CRYPTO 2022. Part III, Lecture Notes in Comput. Sci., vol. 13509, Springer, Cham, [2022] ©2022, pp. 285–314, DOI 10.1007/978-3-031-15982-4_10. MR4539902

[38] Victor Shoup et al., *NTL: A library for doing number theory*, 2001.

[39] A. V. Sutherland, `classpoly`, Software package, version 1.0.3, `https://math.mit.edu/~drew/classpoly.html`, accessed 10 January 2025.

[40] Andrew V. Sutherland, *Computing Hilbert class polynomials with the Chinese remainder theorem*, Math. Comp. **80** (2011), no. 273, 501–538, DOI 10.1090/S0025-5718-2010-02373-7. MR2728992

[41] Andrew V. Sutherland, *On the evaluation of modular polynomials*, ANTS X—Proceedings of the Tenth Algorithmic Number Theory Symposium, Open Book Ser., vol. 1, Math. Sci. Publ., Berkeley, CA, 2013, pp. 531–555, DOI 10.2140/obs.2013.1.531. MR3207430

[42] Andrew V. Sutherland, *Identifying supersingular elliptic curves*, LMS J. Comput. Math. **15** (2012), 317–325, DOI 10.1112/S1461157012001106. MR2988819

[43] The Sage Developers, *SageMath, the Sage Mathematics Software System (Version 10.0)*, 2024, `https://www.sagemath.org`.

[44] Jacques Vélu, *Isogénies entre courbes elliptiques* (French), C. R. Acad. Sci. Paris Sér. A-B **273** (1971), A238–A241. MR294345

[45] John Voight, *Quaternion algebras*, Graduate Texts in Mathematics, vol. 288, Springer, Cham, [2021] ©2021, DOI 10.1007/978-3-030-56694-4. MR4279905

[46] Joachim von zur Gathen and Jürgen Gerhard, *Modern computer algebra*, Cambridge University Press, New York, 1999. MR1689167

ENS de Lyon, CNRS, UMPA, UMR 5669, Lyon, France
Email address: `maria.corte_real_santos@ens-lyon.fr`

COSIC, KU Leuven, Belgium
Email address: `jeriksen@esat.kuleuven.be`

DGA-MI, Bruz, France, IRMAR - UMR 6625, Université de Rennes, France
Email address: `antonin.leroux@polytechnique.org`

University of Regensburg, Germany
Email address: `michael@random-oracles.org`

Technische Universität München, Germany
Email address: `lorenz@yx7.cc`

Contemporary Mathematics
Volume **840**, 2026
https://doi.org/10.1090/conm/840/16824

Explicit supersingular cyclic curves

Marco Streng

Abstract. Li, Mantovan, Pries, and Tang proved the existence of supersingular curves over $\overline{\mathbb{F}_p}$ in each of the special families of curves in Moonen's classification. Their proof does not provide defining equations of these curves. We make some of their results explicit using the reductions modulo p of previously computed curves with complex multiplication.

1. Introduction

Moonen [**10**, Table 1] showed that there are exactly 20 families of cyclic covers of $\mathbb{P}^1$ that give rise to a special subvariety of the moduli space of principally polarised abelian varieties. Denoting these families by $M[1]$–$M[21]$, more recent work of Li, Mantovan, Pries, and Tang considers five of these families and shows that they contain supersingular curves as follows (and as left open in [**7**, Theorem 1.1]).

Theorem 1.1 ([**8**, Theorem 7.1]). *In each of the following families there exists a supersingular smooth curve of genus g defined over $\overline{\mathbb{F}_p}$ for all sufficiently large primes p that satisfy the given condition:*

1. *in $M[6]$ with $g=3$, assuming $p \equiv 2 \pmod 3$;*
2. *in $M[8]$ with $g=3$, assuming $p \equiv 3 \pmod 4$;*
3. *in $M[10]$ with $g=4$, assuming $p \equiv 2 \pmod 3$;*
4. *in $M[14]$ with $g=4$, assuming $p \equiv 5 \pmod 6$; and*
5. *in $M[16]$ with $g=6$, assuming $p \equiv 2$, 3 or $4 \pmod 5$.*

By taking reductions modulo primes of known complex multiplication curves, we prove the following more explicit version of cases 1, 2, and 5.

Theorem 1.2. *In each of the following families there exists a supersingular smooth curve of genus g defined over $\mathbb{F}_p$ for all primes p that satisfy the given condition:*

1. *in $M[6]$ with $g=3$, assuming $p \equiv 2 \pmod 3$;*
2. *in $M[8]$ with $g=3$, assuming $p \equiv 3 \pmod 4$; and*
5. *in $M[16]$ with $g=6$, assuming $p \equiv 2$, 3 or $4 \pmod 5$ and Conjecture 4.1.*

The curves are given explicitly in Theorems 2.1, 3.1, and 4.3 respectively.

2020 *Mathematics Subject Classification.* Primary 14G10, 11G20, 11M38, 14H10, 14K22; Secondary 11G18, 14G10, 14G15, 14H30, 14H40.

The author thanks Edgar Costa, Jeroen Sijsling, and Anna Somoza for helpful discussions.

The improvement that Theorem 1.2 provides over Theorem 1.1 is as follows:

(a) we remove the condition that p be sufficiently large,
(b) we give explicit equations of the curves,
(c) we give curves that are defined not only over $\overline{\mathbb{F}_p}$, but over $\mathbb{F}_p$.

2. Supersingular Picard curves in $M[6]$

The family $M[6]$ consists of the smooth algebraic curves that have a model of the form $y^3 = f(x)$ for a separable polynomial f of degree 4. This is a special family of smooth plane quartic curves known as Picard curves. It features in the LMFDB [**9**] as 3.3-1.0.3-3-3-3-3.1 and 3.3-1.0.3-3-3-3-3.2.

Koike-Weng [**5**] and Lario-Somoza-Vincent [**6**] computed many examples of complex multiplication (CM) curves in this family. However, in order to prove Theorem 1.1.1 we can make do with one curve

$$C : y^3 = x^4 - x.$$

Note that it has an automorphism $[\zeta_9] : (x, y) \mapsto (\zeta_9^3 x, \zeta_9 y)$ over $K = \mathbb{Q}(\zeta_9)$. This induces an automorphism of the Jacobian $J = \mathrm{Jac}(C)$ over K and in fact an embedding $\mathbb{Z}[\zeta_9] \to \mathrm{End}(J_K)$, proving that this curve indeed has CM.

For curves of genus 3 with CM by a cyclic sextic field, Proposition 4.1 of Kılıçer-Labrande-Lercier-Ritzenthaler-Sijsling-Streng [**4**] tells us exactly when the reduction modulo p is supersingular. We get the following result.

THEOREM 2.1. *For all prime numbers* $p \equiv 2 \pmod 3$ *the plane projective curve over* $\mathbb{F}_p$ *given by*

$$y^3 = x^4 - x$$

is smooth and supersingular in the family $M[6]$.

PROOF. The given model of C has good reduction modulo every prime $p \neq 3$. Then Proposition 4.1 of [**4**] shows that the reduction of J modulo p is supersingular if and only if $p\mathcal{O}_K$ has exactly 1 or 3 prime factors. This is the case exactly when $p \not\equiv 1 \pmod 3$. □

For a self-contained proof we can use the following lemma instead of [**4**, Proposition 4.1]. Note that (p mod 9) has even order in $(\mathbb{Z}/9\mathbb{Z})^\times$ exactly if $p \equiv 2 \pmod 3$.

LEMMA 2.2. *Let* K *be a CM field that is cyclic of some degree* $2g$ *over* $\mathbb{Q}$. *Let* A *be an abelian variety of dimension* g *over a number field* k *with endomorphism ring isomorphic to* $\mathcal{O}_K$.

Let p *be a prime number that is unramified in* K *and such that* $\mathrm{Frob}_p \in \mathrm{Gal}(K/\mathbb{Q})$ *has even order. Let* $\mathfrak{P} \mid (p)$ *be a prime of* k *of good reduction for* A.

Then the reduction $(A \bmod \mathfrak{P})$ *is supersingular.*

PROOF. Without loss of generality (that is, by extending k) all endomorphisms of A are defined over k. Then the Shimura-Taniyama formula [**11**, Theorem 1(π2) in Section 13] tells us that the Frobenius endomorphism of $(A \bmod \mathfrak{P})$ is an element $\pi \in \mathcal{O}_K$.

We claim now that for every Galois conjugate π' of π we have $\mathrm{ord}_{\mathfrak{P}}(\pi') = \frac{1}{2}\mathrm{ord}_{\mathfrak{P}}(p)$. Assuming the claim, we get for every i that the coefficient a_{2g-i} of X^{2g-i} in the characteristic polynomial of π satisfies $\mathrm{ord}_{\mathfrak{P}}(a_{2g-j}) \geq i \cdot \frac{1}{2}\mathrm{ord}_{\mathfrak{P}}(p)$. The Newton polygon therefore is a straight line segment from $(0, 0)$ to $(2g, g)$ of slope $\frac{1}{2}$, hence A is supersingular.

Now it remains only to prove the claim, which we do as follows. By assumption, the decomposition group of p in $K/\mathbb{Q}$ contains an element of order 2 in $\mathrm{Gal}(K/\mathbb{Q})$, and since $\mathrm{Gal}(K/\mathbb{Q})$ is cylic, the only such element is complex conjugation. In particular, we get for every $\alpha \in K$:

$$\mathrm{ord}_{\mathfrak{p}}(\overline{\alpha}) = \mathrm{ord}_{\overline{\mathfrak{p}}}(\alpha) = \mathrm{ord}_{\mathfrak{p}}(\alpha).$$

We also have $\pi'\overline{\pi'} = p$ for every conjugate π' of π, hence

$$\mathrm{ord}_{\mathfrak{p}}(p) = \mathrm{ord}_{\mathfrak{p}}(\pi') + \mathrm{ord}_{\mathfrak{p}}(\overline{\pi'}) = 2\mathrm{ord}_{\mathfrak{p}}(\pi'),$$

which proves the claim and finishes the proof of the lemma. □

3. Supersingular hyperelliptic curves in $M[8]$

The family $M[8]$ consists of the quadruple covers of the projective (U, V)-line of the form

$$C : Y^4 = F_2(U, V)\ F_3(U, V)^2,$$

where the F_i are coprime homogeneous separable polynomials of degree i.

Suppose that F_2 splits. A change of variables then gives $F_2 = UV$, or in affine coordinates

$$C : y^4 = uf(u)^2$$

for a separable polynomial f of degree 3 with $f(0) \neq 0$.

Using a further change of variables puts this curve in hyperelliptic Weierstrass form. Indeed, let x be the element $x = y^2/f(u)$ in the function field $k(C)$ of C. Then $k(C)$ is generated by x and y because we have $u = x^2$. Moreover, these new coordinates satisfy

$$y^2 = xf(u) = xf(x^2),$$

hence we write our curve also as

$$C : y^2 = xf(x^2).$$

Weng [**15**] gave a construction for CM curves of this latter Weierstrass form. For example, she constructed the curve

$$y^2 = u^7 + 6u^5 + 9u^3 + u,$$

which has complex multiplication by $\mathbb{Q}(\zeta_9)^+(i)$, as was later proven by Costa-Mascot-Sijsling-Voight [**2, 3**]. We get the following result.

THEOREM 3.1. *For all prime numbers $p \equiv 3 \bmod 4$ except $p = 3$, the curve*

$$y^2 = xf(x^2) \quad \textit{or equivalently} \quad y^4 = uf(u)^2,$$

with

$$f(u) = u^3 + 6u^2 + 9u + 1$$

is a smooth supersingular curve of genus 3 defined over $\mathbb{F}_p$ in the family $M[8]$. For the prime $p = 3$, the curve with $f(u) = u^3 + 7u^2 + 14u + 7$ has the same property.

PROOF. For $p = 3$ this is one curve, so a direct verification by counting points proves the result. We do this using SageMath [**14**], see the accompanying code [**13**].

For $p \neq 3$ the proof is exactly the same as for Theorem 2.1 (using either [**4**, Proposition 4.1] or Lemma 2.2), but with the field $\mathbb{Q}(\zeta_9)^+(i)$. □

REMARK 3.2. The polynomial $u^3 + 7u^2 + 14u + 7$ again comes from Weng [**16**]. Since we only use it over $\mathbb{F}_3$, we could have equivalently written $u^3 + u^2 - u + 1$.

4. Supersingular cyclic plane quintics in $M[16]$

The family $M[16]$ consists of the smooth plane curves $y^5 = f(x)$ for a separable polynomial f of degree 4 or 5. These curves were dubbed *cyclic plane quintics* or *CPQ curves* by Somoza who constructed in [**12**, Section 2.3] the curve

$$C : y^5 = x^4 - 24x^3 + 3x^2 + x \tag{4.1}$$

and conjectured the following.

CONJECTURE 4.1. The Jacobian J of the curve C of (4.1) has $\mathrm{End}(J_{\overline{\mathbb{Q}}}) \cong \mathcal{O}_K$, where $K = \mathbb{Q}(\zeta_9)^+(\zeta_5)$.

REMARK 4.2. I have no doubt about the validity of the conjecture, though rigourous verification seems to be computationally just out of reach using the methods of [**3**], see [**1**].

We have the following evidence for the conjecture. First of all, it is stated in [**12**, Table 4.2] that there exists a curve of the form $y^2 = f(x)$ with $f(x) \in \mathbb{Q}[x]$ such that its Jacobian has CM over $\overline{\mathbb{Q}}$ by K. The method for computing this curve is high-precision numerical approximation, hence C is numerically extremely close to a curve that does have the correct CM.

Next, the curve C that Somoza obtained in this way has discriminant 3^{10}. Such a smooth number seems unlikely to appear by chance.

Finally, we verified that C is supersingular modulo 2, 7, 13, and 17, consistently with Theorem 4.3 below (see [**13**]). This again seems unlikely to be by chance.

THEOREM 4.3. *Assume Conjecture* 4.1. *For all prime numbers* $p \equiv 2, 3,$ *or* 4 (mod 5) *except* $p = 3$ *the curve over* $\mathbb{F}_p$ *given by* (4.1) *is supersingular.*

Over $\mathbb{F}_3$, *the curve given by* $y^5 = x^4 - 7x^2 + 7x$ *is supersingular.*

PROOF. Again we check the curve for $p = 3$ separately. For the other primes, the equation (4.1) has good reduction. Assume Conjecture 4.1 and apply Lemma 2.2.

The primes $p \neq 3$ that are 2, 3, or 4 (mod 5) are unramified in $K/\mathbb{Q}$ and have order 2 or 4 in $(\mathbb{Z}/5\mathbb{Z})^\times$. In particular, the corresponding Frobenius automorphism of $\mathbb{Q}(\zeta_5)$ has even order, and so does that of K. Therefore, Lemma 2.2 gives the result. □

REMARK 4.4. In this case, Proposition 4.1 of [**4**] does not apply directly, as it is stated only for abelian threefolds. The method of proof applies to more general abelian varieties. We condensed it to what we needed and then generalised this, which resulted in Lemma 2.2.

5. Final notes

5.1. Other families. We restricted to the three families $M[6]$, $M[8]$, and $M[16]$ because the literature contains algorithms for reconstructing curves in these families from their period matrices (and in fact even contains the required example curves). Completely new reconstruction algorithms are far beyond the scope of this paper and are in fact a project in development by the author.

5.2. Databases of curves, and verification. The curves that we needed came from three different sources in the literature. More importantly, the proof that they were correct came from yet two other sources or (in the case of $M[16]$) does not even exist yet. A database with CM curves and certificates of their Jacobians' endomorphisms would be very useful.

References

[1] Edgar Costa, *Prove that* $y^2 = x^4 - 24x^3 + 3x^2 + x$ *has CM*, 2025, Issue #84 of [**2**], `https://github.com/edgarcosta/endomorphisms/issues/84`.

[2] Edgar Costa, Nicolas Mascot, Jeroen Sijsling, and John Voight, *Endomorphisms*, Code accompanying "Rigorous computation of the endomorphism ring of a Jacobian" by the same authors, `https://github.com/edgarcosta/endomorphisms/` See especially `https://github.com/edgarcosta/endomorphisms/blob/master/examples/puiseux/HypCorrG3CM.m`, 2017–2024.

[3] Edgar Costa, Nicolas Mascot, Jeroen Sijsling, and John Voight, *Rigorous computation of the endomorphism ring of a Jacobian*, Math. Comp. **88** (2019), no. 317, 1303–1339, DOI 10.1090/mcom/3373. MR3904148

[4] Pınar Kılıçer, Hugo Labrande, Reynald Lercier, Christophe Ritzenthaler, Jeroen Sijsling, and Marco Streng, *Plane quartics over* $\mathbb{Q}$ *with complex multiplication*, Acta Arith. **185** (2018), no. 2, 127–156, DOI 10.4064/aa170227-16-3. MR3856187

[5] Kenji Koike and Annegret Weng, *Construction of CM Picard curves*, Math. Comp. **74** (2005), no. 249, 499–518, DOI 10.1090/S0025-5718-04-01656-4. MR2085904

[6] Joan-C. Lario, Anna Somoza, and Christelle Vincent, *An inverse Jacobian algorithm for Picard curves*, Res. Number Theory **7** (2021), no. 2, Paper No. 32, 23, DOI 10.1007/s40993-021-00253-1. MR4245483

[7] Wanlin Li, Elena Mantovan, Rachel Pries, and Yunqing Tang, *Newton polygons arising from special families of cyclic covers of the projective line*, Res. Number Theory **5** (2019), no. 1, Paper No. 12, 31, DOI 10.1007/s40993-018-0149-3. MR3897613

[8] Wanlin Li, Elena Mantovan, Rachel Pries, and Yunqing Tang, *Newton polygon stratification of the Torelli locus in unitary Shimura varieties*, Int. Math. Res. Not. IMRN **9** (2022), 6464–6511, DOI 10.1093/imrn/rnaa306. MR4411461

[9] The LMFDB Collaboration, *The L-functions and modular forms database*, `https://www.lmfdb.org`, 2025, [Online; accessed 24 January 2025].

[10] Ben Moonen, *Special subvarieties arising from families of cyclic covers of the projective line*, Doc. Math. **15** (2010), 793–819. MR2735989

[11] Goro Shimura and Yutaka Taniyama, *Complex multiplication of abelian varieties and its applications to number theory*, Publications of the Mathematical Society of Japan, vol. 6, Mathematical Society of Japan, Tokyo, 1961. MR125113

[12] Anna Somoza Henares, *Inverse jacobian and related topics for certain superelliptic curves*, Ph.D. thesis, Universiteit Leiden and Universitat Politècnica de Catalunya, 2019, `https://hdl.handle.net/1887/70564`.

[13] Marco Streng, *Sagemath verifications of supersingularity of some curves in Explicit supersingular cyclic curves*, `https://github.com/mstreng/supersingular`, 2025.

[14] The Sage Developers, *Sagemath, the Sage Mathematics Software System (Version 10.2)*, 2023, `https://www.sagemath.org`.

[15] Annegret Weng, *A class of hyperelliptic CM-curves of genus three*, J. Ramanujan Math. Soc. **16** (2001), no. 4, 339–372. MR1877806

[16] Annegret Weng, *Constructing hyperelliptic curves of genus 2 suitable for cryptography*, Math. Comp. **72** (2003), no. 241, 435–458, DOI 10.1090/S0025-5718-02-01422-9. MR1933830

Universiteit Leiden
Email address: `streng@math.leidenuniv.nl`

Contemporary Mathematics
Volume **840**, 2026
https://doi.org/10.1090/conm/840/16825

Curve equations from expansions of 1-forms at a nonrational point

Raymond van Bommel, Edgar Costa, Bjorn Poonen, and Padmavathi Srinivasan

Abstract. We exhibit an algorithm to compute equations of an algebraic curve over a computable characteristic 0 field from the power series expansions of its regular 1-forms at a *nonrational* point of the curve, extending a 2005 algorithm of Baker, González-Jiménez, González, and Poonen for expansions at a rational point. If the curve is hyperelliptic, the equations present it as an explicit double cover of a smooth plane conic, or as a double cover of the projective line when possible. If the curve is nonhyperelliptic, the equations cut out the canonical model. The algorithm has been used to compute equations over $\mathbb{Q}$ for many hyperelliptic modular curves without a rational cusp in the L-functions and Modular Forms Database.

1. Introduction

A curve X is called nice if it is smooth, projective, and geometrically integral. From now on, X is a nice curve of genus $g \geq 2$ over $\mathbb{Q}$, but all our theorems and algorithms work over any ground field F of characteristic 0 if field operations in F are computable. Our goal is to give an algorithm that takes as input the initial terms of the expansions of 1-forms forming a basis of the $\mathbb{Q}$-vector space $\Gamma(X, \Omega^1_{X/\mathbb{Q}})$ at a *nonrational* point and returns equations for X over $\mathbb{Q}$; such input arises naturally in [**Zyw24**, Section 5], for instance. (The analogue for expansions at a *rational* point is covered in [**BGJGP05**, Section 2.1].) These equations will cut out the canonical model if X is nonhyperelliptic, and a double cover of a genus 0 curve if X is hyperelliptic; see Theorem 3.1 for more details. Examples of the nonhyperelliptic case (the easier case) were worked out in [**Bar14**, Section 7] and [**MS20**, Sections 7 and 8]. So the main new work is in the hyperelliptic case.

The algorithm in [**BGJGP05**, Section 2.1] will produce a model over the field of definition of the nonrational point, but there is no easy way to pass from that to the equation over $\mathbb{Q}$. Also, the presence of the rational point in [**BGJGP05**, Section 2.1] meant that in the hyperelliptic case, the image of the canonical map was $\mathbb{P}^1_{\mathbb{Q}}$, whereas in the present article, it could instead be a pointless genus 0 curve instead of $\mathbb{P}^1_{\mathbb{Q}}$, in which case X will need to be given as a double cover of a plane conic. Moreover, there are additional complications in our article coming from the fact that even the expansion of objects defined over $\mathbb{Q}$ have coefficients in a larger number field.

The motivation for our algorithm is the problem of finding equations of modular curves *that have no rational cusp.* The algorithm has been used so far to calculate equations of over 4700 such hyperelliptic modular curves without a rational cusp for the L-Functions and Modular Forms Database [**LMFDB**], among which over 1500 are a double cover of a pointless genus 0 curve.

2. Hyperelliptic curves

The curve X is called hyperelliptic if the canonical map $X \to \mathbb{P}^{g-1}$ is not a closed immersion. Equivalently, X is hyperelliptic if there exists a degree 2 morphism π from X to some genus 0 curve C. Suppose that this is the case. Then C and the morphism π are unique up to isomorphism. In fact, C is the image of the canonical map. The curve C need not be isomorphic to $\mathbb{P}^1$ over $\mathbb{Q}$, but the anticanonical map for C identifies C with a smooth plane conic in $\mathbb{P}^2 = \operatorname{Proj}\mathbb{Q}[a,b,c]$. For any $d \in \mathbb{Z}_{\geq 0}$, let $\mathbb{Q}[a,b,c]_d$ be the space of degree d homogeneous polynomials in $\mathbb{Q}[a,b,c]$.

3. Main theorem

Now return to the general case, in which X is any nice curve of genus $g \geq 2$ over $\mathbb{Q}$. Let $K \supseteq \mathbb{Q}$ be a finite extension. Let $X_K = X \times_{\mathbb{Q}} K$. Let $P \in X(K)$. Assume that $K = \mathbb{Q}(P)$. Let q be a uniformizer of the completed local ring $\widehat{\mathscr{O}}_{X_K,P}$. Let $\omega_1, \ldots, \omega_g$ be a $\mathbb{Q}$-basis of $\mathrm{H}^0(X, \Omega^1)$. For each $i \in \{1, \ldots, g\}$, the Taylor expansion of ω_i at P is $w_i\, dq$ for some $w_i \in K[[q]]$. For $B \in \mathbb{Z}_{>0}$, let $K[q]_{<B} \simeq K[[q]]/(q^B)$ be the vector space of polynomials of degree $< B$. Let $\bar{w}_i := (w_i \bmod q^B) \in K[q]_{<B}$.

THEOREM 3.1. *Let $B = 19g + 48$. There exists an algorithm with*
Input: g, K, and polynomials $\bar{w}_1, \ldots, \bar{w}_g \in K[q]_{<B}$ arising from some nice curve X over $\mathbb{Q}$ and $P \in X(K)$ as above.
Output:

- *If X is nonhyperelliptic, return* `nonhyperelliptic` *and a finite list of homogeneous polynomials over $\mathbb{Q}$ cutting out a curve in $\mathbb{P}^{g-1}$ linearly isomorphic over $\mathbb{Q}$ to the canonical model of X.*
- *If X is hyperelliptic and g is even, return* `hyperelliptic` *and a separable polynomial $f \in \mathbb{Q}[x]$ of degree $2g+1$ or $2g+2$ such that X is birational to the curve $y^2 = f(x)$.*
- *If X is hyperelliptic and g is odd, return* `hyperelliptic` *and homogeneous polynomials $Q \in \mathbb{Q}[a,b,c]_2$ and $H \in \mathbb{Q}[a,b,c]_{g+1}$ such that*

$$C \simeq \operatorname{Proj} \frac{\mathbb{Q}[a,b,c]}{(Q)} \subset \mathbb{P}^2 \quad \text{and} \quad X \simeq \operatorname{Proj} \frac{\mathbb{Q}[a,b,c,y]}{(y^2 - H, Q)} \subset \mathbb{P}\left(1,1,1,\frac{g+1}{2}\right).$$

  In this case, if a rational point on C is given, find a model $y^2 = f(x)$ as in the even genus hyperelliptic case.

REMARK 3.2. In the odd genus hyperelliptic case, we may require the quadratic form Q to be *diagonal*, if desired.

REMARK 3.3. By computing Hilbert symbols, one can determine whether a given smooth plane conic C over $\mathbb{Q}$ is isomorphic to $\mathbb{P}^1_{\mathbb{Q}}$; this is essentially due to Legendre. (More generally, by the Hasse–Minkowski local–global principle for quadratic forms, this can be done over any number field; see, e.g., [**Shi10**, Theorem 26.3]. But it involves more than just field operations, so it is not an algorithm that generalizes to *any* characteristic 0 field.)

4. Theoretical lemmas

Before explaining the algorithm, we prove a few theoretical lemmas. Let $S = \mathbb{Q}[x_1, \ldots, x_g]$ be the homogeneous coordinate ring of $\mathbb{P}^{g-1}$ over $\mathbb{Q}$. Let $I \subset S$ be the homogeneous ideal of the canonical image of X. Let $I_d \subset S_d$ be the degree d parts of $I \subset S$.

LEMMA 4.1. *Let $f \in S_d$. If $f(w_1, \ldots, w_g) \in K[[q]]$ vanishes at $q = 0$ to order $> d(2g-2)/[K : \mathbb{Q}]$, then the corresponding section of $(\Omega^1)^{\otimes d}$ is 0.*

PROOF. The section has more than $d(2g-2) = \deg(\Omega^1)^{\otimes d}$ geometric zeros in total (at P and its conjugates), so it is 0. □

COROLLARY 4.2. *If $B > d(2g-2)/[K : \mathbb{Q}]$, then from the input as in Theorem* 3.1 *one can compute a basis for I_d.*

PROOF. By Lemma 4.1, I_d is the kernel of the $\mathbb{Q}$-linear map $S_d \to K[[q]]/(q^B)$ sending each monomial to its truncated expansion. □

LEMMA 4.3. *The dimension of I_2 is $\binom{g-1}{2}$ if X is hyperelliptic, and $\binom{g-2}{2}$ if not.*

PROOF. We may work over $\mathbb{C}$. Let $\mathbb{C}[x]_{\le n}$ be the space of polynomials of degree at most n. In the hyperelliptic case, X is the smooth projective model of $y^2 = F(x)$ with $\deg F = 2g+1$, and $H^0(X, \Omega^1) = \mathbb{C}[x]_{\le g-1} \frac{dx}{y}$ (see [**ACGH85**, p. 11], for example), so I_2 is isomorphic to the kernel of the surjective map $\ker(\operatorname{Sym}^2 \mathbb{C}[x]_{\le g-1} \to \mathbb{C}[x]_{\le 2g-2})$, so $\dim I_2 = g(g+1)/2 - (2g-1) = \binom{g-1}{2}$. In the nonhyperelliptic case, this follows from Max Noether's theorem [**ACGH85**, p. 117]. □

LEMMA 4.4. *Let X be a hyperelliptic curve over $\mathbb{Q}$. Let L be a finite extension of $\mathbb{Q}$. Let $P' \in X(L)$. Suppose that $\omega'_1, \ldots, \omega'_g$ is an L-basis for $\mathrm{H}^0(X_L, \Omega^1)$ such that $\operatorname{ord}_{P'}(\omega'_1) < \ldots < \operatorname{ord}_{P'}(\omega'_g)$. Let $t = \omega'_{g-1}/\omega'_g \in L(X)$. Then $t \in L(C)$ and is of degree* 1 *(as a rational function on C_L).*

PROOF. We may assume that X_L is the smooth projective model of $y^2 = F(x)$ for some $F \in L[x]$, and P' is at infinity. Then for $i = 0, \ldots, g-1$, we have $\omega'_{g-i} = J_i(x)\, dx/y$ for some $J_i(x) \in L[x]$ of degree exactly i. Then t is a degree 1 polynomial in $L[x]$. □

LEMMA 4.5. *Let C be a genus* 0 *curve over $\mathbb{Q}$. Let $\mathcal{T}$ be the tangent bundle of C. Let $V := \mathrm{H}^0(C, \mathcal{T})$. Let L be a finite extension of $\mathbb{Q}$, with $\mathbb{Q}$-basis $\lambda_1, \ldots, \lambda_\ell$. Let t be a degree* 1 *rational function on C_L.*

(a) *The meromorphic sections $\frac{d}{dt}, t\frac{d}{dt}, t^2\frac{d}{dt}$ of $\mathcal{T}$ form an L-basis of V_L.*
(b) *The elements $\operatorname{Tr}_{L/\mathbb{Q}}(\lambda_j t^i \frac{d}{dt})$ for $0 \le i \le 2$ and $1 \le j \le \ell$ span V.*

PROOF.

(a) Without loss of generality, $C_L = \mathbb{P}^1$ and t is the standard coordinate. Then $\mathcal{T} \simeq \mathscr{O}(2)$, so $\dim V_L = 3$. Also, $\frac{d}{dt}$ has a double zero at ∞, so $\frac{d}{dt}, t\frac{d}{dt}, t^2\frac{d}{dt}$ are independent global sections.

(b) The map $\mathrm{Tr}_{L/\mathbb{Q}}\colon V_L \to V$ is surjective. □

LEMMA 4.6. *Let C be a smooth plane conic in $\mathbb{P}^2$ over a field k. Let $h \in k(C)$ be a rational function of degree d. Then h is given by a ratio of two homogeneous forms on $\mathbb{P}^2$ of degree $\lceil d/2 \rceil$.*

PROOF. Let $L \in \operatorname{Div} C$ be a hyperplane section of $C \subset \mathbb{P}^2$. Write $(h) = (h)_0 - (h)_\infty$, where $(h)_0$ and $(h)_\infty$ are effective and of degree d. Then $\lceil d/2 \rceil L - (h)_\infty$ is of degree ≥ 0, so by Riemann–Roch there exists a section s of $\mathscr{O}_C(\lceil d/2 \rceil)$ vanishing at the poles of h. Then hs is another global section of $\mathscr{O}_C(\lceil d/2 \rceil)$. Both s and hs are restrictions of homogeneous forms on $\mathbb{P}^2$, and h is their ratio. □

LEMMA 4.7. *Let $\pi\colon X \to Y$ be a morphism of nice curves over $\mathbb{C}$. Let $P \in X(\mathbb{C})$. Let $Q = \pi(P)$. Let e be the ramification index of π at P. Let s be a nonzero meromorphic section of $(\Omega^1_Y)^{\otimes n}$ for some $n \in \mathbb{Z}$. Then $\operatorname{ord}_P(\pi^* s) = e \operatorname{ord}_Q s + n(e-1)$.*

PROOF. Let t be a uniformizer at $\pi(P)$ on Y. For any $f \in \mathbb{Q}(Y)^\times$, we have $\operatorname{ord}_P(\pi^* f) = e \operatorname{ord}_Q(f)$ by definition, and $\operatorname{ord}_P(\pi^* dt) = e - 1$ as in the proof of the Hurwitz formula. Since $s = f\, dt^{\otimes n}$ for some $f \in \mathbb{Q}(Y)^\times$, the formula follows. □

5. Proof of main theorem ignoring precision

We now start the proof of Theorem 3.1. Compute a basis for I_2 using Corollary 4.2 and apply Lemma 4.3 to test if X is hyperelliptic. If X is nonhyperelliptic, compute bases for I_2, I_3, I_4 using Corollary 4.2; these are enough to cut out $X \subset \mathbb{P}^{g-1}$, by Petri's theorem [**Pet23**]. *Henceforth, we assume that X is hyperelliptic.*

Steps (1)–(2) below require working over a field L such that X has an L-point P', so that there is an isomorphism $C_L \simeq \mathbb{P}^1_L$ such that P' maps to ∞. We choose L to be an isomorphic copy of K, and let $P' \in X(L)$ be the result of applying the isomorphism to $P \in X(K)$. (We will need to consider $(L \otimes K)/K$-traces, so keeping separate names for L and K will help clarify things.) We will need to take $L/\mathbb{Q}$-traces of elements of $\mathrm{H}^0(C_L, \mathcal{T}_L)$ as in Lemma 4.5 to get elements of $\mathrm{H}^0(C, \mathcal{T})$; these will be computed as $(L \otimes K)/K$-traces of their expansions at P (the tensor product is over $\mathbb{Q}$).

We first explain the algorithm as if we had $w_1, \ldots, w_g \in K[[q]]$ to infinite precision, and later in Section 6 explain what modifications are needed when we have only their truncations $\bar{w}_1, \ldots, \bar{w}_g$.

(1) (Find the expansion of a rational function $t\colon X_L \to C_L \simeq \mathbb{P}^1_L$.) Let $W \subset K[[q]]$ be the $\mathbb{Q}$-span of $w_1, \ldots, w_g$, and let $W_K \subset K[[q]]$ be their K-span. Run Gaussian elimination over K to find a new K-basis $w'_1, \ldots, w'_g$ of W_K such that $\operatorname{ord}_P(w'_1) < \cdots < \operatorname{ord}_P(w'_g)$. Let $M \in \mathrm{GL}_g(K)$ be the change-of-basis matrix sending $w_1, \ldots, w_g$ to $w'_1, \ldots, w'_g$. Applying the isomorphism $K \to L$ yields a matrix $M_L \in \mathrm{GL}_g(L)$. Then M_L sends $\omega_1, \ldots, \omega_g$ to an L-basis $\omega'_1, \ldots, \omega'_g$ of $\mathrm{H}^0(X_L, \Omega^1)$ as in Lemma 4.4. Computing the same L-linear combinations of $w_1, \ldots, w_g \in K[[q]]$ produces elements $w''_1, \ldots, w''_g \in L \otimes W \subset (L \otimes K)[[q]]$ representing the expansions at P of the ω'_i, which have increasing order of vanishing at P'.

Let $t = \omega'_{g-1}/\omega'_g \in L(X)$, as in Lemma 4.4, so t is the "x-coordinate" on a hyperelliptic model. Its expansion at P is in $(L \otimes K)((q))$.

(2) (Find expansions of a $\mathbb{Q}$-basis of $\mathrm{H}^0(C, \mathcal{T})$.) Let $\lambda_1, \ldots, \lambda_\ell$ be a $\mathbb{Q}$-basis of L. The $L/\mathbb{Q}$-traces in Lemma 4.5(b) span $V := \mathrm{H}^0(C, \mathcal{T})$, so three of them form a $\mathbb{Q}$-basis of C. To calculate with them, we start with the expansions of $\lambda_j t^i \frac{d}{dt}$ in $(L \otimes K)((q))$ for $i = 0, 1, 2$ and $j = 1, \ldots, \ell$, calculate $(L \otimes K)/K$-traces (traces are compatible with base change), and find three of them that are K-linearly independent and hence $\mathbb{Q}$-linearly dependent; call them $\partial_0, \partial_1, \partial_2 \in K((q))\frac{d}{dq}$; these are the expansions of a basis of global sections of $\mathcal{T}$ pulled back to X.

(3) (Find the equation $Q = 0$ of the conic C.) There is a unique $Q \in \mathbb{Q}[a, b, c]_2$ up to scalar such that $Q(\partial_0, \partial_1, \partial_2) = 0$ in $K((q)) \left(\frac{d}{dq}\right)^2$. Then $Q = 0$ is the anticanonical model of C in $\mathbb{P}^2$. We find Q by linear algebra.

(4) (Find the expansion of $h \in \mathbb{Q}(C)$ such that $\mathbb{Q}(X) = \mathbb{Q}(C)(\sqrt{h})$.) In this step, we compute $h \in \mathbb{Q}(C)$ such that $\mathbb{Q}(X) = \mathbb{Q}(C)(\sqrt{h})$. Let $f := a/b$, viewed as a rational function on C; its expansion is ∂_0/∂_1. Let $y = df/\omega_1 \in \mathbb{Q}(x)$ and $h = y^2$. The hyperelliptic involution fixes df and acts as -1 on $H^0(X, \Omega^1)$, so it negates y and fixes h; that is, $h \in \mathbb{Q}(C)$. Then $\mathbb{Q}(X) = \mathbb{Q}(C)(y) = \mathbb{Q}(C)(\sqrt{h})$.

(5) (Write h as a ratio of homogeneous forms.) We now show how to write h explicitly as F/G for some $F, G \in \mathbb{Q}[a, b, c]_{g+3}$. Since f is a rational function of degree 2 on C, it has at most 2 poles with multiplicity, so df on C has at most 4 poles with multiplicity (the worst case being when f has two simple poles), so its pullback to X has at most 8 poles. On the other hand, ω_1 has at most $2g - 2$ zeros on X, so y has at most $2g + 6$ poles on X. Then h has at most $2(2g+6)$ poles on X, so its degree on C is at most $2g+6$. By Lemma 4.6, there exist homogeneous forms $F, G \in \mathbb{Q}[a, b, c]_{g+3}$ such that $F/G = h$. To find the coefficients of possible F and G, we solve the linear system $F = hG$ in these unknown coefficients, using expansions of $\partial_0, \partial_1, \partial_2$ and h.

(6) (For even g, find an equation $y^2 = f(x)$ for X.) Suppose that g is even. In this case, $C \simeq \mathbb{P}^1$, and we will describe a method to find a rational parametrization of C, following the strategy of Lemma 4.6. The 1-form ω_1 corresponds to a linear form on $\mathbb{P}^{g-1}$, which cuts out a divisor D of odd degree $g - 1$ on C. Let $\mathcal{S}$ be the space of $S \in \mathbb{Q}[a, b, c]_{g/2}$ that vanish along D. By the Riemann–Roch theorem, $\dim \mathcal{S} = 2$; we next seek an explicit basis of $\mathcal{S}$, which will define an isomorphism $C \to \mathbb{P}^1$. For each $S \in \mathcal{S}$ and for $j = 2, \ldots, g$, the element $R_j := S\omega_j/\omega_1 \in \mathbb{Q}(a, b, c)$ lies in $\mathbb{Q}[a, b, c]_{g/2}$ since S vanishes along D. Thus $\mathcal{S}$ is the projection on the last coordinate of the space $\mathcal{R}$ of g-tuples $(R_2, \ldots, R_g, S)$ of polynomials in $\mathbb{Q}[a, b, c]_{g/2}$ such that

$$\omega_1 R_j = S\omega_j$$

for all $j = 2, \ldots, g$. Using the expansions of $a, b, c, \omega_1, \ldots, \omega_g$ at P, we compute $\mathcal{R}$ by linear algebra. Thus we obtain an isomorphism $C \simeq \mathbb{P}^1$.

Under $C \simeq \mathbb{P}^1$, the function h corresponds to some $f \in \mathbb{Q}(x)^\times$. Now X is birational to the curve $y^2 = f(x)$. Multiply f by a square to make it a polynomial. Remove square factors (by computing $\gcd(f, f')$, etc.) to make f separable. By Riemann–Hurwitz, $\deg f$ is $2g + 1$ or $2g + 2$.

(7) (For odd g, find H.) Now assume that g is odd. Let F, G be as in Step 5. We seek $H \in \mathbb{Q}[a,b,c]_{g+1}$ separable and $J \in \mathbb{Q}[a,b,c]_{(g+5)/2}$ such that $FG \equiv HJ^2 \pmod{Q}$; then the rational function $h = F/G$ equals HJ^2/G^2 on $C\colon Q = 0$, so the function field of the smooth projective curve $X' := \operatorname{Proj} \frac{\mathbb{Q}[a,b,c,y]}{(y^2-H,Q)}$ equals $\mathbb{Q}(C)(\sqrt{HJ^2/G^2}) = \mathbb{Q}(C)(\sqrt{h})$, so $X' \simeq X$; that is, H is as in the statement of the theorem. We cannot simply factor FG to find H and J, since $\mathbb{Q}[a,b,c]/(Q)$ is not a UFD. Instead we will decompose the zero locus $D := Z_C(FG) \in \operatorname{Div} C$ as $U + 2V$ with U, V effective divisors on C and U reduced. First, choose $p \in \mathbb{P}^2(\mathbb{Q})$ not on any line connecting geometric points in D and not on any line tangent to a geometric point in D; then the projection from p restricts to a morphism $\nu\colon C \to \mathbb{P}^1$ that is injective on the geometric points in D and unramified at those points. Write $\nu_* D = U' + 2V'$ with U', V' effective divisors on $\mathbb{P}^1$ and U' reduced, using factorization in the homogeneous coordinate ring of $\mathbb{P}^1$. Let $U = \nu^* U' \cap D$ and $V = \nu^* V' \cap D$; then $D = U + 2V$ by choice of ν. We have $\deg U = \deg U' = 2g+2$, so $\deg V = \deg V' = g+5$. By Riemann–Roch on C, an effective divisor of even degree $2d$ is the zero locus of a form in $\mathbb{Q}[a,b,c]_d$, unique up to scalar and modulo multiples of Q; in particular, there exist $H \in \mathbb{Q}[a,b,c]_{g+1}$ and $J \in \mathbb{Q}[a,b,c]_{(g+5)/2}$ with $Z_C(H) = U$ and $Z_C(J) = V$; we find explicit H and J by linear algebra. Then $FG \equiv \alpha H J^2 \pmod{Q}$ for some $\alpha \in \mathbb{Q}^\times$. Evaluate F, G, H, J at some zero of Q in $\overline{\mathbb{Q}}^3$ to find α, and replace H by αH to get $FG \equiv HJ^2 \pmod{Q}$.

If a rational point on C is given, projection from it defines an isomorphism $C \to \mathbb{P}^1$. Find an equation $y^2 = f(x)$ for X as in the last paragraph of (6).

6. Precision analysis

TABLE 1. Tracking q-adic precision of objects in the proof of the main theorem.

Object	Space	ord_P	Absolute error	Relative error
ω_j	$\mathrm{H}^0(X, \Omega^1)$	$[0, 2g-2]$	$+O(q^B)\,dq$	$\cdot(1+O(q^{B-2g+2}))$
ω'_j	$\mathrm{H}^0(X_L, \Omega^1)$	$[0, 2g-2]$	$+O(q^B)\,dq$	$\cdot(1+O(q^{B-2g+2}))$
t	$L(C)$	$[-2, 2]$	$+O(q^{B-2g})$	$\cdot(1+O(q^{B-2g+2}))$
dt	$(\Omega^1_{C_L})_{\eta_{C_L}}$	$[-3, 1]$	$+O(q^{B-2g-1})\,dq$	$\cdot(1+O(q^{B-2g-2}))$
$t^i \frac{d}{dt}$	$\mathrm{H}^0(C_L, \mathcal{T})$	$[-1, 3]$	$+O(q^{B-2g-3})\,\frac{d}{dq}$	$\cdot(1+O(q^{B-2g-2}))$
∂_i	$\mathrm{H}^0(C, \mathcal{T})$	$[-1, 3]$	$+O(q^{B-2g-3})\,\frac{d}{dq}$	$\cdot(1+O(q^{B-2g-6}))$
$M(\partial_0, \partial_1, \partial_2)$	$\mathrm{H}^0(C, \mathcal{T}^d)$	$[-d, 3d]$	$+O(q^{B-2g-d-2})\left(\frac{d}{dq}\right)^d$	$\cdot(1+O(q^{B-2g-4d-2}))$
f	$\mathbb{Q}(C)$	$[-4, 4]$	$+O(q^{B-2g-10})$	$\cdot(1+O(q^{B-2g-6}))$
df	$(\Omega^1_C)_{\eta_C}$	$[-5, 5]$	$+O(q^{B-2g-11})\,dq$	$\cdot(1+O(q^{B-2g-16}))$
y	$\mathbb{Q}(X)$	$[-2g-3, 5]$	$+O(q^{B-4g-19})$	$\cdot(1+O(q^{B-2g-16}))$
h	$\mathbb{Q}(C)$	$[-4g-6, 10]$	$+O(q^{B-6g-22})$	$\cdot(1+O(q^{B-2g-16}))$
hG	$\mathcal{T}^{g+3}_{\eta_C}$	$[-5g-9, 3g+19]$	$+O(q^{B-11g-23})\left(\frac{d}{dq}\right)^{g+3}$	$\cdot(1+O(q^{B-6g-14}))$
$F-hG$	$\mathcal{T}^{g+3}_{\eta_C}$		$+O(q^{B-11g-23})\left(\frac{d}{dq}\right)^{g+3}$	
$\omega_1 R_j,\ S\omega_j$	$\mathrm{H}^0(X, \Omega^1_X \otimes \pi^* \mathcal{T}^{\frac{g}{2}})$	$[-g/2, 7g/2-2]$	$+O(q^{B-9g/2-2})\left(\frac{d}{dq}\right)^{\frac{g}{2}-1}$	$\cdot(1+O(q^{B-4g-2}))$

In Section 5, we assumed that $w_1, \dots, w_g \in K[[q]]$ were given to infinite precision. Now, in Table 1, we track how much precision we have in the steps if we start only with $w_1, \dots, w_g$ up to addition of $O(q^B)$. For each Laurent series, we bound both absolute error (addition of $O(q^n)$ for some n) and relative error (multiplication by $1 + O(q^n)$ for some n); we can pass between them if the valuation of a power series is controlled; these valuations lie in the range given in the ord_P column of Table 1. For series with coefficients in the étale algebra $L \otimes K$, the bounds apply when projected onto any field factor of $L \otimes K$. Let η_C be the generic point of C, so the stalk $(\Omega^1_C)_{\eta_C}$ is the space of meromorphic 1-forms on C. Define η_{C_L} similarly.

LEMMA 6.1. *Table 1 is correct.*

PROOF. Each ω_j is regular and has $2g-2$ zeros in total, so $\operatorname{ord}_P(\omega_j) \in [0, 2g-2]$. It is given to absolute error $O(q^B)\, dq$. The ω'_j are L-linear combinations of the ω_j, so they have the same absolute error. Since each ω_j and ω'_j vanishes at P to order at most $2g-2$, their relative error is $1 + O(q^{B-2g+2})$ (as usual, big-O notation allows for the possibility that the error could be smaller than specified).

Now $t = \omega'_{g-1}/\omega'_g$, so its relative error is again $1 + O(q^{B-2g+2})$. On the other hand, t is the "x-coordinate" of a hyperelliptic model of X, so $\operatorname{ord}_P(t) \in [-2, 2]$. Since $\operatorname{ord}_P t \geq -2$, the absolute error of t is $O(q^{B-2g})$.

The absolute error of dt is then $O(q^{B-2g-1})\, dq$. Again since t is the "x-coordinate" of a hyperelliptic model of X, we have $\operatorname{ord}_P(dt) \in [-3, 1]$. Since $\operatorname{ord}_P(dt) \leq 1$, the relative error of dt is $1 + O(q^{B-2g-2})$.

Fix $i \in \{0, 1, 2\}$. The relative error of $t^i \frac{d}{dt}$ is the worse of the relative errors of t and dt, which is $1 + O(q^{B-2g-2})$. The section $t^i \frac{d}{dt}$ is regular on C, with 2 zeros, so $\operatorname{ord}_{\pi(P)}(t^i \frac{d}{dt}) \in [0, 2]$, so $\operatorname{ord}_P(t^i \frac{d}{dt})$ is in $[0, 2]$ or $-1 + 2[0, 2] = [-1, 3]$ according to whether π is unramified or ramified at P, by Lemma 4.7 applied with $n = -1$. Since $\operatorname{ord}_P(t^i \frac{d}{dt}) \geq -1$, the absolute error of $t^i \frac{d}{dt}$ is $O(q^{B-2g-3}) \frac{d}{dq}$.

The ∂_i are linear combinations of the $t^i \frac{d}{dt}$, so they have the same absolute error $O(q^{B-2g-3}) \frac{d}{dq}$. As for $t^i \frac{d}{dt}$, we have $\operatorname{ord}_P(\partial_i) \in [-1, 3]$. Since $\operatorname{ord}_P(\partial_i) \leq 3$, the relative error of ∂_i is $1 + O(q^{B-2g-6})$.

We will need to analyze the error in $M(\partial_0, \partial_1, \partial_2)$ for various nonzero forms $M \in \mathrm{H}^0(\mathbb{P}^2, \mathscr{O}(d)) = \mathbb{Q}[a, b, c]_d$, for various $d \geq 1$, so we do a calculation for all of these at once, and later specialize to the particular M we need. Its order of vanishing at $\pi(P)$ is in $[0, 2d]$, since C is a curve of degree 2 in $\mathbb{P}^2$. Then its order of vanishing at P is in $[0, 2d]$ or $-d + 2[0, 2d] = [-d, 3d]$, according to whether π is unramified or ramified at P, by Lemma 4.7 applied with $n = -d$. If M is a monomial, then the absolute error of $M(\partial_0, \partial_1, \partial_2)$ is at worst that of one ∂_i minus $d - 1$ (because all the ∂_j in the monomial have at worst a simple pole), hence at worst $O(q^{B-2g-d-2}) \left(\frac{d}{dq}\right)^d$. Since $\operatorname{ord}_P(M(\partial_0, \partial_1, \partial_2)) \leq 3d$, the relative error is at worst $1 + O(q^{B-2g-4d-2})$.

The rational function $f := a/b$ is of degree 2 on C, and the ramification index of π at P is at most 2, so $\operatorname{ord}_P(f) \in [-4, 4]$. Its relative error is the same as that of $a = \partial_0$ and $b = \partial_1$, which is $1 + O(q^{B-2g-6})$. Since $\operatorname{ord}_P(f) \geq -4$, its absolute error is $O(q^{B-2g-10})$.

Then $\operatorname{ord}_P(df) \geq \operatorname{ord}_P(f) - 1 \geq -5$. Since f on C has at most 2 poles with multiplicity, df on C has at most 4 poles with multiplicity, but the divisor of df on C has degree -2, so df has at most 2 zeros on C, so $\operatorname{ord}_P(df) \leq 2 \cdot 2 + 1 = 5$, the worst case being if π is ramified at P. The absolute error of df is $O(q^{B-2g-11})$, so the relative error is $1 + O(q^{B-2g-16})$.

Since $\operatorname{ord}_P(df) \in [-5, 5]$ and $\operatorname{ord}_P(\omega_1) \in [0, 2g-2]$, we have $\operatorname{ord}_P(y) = \operatorname{ord}_P(df/\omega_1) \in [-2g-3, 5]$. The relative error of y is the worse of the relative errors of df and ω_1, which is $1 + O(q^{B-2g-16})$. Then the absolute error of y is $O(q^{B-4g-19})$.

Squaring gives $\operatorname{ord}_P(h) \in 2[-2g-3, 5] = [-4g-6, 10]$, and h has relative error $1 + O(q^{B-2g-16})$ and absolute error $1 + O(q^{B-6g-22})$.

For hG, we compute ord_P and the relative error from the corresponding numbers for h and $M := G$ of degree $d = g+3$. Since $\operatorname{ord}_P(hG) \geq -5g-9$, the absolute error is then $O(q^{(B-6g-14)+(-5g-9)}) \left(\frac{d}{dq}\right)^{g+3} = O(q^{B-11g-23}) \left(\frac{d}{dq}\right)^{g+3}$.

The absolute error for F, from the $M(\partial_0, \partial_1, \partial_2)$ row with $d = g+3$, is $O(q^{B-2g-(g+3)-2}) \left(\frac{d}{dq}\right)^{g+3}$. Combining these gives $F - hG$ with absolute error $O(q^{B-11g-23}) \left(\frac{d}{dq}\right)^{g+3}$. (We do not need the ord_P and relative error of $F - hG$.)

The calculations for $\omega_1 R_j$ and $S\omega_j$ are analogous to those for hG. □

LEMMA 6.2. *If $B \geq 19g + 48$, then we can perform Steps 1–6 in the proof of Theorem* 3.1. (*Step 7 does not involve expansions; it is carried out exactly.*)

PROOF. In Step 3, Q is determined by $Q(\partial_0, \partial_1, \partial_2) \bmod q^7$ because when $M \in \mathrm{H}^0(\mathbb{P}^2, \mathscr{O}(2))$, we have $\operatorname{ord}_P M(\partial_0, \partial_1, \partial_2) \leq 6$. We computed $Q(\partial_0, \partial_1, \partial_2)$ to absolute error $O(q^{B-2g-d-2}) \left(\frac{d}{dq}\right)^d$ with $d = 2$, which is good enough since $B - 2g - 2 - 2 \geq 7$.

Let $(h)_\infty$ be the polar part of the divisor of h, which has degree at most $2g+6$ as explained in Step 5. For any $F, G \in \mathbb{Q}[a, b, c]_{g+3}$, the expression $F - hG$ is a global section of $\mathcal{T}^{g+3} \otimes \mathscr{O}_C((h)_\infty)$, which is a line bundle of degree at most $2(g+3) + (2g+6) = 4g + 12$. The pullback of this bundle to X has degree at most $2(4g+12) = 8g + 24$. Thus, if $\operatorname{ord}_P(F - hG) > 8g + 24$, then $F - hG = 0$. In other words, it suffices to do the linear algebra in Step 5 to absolute precision $+O(q^{8g+25}) \left(\frac{d}{dq}\right)^{g+3}$. By Table 1, we have this precision if $B - 11g - 23 \geq 8g + 25$, or, equivalently, $B \geq 19g + 48$.

The degree of $\Omega^1_X \otimes \pi^* \mathcal{T}^{g/2}$ on X is $2g - 2 + 2 \cdot 2(g/2) = 4g - 2$, so if $\operatorname{ord}_P(\omega_1 R_j - S\omega_j) > 4g - 2$, then $\omega_1 R_j - S\omega_j = 0$. In other words, it suffices to do the linear algebra in Step 6 to absolute precision $+O(q^{4g-1}) \left(\frac{d}{dq}\right)^{g/2-1}$. By Table 1, we have this precision if $B - 9g/2 - 2 \geq 4g - 1$, or, equivalently, $B \geq 17g/2 + 1$. □

REMARK 6.3. In each of the models found, we have the expansions at P of the new coordinate functions, as Laurent series in q, so we can find the coordinates of P in the new model. Similarly, by linear algebra we can express $\omega_1, \ldots, \omega_g$ in terms of the new coordinates, if desired.

REMARK 6.4. The genus of a modular curve with geometric gonality 2 is at most 17 [**BGJGP05**, Remark 4.5]. So, in running the algorithm of Theorem 3.1 on hyperelliptic modular curves, we always have $g \leq 17$.

Acknowledgments

We thank David Roe for his comments on early versions of the algorithm of this article. His computations using our algorithm were essential in helping us debug and improve it. We thank the anonymous referees for corrections and suggestions.

Van Bommel, Costa, and Poonen were supported by Simons Foundation grant 550033. Van Bommel was additionally supported by Céline Maistret's Royal Society Dorothy Hodgkin Fellowship. Costa was additionally supported by Simons Foundation grant SFI-MPS-Infrastructure-00008651. Poonen was partially supported also by National Science Foundation grant DMS-2101040 and Simons Foundation grant 402472. Srinivasan was supported by National Science Foundation grant DMS-2401547 and Simons Foundation grant 546235.

References

[ACGH85] E. Arbarello, M. Cornalba, P. A. Griffiths, and J. Harris, *Geometry of algebraic curves. Vol. I*, Grundlehren der mathematischen Wissenschaften [Fundamental Principles of Mathematical Sciences], vol. 267, Springer-Verlag, New York, 1985, DOI 10.1007/978-1-4757-5323-3. MR770932

[Bar14] Burcu Baran, *An exceptional isomorphism between modular curves of level 13*, J. Number Theory **145** (2014), 273–300, DOI 10.1016/j.jnt.2014.05.017. MR3253304

[BGJGP05] Matthew H. Baker, Enrique González-Jiménez, Josep González, and Bjorn Poonen, *Finiteness results for modular curves of genus at least 2*, Amer. J. Math. **127** (2005), no. 6, 1325–1387. MR2183527

[LMFDB] The LMFDB Collaboration. The L-functions and modular forms database. `https://www.lmfdb.org`, 2025. [Online; accessed 23 January 2025].

[MS20] Pietro Mercuri and René Schoof, *Modular forms invariant under non-split Cartan subgroups*, Math. Comp. **89** (2020), no. 324, 1969–1991, DOI 10.1090/mcom/3503. MR4081925

[Pet23] K. Petri, *Über die invariante Darstellung algebraischer Funktionen einer Veränderlichen* (German), Math. Ann. **88** (1923), no. 3-4, 242–289, DOI 10.1007/BF01579181. MR1512130

[Shi10] Goro Shimura, *Arithmetic of quadratic forms*, Springer Monographs in Mathematics, Springer, New York, 2010, DOI 10.1007/978-1-4419-1732-4. MR2665139

[Zyw24] David Zywina. Explicit open images for elliptic curves over $\mathbb{Q}$. *preprint*, 2024. `arXiv:2206.14959v2`.

Department of Mathematics, Massachusetts Institute of Technology, Cambridge, MA 02139, USA

School of Mathematics, University of Bristol, Fry Building, Woodland Road, Bristol, BS8 1UG, UK
Email address: r.vanbommel@bristol.ac.uk
URL: https://raymondvanbommel.nl/

Department of Mathematics, Massachusetts Institute of Technology, Cambridge, MA 02139, USA
Email address: edgarc@mit.edu
URL: https://edgarcosta.org

Department of Mathematics, Massachusetts Institute of Technology, Cambridge, MA 02139, USA
Email address: poonen@math.mit.edu
URL: https://math.mit.edu/~poonen/

Boston University, 665 Commonwealth Avenue, Boston, MA 02215, USA
Email address: padmask@bu.edu
URL: https://padmask.github.io/

Selected Published Titles in This Series

840 **John W. Jones, Jennifer Paulhus, Andrew V. Sutherland, and John Voight, Editors,** LuCaNT: Databases, Algorithms, and Computational Number Theory, 2026

839 **Sangita Jha, William Ott, and Mrinal Kanti Roychowdhury, Editors,** Current Trends in Dynamical Systems, Fractal Geometry, and Their Applications, 2026

835 **Mee Seong Im and Tony Shaska, Editors,** Artificial Intelligence and Mathematics Research, 2026

834 **Mee Seong Im, Hanna E. Makaruk, Bach Nguyen, and Robert Owczarek, Editors,** Inverse Problems, Geometric and Topological Aspects of Representation Theory, and Low-Dimensional Topology, 2026

832 **Nils Bruin, David Kohel, and Chloe Martindale, Editors,** Arithmetic, Geometry, Cryptography and Coding Theory, 2026

831 **Irena Swanson and Kei-ichi Watanabe, Editors,** Lectures on Tight Closure and Its Applications, 2025

830 **O. Bezushchak, Yu. Drozd, P. Etingof, V. Futorny, R. Grigorchuk, and E. Zelmanov, Editors,** Modern Algebra Volume 2, 2025

829 **O. Bezushchak, Yu. Drozd, P. Etingof, V. Futorny, R. Grigorchuk, and E. Zelmanov, Editors,** Modern Algebra Volume 1, 2025

828 **Hafedh Herichi, Maria Rosaria Lancia, Therese Basa Landry, Anna Rozanova-Pierrat, and Steffen Winter, Editors,** Fractal Geometry in Pure and Applied Mathematics, 2025

827 **Carmen Caprau, J. Scott Carter, Neslihan Gügümcü, and Sam Nelson, Editors,** Algebraic Structures in Knot Theory, 2025

826 **Ahmed Laghribi and André Leroy, Editors,** Algebraic Structures and Applications, 2025

825 **Saurabh Verma, María A. Navascués, and Amit Priyadarshi, Editors,** Advances in Dimension Theory, Fractal Functions and Measures, 2025

824 **Hanna E. Makaruk and Robert M. Owczarek, Editors,** Inverse Problems: In Memory of Professor Zbigniew Oziewicz, 2025

823 **Mikhail Bershtein, Anton Dzhamay, and Andrei Okounkov, Editors,** The Versatility of Integrability, 2025

822 **Ahmad Barhoumi, Roozbeh Gharakhloo, and Andrei Martinez-Finkelshtein, Editors,** Recent Developments in Orthogonal Polynomials, 2025

821 **Bogdan D. Suceavă, Adara M. Blaga, Marian Ioan Munteanu, Cezar Oniciuc, Shoo Seto, Gabriel-Eduard Vîlcu, and Lihan Wang, Editors,** Recent Advances in Differential Geometry and Related Areas, 2025

820 **Sudeb Mitra, Rakesh Kumar Parmar, Saminathan Ponnusamy, and Toshiyuki Sugawa, Editors,** Geometric Function Theory and Related Topics, 2025

819 **Howard S. Cohl, Roberto S. Costas-Santos, and Robert S. Maier, Editors,** Applications and q-Extensions of Hypergeometric Functions, 2025

818 **Howard S. Cohl, Roberto S. Costas-Santos, and Robert S. Maier, Editors,** Classical Hypergeometric Functions and Generalizations, 2025

817 **Mikhail Khovanov, Joshua Sussan, and Anton Zeitlin, Editors,** From Representation Theory to Mathematical Physics and Back, 2025

816 **Luca F. Di Cerbo and Laurenţiu G. Maxim, Editors,** Geometry and Topology of Aspherical Manifolds, 2025

815 **Daniel Orr and Joshua Jeishing Wen, Editors,** Macdonald Theory and Beyond, 2025

814 **Susan Montgomery and Siu-Hung Ng, Editors,** Quantum Groups, Hopf Algebras, and Applications, 2025

For a complete list of titles in this series, visit the
AMS Bookstore at **www.ams.org/bookstore/conmseries/**.